THE ANATOMIST ANATOMIS’D

The History of Medicine in Context

Series Editors: Andrew Cunningham and Ole Peter Grell

Department of History and Philosophy of Science
University of Cambridge

Department of History
Open University

Titles in this series include

Centres of Medical Excellence?
Medical Travel and Education in Europe, 1500–1789
Edited by Ole Peter Grell, Andrew Cunningham and Jon Arrizabalaga

Crafting Immunity
Working Histories of Clinical Immunology
Edited by Kenton Kroker, Jennifer Keelan and Pauline M.H. Mazumdar

Before My Helpless Sight
Suffering, Dying and Military Medicine on the Western Front, 1914-1918
Leo van Bergen

Negotiating the French Pox in Early Modern Germany
Claudia Stein

Ireland and Medicine in the Seventeenth and Eighteenth Centuries
Edited by James Kelly and Fiona Clark

The Anatomist Anatomis'd
An Experimental Discipline in Enlightenment Europe

ANDREW CUNNINGHAM
University of Cambridge, UK

ASHGATE

Published by
Ashgate Publishing Limited
Wey Court East
Union Road
Farnham
Surrey, GU9 7PT
England

Ashgate Publishing Company
Suite 420
101 Cherry Street
Burlington
VT 05401-4405
USA

www.ashgate.com

British Library Cataloguing in Publication Data
Cunningham, Andrew, Dr.
The Anatomist Anatomis'd: An Experimental Discipline in Enlightenment Europe. – (The History of Medicine in Context)
1. Anatomy – Research – Europe – History – 18th century. 2. Anatomists – Europe – History – 18th century. 3. Enlightenment – Europe.
I. Title II. Series
611'.0072'04-dc22

Library of Congress Cataloging-in-Publication Data
Cunningham, Andrew, Dr.
The Anatomist Anatomis'd: An Experimental Discipline in Enlightenment Europe / Andrew Cunningham.
p. ; cm. – (History of Medicine in Context)
Includes index.
1. Human anatomy – History – 18th century. 2. Enlightenment. I. Title. II. Series: History of medicine in context.
[DNLM: 1. Anatomy – history--Europe. 2. Anatomy – trends – Europe. 3. History, 18th Century – Europe. 4. History, 19th Century – Europe. QS 11 GA1 C973a 2009]
QM11.C863 2009
611.0094–dc22 2009045492

ISBN 9780754663386 (hbk)

Mixed Sources
Product group from well-managed forests and other controlled sources
www.fsc.org Cert no. SA-COC-1565
© 1996 Forest Stewardship Council
FSC

Printed and bound in Great Britain by
MPG Books Group, UK

Contents

List of Figures *vii*
Acknowledgements *xv*
Preface *xix*

Introduction: The Anatomist Anatomis'd **1**

An anatomist anatomised 1
Discipline history 8
The attitude of the Catholic Church 12

1 'This awful subject' **17**

1. The tradition of anatomy 17
2. The identity of anatomy 19
3. The Theatre of Anatomy 25
Giving a good show 25
Getting a good look 29
Invitations to the show 44
Getting the message across 47
4. Practical Matters 55
5. A research culture 63
6. The state of the art 65
7. A community of scholars 72
8. A great tradition 73
Reviving the Moderns 73
Making history 80

2 'Merit is sure of its reward': Careers and Courses **83**

Part 1: Men 83

1. Why anatomy? 83
2. How to become an anatomist 85
Keeping it in the family 86
Making one's way 91
'Rising merit' 96

3. Where to teach and practise anatomy 98
Anatomy in institutions under royal patronage in Paris: The Jardin royal, the Collège royal, the Académie royale des sciences, royal anatomists 98
Royal surgical institutions in Paris 107
Anatomy in University institutions in Paris: The Faculté 109
Hospitals and anatomy 112
Private institutions 114
4. Ways of teaching anatomy 119
Paris, 1673–80: Dionis 119
Edinburgh, 1704: The Incorporation of Surgeon–Apothecaries 121
Leiden, 1713–19: Rau 123
Padua, 1716–71: Morgagni 125
Paris, c. 1731–58: Winslow 131
Edinburgh, 1726–40s: Monro Primus 133
London, 1746–83: William Hunter 135
London, 1773–93: John Hunter 139

Part 2: Women 142

3 Experimental Anatomy and its Sub-Disciplines 149

1. Anatomy, an experimental discipline 149
Experiment 1 149
Experiment 2 150
Experiment 3 151
Experiment 4 152
Experiment 5 152
Experiment 6 154
2. Physiology: A theoretical science and sub-discipline of anatomy 156
3. Generation and man-midwifery: Sub-disciplines of anatomy 165
4. Comparative anatomy: A sub-discipline of anatomy 186
5. Pathology: A sub-discipline of anatomy 186
Why anatomy in pathology? 186
Pathology in the library 190
Pathology in the hospital and anatomy theatre 196
Pathology elsewhere 217

4 Human Bodies: Getting, Keeping, Picturing, Publishing, Arguing 223

1. Acquisition: Getting bodies 223
2. Preservation: Preserving bodies better and longer 231

Preparations dry and wet 231
Injections 235
Wax modelling 244
3. Picturing bodies better 251
Better portrayal 252
The Albinus method 253
The Cheselden method (Camera obscura) 262
The anatomist and the art student 264
4. Better printing 276
5. 'Warm controversy': Discoveries, credit, and the making of anatomical facts 280
A charlatan case: Ruysch vs. Bils, 1665 282
A plagiarism case: Bidloo vs. Cowper 1698 283
Facticity challenged: 'VENITE! ET VIDETE!' Bidloo vs. Ruysch, 1690s 284
A case of theft? Hunters vs. Monros: The testis and the lymphatics 1752–64 288
An internal job? Hunter versus Hunter: The structure of the placenta, 1755–80 291

5 Animal Bodies and Comparative Anatomy 295

1. The animals of the anatomist: Vivisection and dissection 295
2. The source of exotic animals for dissection 300
3. What did 'comparative anatomy' mean in the long eighteenth century? 308
4. George Stubbs and the dissection of the horse 340
5. George Stubbs as comparative anatomist 355

6 The End of Old Anatomy 361

1. 'Dutch' anatomy in Japan 361
2. 'The perfection of anatomy' 367
3. Fields of change 368
Physics and chemistry 369
Experimental physiology 371
Comparative anatomy and the remaking of Natural History 375
Clinical medicine and a new pathology 380
The rise of organisation and the death of the soul 383
Ending the sacred ritual 386

Bibliography of Reference and Illustrative Works *391*
Bibliography of Works Cited *395*
Index *433*

List of Figures

I am most grateful to the relevant institutions for permission to reproduce images in their possession.

Key

Author. By me, or in my collection.
BL. The British Library.
CUL. Reproduced by kind permission of the Syndics of Cambridge University Library.
RCS. The Royal College of Surgeons of England.
Wellcome. Wellcome Library, London.
Whipple. The Whipple Library of the History and Philosophy of Science, University of Cambridge.

Cover

Valverde de Amusco, Jean, *Anatomia del corpo humano*, Rome, 1560, Table to Book 4, p. 108 (detail). Courtesy of the National Library of Medicine, Bethesda, USA.

Frontispiece

Democritus dissecting, visited by Hippocrates. From W. Cheselden, *The anatomy of the human body*, 6th edn, 1741, frontispiece. CUL.

Introduction

0.1 Door protecting the papal praecordia, Church of Saints Vincent and Anastasius, Rome. Author. 2
0.2 Detail of frontispiece to Jean Jacques Manget and Daniel Le Clerc, *Bibliotheca anatomica*, vol. 1, 1685. CUL. 2
0.3 Portrait of Malpighi, sent to the Royal Society from Italy in 1680. Courtesy of the Royal Society of London. 4

0.4 Trevi Fountain and Church of Saints Vincent and Anastasius, Rome. Author. 8

Chapter 1: 'This awful subject'

1.1 Dissection and vivisection table, from Andreas Vesalius, *De humani corporis fabrica, libri septem*, 1543. Author. 24
1.2 Dissection Mondino-style. From Johannes de Ketham, *Fasciculo di Medicina*, 1491 (1925 reprint). Author. 27
1.3 Human anatomy, from G. Reisch, *Margarita philosophica*, 1503 (1512 edition), showing the anatomy of the human trunk and the diaphragm. Whipple. 28
1.4 Paduan anatomical theatre (1594), portrayed still in use in 1842. From P. Tosoni, *Anatomia degli antichi e della scuola anatomica padovana*, 1844. CUL. 31
1.5 Anatomical theatre in the Charité Hospital, Berlin, engraving by Friedrich Wilhelm Schmidt, Junior, *c.* 1730. Courtesy of the Kunstbibliothek, Staatliche Museen zu Berlin. 32
1.6 View of the Cutlerian theatre, 1803. Courtesy of the Royal College of Physicians of London. 33
1.7 Plans for the new university anatomy theatre, Edinburgh University. From D. B. Horn, 'The anatomy classrooms in the present Old College, 1725–1880', *University of Edinburgh Journal* 1965, plate 3. Courtesy of Edinburgh City Council Archives. 35
1.8 Edinburgh University anatomical theatre (completed 1764), detail from Monro class card, 1785. Courtesy of Edinburgh University Special Collections (31860). 36
1.9 From Gunnar Nyström, *Olaus Rudbeck's anatomical theatre in the 'Gustavianum'*, 1955. Whipple. 37
1.10 From Nyström, ibid. Whipple. 38
1.11 Olaus Rudbeck's anatomical theatre, Uppsala University, inside view. Uppsala University website. 38
1.12 Anatomical theatre in De Waag, Amsterdam, 1733. With the kind permission of the Amsterdam Historisch Museum. 39
1.13 Anatomical theatre, Madrid Surgeons Company. From Martín Martínez, *Anatomía completa del hombre*, 1728. Wellcome. 40
1.14 Ingoldstadt anatomy theatre and physic garden, 1723. Figure 7 from S. Hofmann, *Die alte Anatomie in Ingoldstadt*, 1974. Sketch by Melchior Büchner, without further reference. 41

1.15 Anatomical theatre, Halle University. From Georg Daniel Coschwitz, *Theatri anatomici … Natalitia*, 1727. British Library, Shelfmark T514 (1). 42
1.16 Anatomical theatre, Company of Surgeons, London, eighteenth century. From William Jackson, *The new and complete Newgate Calendar, or Malefactor's bloody register*, London, (1779) 1795, facing p. 376. Courtesy Royal College of Surgeons of England. 43
1.17 Anatomical theatre, Royal College of Surgeons, Barcelona. Author. 44
1.18 Invitation to a public dissection by Georgius Francus, Heidelberg, 1683. British Library, Shelfmark T514 (9). 45
1.19 Anatomical theatre of Thomas Bartholin, Copenhagen. Frontispiece to his *Cista medica Hafniensis*, Copenhagen, 1662; he describes the anatomical theatre in his *Domus anatomica Hafniensis brevissime descripta*, which is appended to the *Cista medica*. Reproduced from *Thomae Bartholini Vasa Lymphatica edited on the tercentenary of the birth of Thomas Bartholinus by Vilhelm Maar*, Copenhagen, 1916. Whipple. 49
1.20 Anatomical demonstration by John Banister, 1581, to the London Surgeons. Courtesy of University of Glasgow Library, Department of Special Collections (Hunter 364). 50
1.21 Skeleton from Vesalius, *De humani corporis fabrica libri septem*, 1543. Author. 51
1.22 Dissecting knives and other tools, from Michael Lyser, *The art of dissecting the human body* (1653), English edn 1740. CUL. 56
1.23 Machines for holding and sawing the skull, from Lyser, ibid. CUL. 58
1.24 Revolving table for displaying the body, from Lyser, ibid. CUL. 58
1.25 Tabula V from Book 1 of Thomas Bartholin, *Anatomia*, 1651 edn. Whipple. 61
1.26 Dissectors wearing aprons, from the 1756–60 Lausanne (French) edition of Haller's *Mémoires sur la nature sensible et irritable, des parties du corps animal*. CUL. 62
1.27 Frontispiece to Manget and Le Clerc, *Bibliotheca anatomica*, 1685. CUL. 68
1.28 Eustachius dissecting in 1560s Rome, presumably as reconstructed as an ancient Roman by Lancisi in early eighteenth-century Rome. From title-page of Bartholomaeus Eustachius, *Tabulae anatomicae*, 1711 (Amsterdam 1722 edn). CUL. 74
1.29 Eustachius, ibid., figure 26, showing the veins as they pass through the two lower venters and the inside of the arms. CUL. 76
1.30 The veins and bleeding points (at the bottom, differing angles for using the lancet). Table XXIIII from Pietro Berrettini da Cortona, *Tabulae anatomicae*, 1741. CUL. 77

Chapter 2: 'Merit is sure of its reward'

2.1 Portrait of Haller, 1778. Courtesy of Niedersächsische Staats- und Universitätsbibliothek Göttingen (Gr2 HLP IV, 68/10 RARA, Bl. 32). 92

2.2 Haller's anatomical theatre (left) and house (right) in the new botanic garden of Göttingen University, 1758, engraving by Heumann. Courtesy of Niedersächsische Staats- und Universitätsbibliothek Göttingen (Gr2 H HANN V, 29 RARA, Bl. 8). 93

2.3 The Jardin royal, from Pierre Dionis, *The anatomy of humane bodies improv'd*, 1703, frontispiece. CUL. 99

2.4 Portrait of Pierre Dionis, from his *Anatomie de l'homme*, 1723, frontispiece. CUL. 101

2.5 Hospital dissection scene, Amsterdam. From Bernhard Swalve, *Pancreas Pancrene adornante*, 1668, title-page. CUL. 110

2.6 Hospital dissection scene, Amsterdam. From Stephen Blancard, *Anatomia practica rationalis. Sive rariorum cadaverum morbis denatorum anatomica inspectio*, 1688, title-page. CUL. 112

2.7 William Hunter's proposed museum for the improvement of anatomy, surgery and physic in London, 1763. Printed as a supplement to his *Two introductory lectures*, 1784, opposite p. 114. CUL. 112

2.8 Winslow's anatomical theatre, Paris, 1744, built by the Paris Faculty of Medicine. From William Brockbank, 'Old anatomical theatres' (1968), figure 4, without further reference. 116

2.9 Grilled balcony, Barcelona surgeons' theatre, eighteenth century. Author. 143

Chapter 3: Experimental anatomy and its sub-disciplines

3.1 Anatomy and physiology represented by Muses and putti. From Albrecht von Haller, *Elementa physiologiae corporis humanae*, 1757–66, vol. 2, frontispiece. Whipple. 163

3.2 George Stubbs, illustrations of female pelvis and shape and size of head of fetus. Table 1 from John Burton, *An essay towards a complete new system of midwif*[*e*]*ry*, 1751. Wellcome. 174

3.3 Fetus attached to umbilical cord and placenta. Tab. V from Bernhard Siegfried Albinus, *Tabulae VII uteri mulieris gravidae cum iam parturieret mortuae* …, 1748–51. Wellcome. 175

3.4 Table 10 from William Smellie, *A sett of anatomical tables*, 1754. Wellcome. 178

3.5 Table 17 from Smellie, ibid. Wellcome. 180

3.6 'The child in the womb, in its natural situation', Table VI from William Hunter, *The anatomy of the human gravid uterus*, 1774. Wellcome. 183
3.7 'The external surface of the gravid uterus with its blood vessels', Table 4 from Johann Georg Roederer, *Icones uteri humani observationibus illustrati*, 1759. Wellcome. 185
3.8 Malformed female pelvis. From Roederer, *Élémens de l'art des accouchements*, 1753 (French edn, 1765). CUL. 185
3.9 The obstetric hospital at Göttingen University, 1785. Author. 186
3.10 Frontispiece, Théophile Bonet, *Sepulchretum sive anatomica practica*, 1679. CUL. 191
3.11 Spina bifida, presented by Ignatius Gild at Vienna University, 1754, as reprinted in Haller, ed., *Disputationes ad morborum historiam et curationem facientes*, vol. 7, part 2, p. 500, 1760. CUL. 195
3.12 Hospital of St Mary of Life and Bologna University buildings. Author. 210
3.13 View from rear of the bridge between Hospital of St Mary of Life (right) and Bologna University (left) and its first-floor anatomical theatre. Author. 210
3.14 Morgagni: frontispiece of *Seats and causes of diseases*, 1760 (1769 English edn). Author. 211
3.15 'The most important diseased appearances which are discoverable in apoplexy', from Matthew Baillie, *A series of engravings … intended to illustrate the morbid anatomy of some of the most important parts of the human body*, 1799, tenth fasciclus, Plate VIII, figure 1. A is the flattened surface of the right hemisphere of the cerebrum, and B is a large mass of extravasated and coagulated blood. CUL. 221
3.16 'A portion of the lungs studded with tubercules [internal surface]', from Baillie, ibid., second fasciculus, Plate IV, figure 2. CUL. 221

Chapter 4: Human bodies

4.1 A new jar for wet preparations by Claude Nicholas Le Cat. From *Philosophical transactions of the Royal Society* 46 (1749–50), pp. 6–8. CUL. 235
4.2 De Graaf's new syringe. From R. De Graaf, *Opera omnia*, 1678, p. 387. CUL. 237
4.3 A Ruyschian preserved and injected scene. *Opera*, vol. 2. CUL. 240
4.4 Figure 31 from Anton Nuck's *Adenographia curiosa*, 1691, which 'represents the lymphatics of the kidneys, from a canine subject', injected with mercury. CUL. 241

4.5 Sheldon mercury injecting instrument, made of glass, with a steel collar and injecting pipe. Plate VI, figure 1, from John Sheldon, *History of the absorbent system*, 1784. CUL. Sheldon's mercury injecting table, from Sheldon, ibid. CUL. 243
4.6 Mascagni's mercury injecting instrument. From his *Vasorum lymphaticorum corporis humani historia*, 1787. CUL. 245
4.7 Pole's mercury injecting instruments. From Thomas Pole, *The anatomical instructor*, 1790. CUL. 245
4.8 How to inject the arteries of a human hand with mercury, using Pole's mercury tray. From Pole, ibid. CUL. 245
4.9 Self portraits in wax (before 1755) of and by Giovanni Manzolini and Anna Morandi, Anatomical Museum, Bologna, Italy. Author. 249
4.10 The fourth order of the muscles (front view, with rhinoceros). Albinus, *Tabulae sceleti et musculorum corporis humani*, 1747, Tab. IV. Wellcome. 254
4.11 The fourth order of the muscles (rear view, with rhinoceros). Albinus, ibid., Tab. VIII. Wellcome. 255
4.12 Albinus's tripod. Modern artistic reconstruction by David Lindsay Lee © the artist. 258
4.13 Albinus's viewing frame. Modern artistic reconstruction by David Lindsay Lee © the artist. 259
4.14 Standing ecorché, right arm raised, detailing muscles. From Albinus, *Explicatio tabularum anatomicorum B. Eustachii*, 1744. 260
4.15 Double table from Albinus, ibid. 261
4.16 'The sceleton of a cat frighted at the sceleton of a dog', from Cheselden, *Osteographia, or the anatomy of the bones*, 1733, ch. 2. Wellcome. 263
4.17 'A pocky sceleton in which there is scarce one bone sound', from Cheselden, ibid., ch. 7. Wellcome 263
4.18 Cheselden's camera obscura. From title-page of *The anatomy of the human body*, 6th edn, 1741. CUL. 264
4.19 'A woman's skull who died of the venereal disease', from Cheselden, *Osteographia, or the anatomy of the bones*, 1733, plate 41, demonstrating how the camera obscura makes difficult images clear. Wellcome. 264
4.20 Laocoön, wrestling with the serpents, in the Vatican, Rome. Author. 268
4.21 Laocoön for artists (without serpents or sons), from Bernadino Genga, *Anatomia per uso et intelligenza del disegno*, 1691. CUL. 268
4.22 Camper's facial line. From *The works of the late Professor Camper*, 1821. Whipple. 269
4.23 'Écorché study after Smugglerius'. Sketch by William Linnell, 1840. Fitzwilliam Museum, Cambridge, Ref PD.32-1990. 274

4.24 William Hunter's dissecting room at Great Windmill Street. Cartoon by Rowlandson. Courtesy of the Royal College of Surgeons of England. 280
4.25 Glands of the skin, from G. Bidloo, *Anatomia humani corporis*, 1685. CUL. 285

Chapter 5: Animal bodies

5.1 The Académie royale investigating plants. From C. J. B. Dodart, *Mémoires pour servir à l'histoire naturelle des plantes*, 1676, frontispiece. CUL. 298
5.2 The Académie royale being visited by the king, Louis XIV. From Claude Perrault, *Mémoires pour servir à l'histoire naturelle des animaux*, 1671, 1676. CUL. 298
5.3 The Académie royale investigating animals. From Perrault, ibid. CUL. 299
5.4 Muscles of the orang-outan. From Edward Tyson, *Orang-outang*, 1699, Figure 3. CUL. 301
5.5 Lion. From Perrault, *Mémoires pour servir à l'histoire naturelle des animaux*. CUL. 302
5.6 The Versailles menagerie, late seventeenth century. From Gustave Loisel, *Histoire des ménageries de l'Antiquité à nos jours*, 1912, vol. 2, plate 9. BL. 303
5.7 Skeleton of the friendly elephant in the Versailles menagerie, now in the Muséum d'histoire naturelle, Paris. Author. 306
5.8 Elephant skeleton. From P. Blair, 'Osteographia elephantina: or, A full and exact description of all the bones of an elephant, which dy'd near Dundee, April the 27th, 1706, with their several dimensions', *Philosophical transactions of the Royal Society* 27 (1710), 51–168, Table 2. CUL. 307
5.9 Pietro Longhi, *Rhinoceros at Venice*. Courtesy of The National Gallery of London Picture Gallery. 309
5.10 A comparison of the brains of different animals (human, calf, lamb, pig, spaniel bitch, doe). From Samuel Collins, *A systeme of anatomy*, 1685, Table L. CUL. 315
5.11 Title-page, Gerhard Blasius, *Zootomia, seu anatome hominis brutorumque animalium*, 1676. CUL. 316
5.12 The stomach and guts of a sheep, from Nehemiah Grew, *The comparative anatomy of stomachs and guts begun*, 1681. CUL. 323
5.13 Dissection of a skate by Alexander Monro, *Secundus*. From his *The structure and physiology of fishes*, 1785, p. 92. CUL. 326
5.14 Animal dissection and microscope investigation. From Georges Buffon, *Histoire naturelle*, vol. 1, 1749. CUL. 329

5.15 Colon and external female sexual organs of the lion. From Buffon, ibid., vol. 9, plate V. CUL. 331
5.16 Skulls of hog, deer and horse compared. From Busick Harwood, *A system of comparative anatomy and physiology*, vol. 1, Cambridge, 1796, Table XIV, pp. 70–71. 'In this table the bony structure of the olfactory organs in the herbivorous Quadrupeds, is explained', showing a gradation of structure from the hog to the horse to the deer. CUL. 339
5.17 Horse seen from the side, with the skin etc. removed to show the muscles. From Carlo Ruini, *Anatomia del cavallo, infermita, et suoi rimedii* (Venice, 1598 edn; first publ. Bologna 1598), book 5, Table V, p. 243. CUL. 343
5.18 Table II (1) from George Stubbs, *The anatomy of the horse* (London, 1766), 'The second anatomical table of the skeleton of a horse explained'. CUL. 350
5.19 Table II (2) from Stubbs, ibid., 'The second anatomical table of the skeleton of a horse explained'. CUL. 351
5.20 Table IV (1) from Stubbs, ibid., 'The fourth anatomical table of the muscles, fascias, ligaments, nerves, arteries, veins, glands, and cartilages of a horse explained'. CUL. 352
5.21 Table IV (2) from Stubbs, ibid., 'The fourth anatomical table of the muscles, fascias, ligaments, nerves, arteries, veins, glands, and cartilages of a horse explained'. CUL. 353

Chapter 6: The end of old anatomy

6.1 Page 1 of Sugita Genpaku (and others), *Kaitai shinsho*, Tokyo, 1774, a translation of 1734 Amsterdam Dutch edn of J. A. Kulmus, *Anatomischen Tabellen* (first publ. Danzig, 1722). The title-page is a copy of Juan Valverde de Amusco's *Vivae imagines partium corporis humani*, Antwerp, 1572. Courtesy of the National Library of Medicine, Bethesda, USA. 365
6.2 Page 9 of Genpaku (and others), ibid., trans. of Kulmus, ibid. Courtesy of the National Library of Medicine, Bethesda, USA. 366
6.3 Kassel anatomical theatre, 1776. From F. Sorge, 'Die Kasseler Anatomie von 1779', *Sudhoffs Archiv* 22 (1929), Tafel II. Wellcome. 386
6.4 Bologna anatomical theatre today. Author. 388
6.5 'Russian anatomical institute'. Postcard. Author. 389

Acknowledgements

This book was researched and written while I was the holder of a Wellcome Trust Senior Research Fellowship in the History of Medicine in the Department of the History and Philosophy of Science at Cambridge University. I am deeply grateful to the Wellcome Trustees and their history of medicine advisers for their long-term support of my research. I also thank the shade of Sir Henry Wellcome himself, whose name I was for many years honoured to bear in my professional title.

I was always going to dedicate this book to the memory of Clara (Chapter 5) and of all the millions of animals sacrificed by the anatomists in the long eighteenth century, dissected and vivisected with the best of intentions but with sensibilities different to our own. The recent untimely death of my head of department, the much-loved and greatly missed philosopher of science Professor Peter Lipton, leads me to add him to the dedication. I am much indebted to him for his professional and personal support through difficult times. A head of department who actually nurtured, supported and valued his staff: who could have expected that? I am sure that, with his wonderful sense of humour, he would have said something witty about sharing this dedication with a rhinoceros.

For my research I have had the inestimable privilege of working regularly in three of the world's greatest libraries: Cambridge University Library, the British Library and the Wellcome Library for the History and Understanding of Medicine. There is very little in the way of printed texts I have sought that could not be found in one or other of these fabulous institutions. In addition I have had the facilities of the Whipple Library of the Department of the History and Philosophy of Science to hand, which has recently become the home of many historic volumes from the library of the Anatomy Department of the University. If you keep a scientific library up to date and never throw anything away then it gradually and spontaneously turns into a perfect historical library, which is what had happened to the Anatomy library. I offer my thanks to all these institutions and to their staffs, particularly the librarians of the rare books collections who have fetched so many books for me so promptly and courteously. In the rare books room of Cambridge University Library, where I have sometimes worked for weeks at a time, Helen Hills and Claire Ann Welford in particular have been consistently helpful, patient, professional and cheerful with my many requests. Marianne Smith, the librarian of the Royal College of Surgeons of Edinburgh, and the librarian of the Royal College of Surgeons of England, together with the Senior Curator of the latter, Simon Chaplin, have been very helpful in letting me consult the records of those institutions. I was also allowed to consult the Stubbs manuscripts at the Paul Getty Museum of British Art in New Haven in the United States, and the archives of the

Royal Society in London. I thank the trustees of these institutions for permission to quote from their records.

I have made visits to sites of important past anatomical activity in London, Edinburgh, Leiden, Amsterdam, Rome, Bologna, Padua, Paris and Göttingen. In the course of these I have received generous help from a number of individuals. In Rome Dr Maria Conforti of the Biblioteca di Storia della Medicina, Università di Roma 'La Sapienza', gave me valuable guidance, and Father Pasche Sebastiano of the church of Saints Vincent and Anastasius kindly showed me the pots containing the papal praecordia, while Dr Silvia De Rienzi facilitated my visit. In Göttingen the modern friends of Albrecht von Haller were most helpful: Elisabeth Eck, Marita Nielsen, Wolfgang Böker, and most especially Dr Reimer Eck, who provided me with copies of some rare Haller material. I also thank Dr Rina Knoeff for supplying material on Amsterdam and letting me use her translation of the dispute between Bidloo and Ruysch, and Ralph de Rijke of the Waag Society, Amsterdam. Drs Ole Grell and Silvia De Renzi assisted me in the translations from Italian, and Drs Harmke Kamminga and Rina Knoeff with those from Dutch. Dr Sachiko Kusukawa made a most valuable suggestion that helped me set up Chapter 6. I also want to give special thanks to my friend David Lindsay Lee, who, in a curious recapitulation of the relationship (discussed in this book, Chapter 4) between Albinus the anatomist and Wandelaar the artist, was able to draw for me scenes that I could only envision – of Albinus and Wandelaar at work together (and with acknowledgements to Dr Punt).

Writing a book is always a lonely occupation and every author is in need of open ears and words of encouragement. In particular he or she needs someone in whose encouraging presence they can stumblingly express (or even find out) just what it is they are trying to say. It has to be done out loud; talking to yourself just doesn't do the trick. Dr Yoko Mitsui has in this respect yet again been a great source of strength and support – and counter-arguments, as befits a philosopher – day after day, for which I am once more most grateful. I also thank Dr Brian Glasberg, my companion of the wheel, for listening to and questioning me cheerfully on so much of the work while it was in progress as we rode out together, even remembering what I had said months earlier. Tamara Hug, Administrator of our Department, has been exceptionally supportive of me and the project, for which I thank her. I also thank Steve Kruse and David Thompson, who scanned some of the images for me. Finally it is a pleasure to thank Felicity Teague for her skilful copy-editing of the text, and for her sympathy with Clara.

For the title of my book I am indebted to Sidrach Simpson, an English independent minister who was made master of Pembroke Hall in Cambridge. Simpson published his ill-printed twelve-page pamphlet *The Anatomist Anatomis'd* in 1644. In the *Dictionary of National Biography* Simpson is said to have been 'an extreme advocate for liberty of conscience'. Simpson used the title of his book metaphorically, to protest against a critic who had written *An Anatomy of Independencie*. 'As anatomists dissect one part one day, and then another', Simpson wrote, so he had begun his own metaphorical dissection of his critic 'with the head and tongue'. As will be seen, while I have also used the title metaphorically, I have also taken it more literally. I marginally preferred stealing this title than another attractive candidate, a pamphlet by Jonathan Swift, *The anatomist dissected* of 1727, whose title continues: *or the man-midwife finely brought to bed. Being an examination of the conduct of Mr. St. André touching the late pretended rabbitbearer; as it appears from his own narrative*, which was published in the course of an extraordinary controversy over whether a woman had really given birth to rabbits! Expert opinion was required: it was certainly a case for the anatomists.

Preface

In my previous book on the history of anatomy, *The Anatomical Renaissance* (1997), I argued that the practice of anatomy had not been monolithic: all the anatomists from Aristotle to the Renaissance had not been engaged in a single cumulative investigative enterprise, but there were several different enterprises of anatomical investigation carried out by different people at different times and undertaken for particular reasons. While it would indeed be possible to make the case that there was a comparable set of varied anatomical enterprises in the long eighteenth century,[1] my goal in the present book is somewhat different.

For on this occasion I wanted to explore what it was that anatomists had in common, what it was that united them in a single *discipline* throughout this period. The book could have aptly been called *On the anatomical business*, echoing the title given to their own publications by a couple of past anatomists: *De re anatomica*. Working within the discipline was not of course all harmony and concord, and some of the areas that anatomists argued violently over are also dealt with here, especially in the section on 'Warm controversy' in Chapter 4, and the reader will find other significant topics where the particular investigative project of one investigator was at odds with that of another, and indeed sometimes largely incomprehensible to him, especially across the fault-line that was the boundary between anatomy and physiology. But while there may be conflict within a discipline, the discipline is nevertheless still the discipline.

The main reason I wanted to portray anatomy on this occasion in its identity as a *single* (if not totally unified) discipline, is because anatomy here plays the role of a 'before' picture in a 'before and after' scenario. So here I am taking anatomy as a specimen discipline of the study of nature in the long and enlightened eighteenth century, and looking at both the general and the detailed nature of its practice, in order the better to be able to contrast its practice with its successor disciplines for investigating nature in the nineteenth century. In my view, which has been developing over many years,[2] *all* the disciplines for investigating nature were transformed in the same seismic series of events at the end of the eighteenth century that brought so many other elements of the modern world into being. Some indications as to the nature and importance of this change, and its effects on anatomy, are given in the final chapter of the present book. It is because the present book serves this larger role that I have decided to present the state of anatomy over about a hundred and fifty years in a somewhat static way, rather than tease out all

[1] An excellent preliminary example of how this might be done is Knoeff, 'Moral lessons of perfection', 2007.

[2] See for instance my 'Getting the game right', 1988.

the detectable changes and developments in anatomising over the period, as other historians might have chosen to have done. It is also why I have chosen to look at anatomy as a rounded discipline, with its sub-disciplines, in a way that no historian has hitherto attempted.

With the typical vanity of an academic author I have always imagined that one day this book might actually be read by students. So in order to encourage them to relate to these anatomists as live and passionate people and not as mere ciphers or disembodied names attached to obscure texts or equally obscure parts of the body, I decided that, in order to remove any barriers of strangeness or Latin antiqueness from them, my anatomists would as much as possible all speak English and publish their works with English titles (though for the most part they would still have to work in institutions with foreign names). The price of this decision is that some works long known to scholars by abbreviations of their Latin titles have come out in a new way: the familiar *Fabrica* of Vesalius, for instance, has turned into *Seven books on the fabric of the human body* and the *De motu cordis* of Harvey into *An anatomical exercise on the motion of the heart and blood in animals*. But this might actually also help both scholar and student to read aright the titles that anatomists actually chose to give their works, which were always meant to announce the content and nature of the text within. Full original titles can still be found in the Bibliography.

As for the anatomical images I have used, as printed here in small scale and black-and-white they can only be mere hints of the quality and size of the originals. I urge anyone interested to go and seek out the physical originals in a specialised library – not on the internet – in order to appreciate their sheer size and the artistic and technical skill they embody. As Frederic Ruysch might have said (Chapter 4 below), COME! AND SEE!

The image used on the cover, a merger by Valsalva of two images from the originals in Vesalius's *Seven books on the fabric of the human body*, is my favourite image in the whole of the history of anatomy, as it is both deliberately amusing and also expressive of the fundamentally self-reflexive nature of human anatomising: the anatomist is anatomis'd!

Finally, all I want to say in favour of this book as it goes into the cold world, with all its many faults of omission and commission, of conception and execution, many of which I am aware of, others of which I shall doubtless learn about from my critics, is that it is certainly something like the one that I wished had existed when I started my own studies years ago.

Dedicated to the memory of Clara and of Peter

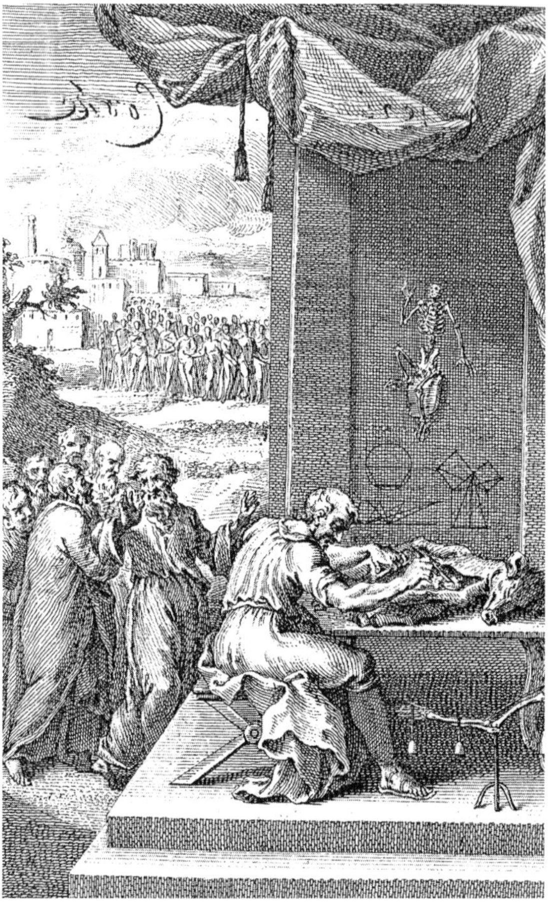

Hippocrates relates at large … how, coming to visit him one day, he found Democritus in his garden at Abdera, in the suburbs, under a shady bower, with a book on his knees, busie at his study, sometime writing, sometime walking. The subject of his book was melancholy and madness: about him lay the carcasses of many several beasts, newly by him cut up and anatomized; not that he did contemn God's creatures, as he told Hippocrates, but to find out the seat of this *atra bilis*, or melancholy, whence it proceeds, and how it was engendred in mens bodies, to the intent he might better cure it in himself …'.

Robert Burton, *The anatomy of melancholy* (first published 1621), 'Democritus junior to the reader'. The full letter can be found in Hippocrates, *Pseudepigraphic Writings*, 1990, pp. 73–9.

Introduction
The Anatomist Anatomis'd

An anatomist anatomised

Come with me. It's December 1694 and we're in Rome. We must push open this heavy door. It's dark and cold inside. We find ourselves in a small Baroque church. The interior is dimly lit by the candles in the three little chapels on each side of the nave. As we approach the altar we know that on the right, behind a screen and beyond the tiny baptistery, there are some shelves where the praecordia – the hearts – of ten popes are preserved, each in its own little pot (Fig. 0.1). For this is the parish church of Quirinale Palace, the summer residence of the popes. If a pope dies in the palace then for a hundred years the tradition has been that his heart is removed prior to the embalming of his body, and is brought here to the Church of Saints Vincent and Anastasius to be retained and venerated.[1]

But as we approach the altar we turn not right but left, and pass through a door to the sacristy. Here the gloom is lifted for there are many more candles that illuminate a strange scene. Half a dozen men are standing around a slab, and on the slab lies the corpse of a man (Fig. 0.2). Two days ago the great Bolognese anatomist Marcello Malpighi, first physician to Pope Innocent XII, died in the Quirinale Palace. His corpse has just been brought from the Palace to this church. It is now lying on the slab.[2] The pupils and followers of Malpighi are gathered round the corpse. They have come to anatomise him.[3] The anatomist is about to be anatomised.

Giovanni Maria Lancisi, the anatomical expert on aneurysms, is there. So is Georgio Baglivi, anatomist of the solids of the body. It's likely too that Antonio Maria Valsalva, explorer of the anatomy of the ear, is in attendance. When their turn comes, in due time, most of these anatomists will also be anatomised by their friends and pupils. The dissection of Lancisi will be performed in the Quirinale in 1720 by his disciple Antonio Volpi, while Valsalva in turn will be dissected by

---

[1] The custom was followed from the death of Sixtus V (1590) to the death of Leo XIII (1903), according to Paravicini-Bagliani, *The pope's body*, 2000, p. 143, and Bradford, *Heart burial*, 1933.

[2] That Malpighi was dissected in the church is attested by both our sources. That he was dissected in the sacristy seems most likely, given the layout of the church.

[3] The dissector was either Lancisi or Alessio Spella; see Adelmann, ed., *Marcello Malpighi and the evolution of embryology*, 5 vols, 1966, vol. 1, p. 659.

Fig. 0.1

Fig. 0.2

his friend and disciple Giovanni Battista Morgagni in Bologna in 1723.[4] Clearly, being an anatomist at the turn of the eighteenth century, at least in Malpighi's circle, demands a strong commitment to one's discipline, a commitment that does not end with death.[5]

In his lifetime Malpighi had, as a contemporary expressed it, 'employed all his time to discover New Worlds by Anatomies', especially by the use of the microscope. Always busy at his researches, Malpighi had made and published completely original discoveries about the fine structure of the brain, tongue, lungs, spleen, liver, kidneys and other internal parts, and he was also one of the first two people to investigate the anatomy of plants. Everyone would have agreed that he

4 For the last illness and post-mortem of Lancisi, see Crescimbeni, *Vita di Monsignor Gio. Maria Lancisi*, 1721, pp. 111–19. Baglivi seems to have escaped this fate: at least there is no mention of it in the 'Life' of him by Angelo Fabronio published in Latin in Kühn, ed., *Georgii Baglivi Opera omnia medica-practica et anatomica*, 2 vols, 1827. Lorenzo Bellini, Malpighi's other great disciple, was in Florence at the time of the present dissection.

5 There are other instances elsewhere of anatomists being anatomised by their pupils: John Hunter, for instance, was dissected by Edward Jenner.

was the greatest anatomist of his age, 'the miracle of Italy', as a contemporary anatomist called him (Fig. 0.3).[6]

Malpighi was 'deservedly celebrated for his great Skill and singular Sagacity in anatomical Researches', covering insects, vegetables, and perfect animals. This account by an eighteenth-century anatomist–historian gives some impression of the extent of Malpighi's many and astonishing discoveries with his microscope (his 'Glasses'), an achievement rivalled by no other anatomist in its range, detail and accuracy:

> Amongst other Discoveries, he found, by his Microscopes, that the cortical Part of the Brain consists of an innumerable Company of very small Glandules, which are all supplied with Blood by the capillary Arteries; and that the animal Spirit, which is separated from the Mass of the Blood in these Glandules, is carried from them into the Medulla Oblongata, thro' little Pipes, one of which belongs to every Gland, whose other End is inserted into the Medulla Oblongata: And that these numberless Pipes, which, in the Brain of some Fishes, look like the teeth of a small Ivory Comb, are properly that which all Anatomists, after Piccolhominus, have called Corpus Callosum, or the medullary Part of the Brain.
>
> Before his Time, the Texture of the Tongue was but guessed at, which occasioned great Disputes concerning the Nature of its Substance; some thinking it to be glandulous, some muscular, and others of a peculiar Nature, not to be matched by any other Part of the Body.
>
> This, therefore, Malpighius examined with his Glasses, and discovered, that it was cloathed with a double Membrane; that, in the inner Membrane, there are abundance of small Papillae, which have Extremities of Nerves inserted into them, by which the Tongue discerns Taste; and that, under this Membrane, it is of a muscular Nature, consisting of numberless Heaps of Fibres, which run, all manner of Ways, over one another, like a Mat.
>
> The Lungs, as most of the other Viscera, were believed to be of a parenchymous Substance, till Malpighius found, by his Glasses, that they consist of innumerable small Bladders, that open into each other, as far as the outermost, which are covered by the outer Membrane that incloses the whole Body of the Lungs; and that the small Branches of the Trachea Aspera are all inserted into these Bladders, about every one of which the Veins and Arteries are entwined, in an almost inconceivable Number of Nets and Mazes, that the inspired Air may press upon, or mix with, the Mass of Blood, in such small Parcels as the Ancients had no Notion of.

---

[6] Schullian, ed., *The Baglivi correspondence*, 1974, p. 56, letter from Guichard Joseph Duverney, 1692, 'miraculum est Italiae'. On Malpighi, see Adelmann, *The correspondence of Marcello Malpighi*, 5 vols, 1975. See also Brown, *The mechanical philosophy and the 'Animal œconomy'*, 1981, chapter 2; Wilkie, *Marcello Malpighi, 1628–1694*, 1969; and the essays in Berteloni Meli, ed., *Marcello Malpighi*, 1998.

Fig. 0.3

Till Malpighius discovered the Texture of the Liver by his Glasses, its Nature was very obscure: But he has found out, that the Substance of the Liver is formed of innumerable Lobules, which are very often of a cubical Figure, and consist of several little Glands, like the Stones of Raisins, so that they look like Bunches of Grapes, and are each of them cloathed with a distinct Membrane; that the whole Bulk of the Liver consists of these Grape-stone-like Glands. From whence it is plain, that the Liver is a glandulous Body, with its proper excretory Vessels, which convey away the Gall that lay before in the Mass of the Blood.

> He also discovered, that the Substance of the Spleen, deducting the numerous Blood-Vessels and Nerves, as also the Fibres that arise from its second Membrane, and which support the other Parts, is composed of innumerable little Cells, like Honey-Combs, in which there are vast Numbers of small Glandules that resemble Bunches of Grapes, and which there hang upon the Fibres, and are fed by Twigs of Arteries and Nerves, and send forth the Blood, there purged, into the Ramus Splenicus, which carries it into the Liver; to what Purpose, is not yet certainly discovered.
>
> The Mechanism of the Reins [kidneys] was wholly unknown till Malpighius found it out. He, by his Glasses, discovered, that the Kidneys are not one uniform Substance, but consist of several small Globules, which are all, like so many several Kidneys, bound about with one common Membrane; that every Globule has small Twigs from the emulgent Arteries that carry Blood to it; Glands, in which the Urine is strained from it; Veins, by which the purified Blood is carried off to the emulgent Veins, thence to go into the Cava; a Pipe to convey the Urine into the great Bason in the Middle of the Kidney; and a Nipple, towards which several of those small Pipes tend, and thro' which the Urine ouzes out of them into the Bason.
>
> This clear Account of the Structure of the Reins, has effectually confuted several Notions, that Men had entertained, of some secondary Uses of those Parts; since hereby it appears, that every Part of the Kidneys is immediately and wholly subservient to that single Work of freeing the Blood from its superfluous Serum and Salt.[7]

This is an astonishing roll-call of achievement on the part of one man. Everywhere in the body that Malpighi had looked, he had found detailed structure hitherto unsuspected, structure that often revealed just how the different organs of the body carry out their roles.

And now after such a lifetime of achievement, having lived by the anatomical sword, Malpighi was also prepared to die by the sword (as it were). For he had long expected his pupils and followers to dissect him after death. However, Malpighi knew 'that some, who seemed dead on a sudden have revived some hours after'.[8] It was of course always embarrassing for everyone concerned if – as occasionally actually happened – the corpse that an anatomist was dissecting should come back to life. There had been a famous case in Oxford in 1650, when Ann Greene came back to life on the slab after being hanged for murdering her new-born baby. So Malpighi had insisted that he was not to be dissected until thirty hours after his death, in case he should spontaneously revive. But, said his follower Lancisi, how vain are the hopes of men! For Malpighi was indeed completely dead, and showed no further signs of life. So the anatomy had to be started.

7 Jenty, *A course of anatomico-physiological lectures*, 1757, pp. cxxxii–cxxxv.

8 Lancisi, 'Extract of a letter', 1697, p. 469.

The point of this particular anatomy was not to demonstrate to students the inner parts of the human body and their location with respect to each other: that is to say, it was not a demonstrating or teaching event in the way that most public dissections were. The point on this occasion was simply to discover what Malpighi had died from. His doctor friends were going to establish whether the symptoms of illness Malpighi had shown during life were accompanied by some deleterious changes to any of the inner organs, and if so, what? By this means they hoped to find the cause of his death. So it was important to first establish the medical history of the now-defunct patient.

For most of his sixty-six years Malpighi had been in fairly good health, though sharp vomiting tormented him for twenty years, and he was troubled with the gravel (pain in passing urine), a haemorrhaging in the kidneys (pain in the kidney region and blood in the urine), and a 'fluxious rheumatism', to which was later added a cruel palpation of the heart. In the final four years of his life he suffered a sharp and biting sweat all summer.

Malpighi had moved to Rome in the last three years of his life, in order to be Papal physician. He lived in the Quirinale Palace. But once arrived there he immediately begun to pass bladder stones through the penis, though relatively painlessly. Then in the third year in Rome,

> … which was the last of his Life, he found himself oppressed, during the Winter with a difficulty of breathing, which troubled him most when he went up Stairs. His Health being thus insensibly undermined, and a bilious Looseness returning ever and anon, he was at length seized with a Vertigo, a loss of Speech and a Contortion of the Mouth (*spasmus cynicus*) and a Palsy of the right side. And though there was appearance that he was out of danger by Bleedings, Purges, Diuretics, and Antapoplectic Medicines, yet one might see, by his Melancholy Countenance, but especially his want of Memory, that there was lodged in his Brain some Melancholy Humour.[9]

Thus far the medical history of the deceased as recalled by Lancisi. These are the conditions to which some originating clue is going to be sought in the body when it is dissected.

And now the dissection must begin. The knife is raised. A vertical cut is made down the front of the body, from the sternum to the pubes, followed by a horizontal one either side of the navel, in a cross-like fashion (as with the corpse in Fig. 0.2), known as 'the crucial incision' (as indeed for the corpse it was). The skin is pulled back and the fat and the fleshy membrane below the skin are pulled off. Then the muscles covering the stomach are cut through, together with the peritoneum and

[9] Lancisi, 'Extract of a letter', 1697, p. 470. An alternate account of Malpighi's post-mortem was given by Baglivi; see Weber, *L'anatomia patologica di Lorenzo Bellini*, 1998, pp. 100–101.

then the fatty omentum. Now the contents of the belly begin to come into view: beginning with the liver, the intestines and the stomach.

The followers conduct the post-mortem in the conventional anatomical sequence – first the belly, then the chest, then the skull – looking for organs that are not in their normal, expected, condition. Here is 'what was found most remarkable' in Malpighi's body:

> The Abdomen being opened, we found the Ventricle [= stomach], with the Guts, the Sweet-bread [= pancreas], the Spleen and the Liver, most sound, both as to colour and bigness; only the Bladder of the Gall abounded with a black Gall. The left Kidney had nothing amiss; but the right was twice as little, and had its *Pelvis* thrice as big; which discovered the cause of the descent of the Stones. We found in the Bladder a little Stone, that seemed to have fallen into it [from the kidney] a few days before.
>
> When the Sternum was taken off, the Lungs appeared withered, with some mark of corruption on the backside. The Heart was bigger than ordinary, and the sides of the left Ventricle felt harder and thicker in some places than in others; yet there was no Polypus found in either of the Ventricles, though there was ground to suspect it.
>
> At last the Skull being cut asunder, the true cause of his death was discovered, for the right Ventricle of the Brain contained almost two Ounces of extravasated Blood [i.e. blood not contained within blood vessels], and the left Ventricle was swelled with a thick and yellow sort of Phlegm, which weighed more than an Ounce. Moreover the *Dura mater* stuck closer to the Skull than is usual.
>
> This proves that the conglobated Glands in the whole Body, had thrown into the Mass of Blood an Acid lymph, and that the conglomerated Glands of the Hypochondria, especially those of the Liver had thrown into it a melancholy Humor, and that these two sorts of Humors being carried into the Vessels of the Brain, had disposed the Blood to coagulate there, and that having there corroded and broken the Tunics which served for a stop to them, they had run into the Cavities, where they caused death without a Remedy.[10]

The 'melancholy humour' suspected in the brain has been found. It is 'the true cause of his death'. Or as Baglivi put it: 'the which extravasation of blood was the cause of the apoplexy and of his death'. After the dissection the body of Malpighi was embalmed, like those of the popes, and the funeral rites were performed. A year later the body was taken in a coffin to Bologna where it was put in a tomb in the Church of San Gregorio. The anatomist had been – literally – anatomised.

Few people today, standing on the church steps trying to get the best view of the Trevi fountain in Rome, would suspect that this particular parish church had such an anatomically interesting history (Fig. 0.4).

---

[10] Lancisi, 'Extract of a letter', 1697, pp. 469–70. The distinction between the conglobated and conglomerated glands had first been made by Malpighi.

Fig. 0.4

Discipline history

In the present book I am trying to perform another act of anatomising, but this time of a discipline, rather than a person: I want to anatomise the practice of anatomists over a period of about a hundred and fifty years, when the discipline was flourishing as never before or since. This meant that it was part of the great intellectual movement known as the 'Enlightenment', in which (supposedly) rationality and facts were to be the guides of human knowledge and conduct, rather than superstition and religion.

So this is a book about a discipline, something rarely attempted. The practice and beliefs, the rules and tacit knowledge of the anatomists are what constitute the core of the discipline of anatomy (just as they do in any other discipline). The practice of a particular discipline involves particular individuals with their natural endowments, their commitment to their subject, their learnt skills and their tacit knowledge, together with what some of my colleagues would refer to as their 'material culture', or what I would call their pertinent stuff and objects. In the present case such stuff included knives and syringes, tables and lenses, human corpses and live animals, waxes and four-colour printing. All these I try to cover.

Discipline history is unusual today. Books that announce in their titles that they are 'a history of astronomy' or of anatomy, are not usually discipline histories in the strict sense at all, but histories of famous men and their particular achievements, or of ideas and how they were developed. Yet all the interesting things about the knowledge of any subject happen within, and sometimes at the edges of, disciplines.

The discipline is what sets the terms for inquiry; it is the context that enables sense to be made of what is found; it is the context in which knowledge is taught. It was as practitioners of particular disciplines that past people made the innovations and discoveries that particularly interest us as historians of the sciences. Understanding the identity of a particular discipline in a particular period will therefore enable us to grasp something of what it was possible for people to think and do – and equally of what is was not possible for them to think or do – as practitioners of those disciplines. Moreover, without some understanding of what constituted a particular past discipline it is too easy for us to project back the concerns and boundaries of modern disciplines inappropriately, and thereby multiply our misunderstandings. An example of such disciplinary misunderstanding, to be explored below, is that between the old discipline of anatomy and the old discipline of physiology, where much modern writing about the history of physiology has simply got hold of the wrong past discipline!

The term 'discipline' comes from the Latin, *disciplina.* It has always had three main meanings. One is discipline as a branch of learning or knowledge, and is the one we shall be concerned with here, with its associated term disciple or *discipulus*, one who learns. The second main meaning is closely related: discipline as a rule or system of rules governing conduct or activity. Hence in the Catholic Church discipline (what one should do, how one should behave) is the counterpart of doctrine (what one should believe). In this sense, for many centuries novitiates entering into monasteries undertook to observe the discipline – the rules – of the religious order they were joining. Physical force might be found necessary as training or corrective, and this is the origin of the third meaning: discipline as punishment. All the meanings centre on the learning and following of rules or systems of rules concerning particular activities. In this sense, anatomy before 1800 was a discipline, and this is how I shall speak of it. Moreover, throughout the period in which I shall be studying it, it was also an academic discipline, taught in universities, though it could also be practised by those who had learnt it, without holding an academic post.

The organisation and preservation of intellectual knowledge, and its development through teaching and innovation, all take place in a disciplinary matrix. This is why disciplinary history is important. A disciplinary history will be the history of practice, of teaching, and of changes in ideas and practices over time (if such occurred). To point to the significance of disciplinary history is of course not to deny the value of cross-disciplinary history, nor to deny that individuals in the past could practise more than one discipline. But that does not mean that each discipline did not have its own persistent identity. The intellectual disciplines that deal with nature are in no privileged or exceptional position. As disciplines, they are no different from other disciplines: physics and theology, mathematics and logic, chemistry and philology – they all have equal status as disciplines, though the subject matter and the particular goals and methodology of any particular discipline are obviously unique to it. Etymologically, according to the *Oxford English Dictionary*,

> *discipline*, as pertaining to the disciple or scholar, is antithetical to *doctrine*, the property of the doctor or teacher; hence, in the history of the words, *doctrine* is more concerned with abstract theory, and *discipline* with practice or exercise.

To put it another way, a discipline is what disciples learn to do; it is what they learn to practise. So when I speak of anatomy as a discipline here I shall be particularly concerned with the fact that it is a *practice*, a putting-into-practice, whether that practice consist in investigation or reflection. But as the etymology of the word also demonstrates, disciplines are also practices that are taught and have to be *learned*, practices into which pupils need to be initiated; an isolated and uninitiated individual cannot be engaged in the practice of a particular discipline. Teachers teach their pupils the discipline. You have to learn it before you can do it. You have to learn it before you can change it.

It is on account of this essential teaching and initiation that the practice of any discipline requires, that we can say that every discipline certainly does have a *core identity*, comprising methods, aims, materials, procedures, topics, questions and results. The core identity of any discipline is readily recognised by its practitioners. We have to keep constantly in mind that the experts on the identity of any particular discipline are the practitioners of that discipline, not outsiders to the discipline (and that includes historians). So in order to grasp the identity of any particular discipline, whether in the past or the present, we need both to listen to its practitioners and to watch what they do in their role as practitioners of this particular discipline. This involves a certain kind of close reading of texts: listening to what their authors do say, and taking care that we do not inadvertently introduce into these writings things their authors did not say – or could not have said or meant.

Of course it would not be possible to give a complete checklist of all the features that constitute the identity of any discipline, nor would it be desirable. As we now know, much of what constitutes correct practice in a discipline is conveyed and absorbed tacitly from one generation to another, and never gets directly expressed or written down. But there is usually enough left from the statements of practitioners and from their actions for us to be able to broadly characterise the domain of any given past discipline and to be highly specific about some features of it, such as its daily practice. It is also possible for the historian to readily recognise some of the features that distinguished one discipline from another.

One of the most striking, if self-evident, things about any discipline is that its practice is shared. Practitioners may even live in different countries, but they will still be reading each others' publications. They may be in competition for discoveries and fame, and even at loggerheads over each others' claims, but they nevertheless constitute a *community* that is in contact. This community is constituted not only by the practitioners living at any one time, but also by the legacy of past practitioners, both the persons and their findings. Past persons and their findings can if necessary be brought back into the arena when a particular matter is being established, or is under dispute, or needs to be celebrated.

So what follows is not intended to be a comprehensive history either of anatomy, or of anatomists, or of anatomising in the 'long eighteenth century', that is, the period from around 1650 to around 1800. I am not concerned with particular anatomical discoveries, nor with evaluating one anatomist against another, nor with the different physiological theories that investigators came up with to explain the hidden workings of the body. Nor am I concerned with variations in ways of seeing and interpreting the body in different cultural and temporal contexts; the history of 'the body' is for other historians to pursue. And even though many of the anatomists I shall discuss were surgeons, and their pupils were preparing to become surgeons in their turn, this is nevertheless not a history of surgeons nor of surgery either. Neither is this a history of collecting and museums, nor of fascination with the monstrous and the curious. Anatomy and anatomists were indeed concerned with these in the eighteenth century, but both these areas have received plenty of attention from other historians in recent years. Finally, I do not explore very far the relations between anatomy and art, or between anatomists and artists; this has received plenty of attention from art historians in recent years.

With all these topics left aside it might sound as though there would be very little history of the discipline of anatomy left to write. But there is plenty. For my intention has been to provide something of a snapshot of the practice of anatomy as one of the major investigative disciplines of the period. Sometimes I am impressionistic in my presentation of certain issues; at others I aim more at completeness, offering extensive quotation and translation from the writings of anatomists and others. Some of the moments I deal with are ones already known; sometimes they are ones that have not been investigated before. It is because I am trying to give this sort of 'snapshot' that I am more concerned with the things about which anatomists agreed – that is, ideas about what the discipline was and how it should be practised – than the things where they differed. So the disputes and debates between anatomists in the long eighteenth century, which have greatly concerned other historians – topics such as who discovered what first, theories of physiology, theories of generation, issues of classification of disease, the great argument over vitalism versus mechanism – make only the most limited appearances here. On the other hand, in looking at the practice of one particular discipline across many of the states of eighteenth-century Europe, one can often see common developments; one can see particular things happening simultaneously across the continent; and one can not only see the discipline 'in the round' as it were, but also begin to recognise which changes and developments require local explanation by the historian, and which ones were on a larger scale.

We need to remember that anatomy was a highly visual subject area, and the images the anatomist deployed said as much as their texts did, so technical matters, particular preparations, and locations of practice are all well represented in images here – more so than in any other work on the history of anatomy, I believe. Martin Rudwick taught us some years ago, with respect to the history of geology, that suppressing or ignoring the visual material that these past investigators used limits our understanding of what they were up to.[11]

Within the history of medicine the history of anatomy has been rather privileged: we don't really have a long-period history of medical theory, or of patient care or even of drugs, but we do have a long-term history of anatomy. For most subjects in the history of medicine the years around 1800 represent a break. Medical theory or treatment before that date seems to have little in common with the more modern views that were created around and after that date. Only in the anatomical domain does it look as though the knowledge is continuous across this break – and indeed continuous from Antiquity to the present. I suspect that this is because anatomy seems to be the one part of old medicine that is about *autopsia*, touching and *seeing for one's self* – that is, it looks most like what we think modern science is like, whereas the rest of the history of medicine before about 1800 frankly looks like fantasy. Moreover, the central truths of 'gross anatomy' have not dated – that is, we take what Galen in the second century AD and Vesalius in the sixteenth century said about the existence, position and interconnections of bones, muscles, blood vessels, as still by and large true. By contrast, we have dispensed with almost all the knowledge and beliefs that constituted the rest of old medicine, both its theory and its practice. Thus, apart from anatomy, modern medicine has completely displaced all western medicine before 1800. So we historians have told a progressive history of anatomy – as the one good thing in old medicine that constantly made progress – presenting discovery upon discovery, improved artistic portrayal upon artistic portrayal. However, even so we have made mis-identifications, and much of what actually was in the history of anatomy we have placed in a (mythical) history of physiology, because it was experimental. So, although the history of anatomy is prominent in the history of medicine, it is still largely un-inspected. And we shall find in fact that for anatomy too there is a break around 1800: there is an *ancien régime* anatomy, just as there was an *ancien régime* political structure. This 'old-' or 'old regime' anatomy is what this book is about.

The attitude of the Catholic Church

One final introductory point remains to be made, which concerns the attitude of the Catholic Church. In the late nineteenth century secular propagandists for science, who had themselves usually been raised in the Protestant traditions of America and Britain, constructed a history of science that presented scientists as the good guys and the Church, especially the Catholic Church, as the bad guys. The most famous books presenting this case were John William Draper's *History of the conflict between religion and science*, published in 1874, and Andrew Dickson White's *A history of the warfare of science with theology in Christendom*,

[11] Rudwick, 'The emergence of a visual language for geological science 1760–1840', 1976.

published in 1896. Although this interpretation was put forward so long ago, the dichotomy it deploys (Christianity/science) still underlies the stories of the history of science that historians tell today, however nuanced our accounts may be about the relation of science to religion or the relation of the Church to the knowledge of nature.[12] According to this story the Catholic Church had supposedly forbidden all investigation into nature that was not directly based upon Biblical and Church teachings. One of the primary areas in which the Catholic Church had exercised this supposed censorship was anatomy.

Andrew Dickson White, first president of Cornell University, and father of this claim, has two sections of his book devoted to this theme. The first is titled 'Theological opposition to anatomical studies':

> One of the main objections developed in the Middle Ages against anatomical studies, was the maxim that 'the Church abhors the shedding of blood'. On this ground, in 1248, the Council of Le Mans forbade surgery to monks. Many other councils did the same, and at the end of the thirteenth century came the most serious blow of all; for then it was that Pope Boniface VIII … issued a decretal forbidding a practice which had come into use during the Crusades, namely, the separation of the flesh from the bones of the dead whose remains it was desired to carry back to their own country … it soon came to be considered as extending to all dissection … it impressed upon the mind of the Church that all dissection is sacrilege ….

Such evidence as is presented by White here in support of this claim is not actually about anatomy or anatomising at all, and the complete lack of evidence for the assertion that the Church considered dissection as sacrilege is quite breathtaking.

White's second section on this theme is 'The scientific struggle for anatomy'. 'More and more', White writes, 'in spite of theological dogmas, came a renewal of anatomical studies by dissection of the human subject … finally [in the sixteenth century] came a far greater champion of scientific truth, Andreas Vesalius, founder of the modern science of anatomy, etc.' Although White claims that Vesalius was attacked using 'weapons theologic', and that a conference of divines was also summoned 'to decide whether dissection of the human body is sacrilege', he presents no evidence or referencing for such claims. It would be difficult to do so, since they are figments of White's imagination. It must be remembered that this is a work of propaganda, and the truth rarely bothers propagandists.[13]

---

[12] The latest in a long line of works on the supposed problematic relation between science and religion is Lindberg and Numbers, eds, *When science and Christianity meet*, 2003.

[13] See White, *A history of the warfare of science with theology in Christendom*, 1896, vol. 2, pp. 31–2, 49–55. Draper does not discuss anatomy as such in his comparable piece of propaganda.

All of this anti-Catholic nonsense is nothing but a lie. It will be noted that our present story began in a church in the heart of Catholic Rome and immediately next door to a papal palace, a church that was being used as a temporary site of dissection. As a life-long evangelical atheist I certainly hold no brief for the Catholic Church. Nevertheless the fact is that the Catholic Church has never been opposed to the practice of anatomy, whether for post-mortem, demonstration, teaching or research purposes. Never, ever, anywhere. But unfortunately the lie has become so widespread that I find not only students but also even some of my fellow historians of science still accepting it. But there were no prohibitions on the grounds that anatomising was interfering in God's creation, nor on the grounds that the blood of humans was somehow sacred. There were no prohibitions at all. The Catholic Church was not in any way, shape or form opposed to the anatomising of dead humans or the anatomising or vivisecting of animals by anatomists. The two Catholic Church fathers who could be quoted as abhorring dissection in the early Christian centuries, Saint Augustine and Tertullian, were both actually opposing the *vivisection* of humans – which probably never took place anyway – not the anatomising of animals and humans.[14]

So not only were there no prohibitions on anatomising, but many anatomists, and especially of course in Rome and the papal states, were friends of popes, were patronised by popes, were even employed by popes as their personal physicians – precisely on the grounds of their reputations as anatomists. We saw this in the case of Malpighi, who was sought out to be his personal physician by Pope Innocent XII *because* he was the greatest living anatomist. Another highly visible instance of the positive attitude that the Catholic Church and its officials held towards anatomical investigation in our period can be seen in the case of the Academy of the Sciences of the Institute of Bologna (the *Accademia delle Scienze dell'Istituto di Bologna*), founded in 1714 with the support of Pope Clement XI, and then given the active personal support and encouragement of the Archbishop of Bologna, Prospero Lambertini, who later became pope himself as Benedict XIV (1740–58). Among the anatomists whose work was supported in this well-equipped research institute was Luigi Galvani, a devout Catholic as well as anatomist, who discovered 'animal electricity' there in the 1780s.

So although it may spoil our customary simplistic story of the development of anatomy and science, which holds that the Catholic Church was – had to be – opposed to the free investigation of nature, it nevertheless needs to be recognised that the opposite is true, and that the Catholic Church was actually virtually a proponent and patron of anatomising. The same is of course the case with all the Protestant churches too: although they may not have been as much in favour of such inquiry as the Catholic Church was, they also never issued any prohibitions against its practice, because they were not opposed to it either.[15]

---

14 See Cunningham, *The anatomical renaissance*, 1997, pp. 22–5 and 35, n.23.

15 It is easy, but mistaken, to read back the outrage of local clerics wherever Professor Von Hagens takes his 'Bodyworlds' exhibition these days. The professor is well aware that

But more than this. Not only were the Catholic and Protestant Christian churches not opposed to human dissection, but such dissection actually carried a message that positively favoured the Christian position and mission. We shall see what this message was in Chapter 1. Andrew White could not have been more wrong!

outrage is good for business and he deliberately provokes it.

Chapter 1
'This awful subject'

It was William Hunter who used this expression in one of his private lectures for students of anatomy in mid-eighteenth-century London. By 'subject' he meant the subject of anatomising, that is to say the human body, and by 'awful' he meant that this subject induces awe in its practitioners 'with a view towards the Creator'. Today we might say 'awesome' to convey something of the same sentiment, although that would not convey the divine sense of the expression as used here by Hunter. Hunter said:

> Who can know and consider the thousand evident proofs of the astonishing art of the Creator in forming and sustaining an animal body such as ours, without feeling the most pleasing enthusiasm? Can we seriously reflect on this awful subject, without being almost lost in admiration? … The man who is really an Anatomist, yet does not see and feel what I have endeavoured to express in words, whatever he may be in other respects, must certainly labour under a dead palsy in one part of his mind.[1]

1. The tradition of anatomy

For the devout philosopher and anatomist such as William Hunter, anatomy was a key to the understanding of the 'whole plan of the Creator',[2] and it had been so since it was first begun as a practice in ancient Greece by the philosopher Aristotle. The books of Aristotle on the parts, movement, generation and functioning of animals are witness to his wide experience and great knowledge of the anatomy of many creatures.

Continuing through Roman times, anatomy was widely pursued and many texts were written on the subject. The works of only one of these classical anatomists have survived in any quantity, those of Galen, who was a Greek physician working in Rome in the second century AD. It is primarily from the writings of Galen that we can see what a very high level of development human anatomical investigation had been taken to by this date, even though the dissection of human corpses was forbidden in the Rome of Galen's day. Galen derived his knowledge primarily from anatomising apes. However, as with almost all other areas of theoretical learning and practical skill, the end of the Roman Empire

---

1 William Hunter, *Two introductory lectures*, 1784, pp. 64–5.

2 Ibid., p. 64.

meant the end of practical anatomising, with the later schools at Byzantium and Alexandria using only diagrams and partial texts for their anatomical knowledge. In the universities of Europe some formal anatomising of the human body was presented from the early fourteenth century, especially at Padua and Bologna, using as textual resources very imperfect texts of Galen. Considerable scholastic attention was given to anatomy and texts were written on the subject by academics in the medical faculties.

As with almost all other areas of learning and skill, anatomising was revived and brought back into full practice in the Renaissance. The most celebrated moment in this revival of anatomy was when the dynamic young Flemish anatomist Andreas Vesal (Vesalius) produced his monumental illustrated volume on anatomy in 1543: *Seven books on the fabric of the human body*. Although his teachers and rivals claimed that Vesalius was being dismissive of Galen in this book, in fact Vesalius was reconstructing Galen's own anatomical project, but using the human body for the first time, where Galen had been restricted to apes and other animals in his own researches.[3] The main centres for anatomical research and teaching came to be Padua, Bologna and Rome in Italy, Paris and Montpellier in France, and Leiden in the Netherlands. Until the 1650s there was little if any anatomical activity to be discerned in the medical faculties of either Oxford or Cambridge, the only two universities in England. As far as England was concerned the first really active anatomist was also the greatest: William Harvey, who worked alone at his dissections and vivisections through the first part of the seventeenth century.

Harvey had studied medicine, including anatomy, at Padua, and when he returned to London in 1602 he continued practising anatomy as a passionate hobby. The gossip John Aubrey wrote that Harvey was 'the first that I hear of that was curious in Anatomy in England'.[4] Harvey's discovery of the circulation of the blood, published in 1628, was highly controversial at first, but was quickly seen as the greatest anatomical discovery since ancient times. Harvey tried to encourage his doctor colleagues in the College of Physicians of London to engage in anatomising and to publish their findings, even leaving money in a Deed for an annual feast and 'an exhortation to the Fellows and members of the College to search and study out the secret of Nature by way of Experiment' (p. 432). He donated a new building to the College in 1651–52 as a library and depository for rarities and simples (i.e. standard medical ingredients before mixing into compound medicines), but this was burnt down in the Great Fire of London. Whether or not Harvey's example was the spur, it is certainly the case that, by the mid-seventeenth century, English physicians and surgeons had become as active in anatomy as any of their colleagues in Europe.

---

3 Cunningham, *The anatomical renaissance*, 1997, chapter 4.

4 Keynes, *The life of William Harvey*, 1978, p. 432.

2. The identity of anatomy

In the long eighteenth century it was the anatomist who was at the forefront of research in medicine and in the whole world of life and living creatures (what we would now call the organic). He – for it was virtually always a man[5] – actively pursued investigation, often deploying experiment, into the structure and functions of both the dead human- and animal body and the live animal body; into the location of disease within the human body (the 'seats of diseases'), and the morbid changes; and into forms of cure insofar as they concerned the structure of the visible parts.

We shall have to keep in mind not only what the discipline of anatomy was, but also what it wasn't. It wasn't just autopsy, though, conversely, autopsy was part of anatomy. As we shall see in Chapter 3, anatomy wasn't physiology. Indeed, most physiologists were totally dependent on anatomy for their facts until around 1800. Nor was anatomy the mere geography of the body. Rather, it was a fundamentally experimental discipline. It was a systematic, methodical, active exploration of the human- and animal body, aimed at acquiring certain kinds of trustworthy knowledge. It was the master discipline for the investigation of life.

Today, however, the anatomist does not have such an exalted rank among the investigators of organic nature, and even within the teaching of medicine his role has become less important with the decline of the centrality of 'gross anatomy' and with the use of methods of learning human anatomy that do not involve the dissection of human or animal corpses. In some ways, indeed, the anatomist is today the poor relation of the molecular biologist or the experimental physiologist.

As is quite typical in writing the history of any of the sciences, the current status of the discipline has tended to be the lens through which its historical position has been assessed. Thus recent historians of anatomy have tended to assume that the shape and importance of the discipline today were ever thus. Hence, for instance, the definition of the discipline offered in the current *Encyclopaedia Britannica* describes it as

> ... a field in the biological sciences concerned with the identification and description of the body structures of living things. Gross anatomy involves the study of major body structures by dissection and observation and in its narrowest sense is concerned only with the human body. 'Gross anatomy' customarily refers to the study of those body structures large enough to be examined without the help of magnifying devices, while microscopic anatomy is concerned with the study of structural units small enough to be seen only with a light microscope.

With respect to 'gross anatomy' the writer says, 'This ancient discipline reached its culmination between 1500 and 1850, by which time its subject matter was firmly established'. With that sentence the writer has effectively defined the old discipline,

5 On women in anatomy, see Chapter 2 below.

at least as practised between 1500 and 1850, as having been concerned only with 'gross anatomy', together with the later addition of 'microscopic anatomy'. By implication, all the other interesting areas that (as we shall see) old anatomy actually did deal with are put to one side, waiting to be credited to the history of other sciences of life. We shall see that, in contrast to this characterisation, anatomy was in fact a dynamic, demanding discipline that set the agenda for the other disciplines and sub-disciplines in which questions about life and living creatures were investigated or discussed.

For a working definition of the discipline of anatomy at the beginning of our period by a contemporary anatomist, we can take the non-controversial and influential definition of the anatomy of man offered by the English physician and anatomist Francis Glisson, who was Professor of Medicine at Cambridge University. This definition also conforms to sixteenth-century practice and was still generally applicable in the late eighteenth century. Glisson drew this up to introduce a series of public anatomies he gave in 1641 at the College of Physicians of London on the liver, and which were published in 1654.

> *Anatome* and *Anatomia*: the words signify as much as a *dissection*. But being taken for an *art* and applied to a certain object, they signify an *artificial dissection* of that object in such manner as may most conduce to the perfect knowledge of the same and all its parts … Now this artificial dissection implies not the *manual* dissection only, but in especial manner the *mental*. For though the manual dissection be first in regard it leads us to the mental, yet the mental is that which mainly denominates the artist an *Anatomist*, and hath use in living as well as dead bodies; and nobody desires the manual dissection of dead bodies but in order to the living, that so by mental assignation of the parts he may give an account of them in the living as well as the dead …
>
> The end of artificial dissection is not to mangle and cut the object it takes in hand rudely into *shreds*, but to gain the *perfect knowledge* of the same and all its parts thereby. Which, that it may the better do, it observes:
>
> 1st. The general *structure* of the whole, with its general proportions, beauties, defects, its general *communities* and *differences* with other creatures, and especially with other individuals of its own kind.
>
> 2ly. It observes the *formation* and texture of every part, with the elements thereof; the *proportion* it hold to other parts and to the whole; the *communities* and *differences* it hath with the same part in other creatures, and with other individuals of the same kind, and other parts of the same creature; also, the several *sympathies* and *antipathies* thereof.
>
> 3ly. And lastly, it observes and discovers the several *faculties*, *actions* and *uses* of every part.
>
> Now, by diligent pursuance of all these scopes [= aims], the anatomist attains at the prime end proposed to his art, towit, the exact knowledge of his object … The most proper object of anatomy is the body of man with all its parts.

> Nevertheless the dissection of other animals in order to this is not excluded, in regard it may be many ways subservient to it …
>
> Now to sum up all that hath been said, the full definition of Anatomy will be: that it is an art which by artificial dissection, commensuration, ponderation, inflation, ligature, and other useful operations, observes and discovers the structures, proportions, perfections, imperfections, elements, communities, differences, sympathies, antipathies, faculties, actions and uses of the body of man and all its parts, as far as human understanding can reach them.[6]

Thus according to Glisson, the anatomist seeks 'perfect knowledge' or 'exact knowledge' of the human or animal body and all its parts; that is to say complete knowledge, as far as human understanding can reach. The anatomist proceeds by cutting up the whole into parts – which is what 'anatomy' means etymologically – and inspecting them from every possible viewpoint, including the use of inflation, ligatures, 'and other useful operations'.

Glisson's definition draws attention to the extent to which anatomising was an integral three-stage investigation into the parts of the body. The anatomist deals first with *structure*, investigated as widely as possible, including connections and other relations between the parts. He then deals with the *action* of the part being investigated. Finally, he makes deductions about the *use* or function of the part. All three of these stages of investigation, taken together, constituted the basis of the practice of anatomising. The investigation of structure is not separated in time or location from the investigation into the action or use of the parts. A proper anatomical account – 'perfect knowledge' – of any part of the body could not be offered until all three of these stages had been investigated by the anatomist. The argument for function is built on structure, not vice versa. The investigation of the use or function of the part is the end-point of such anatomising, not the starting-point. All these operations could usually be successfully investigated on the dead body.

Glisson's definition makes it clear that anatomy was an art, a manual practice. However, he also says that 'this artificial dissection implies not the *manual* dissection only, but in especial manner the *mental*', and it is this mental dissection 'which mainly denominates the artist an *Anatomist*'. So there are complementary roles for hand and mind. The hand-worker, the artist or artisan, is not an anatomist because of his *manual* skills (for butchers also cut up bodies systematically), but because of the *mental* activity that guides his hands. Glisson himself does not here go so far as to call anatomy a science (*scientia*) on the grounds that the anatomist performs mental dissection. But other anatomists did claim that anatomy – or, rather, one aspect of anatomy – was indeed a science because it reached causes,

---

6 Cunningham, *English manuscripts of Francis Glisson (1)*, 1993, pp. 15–21; the English is Glisson's original text, which was translated into Latin by George Ent for publication; emphasis as in Ent's translation; 'definition' (*definitio*) as in Ent's translation, cf. 'description' in Glisson's manuscript.

and hence was superior to a mere art. For instance, the late sixteenth-century Montpellier anatomist André Du Laurens had made this case:

> Now there is amongst Physicians a double acceptation of Anatomy; either it signifyeth the action which is done with the Hand, or the habit of the Mind, that is, the most perfect action of the Intellect. The first is called Practical Anatomy, the later Theoretical or Contemplative: The first is gained by experience, the second by the living voice of a Teacher, or by their learned writings: The first we call Historical Anatomy, the second Scientifical: The first is altogether necessary for the practice of Anatomy, the second is only profitable; but yet this profit is oftentimes more beneficial then the use itself of Anatomy: The first looketh into the structure of the parts, the second into the causes of the structure, and the actions and uses therefrom proceeding. According to the first signification we may define Anatomy thus: *An Artificial Section of the outward and inward parts* … If Anatomy be taken in the latter signification, it is defined *A Science or Art, which searcheth out the Nature of every part, and the causes of the same Nature.* I call it a *Science*, because it hath universal or general Theorems or Maxims, and common Notions, out of which, being the *first*, *true*, *immediate* and *best known*, all demonstrations are framed.[7]

So it seems that, for Du Laurens, exploring the structure and action of the parts was a manual and sensory activity, while exploring the use or purpose of parts was a mental and scientific activity, and therefore superior. There is a further aspect to the identity of old anatomy, which will be discussed below.

The anatomist investigated the parts and the organs of the human- and animal body. The basic meaning of 'organ' from the Greek is 'instrument', and the organs of the body had originally been given that title because they were the *instruments* by which the *soul* carried out its actions in the body, and the body itself, taken as a whole, was also the instrument of the soul. This sense of 'organ' and of its associated terms such as 'organic', 'organised' and 'organisation' (all meaning 'equipped with organs' serving the soul), continued to be the primary ones right through to almost the end of the eighteenth century. For instance, in 1760 the naturalist Charles Bonnet published an *Analytical essay on the faculties of the soul*. He followed this up in 1762 with a work on 'organised bodies': *Thoughts on organised bodies, in which is treated their origin, their development and their reproduction*.[8] Similarly the philosopher Denis Diderot, who almost definitely did not believe in the existence of the soul, nevertheless in the notes he wrote around

---

7 Andreas Laurentius, *Opera anatomica*, 1595, as translated in Crooke, *Mikrokosmographia: A description of the body of man*, 1631, book 1, chapter 15, pp. 26–7. On this see also Wear, 'Harvey and the "Way of the anatomists"', 1983, esp. pp. 227–9.

8 Bonnet, *Essai analytique sur les facultés de l'âme*, 1760; *Considerations sur les corps organisés*, 1762. 'Reproduction' here means the regeneration of lost parts.

1760 for a work on physiology used 'organisation' in the sense of the particular set of organs that an animal has:

> The animal is a being whose form is determined by interior and exterior causes, different ones necessarily producing different animals. The organisation of each animal determines its functions, its needs, and sometimes its needs influence its organisation.[9]

So possession of soul is what made any creature alive, and the origination, maintenance, development and reproduction of organised bodies was one of the functions of the soul, including in the insects and worms that Bonnet was particularly discussing. So to discuss the organs that a particular species possessed, and how those organs functioned, was in effect to discuss the soul of that particular creature, and in this sense the soul was central to all the accounts of animal and human bodies that anatomists gave. The same distinction existed in Latin: the terms 'animate' and 'inanimate', that is, 'possessed-of-soul' (*anima* in Latin) and 'not-possessed-of-soul', distinguished between created things that are alive and created things that are not. The organs were the instruments of the soul.

Anatomists made their investigations on the dissecting table, using the dead bodies of animals and man. But dissecting tables were designed not just for dissecting, but also for vivisecting. The table of Vesalius in the early 1540s had metal rings, and there were holes in the table itself through which rope was threaded, in order to tie down live animals during investigations (Fig. 1.1). This reminds us that the dissector was also always a vivisector: it was another part of the disciplinary role of the anatomist. There were rumours in the sixteenth century that Berengario da Carpi, Vesalius and Juan Valverde had actually broken the greatest of taboos and vivisected men, but I think we can dismiss them as just rumours.[10] Pigs and dogs were the animals usually preferred for vivisection, because of their availability, their convenient size, and their general anatomical similarity to man.

Anatomists certainly had some important uses for vivisection.[11] Vesalius had devoted the final chapter of his *Seven books on the fabric of the human body* to the importance and roles of vivisection in investigating and learning anatomy. He wrote:

> Just as the dissection of the dead teaches us very accurately the number, the site, the figure, the property of the substance, and the composition of each part, so in addition the dissection of the living animal both shows the function of each part very clearly, and also conveniently presents lines of reasoning for discovering

---

9 Diderot, *Élémens de physiologie*, 1964, p. 42; for his views on the contradictions in the concept of the soul, see pp. 54–60.

10 For some of the rumours, see Winslow, *The uncertainty of the signs of death*, 1748.

11 On the tradition of vivisection, see French, *Dissection and vivisection*, 1999.

> function. This is why students rightly start by being trained on dead animals, so that afterwards when they want to investigate the action and use of the parts, they may move straight on to the live animal.[12]

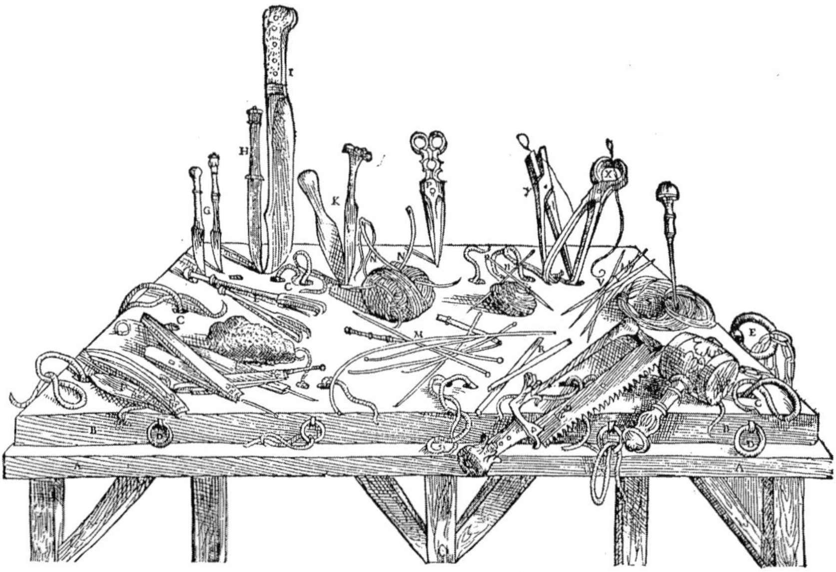

Fig. 1.1

So the primary use of vivisection for the anatomist was to both find and then demonstrate the actions and uses of the parts. However, a teacher like Vesalius had assumed, rightly, that most of the actions and functions of the parts had already been found, so in his public demonstrations he had used vivisection only for demonstrating particular things such as the action of muscles, and the effect of cutting the recurrent laryngeal nerves (the animal's voice is affected), or for clarifying certain issues, such as whether the pulse of the heart and arteries corresponds.[13] We shall see that vivisection played a central role in the approaches that anatomists took to their subject matter in the long eighteenth century.

---

12 Vesalius, *De humani corporis fabrica libri septem*, 1543, book VII, cap. XIX, p. 658. See also Farrington, 'The last chapter of the *De Fabrica*', 1931.

13 Eriksson, ed., *Andreas Vesalius' first public anatomy at Bologna 1540*, 1959; Cunningham, *The anatomical renaissance*, 1997, pp. 114–15.

3. The Theatre of Anatomy

Giving a good show

Anatomy was always theatrical. It took the most fascinating object in the world – ourselves – and performed what in any other circumstance would be a series of taboo or criminal acts on it, by dissecting it – cutting it open slowly, piece by piece, often in public, and presenting each part to the gaze of the onlookers. Then as now people were astonished, delighted and repulsed to see their insides, and enjoyed a frisson of excitement about transgressing the taboo. 'Man is revolted by studying his organisation', according to Jacques Tenon, professor at the Royal College of Anatomy in Paris in 1785.[14] And not everyone found the anatomical theatre appealing. The philosopher Jean-Jacques Rousseau, whose fastidious temperament made him avoid anatomy in favour of the study of plants, found the anatomical theatre overwhelming: 'What a terrible sight an anatomy theatre is! Stinking corpses, livid running flesh, blood, repellent intestines, horrible skeletons, pestilential vapours! Believe me, this is not the place where Jean-Jacques will go looking for amusement'.[15] Similarly, according to some later anatomists, while everyone knows how valuable anatomy is to medicine and surgery, 'everyone also knows how difficult, nauseous and grave a matter it is to look with exacting care into the dissected parts of the human body, parts gushing forth not only blood but sometimes also fetid pus, to investigate every hidden thing, to scrutinise the concealed, and in the same way to be engaged almost for ever in the murky remnants of death'.[16]

In Galen's time, and again in the medieval universities, dissection and vivisection had always been a public show, and this continued to be the case until the end of the eighteenth century.[17] From at least 1316 in the University of Bologna, human dissection was performed in public. Professor Mondino (Mundinus) began teaching anatomy there from a human corpse, and he wrote a dissection guide in Latin, the *Anatomia*, using the anatomical writings of Galen that he was familiar with, to steer one through the sequence of anatomising a human corpse. He used the bodies of those who had been judicially executed by hanging, and which had been made available by the appropriate local authority. There are numerous surviving portrayals of this event as it continued to be practised annually at the best

[14] Tenon, 'Observations sur les obstacles', 1785, p. 19. As just discussed above, 'man's organisation' here means the set of organs serving the soul.

[15] *C.* 1760s, Rousseau, 'Seventh walk', in *Reveries of the solitary walker*, 1979, at p. 114.

[16] Thus the editors of Mascagni, *Anatomia universa* (the Proemium), writing in the early 1820s.

[17] Staden, 'The discovery of the human body', 1992; Staden, 'Anatomy as rhetoric', 1995; Park, 'The criminal and the saintly body', 1994; Park, 'The life of the corpse', 1995.

universities (Fig. 1.2). The *professor*, a man with a book, sits in the professorial chair or *cathedra*, reading out loud the *Anatomia* of Mondino and perhaps adding his own comments. The *ostensor*, a man with a rod, shows the parts that the professor mentions in the course of his lecture. Finally there is a *surgeon*, a man with a knife, who actually does the dissection. Here the relative status of head and hand are clearly visible: the professor works with his brain and does not touch the corpse; the surgeon works with his hands and does not read the book.

The sequence of dissection is according to the 'three bellies' or 'venters' as they were called. First the lower belly is opened, revealing the liver, stomach, intestines, spleen, kidneys, gall-bladder, and sexual organs. Next the 'middle' belly, or chest, is opened, and the lungs and heart are dissected. The diaphragm constitutes the line between the lower and middle bellies; it is clearly visible in the illustration from Gregor Reisch's *Margarita philosophica*, a textbook covering the standard university course, first published in 1503 (Fig. 1.3). Finally the highest 'belly' is opened: the brain and the sense organs located in the head (the eyes, ears, tongue, sense of smell). The dissection of the three bellies each occupied a day. The limbs were not usually systematically dissected until the 1540s.

This order of dissection follows the order of corruption of the parts: the lower belly and its contents rot faster than the heart and lung or the brain. But, following the views of Plato in the fourth century BC, this sequence was also seen as one of relative 'nobility': the organs of the lower belly were less 'noble' (and indeed more lowly and servile) than those of the middle or upper bellies.[18] The brain, and above all the human brain, was thus customarily seen as the noblest organ of all.

Historians used to see this tradition of the public dissection of human bodies in the university, from 1316 until 1543, as exclusively a *teaching* event, and their general verdict was that as such it could have been only moderately successful at best since the professor was so distant from the corpse, not even touching it, and did not see his role as testing the texts of Mondino or Galen against the body lying right in front of his eyes. The text showed the audience what to see in the body. And how well could the facts of human anatomy be conveyed without the students themselves participating in the dissection with their own hands?

As a *research* event, historians have judged this kind of anatomising to have been even worse, since the professors assumed that all the truth was already in the texts of Mondino or Galen. But it was not a research event. It is true that a particular professor might introduce some findings from his own private research into the proceedings, as William Harvey did in his London anatomical lectures in the first half of the seventeenth century.[19] But theatrical, ceremonial, anatomy was not a research context.

Throughout the fourteenth, fifteenth and part of the sixteenth centuries, for a total of almost two-and-a-half centuries (i.e. from 1316 to 1543), this form of anatomising was used in the best Italian universities, especially at Padua

---

[18] Cunningham, *The anatomical renaissance*, 1997, pp. 10–13.

[19] Harvey, *Praelectiones*, 1886.

Fig. 1.2

and Bologna, and sometimes at Paris and one or two other places outside Italy. Ideally the dissection was performed annually in the winter, and it depended of course on the availability of bodies, which were usually supplied by the local government, granting the bodies of strangers or of executed or strangled criminals or of suicides.

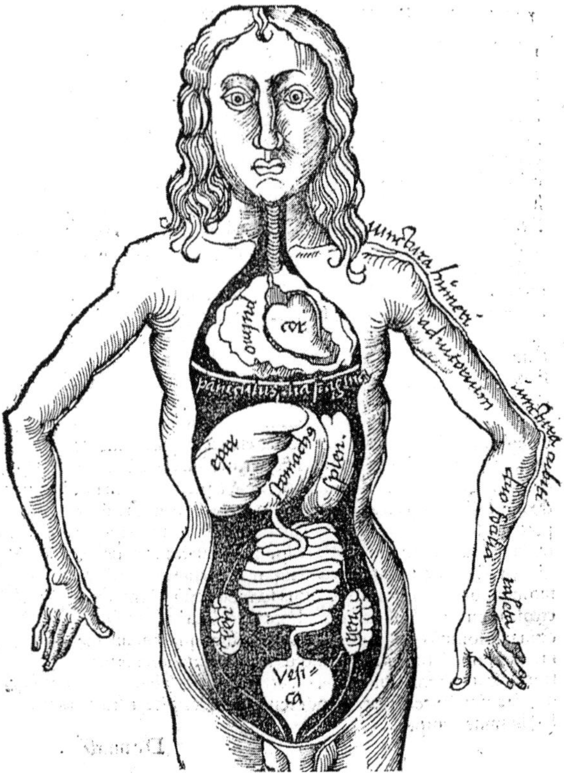

Fig. 1.3

But in the early 1540s Andreas Vesalius, working at Padua and Bologna, began to dispense with the offices of the *ostensor* and the *surgeon*, and began to carry out all three roles himself. In the traditional accounts of the history of anatomy this has been taken as the great revolution in anatomising, the moment when anatomising became modern. For Vesalius, who had been involved with the translation from Greek to Latin of the works of Galen for the great edition by the publisher Guinta, had discovered lots of errors in the anatomical accounts given by Galen, and he now advised – no, much more strongly, *demanded* – that people stop looking at the text of Galen (or Mondino), and instead look at the text of the body, 'the non-lying book of the body' as he called it.[20] So Vesalius has had the reputation among historians of having been the man who rejected textual authority, the authority of the ancients, and replaced this with *autopsia*, 'seeing-for-oneself'.

This new 'Vesalian' pattern of the professor or dissector performing all parts of the dissection – the cutting, the showing and the teaching – was certainly generally adopted by anatomists across Europe and became the standard way of anatomising. In Fig. 1.20 we can see how a later anatomist, John Banister, had himself portrayed similarly in 1581, in that he has a book and a rod and he is 'hands on' the corpse being dissected. Interestingly, the onlookers or students are also provided with probes and knives to participate when appropriate. William Harvey, too, undertook all these roles in the annual series of anatomical lectures with dissections he gave for the apprentices of the London surgeons.

But on the other hand Vesalius was not in fact rejecting the ancient approach, but actually trying to emulate Galen, insisting (like Galen) on personal experience of dissecting; insisting (like Galen) that the evidence of the body was superior to the evidence of any book; and insisting also (which Galen could not do) that the knowledge of the anatomy of the body of man could only properly be acquired from dissecting human (not just animal) bodies.[21]

Getting a good look

Public anatomising was performed in a theatre. Why did anatomising need a special place? There are two reasons. The first is that a specially designed theatre made it possible for more people to see more clearly. Without a proper theatre all was potentially confusion. For instance, in 1627/28 a keen teacher at Cambridge University, presumably the Regius Professor of Medicine, who was officially obliged to perform a dissection every year, publicly dissected an 18-year-old boy over three days. One of those present, young Joseph Mede, complained about it to his father: it took place 'twice a day the last two days. I was once there, but saw it so ill accommodated that I came no more; for it was in the regent house upon a table, when only half a score doctors could come to see anything, standing close by the table, and so hindering others seeing, which was the chief; for I can read as good as they could hear, and with more ease. It will be next time I hope better, for our new doctor will have one every year'.[22] But Cambridge University was not to have a dedicated anatomical lecture room until 1716.[23]

Elsewhere students of medicine were more fortunate. In the latter sixteenth and the seventeenth centuries purpose-built anatomical theatres were created in most of the universities of Europe that had medical faculties. They are usually

---

20 Quoted from the *Fabrica* by Cunningham, *The anatomical renaissance*, 1997, p. 121.

21 Cunningham, *The anatomical renaissance*, 1997, chapter 4.

22 Keynes, *The life of William Harvey*, 1966, p. 16.

23 Fozzard, *Professors of anatomy in the University of Cambridge*, 1983.

quite easy to spot, since they had an unusual shape, size and position. Typically they were round or oval, in order to bring the standing or sitting places as near the corpse as possible, and preferably they had windows above to light the subject, and were placed on the top of buildings.

The earliest, most famous such permanent theatre (replacing seasonal temporary theatres), and which was a model for many others, was that at Padua University, constructed in 1594. It has recently been well described as a 'huge, upside-down telescope' for inspecting the dissected human body.[24] It was in regular use well into the nineteenth century (Fig. 1.4) and is still extant. Others were built or rebuilt at Leiden (1597), Bologna (1649, restored 1733, reconstructed after bombing 1955; see Chapter 6 for illustration), Altdorf (reconstructed 1650), Berlin (1713, see Fig. 1.5), Ferrara (1731), Rome (in the Sapienza, restored in the 1680s), and many other places.[25]

We may take a few instances to see something of the context and extent of anatomical theatre building in Europe. They were not limited to universities: professional colleges of physicians or surgeons had them, as did the occasional city and royal court and even hospitals.

After the Great Fire of London of 1666, the Royal College of Physicians, like the other City companies, had to rebuild their premises. To round off their new building the physicians were given a new anatomical theatre, which was designed by Robert Hooke and opened in 1679. It was the gift of Sir John Cutler, a very wealthy grocer who was simultaneously both a miser and a seemingly generous benefactor. The whole building, seen from the outside here in 1803 in Fig. 1.6, was put up at Cutler's expense, but after his death his executors treated it as if it had been a loan, and tried to get the College to repay the money.[26] It is no longer clear why Cutler decided on this piece of philanthropy, but the Fellows of the College were grateful, and the first lectures in the new theatre were given by Dr Walter Charleton. Although he began with much praise of anatomy, Charleton's lectures were actually on physiology (nutrition, chylification, life, etc.), and no body seems to have been dissected on this initial occasion.[27]

In Edinburgh the beginnings of the anatomy theatre lie in the late 1690s when Archibald Pitcairne wanted to weigh the skin of a human. Pitcairne was a renegade physician who had some ultra-modern views on physiology, believing that the human body acted according to Newtonian physical laws, and knowing

---

24 Semenzato, ed., *The anatomy theatre* [*at Padua*], 1995, p. 123.

25 On this subject in general see Richter, 'Das anatomische Theater', 1936; Brockbank, 'Old anatomical theatres', 1968. On Berlin in particular, see Jaeckel, *Die Charité*, 2001, pp. 20–41; Wendler, '275 Jahre Theatrum anatomicum Berolinense', 1988. On Rome, see Canezza, *Gli Archispedali di Roma*, 1933, pp. 110–15.

26 Clark, *History of the Royal College of Physicians*, 2 vols, 1964, vol. 1, chapter 18; *Dictionary of national biography*, Sir John Cutler.

27 Charleton, *Enquiries into human nature*, 1680.

Fig. 1.4

Fig. 1.5

Fig. 1.6

the weight of the human skin was important for him to develop his views.[28] He made friends with the surgeons of the Corporation of Surgeon-Apothecaries of Edinburgh, siding with them against the new Royal College of Physicians there. He persuaded them to let him apply to the town council on their behalf for the grant of the bodies of executed felons and suicides to dissect. The town council made it a condition of its grant of bodies to dissect that the Corporation should build an anatomical theatre. This was duly ready by 1697.[29] Actually, this demand prompted the Surgeons to build a complete new hall for themselves, with the theatre as its centrepiece. The theatre was used for the regular public and private anatomies of the Corporation, which were conducted initially by the members of the Corporation in collaboration, and then by two of their number, to whom they gave the title 'professor', here meaning public teacher.[30]

The Corporation's anatomical theatre lost its function, however, when Alexander Monro *Primus* was made Professor of Anatomy (1720), as he immediately transferred the teaching to the University of Edinburgh where, again, a purposely designed theatre was created, this time by the town council itself, which was the body that ran the university. While there seems to be no surviving image of the accommodation provided for Monro *Primus*, a new theatre was built in the garden of the university for Monro *Secundus* in 1764 (after much petitioning of the town council by him), and it was he who also paid for its construction in the first instance. It held 300 students, though sometimes, as in 1783, the number of students was well above this figure, at 436, and a gallery had to be built to accommodate them.[31] A view of the theatre is shown here in Fig. 1.8, taken from a student class card, and a sketch of the plans for the theatre in Fig. 1.7.

Our third example will be the University of Uppsala in Sweden.[32] Here in this theatre in 1650 Olaus (Olof) Rudbeck had discovered the lymphatics and the thoracic duct. With royal patronage he had furthered his medical studies at Leiden, and when he returned was made professor. Turning enthusiastically to reorganise the university Rudbeck created both a botanic garden and an anatomical theatre (1662), the latter on the model of the one at Padua. He built it – and he claimed later to have built much of it with his own hands – on top of the Gustavianum, as the main university building was called, within a cupola offering light all round. It has the same steep standing places as Padua, bringing the viewers as close

[28] Cunningham, 'Sydenham versus Newton', 1981.

[29] Minute book of the Royal College of Surgeons of Edinburgh, 17 December 1697.

[30] For the content of the first lectures in the new theatre, see Chapter 3 below. The first two professors were Robert Eliot and Adam Drummond, and on the death of Eliot in 1717 Mr John McGill succeeded him. Minute book of the Royal College of Surgeons of Edinburgh, 28 March 1717.

[31] Horn, 'The anatomy classrooms', 1965; and Wright-St. Clair, *Doctors Monro*, 1964, pp. 78–9.

[32] The information here is from Nyström, *Olaus Rudbeck's anatomical theatre*, 1955; the quotation from Rudbeck is from p. 9.

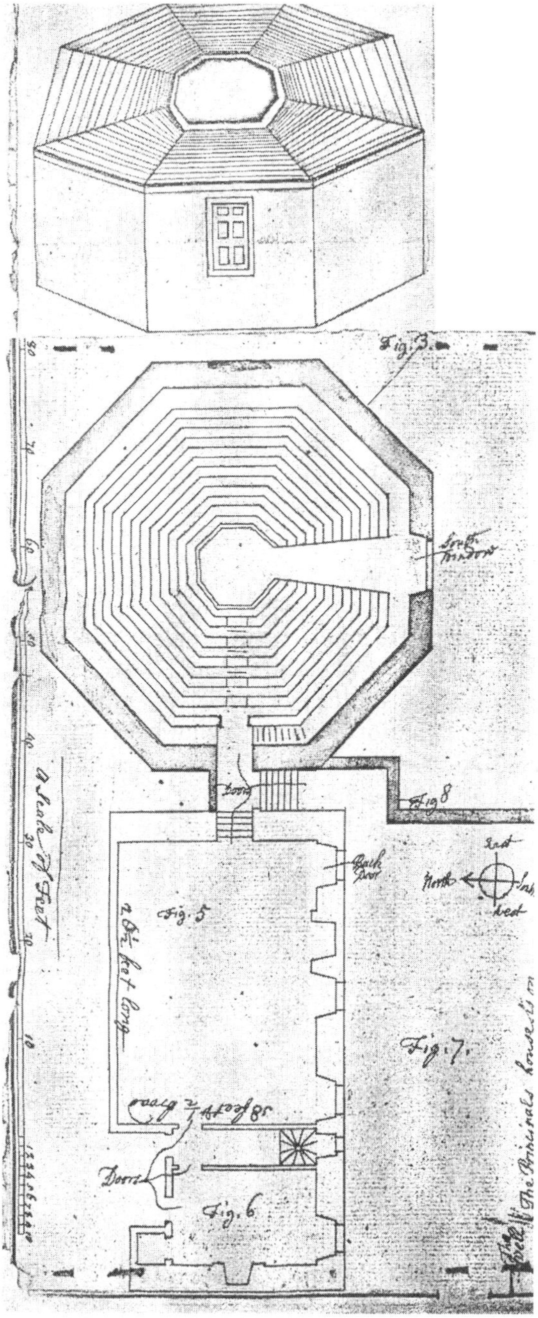

Fig. 1.7

Fig. 1.8

as possible to the dissection, but arranged as an octagon rather than a circle as at Padua, and holding some 200 students. As Rudbeck said in 1684 when defending the cost of the theatre, 'no year passes by without foreigners coming here … from all parts of Europe: when seeing the Theatre, they have to admit that there is not the like of it in Europe as concerns the architecture, the proper disposition of daylight, and the fair accommodation of the public'. However, Rudbeck seems to have lost interest in the theatre once built, and only a couple of public anatomies were held in it in the 1660s. Subsequent anatomies were held only in 1677, 1701 and 1731; not until 1747 did royal decree provide for an adequate supply of corpses, with the last anatomy held probably in 1766. It was restored to its original stunning beauty in the 1950s, being opened again in 1955. Fig. 1.9 shows the theatre built on top of the Gustavianum, Fig. 1.10 shows a bird's-eye view of the dissecting table and the standing arrangements, while Fig. 1.11 takes a cross-sectional view of the whole structure.

In Amsterdam, a capital city without a university, the barber–surgeons had a long tradition of teaching anatomy to their apprentices and to future midwives. From 1638 the guild had had a small anatomical theatre, built at the expense of the city, in their premisses in a disused chapel.[33] In 1688 the guild petitioned the municipal authorities for a new anatomical theatre. They were granted the use of the 'Waaggebouw', the old weigh-house on the Nieuwmarkt.[34] The conversion was completed in 1691 and cost the municipal authorities almost 10,000 guilders; the guild paid 390 guilders a year in rent. This new anatomy theatre in the Waag was described as 'so large and full of rarities, that no other anatomical theatre could compete with it'. The theatre was a tourist attraction, with or without dissections. It had a revolving dissecting table at the centre, illuminated by skylights. Like the Leiden theatre on which it was modeled, around the theatre were marvellous objects such as skeletons of an elephant, a lion, a lioness, of a crocodile, and of a giantess, together with some human flayed skins. This is the theatre in which Frederic Ruysch was *praelector anatomicus* for over sixty years from 1666, and in which he developed his techniques of injections. The praelector was appointed by the city council, in consultation with the senior members of the barber–surgeon guild. Teaching was in Dutch. Sources of corpses were hanged criminals, a few

---

[33] Hansen, 'Resurrecting death', 1996, p. 665, n. 9.

[34] I take all this information from Kurpershoek, *De Waag op de Nieuwmarkt*, 1994; my thanks to Harmke Kamminga for help in translation.

Fig. 1.9

Fig. 1.10

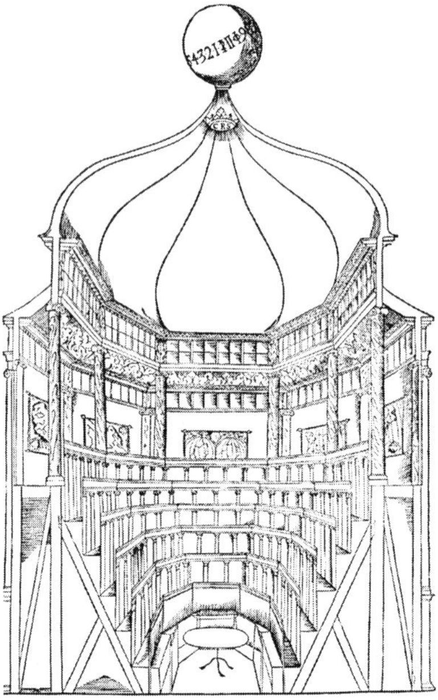

Fig. 1.11

bodies of poor people dying in hospital, and those found dead in the street. The city council was quite generous in allowing the bodies of those who died unclaimed in the hospital to be used for anatomy. Peter Camper once dissected ten corpses in the winter of 1757/58, which constituted something of a record. The theatre was in active use until 1869, and was restored in the 1990s (Fig. 1.12).

Finally as far as the seventeenth century is concerned, we can glance at the anatomical theatre at the royal court in Madrid, built to train surgeons in anatomy and surgery. This was under discussion from the 1680s when the Protomedicato advised the king of 'the need for an anatomical theatre or some other place where the physicians, surgeons and the rest of the audience can comfortably watch the dissections'.[35] An image of it, which appeared in Martín Martínez's *Complete anatomy of man* (*Anatomía completa del hombre*), shows it in 1728 (Fig. 1.13).

Anatomical theatres were being built in the early eighteenth century too. The image of the Ingolstadt university buildings in Fig. 1.14, of 1723, shows the anatomical theatre as part of a whole research institute, an 'Academic–Medical Garden'. The demonstration room is the round building in the centre of the block, with a room dedicated to anatomy and surgery on the right as we look at it, and one to a college of experimental physics on the left, with an astronomical observatory

[35] Quoted in Martínez-Vidal and Pardo-Tomàs, 'Anatomical theatres ... in early modern Spain', 2005, p. 262.

on the roof, and rooms also for a chemical laboratory and a 'hybernaculum' or indoor winter garden, all placed within a decorative garden.

Fig. 1.12

The University of Halle, in Saxony, which had been founded in 1689 by Frederick III, the Elector of Brandenburg, similarly had a new purpose-built anatomical theatre in 1727 (Fig. 1.15). The Professor of Anatomy, Georg Daniel Coschwitz (1679–1729), gave an excited welcome to it:

> At last! At last a good cause triumphs! Finally dawns that auspicious day which we have long hoped for. The day in which Anatomical Study, for a number of years obscure and in the shadows here in our Frederickian university for lack of a large and suitable public place, henceforth begins again to shine and glitter with a greater Light. From the first days of the Frederickian university opportunity was taken of dissecting cadavers and demonstrating them publicly, while the first founder of the University of glorious memory, FREDERICK THE FIRST, most wise king of the Prussians, in the privileges with which he provided this sacred seat of the Muses, by his most Wise Inspiration enacted a perpetual law, that to the Faculty of Medicine, for the study of anatomy, should be given the bodies of those punished by death in the Dukedom of Magdeburg and Hemipolitan.[36]

[36] Coschwitz, *Theatri anatomici … Natalia*, 1727, pp. 3–7. On Coschwitz, see Kaiser, *Zur Geschichte der medizinischen Fakultät der Universität Halle*, 1964.

AMPHITHEATRUM
MATRITENSE
Solerti Naturæ Fugax aperitur ... Arte
Naturæ ingenium dissecta cadavera pandunt
Plus quam vita loquax, mors taciturna docet

Fig. 1.13

Fig. 1.14

Frederick Hoffmann had been very active as an anatomist there, Coschwitz relates, and had had an anatomical theatre erected. 'But the fatal injury of time again destroyed that well-designed Theatre, and hence it was that anatomical study sort of slept for some time. But it did not completely expire (for within the last seven years, notwithstanding the lack of a suitable place, nine anatomical dissections have been publicly made by me)'. Now, however, our glorious king Frederick William I, has heard our pleas and paid for the building of this new theatre. At last a good cause triumphs. Bring out the bunting!

Even in the mid-eighteenth century anatomical theatres were still being built and were centres of anatomical activity. In London, for instance, the new building of the Surgeons' Company, which was being planned in 1746, was to have an anatomical theatre as its central feature, at the insistence of the celebrated anatomist William Cheselden, who was at that time the Master of the Company. The theatre, which was completed in the early 1750s (Fig. 1.16), had its own 'masters', 'demonstrators' and 'stewards of anatomy' (a very old-fashioned arrangement by this date), appointed each year, to perform the public anatomy. It replaced one designed by Inigo Jones that had been completed in 1638.[37]

[37] For the Jones drawings, now at Worcester College, Oxford, see for instance Billing, 'Modelling the anatomy theatre', 2004.

THEATRUM ANATOMICUM FRIDERICIANÆ
EXSTRUCTUM ANN. MDCCXXVII.

Fig. 1.15

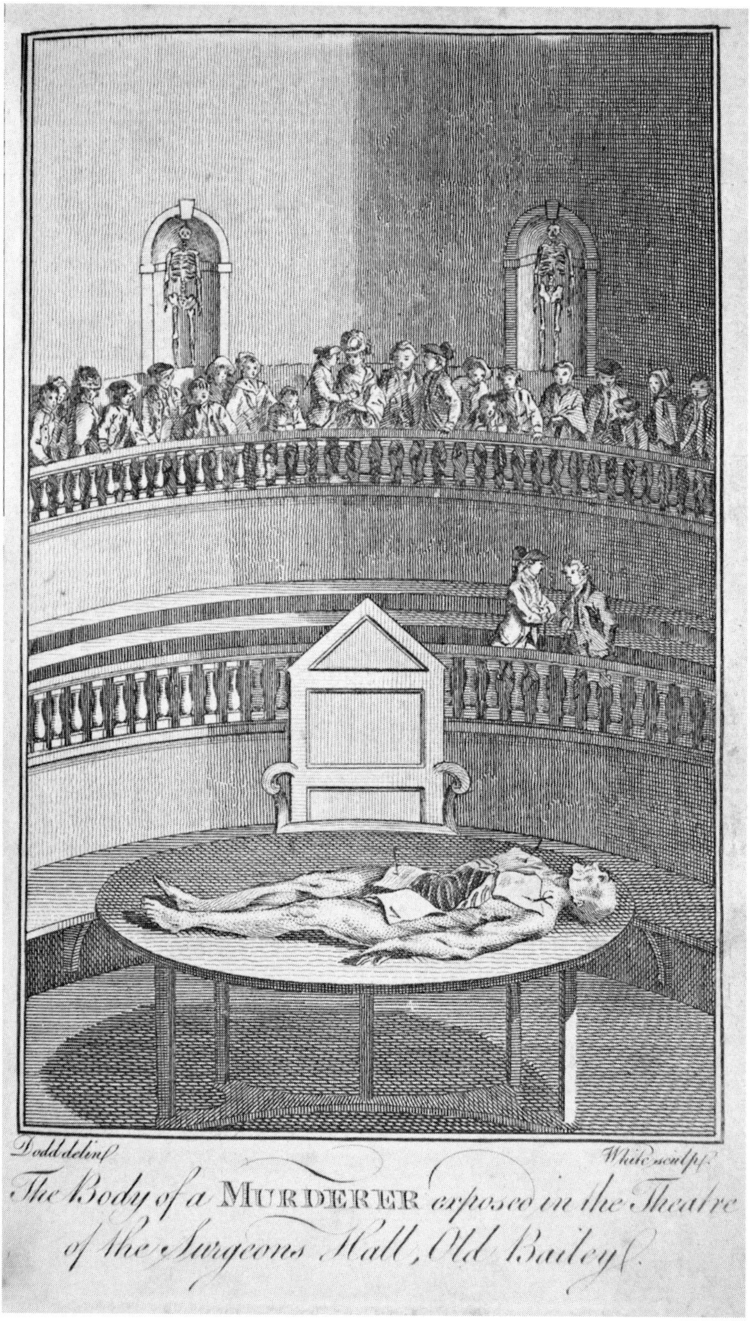
Dodd delin.
White sculp.
The Body of a MURDERER exposed in the Theatre
of the Surgeons Hall, Old Bailey.

Fig. 1.16

Similarly, in Barcelona, a wonderfully baroque new anatomical theatre was built in 1761 for the anatomical demonstrations of the Royal College of Surgery (Fig. 1.17), also effectively replacing an older theatre that had functioned sporadically from 1573 to 1714.[38] The seating here is on such a grand scale and so formal that probably nothing precise could be seen of the body from the seats themselves,[39] though the building itself would help inspire awe in the sacred ritual. It still exists and is used for ceremonial purposes.

Fig. 1.17

Invitations to the show

When a body became available, it seems to have been quite common to notify people of a forthcoming public dissection by printing a pamphlet in the form of 'The professor invites *X* to a public dissection in the anatomical theatre'. Duverney, the Professor of Anatomy at the Jardin royal in Paris, sent out 700 broadsides in the 1680s inviting people to his dissection course, 300 in Latin and 400 in French.[40] Sometimes the professor invited the medical students only. On occasion

[38] On the earlier theatre, see Martínez-Vidal and Pardo-Tomàs, 'Anatomical theatres and the teaching of anatomy in early modern Spain', 2005, pp. 268–72.

[39] For this complaint about such theatres, see Chapter 6 below.

[40] Guerrini, 'Duverney's skeletons', 2004, p. 600.

NOSSE DEUM
NOSSE SE
UNICA SAPIENTIA!
Ergo
ad
ANATOMEN
SVSPENSI
Clementissimo permissu
SERENISSIMI
DN. ELECT. PALAT.
quotidie hora X. & IV. habendam
a XV. Octobr. ꝏ Iↄc XXCIII.
QVOTQVOT CORPORIS SVI ET
SANITATIS COGNITIONE
DVCVNTVR
officiose & peramanter
invitat
GEORGIVS FRANCVS.

HEIDELBERGÆ,
Literis SAMUELIS AMMONII, UNIV. Typogr.

Fig. 1.18

the anatomist also undertook to demonstrate a related topic as well as to anatomise a human corpse, as Albrecht von Haller did twice in 1737 at Göttingen. In the spring he anatomised the body of a girl, while also communicating to his audience many recent observations he had made on the valves of Eustachius, which keep the blood circulation of the embryo diverted from the lungs.[41] Again, in July of that year, while himself dissecting a human body, he invited 'all those interested in anatomy' to hear him also prove that Hippocrates had dissected the human body. Haller could afford to make these varied discourses because, as he said, 'by the most magnificent solicitude of the Curators of the university the numbers of cadavers for dissection is increasing'.[42]

Bernhard Albinus invited people to anatomical demonstrations on a male body at Frankfurt an der Oder, at the opening of the anatomical theatre he founded there as Professor of Anatomy in 1684.[43] In a collection of pamphlets at the British Library invitations of this sort survive from the 1660s to the 1760s from German anatomists active in Wittenberg (J. G. Berger, 1689; P. G. Sperling, 1698), in Altdorf (Moritz Hoffmann, 1662 and 1682; J. N. Weiss, 1733, 1736 and 1745), in Leipzig (A. F. Walther, 1728), and in Helmstadt (P. C. Fabricius, 1750).[44] The one issued in Heidelberg by G. Francus, in 1683, seen here in Fig. 1.18, starts with the message, 'To know God, to know yourself, is the SAME KNOWLEDGE'. There must have been many more such invitations.[45] Sometimes, as in the case of Professor Johannes Travgott Adolph in Helmstadt in November 1762, the invitees were specified as 'the Vice-Rector, the professors with their guests, the eminent and worthy citizens, and all admirers of anatomy'. On this occasion Adolph was dissecting a pregnant 20-year-old woman and an infant that had been killed. The invitations usually specified the nature of the body, male or female, and whether it had been put to death by hanging or decapitation. The custom initially seems to have been to devote the invitation pamphlet to a preface on the antiquity and

---

41 Haller, *Omnes rei anatomicae studiosus ad dissectionem cadaveris humani invitat*, 1737.

42 Haller, *Semestris hyberni anatomen primam in cadavere puellae auditoribus suis indicit*, 1738, p. A2. My thanks to Dr Rymer Eck for providing me with copies of these two invitations from the rich collection of Haller material at Göttingen.

43 Albinus, *Demonstrationibus anatomicis in corpore masculine invitat Bernh. Albinus*, 1684.

44 British Library shelfmark T. 514. Invitations to Leipzig anatomies for 1706 (J. C. Schamberg, female, in 'the new and splendid anatomical theatre' opened in 1704), undated (Walther, infanticide), 1726 (Walther, male), 1728 (Walther, male) and 1734 (J. Z. Platner, male) are also reproduced in Rabl, *Geschichte der Anatomie an der Universität Leipzig*, 1909, pp. 18–20. For the invitation of 1645 by Simon Paulli at Copenhagen, see Bartholin, *Domus anatomica* (in his *Cista medica Hafniensis*), pp. 7–33. For Thomas Bartholin's own invitations, see *Domus anatomica* , pp. 35–7.

45 For an example from Holland, see Peter Camper's invitations to public dissections of rare animals; his invitation to the dissection of a rhinoceros in 1772 is reproduced in Loisel, *Histoire des ménageries*, 2 vols, 1912, vol. 2, pp. 301–2.

desirability of anatomy in general, but later ones are used by the anatomist to report on his findings in other recent public anatomies.

Getting the message across

We come here at last to the second reason why public anatomising needed its own special place. It was because it was a ritual, with a special message. It was in fact a sacred ritual.[46]

First, it was a ritual since it was conducted ritually, that is to say as a formal procedure constantly followed. Since its establishment in the early fourteenth century, the anatomical demonstration was formally conducted, usually at a set place and time of year, with officers to carry out the separate tasks and to maintain crowd control. It was a *calendar* event, taking place in winter, usually over three days, but this could be extended to five or six days. This timing meant that the bodies would not decay too quickly. Moreover, it often had to coincide with the local trials, so that there would be at least one body guaranteed for dissection. It was a *public and social* event. Not only were the masters and students of the medical faculty of the university admitted, but also local dignitaries and even the general public. In some places, especially Bologna, it was a central part of the celebrations of carnival.

Additionally, it had a *programme or sequence*. The dignitaries and lesser folk were ushered into their respective seats. The professor arrived and sat in his proper chair; the assistants, the *ostensor* and the *surgeon*, took up their allotted roles; after the innovations of Vesalius, the professor of course was expected to participate in all three roles, though he usually had someone to help him. The parts of the corpse were dissected always in the same sequence over the three days. The dissected parts were finally gathered together and given a Christian burial. Then there was a feast for the participants.

The dissection had special *officers and instruments*. Alessandro Benedetti, in his *Anatomy*, published in Venice in 1502, had described the situation thus:

> The order of seating is to be allocated according to the rank of the spectators. For this purpose there ought to be one *Praefectus* who takes care of and arranges everything. *Custodes* are to be appointed who should prevent the plebs rushing in. Two trustworthy *Questors* are selected, who should provide the necessaries out of the money collected. In this respect there is need of razors, knives, hooks, drills, trepans, sponges with which the blood may be soaked up in the dissection, paring knives and bowls.

The academic participants wore their academic dress. Thus in its performance, its special celebrants, instruments, location, timing, and so on, the public anatomical demonstration was a ritual.

---

[46] Cunningham, 'The end of the sacred ritual of anatomy', 2001.

But this was not all. For not only was it a ritualised teaching event, but it was also a ritual event on a much higher level, for theatrical, ceremonial, anatomy had a higher message. For this public dissection was a *celebration*. Indeed, this is the Latin term for carrying it out (*celebrare*). For the message that it was set up to convey – something much more important than the relation of particular parts of the body to each other – was about God, the Christian God. The message was that God is good, that God has created us, that God takes care of us, that God's handiwork at its most supreme is pre-eminently visible in mankind, the high-point of creation. In viewing an anatomical demonstration we should marvel at God and His creation. What is being demonstrated is God's power and forethought. Even though it is a piece of low-life being dissected – a criminal – nevertheless God's highest handiwork and forethought is visible in that criminal's body, in the structure, the action and the use of its parts. For God is the Designer of the body. This is the great pleasure – consisting in wonder, amazement, gratitude to God, awe at this awful sight (as William Hunter put it) – that this event provides to the participants and bystanders.

In Christian times the Intelligent Designer of whom Galen spoke was identified as the Christian creator God. Moreover, both for Greeks like Aristotle and Galen, and for medieval Christians, the body is, in one way or another, the vehicle of soul (or the soul).

The evidence for the sacred nature of public anatomising is all over the place and of several kinds. Visually, we can see this message in many illustrations. For instance, Fig. 1.19 is supposed to show the anatomical theatre of Thomas Bartholin in Copenhagen, opened in 1644. The message that floats over the top is 'The gods are also here', which is a quotation ultimately from Aristotle, himself quoting Heraclitus, and it means that evidence of design is not just in the unchanging heavenly bodies, but also amidst all the changeableness and muckiness of living creatures. Best of all such visual representations of this message is this portrayal of John Banister's anatomical demonstrations to the apprentices of the London surgeons in the 1570s, a picture Banister had had drawn for his own purposes in 1581 (Fig. 1.20). We see over the top of the picture a motto; it is, as it were, the title of the whole piece. It says, 'Anatomia scientiae dux est aditumque ad dei agnitionem praebet': 'Anatomy is the leader of knowledge, and shows the way to the knowledge of God'. The portrayal by Rembrandt of the anatomy demonstration given by Dr Tulp in Amsterdam in 1632 to the guild of surgeons there is perhaps the most famous picture of such an event. William Schupbach has brilliantly shown that every element of the picture is aimed at giving this message: that everyone should seek to 'know himself' by recognising that God lies concealed in even the smallest part of the human body.[47] The motto of the Amsterdam anatomical theatre, completed in 1691, was 'God's power is evident even in the smallest part of the body'.[48]

---

47 Schupbach, *The paradox of Rembrandt's 'Anatomy of Dr Tulp'*, 1982.

48 Kurpershoek, *De Waag op de Nieuwmarkt*, 1994.

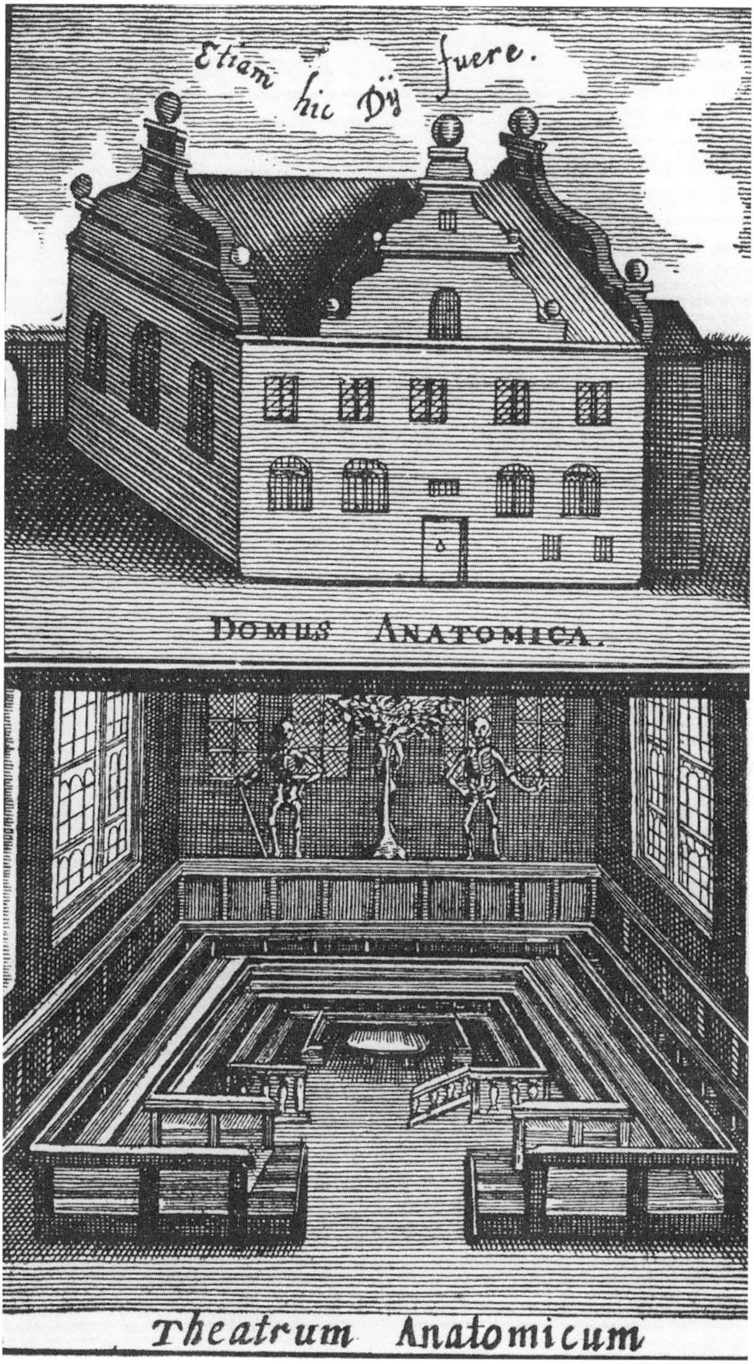

Fig. 1.19

Fig. 1.20

This message that knowledge of one's own anatomy leads one to knowledge of God is similar to, but not the same as, the Delphic Oracle's advice to 'know thyself' and know thine own mortality. Vesalius, in his *Seven books on the fabric of the human body* of 1543, made a conceit of the 'know thyself' advice (Fig. 1.21). A skeleton looks at a skull. Of course, 'know thyself' and thy mortality requires that one of the parties is still alive, so this is clearly a joke!

There are many examples of anatomists giving this message in their texts throughout the long eighteenth century, both in Britain and on the continent. William Harvey, for instance, said around 1650 that 'it hath ever been the delight of my Genius, to make strict Inspection into Animals themselves: And I have constantly been of opinion, that from thence we might acquire not only the knowledge of those less considerable Secrets of Nature; but even a certain Adumbration of that Supreme Essence, the Creator'.[49] Pierre Dionis, another instance almost half a century later in the 1690s, while lecturing on the abdomen, talked about it as the kitchen where all the nourishment is prepared for the rest of the body, and also the

[49] Harvey, *Anatomical exercitations concerning the generation of living creatures*, English version, 1653, Epistle dedicatory (by George Ent), quoting Harvey, A3v.

sink through which all its impurities are drained. Nevertheless it is still admirable because its architect is God:

> An Architect who undertakes a great Edifice, is sometimes as much concerned to dispose the Kitchen, and the Offices [i.e. the toilets], in convenient places, as to order the more noble Apartments; and he lets us see the strength of his Genius in their construction, as well as in his forming a Chamber or a Closet. God has no less made appear his great Wisdom and Power in forming the meanest Parts of a Human Body, than he does in the most noble.[50]

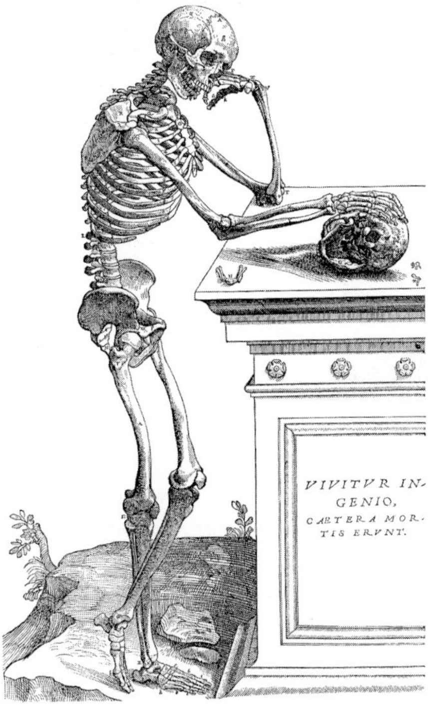

Fig. 1.21

Similarly, here is John Browne, a surgeon to the British king, dedicating his book on the anatomy of the muscles to his monarch in 1697. There is no subject more sacred to monarchs than anatomy –

> which presents to them the Divine Wisdom, and admirable Contrivance of the Supreme Architect in thus framing Man, and the well disposing of all his Parts

[50] Dionis, *The anatomy of humane bodies improv'd*, 1703, p. 113.

> into that excellent Order, Symmetry and Figure, we see they obtain; the true understanding whereof, being not only of great Advantage towards our knowing this Our Divine Benefactor, but of our selves also, who are the *Wonderful Works of His Hands*.[51]

In the course of their anatomical presentations, whether in lectures, books or speeches, the host of anatomists all say such things about how anatomy shows the workings of a providential God. Albrecht von Haller, for instance, a dedicated Calvinist working in a Lutheran world, in a degree-day speech in 1742 he gave at Göttingen on the joy of anatomy, rhetoricised in just this mode:

> In olden times Galen saw how much delight comes from anatomies to someone acquainted with true beauty. We are inquiring into machines made by the hand of God, where you will find nothing except what has been formed precisely for its purposes, except perhaps in mathematics. How much joy there is in this contemplation those men understood who placed the greater part of eternal blessedness in learning, with a blessed soul, the wisdom of God and the laws by which He formed the worlds and the inmost visible forms of things. Anatomy, above all other arts, supplies this intellectual pleasure (as it is called). For nowhere else do we see more clearly both the purposes of the structure in its action, and the structure shaped to those purposes …
>
> But when we add another assistance to human anatomy, the contemplation of brute animals, a new theatre of divine wisdom becomes visible. Everywhere is manifest the AGENT who prepared the individual citizens of his state not only with their proper functions – something that even men can do – but pre-prepared them for the functions to which they were destined, and this Agent is the finger of GOD … This contemplation is unending, and everywhere most beautiful: all things that live praise their common CREATOR and celebrate the SUPREME WISDOM delineated in every limb ….[52]

Luigi Galvani, a devout Catholic working in Bologna, also held such views. Indeed, when the Napoleonic Cesalpine Republic insisted on an oath of allegiance to its constitution in 1798, Galvani refused to take the oath because it was atheistic and contrary to his principles. As a result he lost all his jobs in anatomy.[53]

---

[51] Brown(e), *Myographia nova*, 1697 (first edn 1675). Russell, *British anatomy*, 1987, p. 25, discusses how this work is a complete plagiarism of two other works. However, the passage above is Browne's own work.

[52] Haller, 'De amoenitate anatomes', 1751, pp. 333–5.

[53] Green, *Translation of Luigi Galvani's De viribus electricitatis in motu musculari commentarius*, 1953, p. xix.

Charles Collingnon, the somewhat pathetic professor of anatomy at Cambridge University and a member of the Church of England, maintained in 1763 that

> The first and principal end of the Study of the human body, should be to awaken in us an awful Sense of the amazing power of its Creator ... And yet farther – we shall in the contemplation of the human frame, as strongly see the necessity of Providence to preserve, as we did of Omnipotence to create such a Being.[54]

As late as 1790 another Englishman, John Pole, who was an anatomist as well as a Quaker, was still putting forward the same view:

> To the particular department of natural philosophy, which is the subject of the following pages, I wish most anxiously to draw the attention of rational men; to rouse the activity of the intelligence to a research, which promises the amplest gratification, the sublimest truth; no less than a wide display of the wisdom of Omnipotence in the admirable structure of the first of created beings, a demonstration of the fabric, which serves to convey sentiments emanating from that divine principle which constitutes the essence of human nature.[55]

Thus we can see that this view of what the central message is of anatomising, was independent of particular variants of Christianity. Moreover, this message was independent of what kind of physiology you believed the human body followed. For instance, one could be a mechanist, such as a follower of Descartes, believing that the body functioned entirely as an automatic machine – a position that today looks like a materialist and hence atheistic stance – and nevertheless believe that anatomising still revealed God, the Designer of the machine. Walter Charleton, for instance, anatomical lecturer at the College of Physicians of London, in his lectures published in 1680 could describe the body as nothing more than an automaton:

> My subject is, the most abstruse œconomy of nature in the body of man; a system of innumerable smaller machines or engines, by infinite wisdom framed and compacted into one most beautiful, greater automaton; all whose parts are among themselves different in their sensibilities ... yet all ordained and adjusted to one common end, namely, to compose a living *ergasterium* or work-house, in which a reasonable and immortal soul may ... exercise all her divine faculties.[56]

Further, Charleton could say to the assembled members of the College of Physicians:

---

[54] Collignon, *Tyrocinium anatomicum*, 1758, pp. 9–10.

[55] Pole, *The anatomical instructor*, 1790, p. xviii.

[56] Charleton, *Enquiries into human nature*, 1680, Preface, p. 1.

> Evident it is then, that Your Theatre may be properly enough called a Temple of Natural Theology, where the Perfections of God are studied in the Works of His hands, and his Praises celebrated with understanding. *Deum enim colit, qui novit* [He worships God, who knows the body] ... My PROVINCE is, the *Anatomic Administration* of these ruins of a Man, i.e. to take asunder some few at least of the various Organs of this Master-piece of the Creation, so that we may explore their several conduit-pipes, their springs, movements, actions, communications, offices, in fine, their whole Mechanism or Construction.[57]

It needs to be stressed that this message about knowing God via anatomy was not put forward as an argument against atheists. For although it was always feared that atheists were lurking everywhere, if in fact they existed they very rarely chose to identify themselves, so they were not a clearly visible group in need of conversion. Moreover, as a message, it hardly qualified as an argument at all, for it was no more than a pointing-out of the obvious, but literally pointed-out with the finger and with the rod. Rather than an argument, it was a *demonstration* of the power and foresight of the Christian God.

But even so it was indeed possible for an anatomist to challenge this message. Naturally enough, it required an anatomist who was an atheist or pretty close to being one, which was still something of a rarity. One public challenge of this sort that we know of because it spilled into print was offered by Guillaume Lamy in Paris in the 1670s.[58] Dr Pierre Cressé, the official (deputy) anatomy lecturer at the Jardin royal, made an anatomical discourse (which did not involve performing a dissection). We know that Cressé proposed that,

> ... when we speak of any Part, we ought to assign its End, because it is certain that every one has its End; and that God having made nothing useless, we ought, when we show any part, to say it was made for such or such an Action, since it has such an Action: for Example, that we may certainly say, that the Eye was made for seeing, the Hand for taking, the Foot for going, and so of the rest.[59]

Lamy, himself a doctor and a member of the Paris Faculty, took exception to this and opposed it on the grounds that it was arrogance posing as knowledge. He wrote a series of *Anatomical discourses* against this position. Lamy was a Cartesian (that is, a follower of Descartes) and even an Epicurean, and like many other adherents

---

57 Ibid., pp. a3r, b1v.

58 The event is recounted by Lamy himself, in his *Discours anatomiques*, 1685. The texts have been recently edited as Belgrado, ed., *Guillaume Lamy. Discours anatomiques [1675]*, 1996. See also Dionis, *The anatomy of humane bodies improv'd*, 1703, who witnessed the event.

59 Dionis, *The anatomy of humane bodies improv'd*, 1703, pp. 11–12. For the documents of one side of this dispute, see Belgrado, *Guillaume Lamy, Discours anatomiques [1675]*, 1996.

of the old atomistic message of Epicurus, Lamy claimed that this position was not godless at all but in fact more respectful of the power of God than the standard Christian position as put forward by Cressé. Lamy maintained,

> That it was not our part to determine of the End, for which a part was made; That it was true, the Author of Nature had made nothing in vain, and that he had given some End to every part that compounds a Humane Body: but when we do undertake to assign their several Ends, we do run a risk of imposing upon our selves, because Nature may have intended some other than that which we assert; and that therefore we ought never to say this Part has been made for that purpose, but that this part does so and so.[60]

Rather than claiming that the eyes are made for seeing, the legs for walking, wings for flying, Lamy argued, 'it is sufficient to know that among the different animals that God has made, He wanted to make some of them with eyes, legs, wings, and that He well knew that those which had eyes would see, those which had legs would walk, and those which had wings would fly'.[61] To reason about the parts of the body in terms of final causes – that the eyes were made for seeing, for instance – was, in Lamy's words, 'to enter into a labyrinth from which one could not escape'.[62]

It was thus possible, though extremely unusual, for a bold or brave or foolhardy Epicurean to oppose the familiar claim. Where the traditional anatomist said we can see the wisdom and providence of God in the parts of the body and thus far read the mind of God, Monsieur Lamy said that we cannot read the mind of God: our way of working via final causes gives us no legitimate insight into God's intentions. Yet even here in the case of M. Lamy's protest against the familiar argument, he still employs all the elements of the old argument: parts, God, intentions and final causes.

4. Practical Matters

How did one cope with sorting out and systematically cutting up all the slithery and slippery insides of the body? The practicalities of anatomising were not easy even for professional anatomists undertaking research. Toward the end of the eighteenth century the French anatomist Vicq d'Azyr could state:

> The proper means to learn about the structure and interplay of the organs can be reduced to the following: anatomical dissection, the experiments that one

[60] Dionis, *The anatomy of humane bodies improv'd*, 1703, p. 12.

[61] Quoted from Lamy's *Discours anatomiques*, 1685, in Vartanian, *La Mettrie's L'homme machine*, 1960, p. 46.

[62] Lamy, *Discours anatomiques*, 1685, p. 40.

> attempts on living animals, and the exact observation of the phenomena of the organs, both in the state of health and in that of disease, and the account of the changes that disease brings about in their tissues.

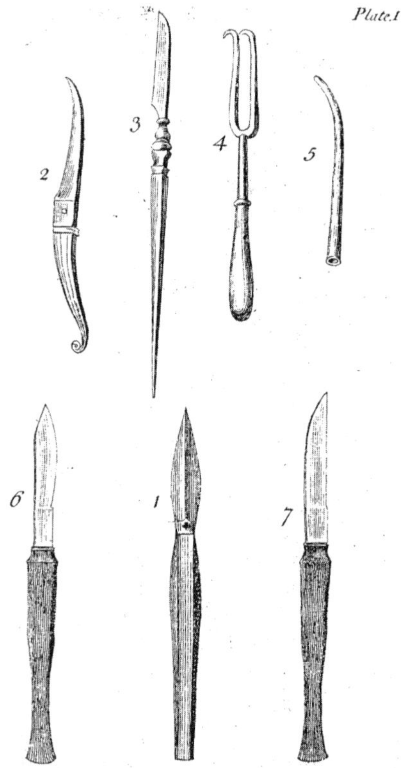

Fig. 1.22

But all these approaches present major problems. In studying the dead corpse, 'all connection, all sympathy' between the parts is lost; and dissection is fraught with difficulties in distinguishing one part from another. If the anatomist decides to interrogate living Nature herself, by vivisection, then he finds himself in the middle of floods of blood and convulsions, of cries of pain and agonies of the animals: 'Who can be certain, among such total confusion, of finding the traces of the natural movements?' Even in the third way of investigating, the exact and assiduous observation of the phenomena that the different organic functions present in normal life, it is difficult to isolate those that belong to each organ, so many are the connections of the parts with each other! And the comparison of diseased viscera with their intact versions has problems too, since the seat of a disease is very often far removed from the location of the pain.[63]

If that is how difficult things were for the anatomist researching on his own, then it is clear that, when it came to a *public* anatomy, the anatomist needed all the guidance he could get. Fortunately there were guides to show him how a public anatomy was conducted or, as the anatomists said (following Galen), how it was 'administered'. What did the anatomist do, and what did the spectators see? In 1653 Michel Lyser (1627–60) published in Copenhagen a Latin practical guide for the Moderns in their dissections, *The anatomical knife (Culter anatomicus*). Lyser was a German who was based in Denmark and as an anatomist has a claim to having, with Thomas Bartholin, the credit of discovering the lymphatics. This book was built on Galen and, of the more recent anatomists, on Jean Fernel and Jean Riolan, as well as on Lyser's own extensive experience. This guide was a very

63 Vicq d'Azyr, *Traité d'anatomie et de physiologie*, 1786, p. 2.

useful work, being published again in Copenhagen in 1665, in Frankfurt in 1679, in Leiden in 1726 and 1731, and in German at Berne in 1735. An abridged English version appeared in London in 1712 as part of the London *Bibliotheca Anatomica*. It still seemed valuable enough almost ninety years after its first appearance to be translated complete into English in 1740 as *The art of dissecting the human body*. The translator, G. Thomson, a physician, said that 'a Person who would be thoroughly acquainted with Anatomy, ought not to be a continual Spectator, but to take hold of the Knife, in order to satisfy himself, and convince others'.[64]

Lyser's ideal anatomist had an assistant to help him, and a set of indispensable tools, shown here in Fig. 1.22. These were a needle; waxed thread for making ligatures; dissecting knives of different shapes, all made of steel; small hooks to take firm hold of the parts; a whet-stone and strap for sharpening the knives; a sponge to wipe up blood and excrements; probes like bodkins; a pair of scissors for cutting the membranes; a small tube for opening the various ducts of the body; a pair of bellows for inflating certain parts to demonstrate them; a saw for opening the skull; and finally, an elevator to lift off the cranium after it has been sawn through. Further to these, Lyser recommended perforators and pins to make holes in the larger bones and remove the marrow if one wanted to articulate a skeleton. Lyser had also invented a couple of machines to hold the head firm while the cranium was sawn off (Fig. 1.23). Such machines could of course not be bought off the shelf: the illustrations served also as a guide for each anatomist to have one built for his own use – a kind of bespoke technology.

Lyser's guide was not intended to be followed in a public anatomy, but to show and tell the anatomist how to rehearse in order to be able to perform the necessary operations when called upon to perform a public anatomy, or to conduct private dissections for research or post-mortem. In the first place he insisted on making the maximum visible, for which purpose the body was to be placed on a special moving – more precisely, a revolving – table (Fig. 1.24). 'This moving Table is absolutely necessary in public Dissections, for every one standing round to have a full View of what Parts are proposed by the Anatomist'.[65] Secondly, the book is full of practical detailed information, about where the anatomist should stand, how he should use his hands, and precisely which finger to use for what purpose. Here, for instance, Lyser describes how to remove, view and prepare the stomach, intestines and mesentery of a human subject:

[64] Lyser, *The art of dissecting the human body*, 1740, pp. iv–v.

[65] Ibid., p. 12.

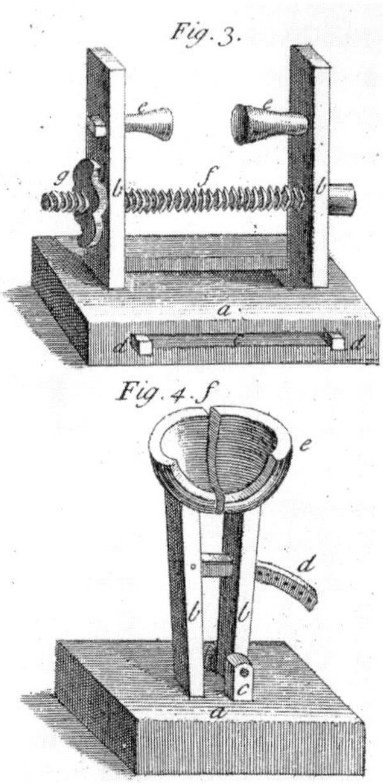

Fig. 1.23

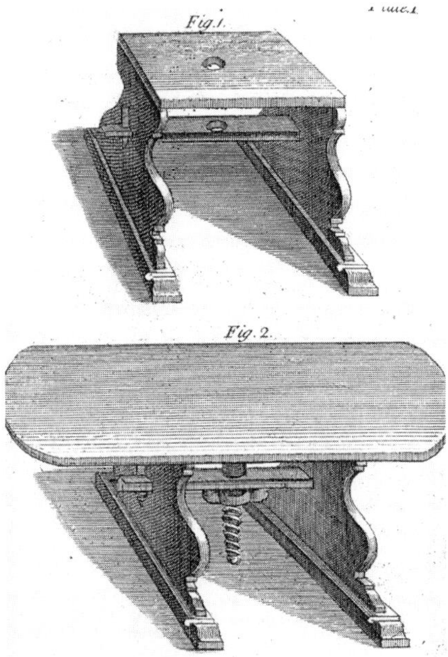

Fig. 1.24

> Anatomists commonly take out the *Stomach*, *Intestines*, and *Mesentery*, at one Operation, which is thus performed: Place yourself on the Right Side of the Body, and turning the small Guts from the left to the right, free the *Colon* from the *Peritoneum* with the subadjacent Membrane to which it adheres; then in like Manner separate the *Spleen* from the *Peritoneum*, taking Care not to lacerate its thin Coat. Thus you will come at the Stomach and its upper Orifice, where it unites with the *Oesophagus*: But because it lies very deep to the *Vertebrae*, you must lay under the Body near the Stomach a Log of Wood, or a Brick, to elevate the Region of the Stomach above the Breast and lower Parts; then slide your Left-hand gently betwixt the Stomach and the left *Hypochondrium*, and with your Finger separate the small Membranes lying between the *Oesophagus* and the *Peritoneum*; afterward fold it round your Middle and Ring-finger, and pull it so far forward as to be able to tie a firm double Ligature; then cut through the *Oesophagus* in the Middle between the two Ligatures.[66]

After doing the same on the left side, separating off the colon and kidney, the anatomist may now take out the mesentery and the intestines with the stomach and spleen. These organs must now be separated the one from the other:

---

[66] Ibid., pp. 36–7.

> The Intestines in all Demonstrations being laid on a Table, and exposed to View, should be cleaned, and afterwards inflated. The first Operation is performed by adapting a Funnel to the Orifice of the Stomach, through which pour a certain Quantity of Water, till it finds its Exit at the *Colon* … The second Operation is performed, first, by making a Ligature upon the lower Extremity of the Guts, then fixing the Mouth of a Pair of Bellows into the Orifice of the Stomach, by firmly tying a Thread round it, and inflating the whole Intestines.[67]

Lyser's guide took the anatomist through the conventional route: first the lower belly with the organs it contains, including the stomach, intestines, liver, spleen, kidney, and the male and female organs of generation; then the thorax to see the heart and lungs; the head to see the brain and nerves; and then the muscles of the arms and legs and hands and feet. Finally the book taught the anatomist how to prepare the bones before re-articulating them as a skeleton.

Subsequently a number of anatomists wrote their own versions (so to speak) of *The anatomical knife*. In 1679 the young Caspar Bartholin produced his own *Specimen of anatomical procedures* to accompany a third edition of Lyser. One of the suggestions he makes here is for the dissection of the trunk. Instead of first investigating the contents of the lower belly, and then moving on to the thorax, Bartholin suggests removing all the muscles and other coverings of the front of the body, and then removing the centre of the rib-cage, the sternum. In this way one can see all the organs of the trunk in their natural places, with the diaphragm separating them. But his main innovation here is with respect to the instruments the anatomist should use, including the microscope, and injections of coloured water, not just to demonstrate the bases of the circulation of the blood, but also to explore the structure and operations of the other organs. Similarly Leonard Tassin, surgeon to the royal hospital at Maastricht, produced his little *Anatomical administrations* in French first in 1678, clearly aimed at surgeons. The sequence followed by Lyser was the standard order for a dissection, but Tassin expounds the anatomical administrations in a very strange order – from top to toe – starting from the head and the eye, descending to the chest and its contents. Then, 'The dissection of the brain and chest having been completed, we must pass to that of the lower belly. It is with this part that one commences the demonstration of the parts, on account of the fact that it is the most susceptible to corruption'.[68]

An anonymous *Method of cutting human cadavers* was published in Latin in 1720 in Frankfurt by Michael Valentinus as an appendix to his Latin *Zoological amphitheatre*.[69] Here eighteen different preparations of the parts of the corpse are given in detail, as well as preparations on live animals in order to show the lacteals, lymphatics, thoracic duct and salivary duct. It is shown how to inject the blood vessels and lymphatics with wax.

---

67 Ibid., pp. 39–40.

68 Tassin, *Les administrations anatomiques*, 1678, p. 56.

69 Valentinus, *Amphitheatrum zootomicum*, 1720.

Johann Friedrich Cassebohm, professor in Berlin, published another such guide for students in 1746, this time in German. Cassebohm, like Lyser, gives very detailed and first-person instructions on where to stand and which hand to use, and is very full on modern discoveries such as, for instance, 'the capsule of Glisson' in the liver. 'There is no shortage of anatomical books … [but] it is not enough for someone to be familiar with the arrangement of the body: it will be especially necessary for him to use his own hand and to work on the body and all its parts. This work is called "anatomical preparations"'.[70]

Finally, a hundred years after Lyser, another such book dedicated to the details of tools, modes of cutting, injecting and embalming was produced by Jean Joseph Sue (Süe), Royal Professor of Anatomy in the Schools of Surgery of Paris and at the Royal Academy of Painting and Sculpture in Paris. Sue's little book, again aimed at conveying the practical knowledge a beginner needed, was first produced in 1750 and then as a separate work in 1765 with the title *The Human-cutter* (*Anthropotomie*). Since Lyser was writing in the early seventeenth century, injection had become a major technique for the anatomist (as we shall see more in Chapter 2), and Sue dedicated much of his advice to this. But here he is describing how to make that all-important first incision, giving detailed instructions as Lyser had done:

> In order to dissect the muscles of the lower belly, the cadaver having been laid on a solid table, one stands on the right; then, taking a scalpel with a back between the middle finger and thumb of the right hand, having the index finger pressing on the back of the scalpel, you stretch the skin of the cadaver with the thumb and index finger of the left hand.
>
> You then hold the instrument between the two fingers of the left [*sic*] hand to cut the skin and the fat from the xiphoid cartilage to the navel, around which you make a circular incision the size of an *écu*. Then you continue the longitudinal incision down to the pubis ….[71]

This table from Thomas Bartholin's *Anatomy* of 1663, Fig. 1.25, shows what the dissector and viewer would see after this first cutting with a cross-like incision. 'In this Table are expressed the common Coverings of the Belly separated, and on one side the Fat besprinkled with its Vessels, and on the other side certain Muscles Detected'.[72] On the right as the viewer looks at it, the lower trunk of the body has had the scarf-skin or epidermis (A) and true skin or dermis (B) pulled back, revealing the fatty layer (E) with its veins (F). On the left, as we look at

70 Cassebohm, *Methodus secandi*, 1746, Vorrede.

71 Sue, *Anthropotomie*, 1765, p. 91.

72 [Thomas Bartholin], *Bartholinus Anatomy; made from the precepts of his father*, 1663, book 1, chapter 3, 'Of fat'.

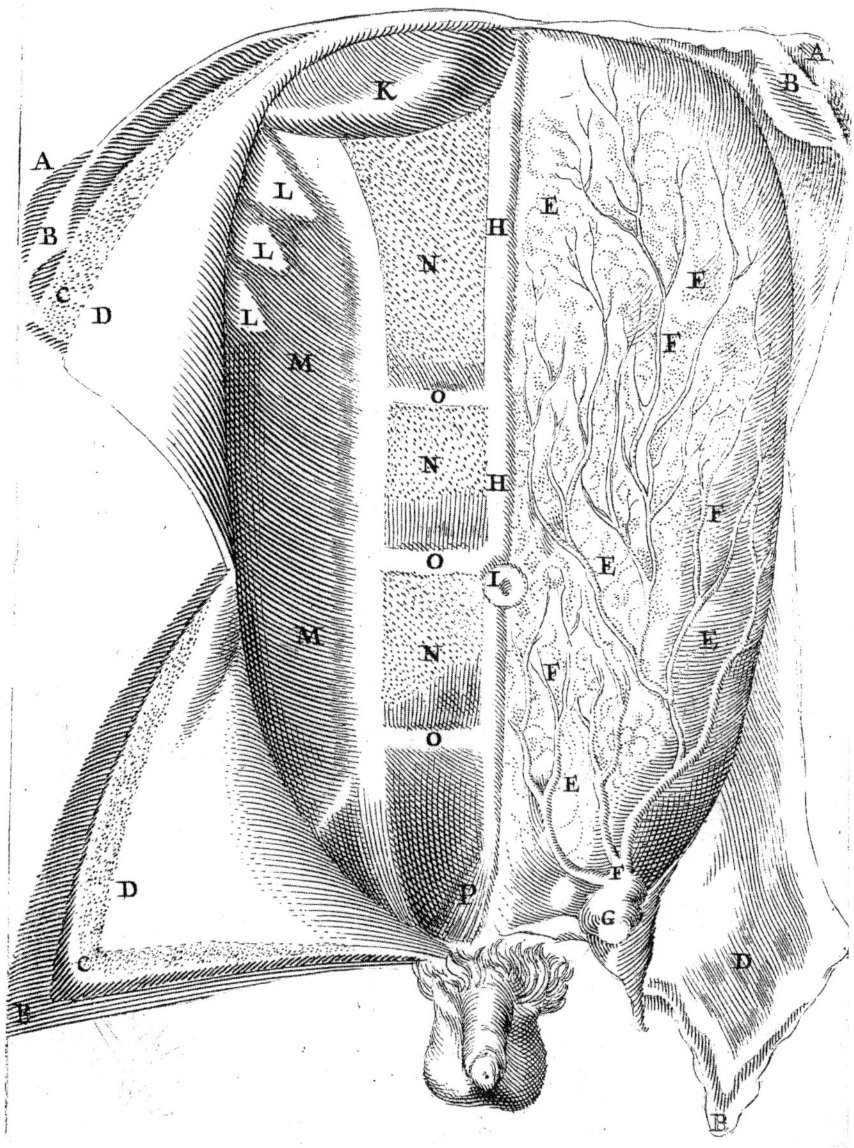

Fig. 1.25

Fig. 1.26

it, the superficial fascia or layer of subcutaneous fat (C and D) has been removed, revealing the abdominal muscles (M the oblique descending, and N the rectus abdominis).

But there was another way of making this first cutting, as Sue writes:

> … or better, when you arrive at four finger-breadths above the umbilicus you make two angular incisions that end in the groins, which preserves complete the strip of skin that is over the pubis, and that one can fold back to cover the parts of generation. This precaution ought to be made, especially when the subject is destined for a public demonstration, or when persons might be present whose modesty could be offended.[73]

From another anatomist, Thomas Pole, we learn how the dissector should dress, and such a scene is illustrated here in Fig. 1.26 in the dissecting rooms of Albrecht von Haller:

> The dress of the dissector consists of a pair of sleeves and a large apron with pockets; these are usually made of chocolate-coloured glazed shalloon [a closely-woven woollen material], and lined with linen … without it, the clothes get soiled, and small pieces of the subjects adhere to them. An inattention to this circumstance would not only be a violation of neatness, but of decency.[74]

In the best anatomical schools attention had been given to ensure that the lighting was satisfactory, and that all the students had the best possible view of the dissection. William Hunter built his Windmill-Street School in London at his own expense in 1768, and it was ideally suited to his purposes. In his introductory lectures to his anatomical course he drew the students' attention to the amenities:

> In our demonstrations here, large and conspicuous objects will be shown to the whole company at once. You may observe that this theatre is particularly well constructed, both for seeing and hearing; a strong sky-light is thrown upon the

73 Sue, *Anthropotomie*, 1765, p. 91.

74 Pole, *The anatomical instructor*, 1790, p. lxix.

> table, and the glass being ground, that is, made rough upon one surface, the glare of sun-shine is not admitted: the circular seats are brought as near the table, as ease in sitting would admit of; and, as they go back, they are a good deal raised, which is a considerable advantage both in seeing and hearing; you may observe another circumstance in this theatre, which has not been sufficiently considered in buildings of that kind, viz. the table, where the object is placed, and by which the demonstrator stands, is not in the centre of the circular room, but about half way between the centre and the circumference; thence the seats make smaller segments of large circles, in proportion as they are farther removed; and the spectators, in proportion as they are at a greater distance, are more directly before the object and speaker, which, both in hearing and seeing, makes some compensation for the greater distance.
>
> Objects that are still more minute, and most of the *preparations*, must be sent round the company; that every student may examine them in his own hand. To prevent confusion, you will please to observe, that, in the first seat, the preparations are to go round from right to left; in the second bench, from left to right; and so alternately, to the farthest seat of all. To prevent loss of time, when you give a preparation to your neighbour, be so good as to point out the *part*, or *circumstance* which is then to be examined; as I shall do, when it is first handed round.[75]

Finally, of course, the anatomist had to ensure that he had all other necessary things at hand, such as 'towels, sponges, and basins, water, coal and wood to heat the dissection chamber and candles to light it', as well as 'copious quantities of *eau de vie* [brandy] or spirit of wine commonly used in human dissection to wash out the putrefying cadaver and arrest the process of decay'.[76] Sometimes the anatomists drank the *eau de vie* to refresh (or perhaps fortify) themselves.[77]

5. A research culture

By the 1650s, quite apart from their study of the writings of the anatomists of Greek antiquity, 'modern' anatomists had long been undertaking their own novel research – that is, undertaking investigations to improve or extend their knowledge of the structure and workings of the human body. This research culture in anatomical investigation is what our anatomists of the long eighteenth century inherited.

---

75 William Hunter, *Two introductory lectures*, 1784, pp. 111–12.

76 Guerrini, 'Duverney's skeletons', 2004, pp. 600–601; Schiller, 'Laboratoires d'anatomie', 1964, p. 109.

77 Guerrini, 'Duverney's skeletons', 2004, p. 601, n. 62, citing Stroup, *Royal funding of the Parisian Académie royale des sciences*, 1987, p. 48. See also Stroup, *A company of scientists*, 1990, p. 41.

Such anatomical research had been being undertaken by a few anatomists from at least the 1490s, and was conducted privately. The first such research anatomists were Italian university professors, working in Bologna and Padua. Such private research might take place in an anatomical theatre, but was not carried out under the gaze of a crowd of onlookers. A favourite student or two of the anatomist might be present, or a colleague or two, but that would probably be all. Or it could be conducted in the anatomist's own home, or elsewhere, as the anatomist Realdo Colombo and the artist Michelangelo in the 1550s together had dissected the body of a young Moor in the Church of Saint Agata in Rome 'because of its being a remote place', where the smell would not offend anyone.[78] Or it could be carried out in a hospital, where a patient had died, almost all hospital patients being poor and their bodies often available to the anatomist without formality.

Many dozens of printed books bear witness to such original research, from that of Gabriele de Zerbis, whose enormously long *Book of the anatomy of the human body and of its individual members* had appeared in print in 1502 in Venice, to the final book of William Harvey, *Anatomical exercitations, concerning the generation of living creatures*, which was published in London in Latin in 1651. Some of these books simply went over the anatomical ground in more detail than before; some offered short ways of learning anatomy; some put forward new (or revived) research programmes and their findings, such as the Aristotelian approach advocated by Fabricius ab Aquapendente of Padua, and also by his pupil William Harvey – whose research in this tradition culminated sometime in the 1610s in the unexpected discovery of the circulation of the blood.[79] Other anatomical books reported on particular investigations of a few months or years, or on chance novel discoveries. The seventeenth century was particularly rich in such new investigations and discoveries, such that in some people's eyes it looked as though anatomy had been completely revised since Galen in the second century AD, or since Andreas Vesalius, his great emulator in the mid-sixteenth century.

The audience for these works, and for discussion of the triumphs and travails of anatomy in general, was strikingly wide, especially if compared to the position today, where scientists rarely speak to the general educated public. 'Gentlemen' and their continental equivalents, that is to say literate male products of the schools and universities, expected themselves to keep up with the latest discoveries and to follow the latest controversies in the whole domain of natural philosophy. Because anatomy was taken to be part of natural philosophy, the implications of anatomical findings for wider philosophy were quickly remarked on, and equally any new approaches to physics, and later also to chemistry, had important implications for how the factual information derived from anatomy was to be interpreted in understanding the body's hidden workings. With the founding in the mid-seventeenth century of learned societies, with extensive lay, amateur memberships and audiences, such as the Royal Society of London (1660), or

[78] Condivi, *The life of Michelangelo*, 1976, p. 99.

[79] Cunningham, *The anatomical renaissance*, 1997, chapter 6.

the French equivalent, the Académie royale des sciences (1666), and with the publication of *Philosophical Transactions* and other newsletters for the 'virtuosi' (as they called themselves), anatomical discoveries and discussions reached a large and widely spread community of educated amateurs across Europe. Many professional anatomists took this Europe-wide amateur audience of gentlemen as their main audience for their books and lectures. For instance, William Cheselden offered a course in anatomy in London in 1721, of which he said in a newspaper advertisement, 'This Course being chiefly intended for gentlemen, such things only will be omitted as are neither instructive nor entertaining, and care will be taken to have nothing offensive'.[80] Similarly the Rouen surgeon and anatomist Claude Le Cat showed in his course on anatomy and surgical operations in 1737 that anatomy was pertinent and of value to everybody: 'a great number of persons, even of the highest distinction, regularly attend his lessons'.[81]

So while some publication on anatomical matters was certainly directed either towards pupils in anatomy or surgery, or to other professional anatomists, most anatomical writers could assume a larger informed audience for their writings and observations. The situation was far from the one today where specialist writes only for specialist in dry technical prose in professional journals of very limited circulation. By contrast, in the long eighteenth century, anatomical discovery was news.

6. The state of the art

What was the state of affairs in anatomical research at the beginning of our period, in the middle- and late seventeenth century? In the first place the excitement of the hope of making new discoveries was widespread. Nicolaus Steno (Niels Stensen) from Denmark gave a celebrated lecture in Paris in 1665, at the house of Monsieur Thévenot, taking the current situation in the anatomical investigation of the brain as his subject.[82] His lecture was both cautionary about the present state of knowledge and positive about the prospect of future discoveries, if anatomists would only think more deeply about what they were doing, if they were to vary their procedures more, and if they were to see the making of discoveries as one of their major roles. To reach certainty in one's present knowledge, and to direct one's investigations at new discoveries, less reliance should be placed on the ancients, according to Steno, and more on one's own findings. The brain was, one

---

[80] Peachey, *A memoir of William and John Hunter*, 1924, p. 18.

[81] Le Cat, 'Discours', 1737, p. 248.

[82] Steno, *Discours … sur l'anatomie du cerveau*, 1669 (1950); the English translation provided in this edition, made by G. Douglas and first published in the English version of the works of Winslow in 1733 (Winslow, *An anatomical exposition of the structure of the human body*, 2 vols, 1733, vol. 2, p. 56ff.), is somewhat free, and the French text needs to be consulted.

might say, something of a soft target on which to base one's recommendations for change, since the form of the parts of which it is composed are so unrevealing of their functions or uses, and its soft and malleable texture meant that it was difficult for the anatomist to handle it without creating unintentional changes in the form and position of the parts, which then misled his observations. However, Steno's advice was intended to apply not just to the anatomising of the human brain, but to the exploration of all the parts.

Steno – following the spirit but not the detail of Descartes – started from the assumption that the body was a machine, and thus there were only two ways of coming to a knowledge of it: either to be told by the Maker directly, or 'to examine each Piece by itself, and as it stands in relation to the rest'.[83] One of the reasons why such investigation had not been adequately performed hitherto, Steno argued, was that neither physicians nor surgeons, nor their teachers of anatomy, could spend enough time at anatomy, since even the professors of anatomy thought they had done their duty in their public demonstrations when they had expounded some works of the ancients on a human corpse. That is, they saw their role simply as teaching, not as exploring, the anatomy of the body. But 'to make the necessary Inquiries for the discovery of the Truth, a Man's whole time must be taken up … everything else must be laid aside, and we must mind nothing but that … Whole Years may sometimes be necessary to discover what may afterwards be demonstrated to others in the space of an Hour'.[84] And 'though Anatomists open a thousand Bodies in the Schools, it is by mere accident if they ever discover anything new. They are obliged to demonstrate the Parts as described by the Ancients, and in doing this it is necessary that they should follow a certain Method; whereas Inquiries (*les recherches*) admit of no settled Method, but must be pursued in every manner that can be thought of'.[85]

A short while later Steno issued more advice to his fellow anatomists. He had moved to Florence, where he received the patronage of Grand Duke Ferdinand II of Tuscany. He had been investigating how muscles work and, as with the brain, he claimed that anatomists still knew almost nothing about the functioning of muscles, even after two thousand years of inquiry. And here he claims that mathematical knowledge is indispensable if anatomists are going to make progress:

> The structure of the muscles requires almost necessarily that they be explained mathematically. But why do I claim for the muscles alone what is due to the entire body? Our body is an organ composed of a thousand organs. Whoever thinks that its true understanding can be sought without mathematical assistance must also think that there is matter without extension, and body without figure. Nor is there any other cause of the many errors which have foully defiled the

[83] Steno, *Discours … sur l'anatomie du cerveau*, 1669 (1950), p. 32; p. 23 of the English version.

[84] Ibid., pp. 36–7; pp. 26–7 of the English version.

[85] Ibid., p. 37; p. 27 of the English version.

> description of the human body than that anatomy has hitherto disdained the laws of mathematics … Anatomy … has thrust on us the dubious for the certain, the false for the true, the unknown for the known. Anatomy finally has brought the matter to such a point that nothing remains more unknown to man than man himself.[86]

In fact, anatomising in the mid- to late seventeenth century was in a flourishing state in Europe as a whole, and there were many anatomists who were already following the injunctions of Steno that 'Anatomy ought to be confined by no Rules, every new Dissection requiring a different Method', though fewer who put so much faith in mathematics as the key to progress. No better starting-point can be found for characterising the state of anatomy in the second half of the seventeenth century and its recent achievements than the enormous two-volume work edited by the doctor and historian of medicine Daniel Le Clerc and his collaborator, that jackdaw of medical knowledge, Jean Jacques Manget, chief physician to the ruler of Prussia: *The anatomical library* (Geneva, 1685). The full title gives a fair indication of their aims: *The anatomical library or a new and very rich treasury of discoveries in anatomy, in which is shown a full and complete description of the human body, and its œconomy* [i.e. physiology] *assembled from the individual treatises, both already printed and even some not yet published, of the most outstanding anatomists; to which is added the anatomical administration of all the parts, with various most interesting preparations of the parts … with the necessary Indexes and engraved figures*.[87] It was a brilliant idea. It was meant to be the one book you needed in order to be up to date with anatomical knowledge, all arranged in the most logical way, a bit like a *Reader's Digest* of anatomical progress, but with most of the eighty or so texts given at full length.

The title-page (Fig. 1.27) has a number of messages. In the first place it is called a library, but no book seems to be visible. Instead we have a human corpse laid out on a table. However, as Vesalius himself had taught, for the true anatomist the body *is* a book; he called it 'the true book of ours, the human body, man himself'.[88] Second, obelisks, with their hieroglyphic inscriptions, were known to have been used by ancient Egyptians in pairs at the entrance to a temple, so here the reader is being invited to enter a sacred arena. Third, the editors quote that familiar encouraging passage from the sixty-fourth of Seneca's *Moral letters*: 'Our predecessors have worked much improvement, but have not exhausted the

---

86 Kardel, ed., *Steno on muscles*, 1994, p. 85, dedication to *Elementorum myologiae specimen* (1667).

87 Manget and Le Clerc, *Bibliotheca anatomica*, 1685. This was only one of the great medical *Bibliothecae* that Manget edited: there were others on practical medicine (1695–98), chemical medicine (1702), pharmaceutical medicine (1703), surgery (1721), and on the writings of the ancient and modern physicians (1731).

88 As cited from the *Fabrica*, in Cunningham, *The anatomical renaissance*, 1997, p. 121.

BIBLIOTHECA ANATOMICA
NEC VL
LI NATO
POST MI
LLE SECV
LA PRAECI
DETVR OC
CASIO ALI
QUID AD
HVC ADII
CIENDI.
SENECA

Fig. 1.27

problem. Much still remains to do, and much will always remain, and he who shall be born a thousand ages hence will not be barred from his opportunity of adding something further'.[89] The Moderns of anatomy had certainly been busy adding their contributions. New work had been performed on just about all of the parts and organs in recent years, and much of it was included in the *Anatomical library*.

The work was composed with the willing co-operation of many of the active anatomists of the day, such as Redi, Malpighi, Bontekoë, Drelincourt, Ruysch, Schelhammer and Peyer, many of whom we shall be meeting later, and some of whom sent their own unpublished treatises to be included. It contained most of the important works of the earlier seventeenth century such as William Harvey's *Anatomical exercise on the motion of the heart and blood in animals* (1628) and his *Anatomical exercises on the generation of animals* (1651), together with Gaspar Aselli's announcement of his discovery of the lacteal veins (1622) and Jean Pecquet's of their destination in the thoracic duct (*New anatomical experiments*, 1651), and Thomas Wharton's *Adenographia* of 1656, the first book dedicated to reporting research on the different glands of the body.[90] It also included more recent work and speculations, such as those of Lorenzo Bellini on the structure and use of the kidneys (1662), Jan Swammerdam's *Miracle of Nature, or the fabric of the female uterus* (1672), and Nicolaus Steno's geometrical treatment of the movement of the muscle fibres: *An example of observations on the muscles and glands* (1664). All treatises had their illustrations duly redrawn and re-engraved.

For the structure of their *Anatomical library* the editors chose to follow the traditional sequence of dissection of the human body; that is, starting from the lower belly, then dealing with the contents of the thorax, then the head, and finally the bones and muscles. The whole of the first enormous volume is on the lower venter or belly, that is, the organs and other parts that are located below the diaphragm: the outer skin, the true skin, the muscles of the abdomen, the peritoneum, the omentum, the stomach and intestines, the glands, the pancreas, the liver, the spleen, the kidneys, the organs of generation, both male and female, and the process of generation from the semen to the formed fetus. Volume two, equally large, deals in sequence with everything else, including the thorax and brain, together with the sense organs, the bones, the muscles, the vessels that run through the body, the physiology (or 'animal œconomy'), and finishing with Lyser's *Anatomical knife*, the key to practical dissection.

The sequence that the editors of the *Anatomical library* adopted meant that some of the treatises they reprinted were interwoven, as it were, with others. The anatomist began his work on the human from the skin of the belly, cutting through it to the organs underneath. For this the editors of the *Anatomical library* chose to start with the first ten chapters of an essay that Francis Glisson, Professor

89 Seneca, *Ad Lucilium Epistolae morales*, 1967, modified.

90 For which see today Freer and Cunningham, *Thomas Wharton's Adenographia*, 1996.

of Medicine at Cambridge University, prefaced to his *Treatise on the stomach and intestines* of 1677: 'On the containing parts in general and in particular of Abdomens', which covers the skin and its nerves, nails and hairs.[91] Then follow two treatises by Marcello Malpighi: 'On the external organ of touch' (i.e. the skin as a sensitive organ), and 'On the growth of horns', as related to nails. The editors then switch back to the rest of Glisson's preliminary treatise, which concludes with the omentum. Another short treatise by Malpighi on the omentum is then inserted. Then follow the first eleven chapters of Glisson's *Treatise on the stomach and intestines* proper, up to and including his treatment of the intestines, and including his important discussion of the 'irritability' of the fibres of the stomach (a theme to be built on and more widely generalised in the next century by Albrecht von Haller). The other fourteen chapters of this work are discarded in favour of other writers, of whom the first is Glisson's own physician colleague Thomas Willis, Sedleian Professor of Natural Philosophy at Oxford from 1660. Willis's account of the 'first ways' in which food and medicines enter the body, via the oesophagus, the stomach, through to the guts, is taken, complete with its several illustrations of the human stomach, from his work on the operations of medicines in man's body, his *Rational pharmaceutics, or a Diatribe on the operation of medicines in the human body* (first published in Oxford in 1674), and was built on his own dissections. Then comes a work by Johan Conrad Peyer of Schaffhausen in Switzerland, his *Anatomico-medical exercise on the glands of the intestines and their use and diseases* (1681), followed by his *Anatomy of the stomach of the hen* of the same date.[92] And so forth throughout the two large volumes and all the parts of the human body.

Some of the anatomical findings in the *Anatomical library* were of course built on premises directly contradictory to others. For example, Harvey's Aristotelian work on the heart and on generation is incompatible in its assumptions with almost anything by Malpighi, with his thorough-going mechanistic interpretations. But at this date mechanism had not yet fully triumphed as the superior explanatory model for anatomy or physiology. Some anatomists and physiologists still spoke in terms of faculties and in qualitative rather than in mechanistic and geometric terms about the body. But being up to date seems to have been more important to the editors than presenting one consistent story.

The *Anatomical library* had at least two further editions in Latin (Geneva 1716, 1717), but it was also imitated in English. From 1711 to 1714 a printer and a number of London booksellers collaborated in producing forty numbers (like a periodical) of a *Bibliotheca anatomica, medica, chirurgica, &c, containing a description of the several parts of the body*, which together built up into three volumes, issued over three years. The audience for this work were primarily 'the

---

[91] Glisson, *Tractatus de ventriculo et intestines*, 1677. On this work see also Cunningham, *English manuscripts of Francis Glisson (2)*, 1998.

[92] Peyer, *Exercitatio anatomico-medico de glandulis intestinorum earumque usu et affectionibus*, 1682.

Gentleman who is pleas'd to study Anatomy or Physick as an Accomplishment, and not as a Profession … In the next Place, we shall address the Student in Physick and Anatomy'. Some people argued that all these recent anatomical discoveries had made no improvement in our knowledge of diseases and their cures. The editor vigorously defended the discoveries:

> Now if we consider what vast discoveries have been made by our modern Anatomists, in relation to the structure of the heart, and its use for the circulation of the blood; the motion of the arteries; the uses of the veins; the existency of the lymphatic vessels; the liquor they contain; their propagation and termination; the lacteal veins, and their progress; the structure of the liver, together with its uses, and the separation of the bile; the structure and uses of the parotids; the existency and structure of the glandules of the palate, and the tracheae; the structure of the brain, which separates that liquid matter that passes into the nerves; the structure of the kidneys, with the succenturiate glandules; the structure of the glandules of the stomach; the structure and uses of the conglomerated glandules; the way by which sweat and perspiration is excerned; the generation of eggs; the observations regarding respiration in all living creatures; the structure of the lungs; the structure and uses of the bones, and their medullary oil; the structure and uses of the cartilages; the structure and uses of the ligaments; the structure and uses of the mucilaginous glands between the muscles and interstices, or joints of the articulated bones, &c … who would be so far beyond himself, as still to insist upon those trifles so much extolled by the ancient physicians, which have scarce any foundation elsewhere than in their own brains?[93]

Rather than write entire anatomical systems, modern scholars have divided the task between them, 'which hath produced this happy effect, that within the space of forty or fifty years past, there have been more discoveries made in anatomy, than all the preceding ages could pretend to'.[94]

This was the atmosphere and research culture in which our anatomists worked, a culture that knew no political or geographical boundaries. They were practitioners of a very successful discipline that was going from strength to strength. There was still more to discover, and the opportunity of immortality awaited those who discovered the most important novelties. It was a glorious field, and it was wide open to men of talents!

---

[93] Nutt and Morphew [Manget and Le Clerc], *Bibliotheca anatomica*, 3 vols, 1711–14, vol. 1, pp. 10–11. Because it was issued in parts this work is now very rare; a complete copy is held by the Wellcome Library in London. The editors claim to be following the second edition of the *Bibliotheca*, but I have been unable to find an edition of the right date, i.e. in the 1710s.

[94] Ibid., vol. 1, p. 12.

7. A community of scholars

It will already have been noticed how many anatomists there were across Europe willing to engage in helping to produce the *Anatomical library*, and to buy the resulting volumes. As we shall see in the next chapter, many of Europe's universities had anatomical teaching and professorships of the subject, while there were many private anatomical teachers too. There were also professional anatomists in several other kinds of institution. If I might venture a rough guess, I would imagine that there were maybe about two hundred men, perhaps more, professionally engaged, mostly part time, in anatomical teaching and research at any one time in the long eighteenth century across Europe. Around them was an extensive penumbra of students; if we are to accept the figures of the anatomists themselves, five- or six hundred students at a time would attend in the largest teaching venues, suggesting that something like two- to three thousand or more men were studying anatomy as students at any one time. Of the active teaching anatomists only a few wrote up and published their findings or opinions, but those who did so could be sure that most of what they wrote would be taken up and eagerly scrutinised by their fellow anatomists across Europe. The periodical publications of the many learned societies of Europe provided another way of keeping up to date and joining in the debate. It took only a few weeks for learned publications to cross Europe, and scholars also engaged in networks of correspondence with each other.

So it would seem reasonable to speak about a *community* of practitioners of the discipline of anatomy, even though any particular anatomist might be the only practitioner of the discipline for hundreds of miles around. This community of scholars shared a set of understandings that (even if they were not written down) enabled them to recognise important new discoveries or claims within the discipline. All academic disciplines today work as communities in this way, even if the practitioners are separated by continents and never meet face to face. So it was too in the eighteenth century. An anatomist who published his findings would wait with baited breath to hear what reception his work was given across Europe. And there were certain practitioners whom everyone in the community recognised as being currently the greatest anatomists – the leaders of the discipline, the key players – and their fellow anatomists would sometimes call a particular practitioner 'great' even in his lifetime, as in the case of Guichard-Joseph Duverney. Similarly they would praise each other in print and acknowledge each other's achievements. The 'mutual admiration society' – doing honour to each other by recognising their abilities – is one of the things that makes disciplines thrive. So it was too in the eighteenth century, but with considerably more effusion than is common today. Hence (for instance) the Italian anatomist Morgagni, living in Padua, and whom we shall later meet as the greatest anatomical pathologist of the age, could praise the equally eminent Swiss anatomist Haller, while co-opting the latter's approval for his own opinions: 'Nor, indeed, did the illustrious Haller judge that those cases [of hydrocephalus] could be otherwise explained', wrote Morgagni, 'not only in other places, but in his *Opuscula anatomica*; in which, from his great humanity to me, he was willing to make my name, little as it is, honourable, when he also gave an observation of this kind, on a female foetus, very accurately made, according to his custom, and took notice of many, from others, in consequence of that great erudition, of which he is master'.[95] The arguments and disputes between anatomists, of which there were many, were also keenly followed by the community.

8. A great tradition

Reviving the Moderns

It is a commonplace among historians today that history is made by the historian, at least as much as it is discovered by the research he or she conducts. People write histories for reasons, reasons that tend to shape what they find in the past and how they interpret it, and our anatomists in the long eighteenth century were no different in this respect and had good reasons to be writing histories of anatomy. They were well aware that they had a place in history, viz. that they were participating in the period that Haller called 'the perfection of anatomy' (see Chapter 6 below). The celebration of the glorious history of which they were part was undertaken by several anatomists. In other words, they wrote the history of anatomy as leading up to themselves, and physically reconstructed that history by editing and reissuing classic texts by their heroes, the Moderns – that is, the anatomists of the Renaissance and since. These were acts of obeisance to their forefathers by the practitioners of a flourishing discipline. The number of such homages in the century is quite remarkable.

One of the first great acts of homage paid by early eighteenth-century anatomists to their modern forebears was the finding and publishing of the forty-six (sometimes calculated as forty-seven or forty-eight) anatomical tables drawn and engraved on copper for Bartolomeus Eustachius in Rome in the 1560s. They had been intended by Eustachius for a work that never appeared, and whose text remains lost: *On anatomical dissensions and controversies*. Most probably it would have been a major attack on Vesalius, and a defence of the doctrines of Galen. Eustachius had been Professor of Anatomy at the papal university in Rome, the Sapienza. When the plates were rediscovered in the early years of the eighteenth century, Giovanni Lancisi persuaded Pope Clement XI to buy them and give them to the library of the Santo Spirito Hospital in Rome, which Lancisi was in the process of creating. Lancisi then published the engravings at his own expense in 1712, opening the work with an imaginative image of Eustachius dissecting, seen here in Fig. 1.28, portrayed significantly as an ancient Roman.

[95] Morgagni, *The seats and causes of diseases*, 4 vols, 1769, vol. 1, p. 249.

Fig. 1.28

Lancisi was very conscious of the fact that Eustachius was one of his predecessors as Professor of Anatomy at the Sapienza, and for him Eustachius was 'Eustachius noster', 'our Eustachius'. He had more than once heard his own master, Malpighi, 'that light of the Anatomists' and investigator of the kidneys among so many other organs, saying about Eustachius that 'if he had investigated the structures of the other vessels and viscera not only with the knife but also with the microscope and the injection of liquids (which he used only in the kidneys), then without doubt he would have deterred all later people from investigating Anatomy'.[96] Lancisi was even willing to credit Eustachius with having discovered the thoracic duct (a discovery usually ascribed to Pecquet in 1642), so skilled and observant an anatomist had he been. Moreover, Lancisi had in 1676 read a particular passage of Eustachius on the kidneys, where Eustachius expressed the wish that he had spent more time when young on dissecting people dead of disease;[97] the young Lancisi took this advice and thereafter devoted himself to pathological investigations. So Eustachius was 'our' anatomist in many ways for Lancisi. Morgagni too, in the letter he wrote to Lancisi, published with the Tables, could see from the Tables that 'our Eustachius' had been the first to note many things that were credited to later anatomists.

In addition to having been among the earliest anatomical illustrations produced by using copper engraving (Vesalius by contrast, just twenty years before, had used woodcuts), Eustachius's set of tables has also been judged by historians as

[96] Eustachius, *Tabulae anatomicae*, 1722, p. x.

[97] Eustachius, *Opuscula anatomica*, 1563, p. xiii.

comprising some of the best anatomical illustrations ever produced. Eustachius gave his illustrations an ingenious referencing system, by taking the longitude and latitude markings from maps. Eight of the tables Eustachius had published in his lifetime, in his writings on the kidneys, veins of the inner arm, and the ear. These Lancisi republished with Eustachius's original explanations. Lancisi wrote the explanations for all the other tables that Eustachius had not published himself (Fig. 1.29).

The eighteenth-century history of these Tables, once rediscovered, is impressive. After the first printing in Rome in 1712 the Tables were republished in Amsterdam in 1722 using copies of the original plates, then again in Rome in 1728 using the original plates once more, then in Leiden in 1744 using newly engraved plates, then in Venice in 1769, in Amsterdam again in 1798 (in German), and finally in Amsterdam again in 1800.

Not only were the Tables of Eustachius brought into the public domain initially by that great Italian anatomist Lancisi, but the equally great Leiden anatomist Bernhard Siegfried Albinus edited the version published in Leiden in 1744. In the years of his teaching before he had completed his own anatomical atlas, Albinus taught publicly from Eustachius's Tables, and indeed his students urged him to publish the Tables with his own notes.[98] The 'map-grid' reference system that Eustachius used on the Tables may have been an origin for Albinus's own system of representing anatomy by visual squares (see Chapter 4 below). Albinus supplemented this reference system by having an outline sketch of each of Eustachius's plates printed opposite, on which each part was lettered. This made for extremely clear reference to the parts, and was the system Albinus later used in his own atlas on the skeleton and muscles of the human body. Albinus produced a second, corrected, edition in 1761.[99]

Another set of 'classic' views of the anatomy of the human body were the plates recovered and published in 1741 in Rome, originally engraved probably about 1618. These were drawings, now 'outstandingly engraved', by a Roman artist of the previous century: *The anatomical tables drawn by the celebrated painter Pietro da Cortona*. Pietro (1596–1669) is also known as Pietro Berrettini. The plates were published by Fausto Amideo, with the notes of the anatomist Gaetano Petrioli.[100] An example, seen here in Fig. 1.30, was used to show the bleeding-points.

---

98 Albinus, *Explicatio tabularum anatomicarum Bartholomaei Eustachii, anatomici summi*, 1744, Epistola Auditoribus Suis, p. *1 [*sic*].

99 Four of Eustachius's plates were used in Winslow, *Exposition anatomique de la structure du corps humain*, 4 vols, 1732; Winslow proposes an alternative way to use the map-grid precisely – see vol. 1, p. viii.

100 Pietro, *The anatomical plates of Pietro da Cortona*, 1986. The original drawings came into the possession of William Hunter, and are now in the Hunterian Collection at the University of Glasgow.

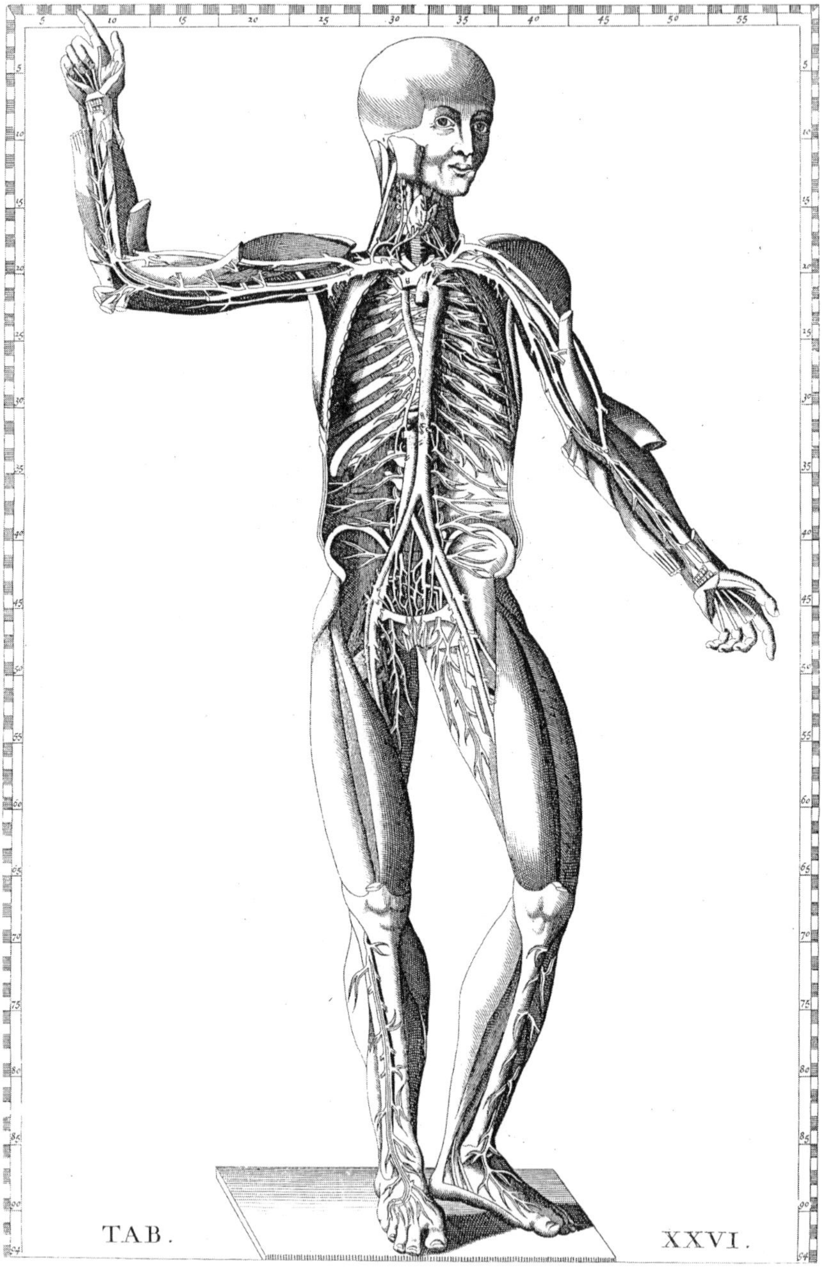

Fig. 1.29

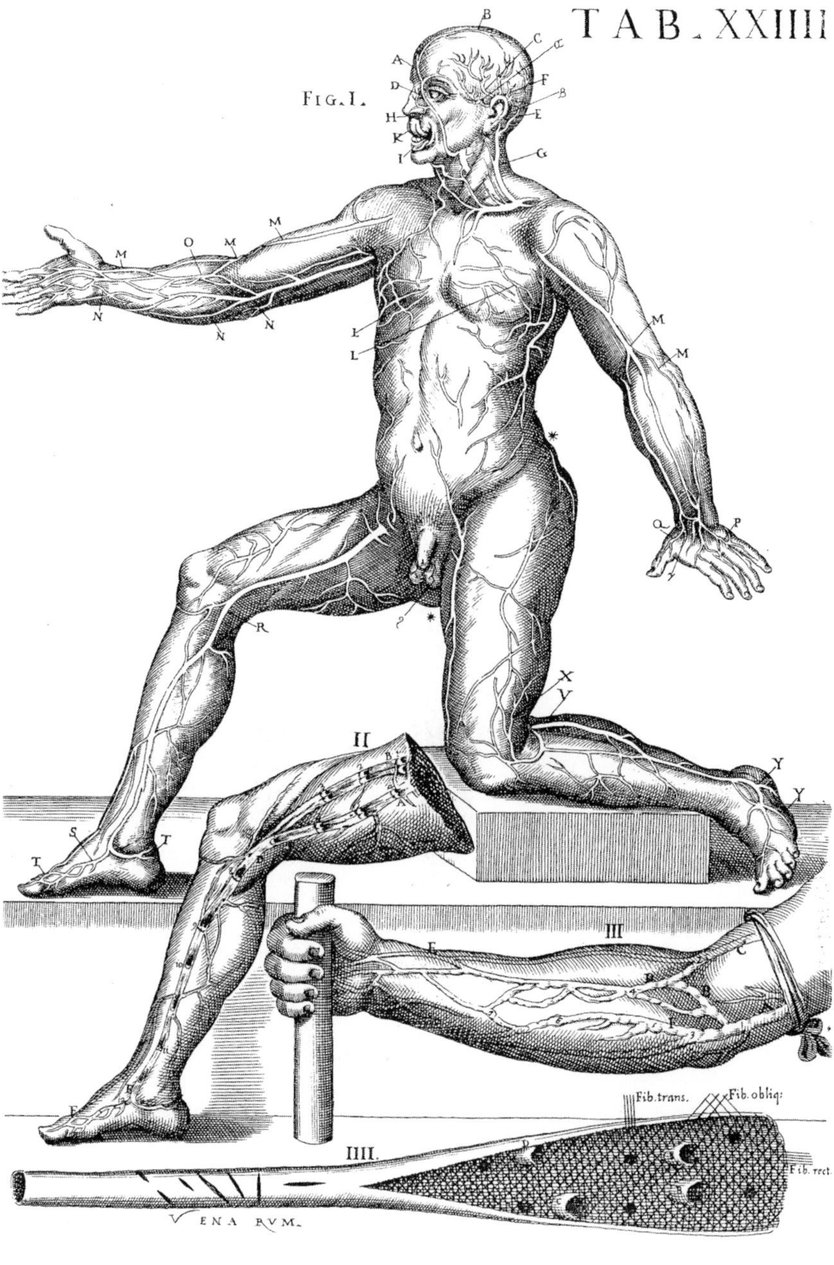
TAB. XXIIII
FIG. I.
II
III
IIII.
Fib. trans.
Fib. obliq:
Fib. rect.
VENA RVM.

Fig. 1.30

William Hunter had a plan to publish a very interesting work, but one that did not actually appear: the anatomical drawings of Leonardo da Vinci. He mentioned his plan in his introductory lectures where he also outlined the history of the discipline:

> In tracing the great revolution of learning, which happened in the fifteenth century, I am enabled to carry the history of the improvement of Anatomy farther back than has been generally done by our own writers; and to introduce into the annals of our art, a genius of the first rate, Leonardo da Vinci, who has been overlooked, because he was of another profession, and because he published nothing upon the subject. I believe he was, by far, the best Anatomist and physiologist of his time; and that his master and he, were the very first who raised a spirit for anatomical study, and gave it credit; and Leonardo was certainly the first man we know who introduced the practice of making anatomical drawings.

William Hunter then mentioned his plan at some time 'to engrave and publish the principal of Leonardo's anatomical designs' in possession of the King.[101]

Other 'classic' texts were republished by the Leiden anatomist Bernhard Siegfried Albinus. Albinus was quite indefatigable in trying to bring back into circulation the works of his great predecessors. In 1725, together with his teacher, the great Boerhaave of Leiden, Albinus produced *All the anatomical and surgical works of Andreas Vesalius* in two volumes, published in Leiden. The first volume contained the *Seven books on the fabric of the human body* of 1543, while volume two contained the *Epitome* of this, the *China-root letter*, Vesalius's replies to Fallopius, together with a work here called Vesalius's *Great surgery in seven books*, which had been originally published in 1568, after Vesalius's death, by Prospero Bogarucci, a work not otherwise attributed to Vesalius. All the images here had been freshly engraved (that is, not woodcuts as in the original), and J. Wandelaar is credited on the title-page picture as the engraver. Strangely, in a 23-page Preface by the editors, they say almost nothing about why they are producing this new book. They give an outline history of anatomy, and then a life of Vesalius, followed by the texts.

A '*Vesalius renovatus*' had been planned by J.-B. Winslow, a Danish anatomist who was Professor of Anatomy in the University of Paris and who taught in the Jardin royal on behalf of Duverney for twelve years (see Chapter 3 below). It is possible that he was diverted from this (instead publishing a different book on human anatomy in 1732, on which see Chapter 3 below) by the appearance of the Boerhaave–Albinus edition of Vesalius. However, a further edition of Vesalius was indeed later produced, which involved Winslow in a different way. The original woodblocks of Vesalius's book had surfaced in Augsburg in 1774, and were bought by Dr von Woltter, chief physician to the Bavarian Elector. He encouraged the physician Heinrich Palmaz Leveling to publish them, with descriptions of the parts taken from Winslow's works. Thus appeared in German at Ingolstadt in 1783 the

101 William Hunter, *Two introductory lectures*, 1784, pp. 37–9.

Anatomical explanation of the original figures of Andreas Vesal, together with the use of the Winslowian dissection doctrine, in seven books.[102]

The next of Albinus's anatomical heroes was William Harvey. An edition containing all of Harvey's published works was put out by a Leiden publisher, Johannes van Kerckhem, who had noticed how hard Leiden medical students had found it to obtain copies, and he asked Albinus to write a preface for it in praise of Harvey. This appeared in 1737. In his Preface Albinus put forward Harvey – 'the immortal glory of Britain' – as a model (*exemplum*) for medical students, as someone who, unlike current physiologists, never mistook the false for the true, and could prove all his claims by experiment. Moreover, Harvey's findings, especially about the heart and circulation, are important in every area of physiology, 'and whoever is not familiar with it, understands nothing of the life of our body, and its causes, nothing of health, nothing of all its functions'.[103]

The next year, in 1738, Albinus, again in association with the publisher Johannes van Kerckhem, produced *All the anatomical and physiological works of Fabricius ab Aquapendente*. Fabricius had been the teacher of William Harvey at Padua. Kerckhem dedicated the work to Morgagni as Fabricius's successor, teaching 'from the same professorial chair (*cathedra*) in which at one time studious young people drank in with avid ears the voice of Fabricius investigating the hidden things of human nature and the amazing structure of the human body'.[104] Albinus wrote a Preface on Fabricius, which is a Life. The images were all new-done engravings, faithfully copying the original woodcuts, though no artist is credited.

Finally, Harvey was again celebrated with an edition of all his published anatomical writings, with *An anatomical exercitation on the motion of the heart* rescued from the many typographical errors of the first edition. However, this edition, put out by the Royal College of Physicians of London in 1766 and edited by the poet and physician Dr Mark Akenside, was more about celebrating Harvey as the College's most illustrious member than with celebrating his work as an anatomist as such. The address to the Reader merely says: 'Since it has been for a long time judged unworthy by those who follow the studies of medicine and natural things that the books of William Harvey, in which he first announced the use of the heart and the motion of the blood, lay so distorted in such great silence, we reckoned that the obligation lay most strongly on us that all his works should be put into your hands in a more decent garb'.[105] It was in these very years, the 1760s, that the fellows of the Royal College were under attack from the licentiates of their own College, who were demanding equal rights in running the College. One of the things that marked

[102] Leveling, ed., *Anatomische Erklärung der Original-Figuren von Andreas Vesal*, 1783.

[103] Harvey, *Opera sive exercitatio anatomica de motu cordis et sanguinis in animalibus* , 1737, Praefatio.

[104] Kerckham and Albinus, eds, *Hieronymi Fabricii ab Aquapendente Opera Omnia anatomica et physiologica*, 1738, Dedication.

[105] Harvey, *Opera omnia*, 1766, Praefatio.

off the gentlemen physicians of the College from the licentiates was their classical culture, and Latin examination was used as a barrier to keep out the unclassical (and especially the Scottish-trained) physicians from the fellowship.[106] And the greatest former member of the College, Harvey, had published in Latin.[107]

Making history

Works of the period that are called 'anatomical history' are mostly using 'history' in the sense of 'account', meaning an account of the parts of the body, and with no reference to the past at all, such as Joseph Lieutaud's *Historia anatomico-medica, sistens numerosissima cadaverum humanorum extispicia*, published in Paris in 1767 ('extispicia' is a fancy word for dissections, originally meaning inspecting the entrails for prophecy).

But our anatomists were very aware of their place in history: aware that their age had ushered in a new epoch in anatomical investigation, and that they were rapidly approaching a state of perfection of anatomical knowledge. Thomas Bartholin, for instance, writing at the beginning of this period, engagingly put it like this:

> Man formerly knew himself by the Face, which he admired as a painted shadow, before he searched into his own Bowels, by courage. Mangling the dead that he might know the living. He first began with Apes, Dogs and Swine and other Creatures, which he long practised upon without satisfaction, and at last came to dissect men that thereby he might better understand himself. And in this famous Art, there are many in every Age, that have immortal Names. But no Age hath been more happy than ours, for Curiosity and Invention, in which men seems [*sic*] to be restored and to flourish, and to know his own Parts, which were before unknown by the Negligence of Anatomists, and to come to perfection in this Art.[108]

In fact it was very common for an anatomist to begin his series of lectures with a short account of history in the sense of 'rise and progress' of the discipline, as Monro *Primus* did every year, or open his lectures with an oration on the topic, as Pieter Luchtmans did in 1760 at Utrecht.[109] These tend to be very platitudinous. By contrast Lorenz Heister in his inaugural oration as Professor of Anatomy, Surgery and Physiology at the Royal and Ducal Julius-University of Helmstadt, chose to speak 'On the increase of Anatomy in this eighteenth century'. Heister claimed that work in anatomy in the eighteenth century exceeded that accomplished in the sixteenth and the seventeenth centuries in every way. As this was still only 1720,

106 Waddington, 'The struggle to reform the Royal College of Physicians', 1973.

107 It is still an open question as to what language Harvey had used to write his books, English or Latin, and whether they had been translated into Latin by George Ent.

108 [Thomas Bartholin], *Bartholinus Anatomy; made from the precepts of his father*, 1663, To the Reader.

109 Luchtmans, *Oratio inauguralis*, 1760.

this was quite a bold claim! Heister supported it by, in effect, going through his library and listing the anatomical publications for each year since 1700 produced 'by outstanding men of various regions of Europe in various languages, Latin, French, Italian, English, Belgian [*sic*] and German'.[110] Heister gave first place among these heroes to Frederick Ruysch, his own teacher, but nevertheless presents the discipline of anatomy as one that knows no borders. Anatomy had been brought 'ad summam perfectionem', to the highest perfection. After giving this extensive oration, Heister publicly dissected a female who had met a violent death.

On a grander scale, there were full-blown histories of anatomy being produced.[111] Charles Jenty, working and teaching in London in the 1750s, complains about the number of big volumes he has had to read in order to find out the tiny discoveries or novelties of most anatomists. So, given the amount of original research that was required to write a history of anatomy or anatomists or anatomising, there are a surprising number of such works. Jenty's three-volume work is basically *A course of anatomico-physiological lectures*, but which was *Illustrated with an historical compendium of the rise, progress and discoveries which have been made in the animal œconomy, gradually traced to the present time*. 'Rise and progress' histories of disciplines and systems was a very eighteenth-century formula of historical writing.

In 1772 the London surgeon William Northcote produced *A concise history of anatomy. From the earliest ages of antiquity. To which are annexed a few thoughts on the uses of anatomy, and rules for giving a course of anatomical lectures*. The work has some good indications of what (from a surgeon's point of view) people had discovered, and what some of their arguments are about. But it has been shown that this was simply a plagiarism of unpublished lectures by Monro *Primus*.[112] Finally another useful history of anatomy was published, this time in French, by Pierre Lassus in 1783 in Paris: *Essay or historical and critical discourse on the anatomical discoveries made by the ancients and by the moderns*.

Another important form of history writing was bibliographic.[113] In the late 1760s, two anatomists, one old and one young, were, unknown to each other, writing great bibliographic surveys of the literature produced over the course of the history of anatomy. In Bern, in Switzerland, Professor Albrecht von Haller (b. 1708), previously the great luminary of the University of Göttingen, a medical commentator and teacher of towering status, a bibliophile and anatomist, was writing in Latin his two-volume *Anatomical library* (*Bibliotheca anatomica*), 'in which are Reviewed the Writings Contributing to Anatomy and Physiology since the Beginnings', as the title says. The first volume covers everything from Antiquity until the year 1700.

110 Heister, *Oratio de incrementis anatomiae in hoc seculo XVIII*, 1720, p. 15.

111 For a listing of such works see Choulant, *Bibliotheca medico-historica*, 1842, pp. 141–9.

112 Russell, *British anatomy*, 1987, pp. 151–2.

113 For instance, Goelicke's book, *Historia anatomicae nova aeque antiqua*, first published in 1713, and republished in a much fuller edition in 1738, in Frankfurt.

The second covers just the period 1701 to 1776. Haller distinguishes a number of stages in the development of anatomy, each represented by a section of his book. The eighteenth century opens, in his account, with a period of 'More Learned Anatomy', covering 1701 to 1718. But then anatomy reaches its peak. The period 1719 to 1775 Haller calls nothing less than 'The Perfection of Anatomy'. It begins (he says) with Bernhard-Siegfried Albinus, that 'summus anatomicus', and includes Antoine Petit, Georges Buffon, the abbé Spallanzani, Antoine Portal, Alexander Monro *Primus*, the Hunter brothers, himself, and others.

Meanwhile in Paris, Antoine Portal (b. 1742), anatomy teacher to the Dauphin, still in his twenties, was preparing his six-volume *History of anatomy and surgery* (*Histoire de l'anatomie et de la chirurgie*), 'Containing the Origin and Progresses of these Sciences, with a Chronological Tableau of the Principal Discoveries', which (like Haller's work) offered a review of just about everything ever written on anatomy, starting in this case from Noah's Flood! Portal surveys almost twelve-and-a-half thousand separate writings, giving his opinion on the merits of each, and on the talents of their writers. Portal's focus is slightly different than Haller's (he is interested in the connection of anatomy with surgery, rather than with physiology), his periodisation of the past of anatomy is somewhat different, and he has different modern heroes. But nevertheless Portal is in total agreement with Haller about the historical importance of the eighteenth century in anatomy:

> Anatomy and surgery have made incomparably more progress in the space of one century than in the previous two thousand years: reading the works of Malpighi, Ruysch, Duverney, Winslow, Lieutaud etc., one would think Anatomy was a new science, and Surgery too has changed its face since the foundation of the Académie de Chirurgie of Paris [in 1731].[114]

Both works are valuable for giving an anatomist's viewpoint on the contribution towards anatomical knowledge that each of their predecessors had made, or were still making. In a supplementary volume Portal also gives a chronological view of discoveries about each part of the human body and where they were first published. Thus for both Haller and Portal the eighteenth century is clearly the high-point of a long history of anatomy and anatomising, a history that has been leading up to this moment.[115]

---

114 Portal, *Histoire de l'anatomie et de la chirurgie*, 6 vols, 1770–73, vol. 5, p. 1.

115 Monro *Secundus* also wrote a history of anatomy that follows the distinction of dividing the history at 1700: *Of the rise and progress of anatomy in two volumes*, consisting of 'Notes concerning the rise and progress of anatomy to the end of the 17th century after Christ', dated 1759, and 'Notes concerning the progress of anatomy in the present 18th century', not dated. This was not printed, but survives in manuscript in New Zealand. See Taylor, *The Monro collection*, 1979, pp. 18, 101–2, and reproduction opposite p. 120. I have not seen this.

Chapter 2
'Merit is sure of its reward': Careers and Courses

Part 1: Men

In this chapter I shall deal with the why, how, where and what of some of the anatomical teaching and research in western Europe, as it applied to men. At the end I ask about the roles of women in the world of anatomising. As usual my account is intended to be impressionistic rather than exhaustive.

1. Why anatomy?

We should not be surprised to find that the greatest centres of activity in anatomy in our period are in some of the richest towns of the wealthiest and most commercial of nations. For anatomical knowledge was a commodity to be bought and sold, and as the customers for it – primarily surgeon apprentices training for the army and navy, and medical students wanting to become doctors – were in the big rich cities, so the purveyors of anatomical knowledge gravitated there as well. This of course is a truism, that intellectual innovation follows trade, commerce and industry, and it applies to all forms of intellectual activity in the modern- and early modern world. Hence for anatomy, as for other subjects, we shall find that the most important cities were Paris, Amsterdam, Leiden, London and Edinburgh. This list of cities shows that the developing commercial north of Europe was coming to dominate the old south. The traditional importance of Italy was on the decline, though it was still strongly represented and we still find much significant anatomical work being undertaken in the traditional home-towns of anatomy: Bologna and Padua, together with Rome. In addition to these, a few towns are important for our story because of the presence of a particular individual, such as Göttingen, where Albrecht von Haller was the highly industrious sole Professor of Anatomy, or Rouen, where Claude Nicholas Le Cat was a one-man band in anatomical teaching and research.

As we shall see, the primary reason why anatomical knowledge was sellable was that it was required for the complete training of a surgeon. As a careers advisor announced in the 1740s:

> The young surgeon must be an accurate anatomist, not only a speculative but a practical anatomist, without which he must turn out a mere bungler. It is not

> sufficient for him to attend anatomical lectures, and see two or three subjects cursorily dissected; but he must put his hand to it himself, and be able to dissect every part, with the same accuracy that the professor performs.[1]

And the demand for surgeons was enormous in the long eighteenth century. Of the many causes of this unprecedented demand, two are particularly important. The first was the creation and maintenance of standing armies and navies. Here the states most responsible (or culpable) were France and Britain. France sustained an enormous army, and surgeons were required in every regiment. From the 1680s every French ship with more than twenty men on board had had to have one surgeon on board. The British army had a surgeon and a surgeon's mate in every regiment of the standing army, with many more recruited in time of war.[2] Britain also maintained an enormous navy for its defence and aggrandisement, and surgeons were required on every fighting ship, from eight to a hundred guns.[3] As far as the British navy was concerned it is very hard to find precise numbers, but in the war years of 1739–44 the numbers of surgeons range between 140 and 440,[4] most of whom would only be on board for a few years, so there was need of a continual supply of trained men. In 1780, during the American war, there were 370 sea surgeons, and 450 in 1783, though most left at the end of the war.

The other main reason for the new importance of surgeons was related to the population increase of the period and its related commercialism. For the increasing population was available for the medical attention of surgeons or surgeon–apothecaries, or their equivalents in the various countries, especially since (at least in Britain) any surgeon who had practised in the army or navy for three years could freely enter medical practice without further examination.[5] Thus the progress – the perfection – indeed, even the continued regular existence of anatomy teaching and research, including within universities, depended not on the abstract existence of problems within the discipline that needed to be solved, but on the commercial and military markets for the very specialised knowledge and practical experience that the discipline provided.

2. How to become an anatomist

'Merit is sure of its reward in this world', said William Hunter to his students: application was all that was needed for the young man of merit to accomplish great

---

1 Campbell, *London tradesman*, 1747, p. 50.

2 Johnstone, *Roll of Commissioned Officers*, 1917, Introduction.

3 Rodger, *The wooden world*, 1986, pp. 348–9.

4 Coulter, with Keevil and Lloyd, *Medicine and the Navy*, 4 vols, 1957–63, vol. 3, pp. 19–20; see also p. 104 for numbers of new surgeons paying into the poor box.

5 Stevenson, 'A note on the relation of military service to licensing in the history of British surgery', 1953, pp. 420–27.

things.[6] Roy Porter has shown how William Hunter lived out this motto in his own life.[7] The long eighteenth century had indeed the greatest number of employments for anatomists, whether in traditional institutions or in private business, and the best anatomists went to the top.

Among the many skills, accomplishments and qualities an anatomist needed if he was to make his name were: manual dexterity; a certain degree of showmanship; confidence in himself and in the veracity of his findings; persistence through long series of experiments and dissections; a thick skin to repel the criticisms of others; and an inurement or habituation both to the cries of the vivisected animals and to the smells and sheer messiness of the dissecting business, and of course to the insects who flourished in the dissection room. But how to become an anatomist?

Keeping it in the family

The easiest route to becoming a professor of anatomy was to follow one's father in the job. To a modern sensibility the succession of sons to the academic chairs of their fathers or uncles smacks of the corruption that the term 'nepotism' conveys, for it seems to assume that blood-ties or connections are more important than the actual merit of the individual. But the 'meritocratic' view is one that in the eighteenth century was only just coming in. More usually the view in the eighteenth century was that the ideal person to succeed one was indeed one's own son, for he had been fully trained up in the business. Family succession was the norm in all sorts of businesses in the eighteenth century, and continues to be so today, with the one exception of the academic world, where merit is all (at least in theory). And it was taken for granted that brainy and able fathers of course had brainy and able sons. Why should the son not be as good as, or better than, the father, and thus be the father's natural successor? We need only look at the serial successes of families of professional musicians of the period, such as the Bachs or the Mozarts, to see how natural a view this was to hold, and how regularly it was confirmed in practice. What held in music also held widely in anatomy.

The first such anatomical dynasty I shall deal with was that of the **Bartholins** in the University of Copenhagen. Caspar (or Kaspar, 1585–1629) Bartholin was the holder of one of the two medical chairs from 1613. In fact this dynasty was one begun by a marriage connection: Thomas Fincke became Professor of Medicine in Copenhagen in 1603, and one of his daughters married Caspar Bartholin, who then became Professor of Medicine in 1613.

His son recalled that Caspar 'spent most of his youth in anatomical experiments (*Magnam aetatis partem inter experimenta anatomica consumpsit*), and had the best tutors that could be found in Europe, travelling to all universities, and these he imitated and partly transcended. He opened many bodies in Padua with Casser. And he consulted with Aquapendent [in Padua], Bauhin and Platter [in Basle], and

6 William Hunter, *Two introductory lectures*, 1784, p. 103.

7 Porter, 'William Hunter: A surgeon and a gentleman', 1985.

also lent them his knife [i.e. assisted them in dissections]'.[8] Caspar's famous book on anatomy, first published in 1611 in Latin, was *Institutions of anatomy showing the account and state of the human body of both sexes*, originally assembled (he says) for his private pupils, and built on all he had learnt from his teachers on his travels. But after eleven years as Professor of Medicine in Copenhagen, Caspar had a severe illness and he vowed that if he recovered (which he did) he would spend the rest of his life studying theology, and so, when he was well again, he was made Professor of Theology in 1624. He produced an ethical handbook or *Epitome of moral philosophy*, published in 1633 after his death. Ole Worm, his brother-in-law, succeeded him when Caspar moved to theology in 1621. Worm died as Professor of Medicine in 1654.

Caspar had six sons and a daughter, and at least four of the sons became a professor of one subject or another, while another was a learned antiquary.[9] Of these the anatomist son was the second son, Thomas (1616–80). Like his father, Thomas spent several years touring the medical schools of Europe, in Leiden, Paris, Montpellier, Padua and Basle. On his return to Copenhagen in 1647 he was made Professor of Mathematics, and in the next year Professor of Anatomy, a post he held until he retired in 1661 (he died in 1680). Thomas's most famous book on anatomy was a reworking of his father's *Institutions of anatomy*, the first edition of which he published at Leiden in 1641 in Latin, even before he had returned to Copenhagen. 'I was required by the Dutch [publishers] to make an addition to them: This I could not refuse, though I had many cares upon me, being a Traveller'.[10] This revised work was one of the most popular anatomical handbooks of the seventeenth century, receiving at least eight editions in the Netherlands and two in Lyons before 1686, as well as translations into English and French. This work, under the title *Anatomy, from the Institutes of his father Caspar Bartholin, and from the observations of all the recent writers, and his own, reformed for the third time according to the circulation of the blood*, had two distinctive features.[11] One was that, unlike Caspar's handbook, it was profusely illustrated with engravings copied from the best anatomists. The other, 'which is the chief, I have in few words set forth the circulation of the blood from Harvey and Wall [Walaeus], which is now received by many'.[12] In later editions he also added his own discovery of the lacteals.

And finally in this anatomical dynasty there was the younger Caspar (1655–1738), son of Thomas, and who studied in Holland under Ruysch and Swammerdam, in Italy where he met Malpighi in Bologna, and in Paris under Duverney. When

---

8 This is according to his son, Thomas: *Bartholinus Anatomy*, 1663, To the Reader.

9 Portal, *Histoire de l'anatomie*, 6 vols, 1970–73, vol. 2, p. 364.

10 *Bartholinus Anatomy*, 1663, To the Reader.

11 The original work was published five times by Caspar before his death. 'For the third time' seems to mean that the edition by Thomas published in Leiden in 1641 is the second.

12 *Bartholinus Anatomy*, 1663, To the Reader.

he returned to Denmark he began making dissections in the university in 1677 at the age of 22, and in 1680 was made Professor of Anatomy. He too made his discoveries and published anatomical works like his father and grandfather before him, and then became a royal administrator from 1691.

So the Bartholin dynasty of father, son and grandson taught anatomy in Copenhagen for about thirty-five years, though not in continuous succession. They did not hold a specifically anatomical chair initially, though the 'new' university statutes of 1621 did emphasise that the holder of the second chair of medicine should give anatomical instruction through dissections. Public – as opposed to private – anatomical demonstrations did not take place until the anatomical theatre was built in 1644, and thereafter the second chair was largely dedicated to anatomy.[13]

Our second anatomical dynasty is that of the **Albinus** family in Leiden. The original German name of the family was Weiss ('White'), but these members of it are always known by the Latin version of the name, as they all taught and published in Latin. In this dynasty the founding father was succeeded by two of his sons in succession in the professorship in Leiden University, while a third son became Professor of Anatomy at another university.

Bernhard (1653–1721), the first of the sequence, was skilled in anatomy, and had been Professor of Medicine at the University of Frankfurt an der Oder before being invited in 1702 to take up the Leiden Chair of Medicine.[14] After his death he was succeeded in 1722 by his 24-year-old Leiden-educated son Bernhard Siegfried (1697–1770) – though this time in a chair of anatomy – in response to the special plea of Boerhaave to the curators of the university. This is the 'great' Albinus, who will play several significant roles in our story. When 'our' Albinus, Bernhard Siegfried, retired from his Chair of Anatomy in 1747 to take up the Chair of Medicine, he was succeeded by one of his younger brothers, Frederick Bernhard (1715–78). This meant that an Albinus was teaching anatomy at Leiden for some 56 years without a break. Moreover another middle son of Bernhard (there had been eleven children in all), Christian Bernhard (1696–1752), had been appointed Professor of Anatomy at the University of Utrecht in 1723.

The success of this dynasty of anatomical professors was equalled by another one, which prevailed even longer. This was the great academic anatomical dynasty of the **Monros**, who occupied the Chair of Anatomy at Edinburgh University in succession for 126 years consecutively. All were called Alexander (though surprisingly the latter two were neither of them first sons), so inevitably they were and are known by numbers: Alexander Monro *Primus* (1697–1767, professor 1726–67), his third son Alexander Monro *Secundus* (1733–1817, professor first jointly with his father and then on his own, 1754–1808), and *Secundus*'s second son Alexander Monro *Tertius* (1773–1859, professor first jointly with his father and then on his own, 1798–1846). Of these three, *Primus* is generally thought

13 My thanks to my colleague Ole Grell for assistance in sorting out the Bartholins.

14 The spellings Bernard and Bernhard are both used in every case.

to have been one of the greatest anatomists who ever lived, while *Tertius* is commonly regarded as one of the worst. There was a nasty rumour current in his lifetime that *Tertius* in the 1820s was still lecturing from his grandfather's notes and talking about his time in Leiden in 1719! However, this is almost certainly apocryphal.[15] In Edinburgh the right of appointment of the professor was in the hands of the town council, which ran the whole university, and its support was therefore needed for each Alexander to take up the post or to succeed his father by first being appointed as joint professor. But this was not just a formality: the petitioners felt they had to demonstrate that each son was a highly competent anatomist who would continue to attract students to the medical school, and thus continue to bring valuable student money to Edinburgh. The first two Monros will appear frequently in the following pages.

To these actual blood-ties in anatomical careers, one might add 'spiritual' dynasties of anatomical professors and investigators, who came from particular schools. There was, for instance, the Anatomical Cabinet of Professor Johannes **van Horne** in Leiden, which was a place where he encouraged anatomical investigation among its pupils. In the early 1660s there was a veritable galaxy of anatomical talent there, not only the future developer of anatomical injections, Frederic Ruysch, but also Johannes Swammerdam, already noted for his discovery and description of the red blood cells (1658) and later a pioneer investigator of insects, and the young Dane Nicolaas Steno, later the discoverer of the parotid duct (which brings saliva into the mouth) and a celebrated teacher of anatomy in Paris.

Another such 'spiritual' dynasty is that of Marcello **Malpighi** and his pupils and grand-pupils over the course of eighty years or more in the great cities of northern Italy. In the Introduction we saw some of the followers gathering around Malpighi's body after his death and assisting in his dissection, especially Giovanni Maria Lancisi and Antonio Maria Valsalva. And then there was Morgagni, still in the 1760s proudly calling himself the 'grandson' of Malpighi. We shall hear more of this group in Chapter 3.

Anatomists were not usually just bred up in one place and one school. They expected to travel to the great centres of anatomical excellence too, such as Amsterdam, Paris or Göttingen, in order to learn from other celebrated anatomists. We have already seen this in the case of the Bartholins, who were prepared to spend years abroad drinking at the founts of anatomical knowledge. Sons were even sent around to learn from their fathers' rivals.

The Monros, for instance, travelled widely to learn from the greatest masters. In the first place young Alexander Monro (later to be called *Primus*) was sent by his father John, who was an Edinburgh surgeon, first to London in 1717 where 'his principal employment was attending … Mr Cheselden's anatomical demonstrations and assiduously dissecting human bodies … This he did without any instructor,

---

15 Wright-St. Clair, *Doctors Monro*, 1964, chapter 13.

except [the books of] Vesalius and Lyser …'.[16] It was to a discussion group run by Cheselden that Monro first gave a disquisition on the bones, which was to become his research speciality. He made some preparations of the bones that he sent home to his father, who used them in Edinburgh to talk up his son's abilities to the town council. Then Monro went to Paris in 1718 where he attended the botanical lectures in the Jardin royal. 'He attended the discourse in the Ecole des Médecins and at the visits of the physicians and surgeons in the Hospitals La Charité and L'Hôtel Dieu in which last place he had a course of anatomy from Mr Bourquet … He likewise was a pupil of M. Grégoire for accouchemens or delivery of women … but always regretted that Mr Winslow, to whose acquaintance he was introduced, gave no course on anatomy while he stayed in Paris'.[17] After some six months in Paris young Monro went to Leiden, 'where he was a student under Boerhaave … Here he met with another misfortune as had happened to him at Paris: Professor Rau was in such a state of health that he could not teach anatomy'. But armed with a letter of recommendation from Boerhaave, Monro visited Professor Frederic Ruysch in Amsterdam and saw his anatomical preparations. In two-and-a-half years the young Monro had met most anatomists worth meeting, and profited from the teaching of some of them. On his return to Edinburgh he was made Professor of Anatomy 'in this citie and colledge' with a salary of £15. But having seen the great centres of anatomising abroad, young Monro had great ambitions for his own teaching in Edinburgh, and soon after he arrived back in Edinburgh he petitioned the town council 'to take into their consideration the properest method of making this place as famous a school of anatomy as any of these to which our youth is sent with so great expence for their education', and urged them to achieve this by making his own appointment for life.

Donald Monro was the second son of Monro *Primus*, and studied medicine at Edinburgh. That he was initially destined for anatomy appears from the fact that in 1747 his father wrote for him 'A treatise on the anatomical encheireses or manual part of anatomy wherein the arts of dissecting, preparing, preserving etc. all the organs of the human body are described', a work that remains in manuscript.[18] He was sent to train in anatomy with Haller in Göttingen and Le Cat in Rouen, and seized every opportunity that came his way of dissecting women who had died just before or just after childbirth.[19] However, by 1754 he was styling himself 'physician at London', where he settled to practise as a domestic and military physician.

---

[16] Monro's autobiography (in the third person), as cited in Wright-St. Clair, *Doctors Monro*, 1964, p. 28.

[17] Ibid., pp. 29–30.

[18] This is discussed further below, in connection with the teaching of Monro *Primus*.

[19] Donald Monro, 'The dissection of a woman with child', 1754, pp. 405, 416: Haller 'when I was his scholar at Gottingen'; a woman who died a few hours after childbirth 'whose body I dissected while I was in Mr. Le Cat's house at Rouen'.

His younger brother, Alexander Monro (later to be styled Monro *Secundus*), also being trained up to anatomy, went on a similar trip, which was to have fateful consequences. For he went to William Hunter in London (who had himself studied for a while in Edinburgh under Monro *Primus*), presenting his father's compliments, and having sent in advance to William a copy of his own Edinburgh MD thesis (in Latin) 'On the testicles and semen in different animals'. William, out of respect for his old teacher (or so he said), offered young Alexander every assistance: 'If you can make any use of us, you will do us a pleasure'.[20] But they immediately began to suspect each other of stealing their ideas and discoveries (Chapter 4 below). His attendance at William's course in London was one stop on Alexander's scholarly travels around Europe to learn what he could from the best masters of the subject; he was then off on his way to Paris, to Berlin where he studied under Johann Friedrich Meckel the elder, and to Leiden where he consulted with Albinus and Peter Camper before returning to Edinburgh via London.

Even Alexander Monro *Tertius* went travelling for anatomy after getting his medical degree in Edinburgh in 1797, 'attending the anatomical and other medical classes in London, and the practice of the London hospitals', and he ventured as far abroad as Paris, returning in 1800.[21]

This kind of education was expensive, both for the travel itself and for the tuition and maintenance costs for long periods abroad. Monro *Primus* was away for two-and-a-half years, for instance. It was therefore an investment in one's career. This was the point of such travels, over and above the obvious aim of learning what other anatomists were up to and what was the latest thing in the subject. In particular it was valuable watching senior anatomists at work and learning their techniques. And having met other famous anatomists one could later ask them for favours; exchange letters, observations and preparations; recruit them on your side in public disputes; and generally build the research community and establish anatomical discussion with one's colleagues and rivals. In the case of the Monros, each successive father could also use the support of famous men abroad to impress people at home, and thus help get his son appointed to a full-time post at Edinburgh. One can see also that, although the eighteenth century was the last one in which intellectual Europe – the 'common-wealth of learning', as it was sometimes called – had a *lingua franca* in Latin, that language was still invaluable for communication in person as well as in print.

Making one's way

Beyond these dynasties, in which it was usually assumed that a son would take up the profession of his real or 'spiritual' father, other young men who became anatomists had usually been on the route to becoming physicians or surgeons,

---

20 William Hunter, *Medical commentaries*, 1777, p. 20.

21 Wright-St. Clair, *Doctors Monro*, 1964, p. 97, quoting a manuscript of Alexander Monro *Secundus*. It is not clear from this account whom *Tertius* studied under at Paris.

and had come across the subject in their own training. They were then diverted into this discipline for reasons of enthusiasm, calling or accident, which would have differed from one individual to another. Some performed anatomy as part of their larger duties as medical teachers, or as one step in the sequence of medical professorships. Others, the ones we are most interested in here, performed anatomy on a full-time basis as professionals.

For an outstanding example of the kind of anatomist who had to make their own way on merit and reputation, we can take Albrecht von Haller, seen here in his maturity in Fig. 2.1. In the early 1730s Baron Gerlach Adolph von Münchausen was creating the new university of Göttingen, the Georg-August-Universität (or the Georgia Augusta), in the Electorate of Hanover, on behalf of its ruler Georg August, who under another crown was also King George II of Great Britain. Münchausen wanted professors whose teaching would 'lead neither to atheism nor naturalism, neither assail the fundamental articles of evangelical religion, nor introduce enthusiasm, nor any evangelical Papacy ... but sober diligent scholars'.[22] The aim was to create a university with a high through-put of students, and hence to turn a profit. Research had not been on the agenda for prospective professors. However, Göttingen has since become one of the two or three most successful universities in the world in terms of original and influential research both in the sciences and in the arts – and even in theology. That astonishing research tradition all started with Münchausen's appointment of the first anatomy teacher, since he wanted an outstanding anatomist to begin the medical teaching there. He asked the advice of the medical king-maker himself, Professor Hermann Boerhaave of Leiden. Boerhaave recommended his own former pupil, the young Albrecht von Haller of Bern. So important was it for Münchausen to acquire Haller for Göttingen and to keep him there, that not only was a splendid professorial house immediately built for him, but also an anatomical theatre on the same scale – shown in Fig. 2.2; the theatre is on the left, the house on the right – both situated in a new botanical garden. Moreover since, as a Calvinist, Haller had no place to worship in Lutheran Göttingen, Münchausen had a church built nearby specially for him and his co-religionists, a church that for theological reasons – there is no altar, and the preacher sees everyone – itself looks somewhat like an anatomical theatre. Münchausen had extremely high expectations of Haller, and Haller actually exceeded all of them, becoming one of the most industrious, innovative and famous anatomists in the world, fertile in discoveries, indefatigable in publication of his results, cajoling his colleagues to engage in research, and attracting many students to study anatomy and medicine at Göttingen.

22 As quoted in Roberts and Tomlinson, *The fabric of the body*, 1992, p. 350.

Fig. 2.1

Fig. 2.2

How had Haller trained himself to fulfil this role for the moment when it was offered to him?[23] Haller's own long-term ambition had actually been to be a *physiologist* and to reform physiology,[24] which he may be said to have accomplished in 1766 with the publication of the eighth and last volume of his *Elements of the physiology of the human body*. But as a young man he knew that in order to become a physiologist he first had to be a skilful anatomist, and that this would be a long job:

> I am convinced that we know scarcely a thing in physiological matters unless we have learnt it through anatomy … But anatomy is a long study, and of many parts: muscles, bones, the exterior limit of the viscera, and the positions and the relations of the parts to each other; then there is the description, new only in our time, of the arteries and veins, the labyrinth of the nerves, still not unwoven, then the more subtle dissection of the smaller parts, out of which the viscera and the glands, and the other visible parts of the animal body are constructed.
>
> This work is so extensive that great men of the past have received praise who have given a full account of one particular part, even a minute or easy part. The circulation of the blood gave Harvey immortal fame; a gland has made famous the name of a Wirsung or a Wharton. So to go through everything, and to have fully seen all the regions of the human body, is as difficult and as rare as a full account of all the immense districts, rivers, valleys and hills. Life is brief, and the life of the anatomist briefer still, which death cuts off or other ways of life

[23] Typically, Haller kept a diary of his student travels, which has been published as his *Tagebuch* (1942).

[24] Haller, *Bibliotheca anatomica*, 2 vols, 1774, 1777, vol. 2, p. 195: 'Nimis gnarus, quam impar fuerim magno a me suscepto labori, constituendae de integro Physiologiae …'.

> interrupt, or some other office of civic life disturbs. It is possible for it to be shown … that all the parts of the animal body cannot be investigated within twenty years; if you take into your calculation the variety there is in the fabric of the parts, and how it is necessary to persist in repeated experiments – unless you want to stop short in part of the truth and be in perpetual danger of being stuck in that oh-so-familiar error into which a single observation of one instance throws us. For, from this fault, we frequently take for true what is really very rare … Of the study of the nerves, after a hundred years, in so many anatomical theatres, by conjoined labours, scarcely anything is known worthily and fully ….[25]

So, with this due appreciation of the enormity of the task he had set himself of becoming an anatomist in order to become a physiologist, Haller carefully planned the steps of his education:

> At Tübingen while still a boy I exercised myself in dissections of brute animals, taught by J. Georgius Duvernoy, so that I had at that stage some knowledge of the viscera and muscles. Duvernoy, that excellent man, used to demonstrate with much labour the whole round of anatomy in dogs, since he lacked human bodies.

Duvernoy[26] lectured on the *Institutes* of Boerhaave. So while still just a boy, Haller went to Leiden so that, 'out of the mouth of such a great man, I should return more learned. I heard Boerhaave's lectures with incredible pleasure'. In fact, Haller recalled, 'In our studies we students had all that we could wish for. Boerhaave, who lectured daily from three to four o'clock, drew people from all over the world to himself, and Albinus could introduce one to dissection better than anyone else'.[27] Haller became familiar with the anatomical theatre, which was within a church, and could accommodate 400 people at a time.[28] Then Haller had to prepare his thesis, in which he denied the existence of a salivary duct that someone had recently claimed to have discovered: 'I began to make my dissections about the Coschwizian duct'. Crossing to England, Haller became acquainted with the excellent anatomist James Douglas ('Der Man is gewiß Anatomicus primi Ord.'),[29] from whom he learnt much about the bones. In Paris Haller attended the teaching of Le Dran, and Winslow showed him the minutest parts of the human

25 Haller, *Elementa physiologiae corporis humanae*, 8 vols, 1757–66, vol. 1, pp. ii–iii.

26 Or Duvernoi. Not to be confused with the much more famous Joseph-Guichard Duverney the Professor of Anatomy at Paris or his many male relatives involved in anatomy there, nor with Georges Louis Duvernoy, who later worked on animals with Cuvier in Paris.

27 Haller, *Tagebuch*, 1942, p. 37.

28 Ibid., p. 40.

29 Haller, *Tagebuch*, 1883, p. 19.

body. But still not satisfied, Haller and a friend acquired a corpse that had been buried, and on it they furtively explored the nerves, the vessels of the heart, the muscles of the hand, and other parts, in imitation of the subtle Winslow. But their theft became known, and Haller felt he had to flee Paris to avoid punishment, so he returned to Basle.

> At Basle, beginning to learn mathematics [with Jean Bernoulli], I fortunately arrived in a period when J. Rudolf Mieg, the Professor of Anatomy, was not able to carry out his duties satisfactorily due to poor health. He therefore entrusted me with the dissections of the bodies, and a little while later with the demonstrations. Skilled in cutting, describing, delineating the parts of the body as accurately as I could, I laid down some rudiments of the art, nor did I lack corpses. At Bern with good fortune I succeeded in getting a place given to me for dissecting and demonstrating bodies publicly.[30]

In fact it was in February 1734 that Haller petitioned the town council for permission to dissect in public. The council's minutes record it thus:

> Doctor Haller presented himself in front of the City Council, and he said that he would like to give public lectures on anatomy and botany to those who were interested, and he was granted permission. In connection with this he would be allowed to dissect the corpses of people who died in Spallenwerk [the poor house] and executed criminals. He furthermore applied to get a place for dissections and lectures. He will be given a place in the Insel hospital [the Inselspitel, founded in 1354], if available, otherwise he should return and he would get a place elsewhere.[31]

Haller's account of this moment continues: 'I worked on a two-headed fetus and then on other corpses, and I learnt much that I had not known'.

So it can be seen that Haller had been deliberately acquiring the necessary knowledge, skills and contacts to build a career in anatomy if the opportunity offered, as it did when Münchausen head-hunted him for the chair at Göttingen in 1736. 'Called to Göttingen, there was nothing more important for me to do than to build an anatomical theatre, and to supply it with corpses'.[32]

---

30 Haller, *Bibliotheca anatomica*, 2 vols, 1774, 1777, vol. 2, p. 196. Haller 'laid down the rudiments' in a manuscript work, still in Bern.

31 I am grateful to the Haller archives for providing a photograph of this entry, and to Dr Tatjana Buklijas for this translation.

32 Haller, *Bibliotheca anatomica*, 2 vols, 1774, 1777, vol. 2, p. 196.

'Rising merit'[33]

As we have seen, there were no apprenticeships in anatomy anywhere as such. Becoming an anatomist was not like becoming a physician or surgeon. There was no direct entry into the profession. Usually one would have received anatomical training as an apprentice surgeon or, on the continent, as an 'apprentice' physician. This is why those hoping to become anatomists sometimes married the daughter of their teacher or employer, in order to inherit the business – keeping it 'in the family' in the most direct way!

So finally we look briefly at how careers were built in the world of private anatomical teaching. Here the situation was quite different from the universities or other formal institutions. If the private teacher did not attract enough pupils, then he simply went out of business. Such private anatomy teachers are very visible in London and Edinburgh, but only occasionally on the continent. So here merit, or at least commercial guile, was crucial to the career of the anatomist.

For our example here we may look at a family business where a brother followed a brother, and then a nephew followed them both: the **Hunter** brothers, William (1718–83) and John (1728–93), and their nephew Matthew Baillie (1761–1823). Here we find the nearest thing to an apprenticeship system, where one anatomist 'trains up' some novice, as William Hunter did on more than one occasion, in order to equip his school with assistant teachers and people able to prepare dissections and guide students. Such 'training up' consisted of seeing what the pupil could do with their natural aptitude, and then either keeping him on and presenting him with new challenges, or dispensing with his services.

Both William and John were very much in favour of merit as the criterion of advancement, in every sphere, not just anatomising, and were against the traditional system in Britain of advancement by family ties or connections, or even by the purchase of positions.

William Hunter had been educated at Glasgow University, and it was initially intended that he become a church minister.[34] Exposure to the teaching of theology, however, turned him very much against this as a career. Instead William trained for medicine, first spending three years following the practice of William Cullen, and then three more years in Edinburgh and London. In Edinburgh he attended the anatomical teaching of Monro *Primus*. He then went to London, arriving in late 1740. Here William introduced himself to another expatriate Scot, William Smellie, with whom he was to live for a while, and studied midwifery with him. He also attended the private anatomy lectures of Dr Frank Nicholls, which may not have included dissections at that time. William was favoured by James Douglas,

---

33 'Gunning's Philippic', as quoted in Cope, *History of the Royal College of Surgeons of England*, 1959, p. 14. This is dealt with further in Chapter 6 below, section 3, 'Ending the sacred ritual'.

34 For William Hunter's career I follow Peachey, *A memoir of William and John Hunter*, 1924, chapter 2, who corrects many errors in earlier accounts.

another Scottish anatomist in London, who arranged for William to be entered as a surgical pupil at St George's Hospital. Then shortly afterwards, in 1742, 'On his death-bed', as William wrote to his mother, Douglas 'acquainted the family that I must go to Paris … After I come from Paris I have a scheme laid out of settling here, and certainly nobody can say that it will not succeed'.[35] On his return from Paris William advertised his course of anatomical lectures in October 1746. Private courses such as this were given at the home of the anatomist or in rented properties, not in institutions. William had about twenty students in 1748, and about a hundred in 1756.[36]

At the age of 21, John Hunter came down from Scotland to London, where William,

> … anxious to form an opinion of his brother's talents for anatomy, gave him an arm to dissect for the muscles, and found the performance greatly exceeded his expectations. He was then given an arm in which all the arteries were injected, and these as well as the muscles were to be exposed and preserved; and in the result William did not scruple to say that he would become a good anatomist, and that he would not want for employment.[37]

William employed John as his dissector and prosector for eleven years, after which John set up business on his own.

In time these two anatomical brothers fell out over who discovered what first, so grievously indeed that they ceased to speak to each other. Part of this dispute will be told in Chapter 4. William never married and died without issue, and left his anatomy theatre and house to his nephew Matthew Baillie, together with the use of his collection for twenty years, which gave Baillie a considerable advantage when teaching.

3. Where to teach and practise anatomy

There were many different places and institutions in Europe during the long eighteenth century where one could earn a living teaching anatomy and related subjects. In most cases the students had to pay for the tuition, but in some cases the teaching was free, including one of the very best places. My aim is not to give a comprehensive account of all possible institutions, but to deal with some of the

---

35 Ibid., p. 74.

36 Ibid., p. 98; John Hunter claims William 'had about 60 the first course and they increased as his course lengthened and became a course of above 100 lectures' – Brock, ed., *William Hunter … A memoir*, 1983, p. 5.

37 Everard Hume, as quoted in Peachey, *A memoir of William and John Hunter*, 1924, p. 101.

most important ones and to offer a sketch of a few of the others, in order to provide an overview of the range and kind of facility available.

I shall put France at the centre of the picture, and even more specifically Paris, since in the long eighteenth century the single most important institution in Europe for teaching and research in anatomy, as for other subjects concerning research on the phenomena of life, was in Paris. Moreover, in Paris almost every possible institutional – and non-institutional – method by which anatomy could be taught and researched existed; that is to say, more different ways than in any other single location in the world.

Anatomy in institutions under royal patronage in Paris: The Jardin royal, the Collège royal, the Académie royale des sciences, royal anatomists

The most important institution for anatomical research and teaching in the world through the long eighteenth century was the Royal Garden, the *Jardin royal* or the *Jardin du roi*, also known as the *Jardin des plantes* or the *Jardin royal des plantes* in Paris, seen here in Fig. 2.3. As its name announces, it was an institution entirely supported on royal funds. The reasons for its importance to anatomy were several. In the first place it maintained at least one full-time anatomist, who was expected to perform research as well as teaching. Second, this anatomist lectured and demonstrated publicly for free, which attracted not just Frenchmen but also foreigners from the Netherlands, England, Scotland, and the German states – and even (it was claimed for the 1790s) Peruvians, Brazilians, Anglo-Americans and Asians.[38] Third, after a quiet start, a series of really excellent anatomists, skilled, dedicated and innovative, were chosen for this post; they were mostly also very good teachers. From 1727 this lecturer was complemented by a demonstrator in anatomy and surgery.[39] Fourth, from 1692 the Jardin had an excellent purpose-built anatomical theatre seating up to 600 people; it is visible as the round building at the back, right, of the picture. And finally, the lecturing was associated with a fine and constantly updated collection of anatomical preparations, both animal and human, which were open to free public view in the Royal Cabinet, and Louis-Jean-Marie Daubenton, no mean anatomist himself, filled this curatorial role for almost fifty years from 1745.

The garden had originally been founded to provide free teaching of the medicinal properties of plants. It had been established at the behest of an apothecary

[38] Hamy, 'Les derniers jours du Jardin du roi', 1893, p. 100.

[39] Laissus, 'Le Jardin du roi', 1964, pp. 338–9. Here too things were very much kept in the family, for the post was held successively by Christophe Duverney, 1727–48 (brother of Joseph-Guichard Duverney, the anatomy lecturer); and then by members of the Mertrud family: Antoine Mertrud, 1748–64; then his nephew Jean-Claude Mertrud, 1764–87; then Jean-Claude's son Antoine-Louis-François Mertrud, 1787–1802. During the Revolution this position was transformed into a professorship of the anatomy of animals (see Chapter 6 below).

Fig. 2.3

and physician in ordinary to the French king, Guy de la Brosse.[40] The founding edict was issued in 1626, and the garden was functioning by 1634. From the very beginning there were three 'demonstrateurs et operateurs pharmaceutiques', whose role was to demonstrate – in French – the exterior and interior of plants and all other medicaments. In time, through various little sleights-of-hand by interested parties, these three initial pharmaceutical demonstrators became professors of botany, chemistry and human anatomy respectively.[41] From 1635, supposedly for just one tenure, a certain Marin Cureau de la Chambre was given one of these positions and was obliged by ordinance 'to make ocular and manual demonstrations of each and every of the operations of surgery, of all possible kinds'. Thereafter this surgeon–anatomist was officially 'Demonstrator–operator of the interior of plants with the title of professor of anatomy and surgery'! While La Chambre's published works were on the passions and on the soul, it is nevertheless likely that he carried out his prescribed teaching in the Jardin, since later anatomists refer to a 'revival' of teaching after his time. He was succeeded in 1671 by his son François, who was

[40] Howard, *Guy de la Brosse*, 1974.

[41] I follow Hamy, 'Recherches sur les origines de l'enseignement de l'anatomie humaine', 1895, closely here, who reproduces the pertinent documents.

also a royal physician. By royal decree of 1673 the professors in the Jardin were granted 'the liberty of performing surgical operations, dissections, and anatomical demonstrations'. The professors of the Jardin were to have delivered to them 'the first corpse executed', in preference to the claims of all others, including the Dean and Faculty of Medicine of Paris, on condition that the courses and demonstrations of anatomy be given to anyone who wanted to attend, for free – 'à portes ouvertes et gratuitement' – and given 'in the customary manner'. François Cureau de La Chambre deputed another doctor of the faculty, Pierre Cressé, to undertake the anatomical discourses on his behalf, and a royal surgeon, Pierre Dionis, to undertake the dissections and demonstrations.

A better choice of deputy could not have been made. For it was Dionis (Fig. 2.4), with his remarkable application and success with the public, who first made the international reputation of the Jardin as an anatomical teaching centre. He claimed that the number of spectators at his dissections and demonstrations was regularly as high as four- or five hundred persons. The numbers were always highest when he demonstrated the organs of generation in the corpse of a woman, since (as he said knowingly to his exclusively male audience and with, one imagines, a wink) 'their curiosity is inflamed by the sight of those parts', and 'nothing can be more natural to a Man, than a Desire of knowing where and how he was formed'.[42] Dionis was a very engaging lecturer and demonstrator: his teaching was laced with humour and anecdotes, and he was extremely clear in his explanations. He carried out his duties, as he himself said, 'avec toute l'ardeur et l'exactitude qui sont dûes' to the orders of His Majesty.[43] Dionis taught for eight years in the Jardin. His teaching came to an end with the death of the titular holder of the post in 1680. This coincided with Dionis's appointment as first surgeon to the Dauphin, which meant he would have been too busy to continue with the teaching.

Why, one might ask, did the French monarchy support this free teaching in the Jardin, which was so eagerly received every year by crowds of young men? Dionis correctly linked it to the need for surgeons for the French army:

> For to how great Dangers would wounded Men be exposed, if the Surgeon, who is to make an Incision, or to use the Trepan, or to extract Bullets, or a Splinter of a Granado [i.e. a grenade], did not know how these Parts are formed? Could he, without this Knowledge, cure so many wounded Men, and perform so many fine Cures in the Army, where strange and amazing Wounds do happen every Day?
>
> And it is for this Reason, GENTLEMEN, that the King, who well knows how useful able Surgeons are, has pleas'd to Order these Exercises in the *Royal Garden* to be again renewed, they having suffered an Interruption for divers Years; to the end that Publick Anatomies should here be made *gratis*, and all the Operations of Surgery should here be shewn, for the giving a free Opportunity

42 Dionis, *The anatomy of humane bodies improv'd*, 1703, p. 178.

43 Dionis, *Cours d'operations de chirurgie*, 1740, p. v.

> to Students to make themselves perfect in an Art, unto which His Majesty does owe the Preservation of his greatest Captains.[44]

In Dionis's own practice, the anatomy teaching was presented as being essential for the proper training of a surgeon: 'the surgeon ought to know Man in his perfect state', Dionis proclaims, 'that he may be the better able to re-establish his Health, when it comes to be altered and destroyed by some Disease, or ill Accident. It is by Anatomy, Gentlemen, that he must acquire this Knowledge, seeing that it is the *basis* and foundation of Surgery'.[45] However, the coverage of anatomy that Dionis presents is in no way restricted to just what a surgeon needed to know for practice, but gave a complete coverage of the subject. Indeed, the present arrangements, under the oversight of the First Physician to the King, M. Fagon, were so successful, Dionis claimed, 'that this School may now be said to be one of the most flourishing in the World'.[46] We shall deal with the content of Dionis's teaching later in this chapter.

Fig. 2.4

Dionis was succeeded by Joseph-Guichard du Verney, more commonly spelled Duverney (1648–1730), known even in his lifetime as 'the great Duverney'.[47] Trained in anatomy at Avignon, he had come to Paris to pursue his fortune, and was given admittance to the Académie royale des sciences in 1676 as an anatomist. Of his early private lectures there Fontenelle, the writer of his Éloge, recorded: 'He demonstrated what had been discovered by Steno, Swammerdam, de Graaf and the other great anatomists, and he soon had a reputation'. But he also had remarkable eloquence:

> This eloquence was not just that of clarity, judgment and order … but it was a fire in his expressions, his turns of phrase, and even in his pronunciation, which would have sufficed in an orator. He could not announce with indifference the discovery of a vessel, or a new use of a part, his eyes shining with joy, and his whole person animated … As he became increasingly fashionable, so he made Anatomy, which hitherto had been enclosed in the Schools of Medicine or at

44 Dionis, *The anatomy of humane bodies improv'd*, 1703, p. 92.

45 Ibid., p. 1.

46 Ibid., p. 93.

47 Duverney, *Traité des maladies des os*, 2 vols, 1751, vol. 1, p. v. Duverney's first names are sometimes given in the opposite order, viz., Guichard-Joseph. On Duverney see in particular Guerrini, 'Duverney's skeletons', 2004.

> [the surgeons' hall at] Saint-Côme, appear in the beau monde, presented by his hand. I recall seeing members of the beau monde who were carrying away with them some dried anatomical preparations of his, to have the pleasure of showing them in company ....[48]

Antoine Portal, anatomist and historian of anatomy, later said of him that 'the obstacles that he had to surmount added to his celebrity: he was a great observer and the best Professor of Anatomy of his century'.[49] And according to Cuvier, writing much later, although Duverney published hardly anything except a treatise on the ear, he 'carried out an infinite amount of research on all parts of the body, compared with the corresponding parts in animals'. In a dispute with Méry over the direction of the circulation of the blood in the human fetus, Duverney 'to clarify the question had recourse to reptiles where the circulation is similar to in the fetus'.[50]

In 1679 Duverney was made Professor of Anatomy in the Jardin, where he attracted crowds of foreign students – 140 at one count – to the anatomical courses who would return to their countries and make the name of Duverney celebrated. However, in time a 'weakness of the chest' led him to pass the demonstrating duties to a surgeon, while continuing to present the discourses.

Just how free the anatomical teaching at the Jardin was during the tenure of Duverney, was witnessed in 1698 by Martin Lister, a Welsh physician visiting Paris for six months.

> I was to see Monsieur *Verney* at his Apartment at the upper-end of the Royal Physick Garden; but missing my Visit, went up, with a young Gentleman of my Lord Ambassador's Retinue, to see Mr. *Bennis* [a visiting Englishman], who was in the Dissecting Room, working by himself upon a Dead Body, with his Breast and Belly gutted; There were very odd things to be seen in the Room. My Companion, it being morning, and his Senses very quick and vigorous, was strangely surprised and offended; and retired down the Stairs much faster than he came up: And indeed, a private Anatomy Room, to one not accustomed to this kind of Manufacture, is very irksome, if not frightful: Here a Basket of Dissecting Instruments, as Knives, Saws, &c. And there a Form [i.e. a table] with a Thigh and Leg flayed, and the Muscles parted asunder: On another Form an Arm severed after the same manner: Here a Trey full of Bits of Flesh, for the more minute Discovery of the Veins and Nerves; and every where such discouraging Objects. So, if Reason, and the good of Mankind did not put Men

48 Fontenelle, *Suite des Éloges*, 1733, pp. 374–97, see pp. 375–7. The accounts of Duverney's life given by Portal and Cuvier depend completely on Fontenelle's *Éloge*.

49 Portal, *L'histoire de l'anatomie*, 6 vols, 1770–73, vol. 3, p. 2.

50 Cuvier, *Histoire des sciences naturelles*, 5 vols, 1841–45, vol. 3, p. 251.

> upon this Study, it could not be endured: for Instinct and Nature most certainly abhors the Employment.[51]

After Duverney's retirement in 1718 the post of Professor of Anatomy was given to his son, Emmanuel-Maurice Duverney, and then in 1729 to François Hunauld (1701–42), who had studied under both the elder Duverney and Winslow. While Portal (who could not have heard him) says his lectures were very instructive because of the sequence he adopted, he is generally contemptuous of Hunauld's publications,[52] and Hunauld never published his lectures.

Jacques Bénigne Winslow was the next holder of the professorship, from 1743, though his age – he was already 74 years old – meant that he could only fulfil the role for a few years, before petitioning the king to name a successor who could teach in his place. His Majesty favoured Antoine Ferrein (1693–1769), who succeeded to the chair in 1751. To complete the listing of the sequence of anatomy teachers in the Jardin: Antoine Petit was appointed in 1769, and when he fell ill in 1775 Vicq d'Azyr substituted for him for three years. But after Petit's withdrawal to the country it was not Vicq d'Azyr who succeeded him in 1778, but Antoine Portal, who kept the post until his death in 1832. Partisans of Vicq d'Azyr portrayed this as a disaster, with the customary eight hundred students reduced to a rump of a dozen or so in an amphitheatre built for twelve hundred.[53] But whatever one might think of this criticism, taken altogether this is an impressive roll-call of anatomists at the very heart of anatomising in Europe throughout most of the long eighteenth century.

There was no equivalent of this institution in any other country or state of early modern Europe – nothing remotely like it in its size and facilities – though rulers did sometimes support anatomy through the creation and maintenance of medical faculties in universities, as the elector Georg August did in Göttingen (as with Albrecht von Haller, discussed above).

But this was not all. For there were other royal institutions in Paris in which anatomy was taught beside the Jardin royal. There was the Collège royal, founded by the crown in 1530 with a dozen 'royal readers' or professors in the arts and sciences. Again, there was no equivalent of this institution anywhere else in the world. The teaching was notionally free to all who wished to attend. In the chairs of medicine the holders would sometimes teach anatomy. Antoine Ferrein (professor 1742–69) did so, and William Hunter attended his lectures in 1743–44.[54] Regulations of this period specified that there should be two teaching terms and that the teachers should give three lessons a week, each lasting an hour-and-a-

51 Lister, *A journey to Paris in the year 1698*, 1699, pp. 63–4.

52 Portal, *L'histoire de l'anatomie*, 6 vols, 1770–73, vol. 4, pp. 668–73.

53 Hannaway, 'Vicq d'Azyr', 1994, p. 284.

54 On Ferrein see Saban, 'Deux anatomistes Monpelliérains', 1985.

half, with the first hour being devoted to the lecture and the half-hour to questions and discussion.[55]

By 1772, when it was placed under the University, the Collège had nineteen such chairs. The Chair of Surgery was now assigned to Anatomy, and a theatre for anatomical and chemical demonstrations was built jointly by the two institutions at the Collège, opening in 1781.[56] There was only one holder of this anatomy post in the eighteenth century and beyond, through all the changes made to this institution during the Revolution: Antoine Portal. He was Royal Professor of Medicine from 1769 and became the Professor of Anatomy in 1772, and he held the post until his death in 1832. It was said by Jean Darcet, Portal's colleague there as Professor of Chemistry, that the Collège royal 'begins where all other education establishments end; it is not the elements of the sciences that is taught there, but the sciences themselves … in a word, what is most profound, most *recherché*, most useful, most direct in each area'.[57] For his part Portal seems to have used his teaching here to develop his research on pathological anatomy.[58]

The third of the royal foundations in which anatomy was practised in Paris in the long eighteenth century was the **Académie royale des sciences**, founded in 1666, which paralleled the Royal Society of London, but initially with some twenty or so pensioned members who were expected to engage in communal research.[59] Here the audience was private rather than public, and the work was aimed at research rather than demonstration. A great joint project on the anatomy of many rare and exotic animals resulted in *Memoirs to aid the natural history of animals*, which was published in 1676 (Chapter 5 below). The bodies for this research were animal bodies acquired from the royal menagerie. Duverney was the chief anatomist at the Académie from 1688 to 1725, where he assembled a huge collection of skeletons.[60] Later, Ferrein was the academician who specialised in anatomy (1743–68), and he was succeeded by Portal, who held the position well into the next century.

In other states of early modern Europe there were similar institutions to the Academy, though none in which the members received pensions, and anatomy was sometimes studied in, or researches were reported to, such institutions as the Royal Society in London.

Finally among the ways of making it as an anatomist and receiving royal pay in Paris, there was also an occasional *royal anatomist*, whose role was to teach some anatomy to the king, or to the heir apparent, and to the courtiers both male

55 Torlais, 'Le Collège royal', 1964, pp. 267–8.

56 Saban, 'Deux anatomistes Monpelliérains', 1985, p. 155.

57 Quoted from Jean Darcet (1725–1801) without reference in Torlais, 'Le Collège royal', 1964, p. 274.

58 Torlais, 'Le Collège royal', 1964, p. 284; cf. p. 282 for what Portal taught before the chair was reassigned.

59 On which see Stroup, *A company of scientists*, 1990, and *Royal funding*, 1987.

60 Guerrini, 'Duverney's skeletons', 2004.

and female. In France there was the position of anatomist to the Dauphin (the heir apparent) from the 1670s, and Duverney was the first to be given this role. It was no sinecure, as Duverney's obituarist reveals:

> Duverney prepared the anatomical parts at Paris, and transported them to Saint-Germain or Versailles. There he found a formidable audience: the Dauphin surrounded by the Duc de Montausier, the Bishop of Meaux [i.e. Jacques-Bénigne Bossuet], and Monsieur Huet, then Bishop of Avranches, and Monsieur de Cordemoi, all of whom were very knowledgeable and very capable of making judgements even on matters new to them. The anatomical demonstrations succeeded so well in front of the young prince, that he sometimes offered not to go hunting if the demonstrations could continue after his dinner.[61]

There were no successors to this honorary role until Antoine Portal held it in the 1760s. In his case it was a ruse used by his allies to get round the prohibition on non-Paris graduated physicians from practising in Paris. Physicians connected to the royal household were exempt from this prohibition, so Montpellier-graduated Portal could use this appointment to build a practice, to set up private anatomical teaching, and even to establish his own anatomical theatre.

A title such as this, of course, shows more about the ambition of the anatomist in question than it does about the wisdom and forethought of the king or his ministers. We can see this particularly in the case of Claude-Nicholas Le Cat (or Lecat, 1700–1768), a Rouen lad who had first trained in surgery and then earned a doctorate in medicine in Paris, before returning to Rouen to practise. He taught surgery and anatomy privately, while also holding the position of surgeon-in-chief to the Rouen hospital. In 1736 he successfully applied to the French crown to elevate his private teaching in Rouen into a public anatomy school, complete with amphitheatre. Le Cat received the title of Royal Professor–Demonstrator in Anatomy and Surgery; and, after offering free teaching in this role for a decade, he also received a royal pension.[62] 'This establishment, so useful to the province, was favoured by the municipal authority, which placed it in a building on top of one of the gates of the town, and M. Lecat contributed to constructing an amphitheatre inside it … M. Lecat enjoyed a great reputation, especially with foreigners. Many came to France to profit from his teaching … Keen students found with him all the necessaries for their education: head of a great hospital, at the head of an anatomical school … his house was a true lycée, where could be found together an excellent library full of choice works, a cabinet of anatomical preparations, another full of machines of physics and a valuable collection of natural history'.[63]

As with so much that happened in France, the initiative of having a royal anatomist was followed occasionally elsewhere. Nicolas Steno appears to have held

61 Duverney, *Traité des maladies des os*, 1751, Éloge de M. Du Verney, p. cxci.

62 Portal, *L'histoire de l'anatomie*, 6 vols, 1770–73, vol. 5, pp. 167–8.

63 *Éloge* of Lecat, by Louis, *Éloges*, 1859, pp. 133, 148.

a similar position as royal anatomist to the Danish crown for a short time. Steno returned to Copenhagen in 1672 after eight years abroad, and having converted to Catholicism in 1667. This position was given to him in response to a request from the king, Christian V (1646–99), via his chancellor. He was promised religious toleration, and a salary of 400 thalers, but apart from serving as an anatomist the precise nature of his position was unclear. Considering the University's Lutheran base and the insistence that all professors should be Lutheran, a university position would have been problematic. Thomas Bartholin described Steno as royal anatomist, as did Matthias Jacobaeus, according to Winslow in his autobiography. His only public dissection, of a female body, took place in January/February 1673. Steno left Copenhagen again in spring 1674.[64]

The position of royal anatomist also existed in Britain, at least in name. Nathaneal St André, by birth Swiss, was appointed anatomist to the royal household in 1723 by George I, primarily it seems because the German-speaking king appreciated his fluency in German. He gave some public lectures in this role, and held it until his death in 1776, though without drawing the salary. When Haller met him in London in 1727 he thought he was a big windbag ('ein großer Windmacher').[65] Unfortunately St André was also gullible and liable to disasters. Not only did he confirm that the fraudster Mary Toft had given birth to eighteen rabbits, but he was also suspected of having poisoned his wife's first husband while giving him medical attention.[66]

We have seen that Paris was well equipped with royal institutions for anatomical teaching, usually offered by people who did the job they were being paid for, rather than through deputies. There was always someone ready to solicit for the posts if the holders showed any lack of enthusiasm in carrying out their duties. But – then as now in France – there was often *cumul*, the accumulation of posts by one person, technically improper but overlooked when the holder was assiduous and energetic in their teaching. Ferrein, for instance, was simultaneously a royal professor at the Collège royal from 1742, and professor in the Jardin from 1751. Similarly, the ambitious Antoine Portal was able to hold several top positions at once and for many years – and all this in addition to being an expensive private physician and prolific author. He was anatomist to the Dauphin shortly after he arrived in Paris in 1766; he was named adjunct to the anatomy professor at the Académie royale in 1768, when he was only 26 years old; in 1778 he succeeded Petit as Professor of Anatomy in the Jardin royal. He held the last two positions until his death in 1832 – for sixty-four years in one instance, forty-four in the other. Yet reportedly he continued to draw crowds of students to his teaching in the Jardin. Importantly, Portal was not constrained by political principle nor fazed

64 My thanks to my colleague Dr Ole Grell for guidance here.

65 Haller, *Tagebuch*, 1942, p. 18.

66 *Dictionary of National Biography*.

by political changes: he swore allegiance to five kings, a republic and an emperor in order to keep his jobs![67]

Royal surgical institutions in Paris

Because of the centralising tendency of the French monarchy, there were more opportunities in France than anywhere else for improving surgical education in general, including anatomy teaching, under the control of the *premier* surgeon of the king. The historian Toby Gelfand has called the surgical profession in eighteenth-century France 'a "monarchical profession", as a way of indicating the extent to which its organisation depended upon and reflected the royal state. The king's *premier* surgeon played a pivotal role, using his position of intimate access to the sovereign at Versailles to obtain royal patronage and legislation for his profession'.[68]

The surgeons' guild dated back to the medieval period, and was originally a licensing body rather than an educational one, and associated with the parish church of Saint-Côme. A royal ordinance of 1533 allowed this organisation to act as if it were an academic institution, including wearing the 'long gown' and examining future members in Latin. There was a similarly ancient barbers' guild, which had progressively taken over much of the manual work of surgery, and these 'barber–surgeons' had taken to wearing the 'short gown'. With the inevitable conflicts between the long gown surgeons and the university faculty of medicine, the physicians came to favour the short gown surgeons and sought to control them by demanding an oath of allegiance to the faculty and for them to pay the faculty annual dues. The two kinds of Parisian surgeon were united into one company, of about fifty members, only at the beginning of our period, in 1655. The resultant organisation thus had two opposed traditions united in one: literacy and illiteracy, theory and practice. Unified, they were put under the control of the first royal surgeon, and while this was an office of profit, the holders of this position through to the end of the eighteenth century treated the role as giving them the authority to turn French surgery into 'a new kind of profession'.[69]

An anatomical/surgical amphitheatre was built in 1691–-95 for Saint-Côme, designed by Blondel, with special inspiration from two surgeons, MM. Bienaise and Roberdeau. It held almost eight hundred students, and around 1700 there was a regular public course of anatomy, as well as courses in surgery.[70] Bodies for dissection were supplied from the Paris hospitals.

The *premier* surgeon established in Paris both a school of surgery in 1724 and an academy in 1731. This was a very serious exercise to raise and maintain the

---

67 The point about Portal's loyalty comes from Michel Dupont, *Dictionnaire historique des médecins dans et hors de la médecine*, 1999, pp. 481–2.

68 Gelfand, 'Monarchical profession', 1984, p. 166.

69 Gelfand, *Professionalizing modern medicine*, 1980, p. 27.

70 Ibid., p. 33.

quality of surgery and the skills of surgeons throughout France. Initially the first royal surgeon – Georges Mareschal, succeeded by François de la Peyronie – was concerned with theoretical knowledge in establishing now five teaching chairs, including anatomy and surgery, at Saint-Côme.[71]

Then in 1750, by royal edict, the first royal surgeon established a practical school of anatomy and operations. The first article of the edict specifies it as a place

> … where … the students can practise daily under the direction of the masters of the school what they have been taught in the surgical courses, and to make the students more assiduous it will be essential to oblige them to sign on at the beginning of each course in the registers of the demonstrators or masters in charge, as well as in the registers of the said Practical School.[72]

The second article specified that 'there should be established forever a practical school of anatomy and surgical operations in the college of Saint Côme in Paris, in which all the parts of anatomy will be demonstrated for free'. Both the masters of surgery in the school and the students had to make 'all the necessary dissections' themselves. Nothing was here specified about the source of bodies, but in practice they came from the hospitals. Students were expected to attend these courses for three years. Then in 1760 an elite corps of young surgeons was envisaged: a maximum of twenty of the very best students from the provinces (and who would return to the provinces) were to be admitted each year, and each of the ten professors of the College of Surgery – that is, of anatomy, operations, pathology, hygiene and therapeutics, each with two holders, that is, a professor and an adjoint – could choose two of these students to attend assiduously for a one-year course. Teaching was to be given by two (later four) demonstrators chosen by the royal surgeon from the professors of the college or other master surgeons, and paid 300 livres a year, and each with a student prosector or *prévôt* who had to purchase corpses, wood for heating, candles, and materials for injections and cleaning.

This was a great success. The only problems with this elite teaching revolved around the acquisition of corpses. Letters patent of 1768 specified that corpses were to officially be provided by the Hôpital Général (that is, the Salpêtrière) for free. But the hospital administrators were loathe to supply them, so there was grave-robbing. In 1760 there is a document indicating that the school was buying two corpses every ten days.

---

71 Gelfand, 'The "Paris manner" of dissection', 1972, p. 108.

72 The edict is quoted here from Imbault Huart, 'L'ecole practique', 1975, pp. 38–53.

Anatomy in University institutions in Paris: The Faculté

Older than any of these royal institutions in Paris was the Paris **Faculty of Medicine**, part of the university. Unlike the royal institutions we have been dealing with so far, the Faculty of Medicine dated from the medieval period, and still continued its ancient practices, which antedated the system of dedicated professors for teaching. Instead it preserved the system in which all doctors graduate of the faculty had the right (and sometimes the obligation) to provide teaching for the next generation of students. In practice this meant that the anatomical teaching – when it actually took place – consisted of a course by one of their number, chosen by lot, teaching both anatomy and physiology over two years,[73] teaching at 6 o'clock in the morning in summer and 7 o'clock in winter. Any member of the faculty might be chosen, and in this sense every member was like a professor or potential professor. The difference in quality in teaching between this system and one where the anatomist was a professional is easy to see, especially since so many of those appointed tried to evade their duties.

In 1733 a strong initiative was taken in the faculty to insist that candidates for the Bachelor's degree in medicine must have direct experience of anatomy and dissection. On the proposal of M. Nicolas Andry the faculty passed a decree specifying a one-week practical course every two years, and attendance at this course – including evidence of having worked with one's own hands at the anatomy – would be required as part of the qualification for a Paris MD.[74] No Bachelor was to be admitted to the licence in medicine in future unless, in the customary biennial curriculum, he shall have given evidence of his skill in anatomy and surgical operations, by his own hand, on the body, in front of the faculty. All the Bachelors, at the same time, should follow a course, in the first year on the human body, in the second year on surgical operations, through seven consecutive days, between the first day of October and the last day of April, as the occasion of obtaining cadavers offers.

> Article 4: Of the anatomical exercises. On the first and second day the abdomen and its viscera are to be demonstrated; on the third day the parts of the chest are to be dealt with; on the fourth the parts of the head; on the fifth all the muscles of the body are to treated; on the sixth the distributions of the [blood] vessels and nerves is to be explained; and on the seventh the bones are to be inspected; the Bachelors should respond to questions about the site, connection, structure and uses of the parts by pointing them out (*demonstrando respondeant*).

All the members of the faculty could participate in examining the candidates.

The faculty did have an anatomical theatre. In the early 1690s they rebuilt it, and then they rebuilt it again in 1744 and engaged Winslow to give the first course

73 Delaunay, *Le monde médical Parisien*, 1906, p. 13.

74 *Decreta Facultatis Medicinae Parisiensis*, 1733.

of anatomy in it in February 1745, which he did give, on a male corpse – and, for the first time in the faculty, in French.[75] This 'Winslow theatre' (Fig. 2.5) had to be abandoned in the 1770s when the faculty was obliged to move premises. However, anatomical demonstrations continued at the old site in the rue de la Bûcherie.[76] For instance, among the 'Cursus publici' for the years 1773–86 was advertised a course on the structure and use of the parts of the human body, by the faculty's Professor of Pathology, M. Stephan Pourfour Du Petit, listed for February 1780. The parts would be demonstrated by M. Claude André Goubelly, Doctor Regent of the faculty. Du Petit was paid 590 livres as professor; the expenses of the anatomy came to 30 livres, with the prosector receiving 30 livres.[77]

Fig. 2.5

[75] Delaunay, *Le monde médical Parisien*, 1906, pp. 14–15.

[76] Sabatier, *Recherches historiques sur la Faculté de Médecine*, 1837, p. 83; however, Sabatier nowhere mentions the anatomical theatre of the faculty.

[77] *Commentaires de la Faculté de Médecine de Paris, 1777 à 1786*, 1903, pp. 427, 680.

Hospitals and anatomy

In a number of old and new hospitals there were facilities for anatomising, both for performing post-mortems on patients, and for teaching students who were studying and working in the hospitals. Hospitals throughout this period were charitable institutions dedicated to the care of the poor. The rich and 'middling sort' did not enter hospitals as patients until the beginning of the twentieth century. Depending on their size and wealth, hospitals would have one or more physician, surgeon or apothecary to serve the needs of the poor patients. Sometimes these were paid; sometimes they were honorary positions (as in the English voluntary hospitals in the eighteenth century) whose holders used them as stepping-stones to the private medical care of the rich. At all events, hospitals meant medical failure as well as medical success, and medical failure meant that human bodies were available to the medical staff for dissection without restriction. Here, when they conceived a need for them, the surgeons and physicians could find ample raw material for their anatomical or pathological researches (on which see Chapter 3 below). As represented here in two frontispieces, the first to Bernhard Swalve's *Pancreas pancrene* of 1668 (Fig. 2.6) and the second to Stephen Blankaart (or Blancard)'s little book *Rational anatomy* of 1688 (Fig. 2.7), such post-mortem anatomising appears to have taken place in the wards in the presence of the patients who were still alive. However though portrayed in one location, these activities may have taken place at different times. Both these authors were based in Amsterdam.

Sometimes there appears to have been a dedicated anatomical theatre or space for post-mortems and teaching. Up until the end of the eighteenth century, the most important region for establishing anatomical spaces in hospitals was Italy. Grace Goldin has photographed a sixteenth-century such anatomical theatre in an Italian hospital, that at Pistoia, at the Hospital del Ceppo. She says: 'We can imagine the lesson taking place in this, the smallest anatomy theatre in Europe, a sixteenth-century artistic treasure which still stands three feet from the rear door of a ward of the Ospedale del Ceppo. There was room for only twelve students on the curving benches'.[78] There certainly were many such anatomical sites in hospitals created in our period, with new ones being founded (for instance) in hospitals in Milan in the Grande Spedale, while in Parma an 'anatomical academy' was set up in the Misericordia in 1751. In Naples the hospital of S. Giacomo had an anatomical theatre in the 1780s.[79]

---

[78] Goldin, *Work of mercy*, 1994, p. 124–5.

[79] All these examples are from Grell, Cunningham and Roeck, *Health care and poor relief in 18th and 19th century Southern Europe*, 2005.

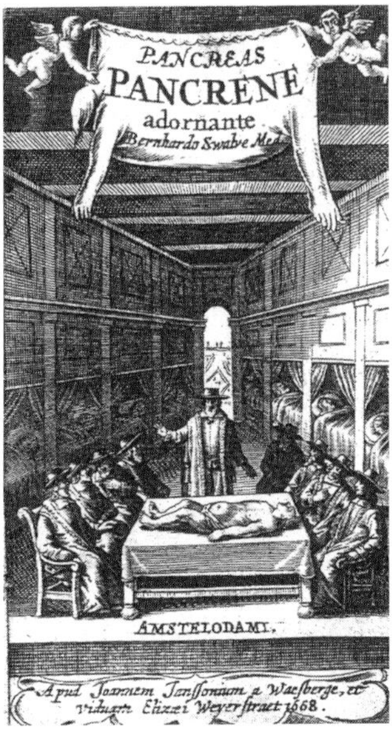

Fig. 2.6

Spain too had several such hospital-based anatomical sites. For instance, the crown founded a professorship and an anatomical theatre in the General Hospital of Madrid in 1689, outside of any university context, but within the orbit of the royal court.[80] This theatre (illustrated in Fig. 1.13) was for regular dissections as well as post-mortems. There was another one, long established in the General Hospital in Valencia, that was rebuilt in 1714, paid for by a tax on wine, 'it being compulsory for the Professor of Anatomy, according to the University regulations, to perform twenty-five anatomies a year … the greatest precision is required because the wards for convalescence having recently been rebuilt these obstruct the entrance to the said classroom'.[81] Again in Barcelona there was an anatomy classroom established right next door to the Hospital de la Santa Creu, functioning intermittently from the 1570s, restored in the 1670s, and in time replaced by the anatomical theatre of the Royal College of Surgery.

Fig. 2.7

In addition to the great Paris hospitals, such as the Invalides, and the Brothers of Charity giving anatomical teaching in an anatomical theatre at La Charité hospital, occasional information indicating hospital-based anatomical sites across France can be found. For instance, 'at Angers, under the initiative of the great dean Pierre Hunauld (1664–1728) the hospital became a veritable annex of the medical faculty and eight to ten regent doctors were teaching there in turn in 1757'.[82] Similarly, a Monsieur Aurran was at some time 'Surgeon and Demonstrator of Anatomy at the Royal Hospital at Strasbourg'.[83]

Finally, there was one way of being a paid anatomist that was not practised in Paris: viz. as city anatomist. I have come across two such instances. One was in Amsterdam, where Ruysch held the post for so long.

[80] Martínez-Vidal and Pardo-Tomàs, 'Anatomical theatres in early modern Spain', 2005, pp. 261–4.

[81] Ibid., pp. 266–7

[82] Huard, 'L'enseignement medico-chirurgical', 1964, p. 183.

[83] Portal, *Histoire de l'anatomie*, 6 vols, 1770–73, vol. 5, p. 178.

The other was in Venice. Santorini was described as 'in Veneto Lyceo Anatomes Professor' on the title-page of his *Anatomical observations*, published in 1739 after his death.

Private institutions

London presents a sorry contrast with Paris for anatomical teaching. Seen from London, the Jardin royal was a model of all that was desirable in the training of surgeons. There was no state-sponsored teaching of medicine or surgery in Britain, as there was at Paris. Nor were the only two English universities, Oxford and Cambridge, with their minuscule medical faculties, very enthusiastic about offering anatomical dissections, though the story was later different with respect to one (perhaps two) of the Scottish universities. There was no British equivalent of the Collège royal. The Royal Society, unlike its Parisian counterpart, had no pensioned researchers. The Royal College of Physicians of London and the Company of Barber Surgeons (from 1745 the Incorporation of Surgeons) had very limited teaching roles when it came to anatomy.[84] It was a consequence of this state of affairs that so many private anatomical teachers set up in London – for the British navy and army had as great a need of skilled surgeons, trained in anatomy, as did the French ones. The historian George Peachey traced some twenty-eight private teachers in London, Oxford and Cambridge working at some time between 1701 and 1744.[85] Some of them used the theatre of the surgeons for this purpose.[86]

This general inferiority of provision of anatomical and surgical teaching in London was partly due to the lax way in which the London Company of Barber–Surgeons fulfilled its historic obligations in this respect. An eminent member of the Company, William Cheselden, complained that it positively suppressed anatomical teaching by forbidding any of its members to demonstrate anatomy outwith its premises. But the Jardin royal (and other institutions) in Paris was certainly visible to some members of the company as a model to be emulated. One of the roles of the company was to examine and approve surgeons intending for the navy, and some petitions were drawn up to be presented to the crown in order to provide in London anatomical and surgical teaching comparable to that available in Paris. Sometime before 1732 the surgeons planned this appeal to the crown –

---

84 However, the College of Physicians of London had had a lively if short-lived period of anatomical research in the 1640s and '50s, following the exhortations of William Harvey. See Cunningham, 'The historical context of Wharton's work on the glands', 1996.

85 Peachey, *A memoir of William and John Hunter*, 1924, p. 8.

86 See [Anon.], *Syllabus musculorum corporis humani*, 1711; [Anon.], *Syllabus musculorum humani corporis*, 1729; Nourse, *Syllabus totam rem anatomicam humanam complectens*, 1729; [Frank Nicholls], *Syllabus partium corporis … Praelectionibus aptatus mox habendis in Theatro Anatomico Chirurgorum Londinensium*, 1733.

> The anatomy and surgery schools endowed by the French King at Paris has rendered that city famous for the attaining of this knowledge and the want of these in London daily obliges several of the British surgeons to seek and purchase these sciences abroad. This is a great evil as well as a great reproach to us, to remedy which and to establish these useful schools among us, the Company of Barbers and Surgeons of London and especially that part of it the Court of Examiners appointed by law to inspect into, examine, qualify and approve all surgeons serving on board His Majesty's fleet as [also] those practising within the Bills of Mortality [i.e. within London], being highly sensible of the great use of these schools and that His Majesty has nothing more at heart than the welfare of his subjects and the good of mankind humbly beg that his Majesty in his great goodness would be graciously pleased to assist them in promoting sciences so beneficial to mankind and enable them to appoint and settle five professors teachers and demonstrators of anatomy and surgery for the instruction of young surgeons and others and to settle upon each of the said professors a salary of about 30lb per ann. for their encouragement ....[87]

The professors and teachers were to be members of the company, rotating every three years, and the teaching was to be offered in the company's hall. The four bodies of malefactors allowed to the company every year should be used by these professors; two of them 'shall by one of the said professors, teachers and demonstrator, be anathomised, its vessels, or the most considerable part of them injected with coloured wax and the parts of these bodies dissected and shown and read upon in the public theatre of the said Company for two hours every day at least so long as those bodies can conveniently be preserved for the instruction of the members of the said Company'. Another professor was to use the other two bodies to demonstrate all the surgical operations. Yet another fresh body was to be used by the third professor to show the bones and ligaments 'as an insight in the cure of fractures and dislocations'. The fourth professor would read on surgery in the summer, excepting midwifery. The fifth professor would deal with all other aspects of surgery, while the sixth one would demonstrate all the surgical instruments and the materia medica. The proposal even included a seventh professor to teach midwifery, 'for this end only who shall every year give two complete courses of midwifery at the said public theatre at Surgeons Hall for the instruction of young surgeons', and he should also 'give two other complete courses every year the one in the city of London and another in the Liberties of Westminster as shall be agreed on for the instruction of women practising midwifery … in which demonstrations these women shall be showed every thing necessary to guide and direct them in this practice'. Nothing seems to have come of it.

---

[87] Undated transcript in Royal College of Surgeons library, Nc5a, 'Barber surgeons 1708–32 Anatomy', unpaginated. It is not mentioned in Flint South's *Memorials*, 1886, nor in Young's *Annals of the barber–surgeons*, 1890.

Years later things had not improved in London with respect to state-sponsored anatomical teaching, and towards the end of his own career as a private anatomy teacher, in 1763, William Hunter put forward another proposal for an anatomy school on royal funds: a museum in London, for the improvement of Anatomy, Surgery and Physic. The teaching of anatomy, crucial as it is for medicine and surgery, is not well rewarded, wrote Hunter. So, while Hunter's own private teaching in London since 1746 has made London 'one of the best schools for Anatomy'[88], yet conditions have not been favourable to breed up new teachers in anatomy: 'though we have had many professors, or teachers, in this great town, we have not had one Winslow, Morgagni, or Albinus: nor can it be expected, that a Briton should be able to do, in a few years, what is done by the labour of a long life in other countries: especially too, when we consider that there is no provision made by our government for supplying him with subjects [viz. corpses for dissection], and that in other countries this article is amply provided for' (p. 120).

Hunter argued that 'Scarce any science or art requires the protection of a prince more than Anatomy, as well on account of its great use to mankind, as because it is persecuted by the prejudices, both natural and religious, of the multitude in all nations' (p. 117). His petition was to the Earl of Bute, who was the first lord of the Treasury and practically prime minister. Hunter wrote of himself in the third person:

> What he at present wishes is this; to be allowed a proper piece of ground, that he may forthwith lay out six, or even seven thousand pounds, in erecting a building fit for the purpose, under any condition that may be agreeable to the King.
>
> Or, if his Majesty's known love of the polite arts, and his benevolence to mankind, should suggest to him a design of establishing an academy on a more extensive plan, Dr. Hunter would be still more happy, if what he now proposes for the advancement of one science, might be made a small part of an institution, worthy of the British nation, and a British king. (p. 121)

Hunter's scheme (Fig. 2.8) had living quarters (bottom right), a museum (bottom left), a yard and graveyard combined (right), and at the top a circular anatomical theatre. It was clearly meant to be double-story in parts, as Hunter has included staircases in his sketch, but presumably the theatre and the museum would have been double-height. But nothing came of this scheme and Hunter eventually withdrew his petition.

This comparison of the situation between France and Britain certainly shows that it was much simpler to initiate and provide anatomical teaching in an absolute monarchy such as France, or in an absolute empire such as the Hapsburg Empire run from Vienna, than it was in a 'free' country such as Britain. The squeamishness about the provision of dead bodies, so frequently shown by the poor or criminal

---

88 William Hunter, *Two introductory lectures*, 1784, p. 119.

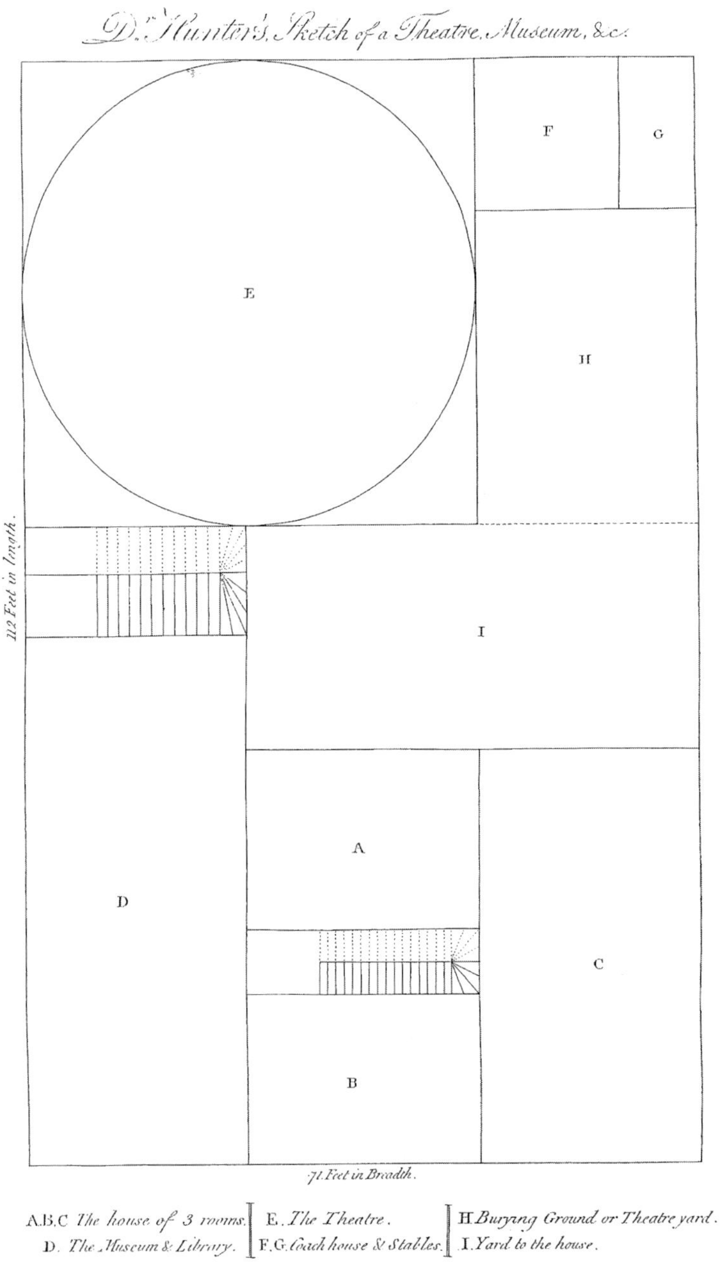
Dr. Hunter's Sketch of a Theatre, Museum, &c.
E
F
G
H
I
A
B
C
D
112 Feet in length.
71 Feet in Breadth.
A.B.C The house of 3 rooms.
D. The Museum & Library.
E. The Theatre.
F.G. Coach house & Stables.
H Burying Ground or Theatre yard.
I. Yard to the house.

Fig. 2.8

(whose bodies they were) or by the administrators of hospitals, could be simply dispensed with by an edict. Equally, as we have seen, absolute monarchies provided the conditions under which anatomical and surgical conditions could be radically improved, if they were properly manipulated by a royal surgeon.

In a comparatively 'free' country like Scotland, the improvement of facilities depended on the initiative and concurrence of local government. In Edinburgh, the capital of Scotland, the anatomical teaching originally arose out of a decades-long conflict between the physicians and the surgeon–apothecaries of the town that went back to at least the 1650s.[89] But it all continued and flourished through the eighteenth century because of the link in Britain between surgery and a standing navy. For after the Union of Scotland and England in 1707, the highway to England lay open for young Scotsmen and many of them took it. One of the best career options open to a young Scot was service in the navy as a ship's surgeon, especially since after three years' service such a surgeon was able to set up in practice wherever he might want, irrespective of the regulations of local guilds or councils. Many young Scotsmen therefore chose to acquire their anatomical education at home, that is, at the universities of Edinburgh or Glasgow or with private anatomy teachers there. Here they were taught by Scotsmen. But if they preferred, as many did, to go to London for their anatomical education, there too they were usually taught by fellow Scots. Anatomy – teaching, researching, learning – was overwhelmingly a Scottish enterprise in eighteenth-century Britain, wherever it was conducted. But anatomical learning in Scotland (and medical learning as a whole) was also attractive to a particular class of Englishman, who trekked north to Edinburgh or Glasgow. These were the non-conformist Protestants and the Catholics, who were not allowed to go to the English universities because of their religion, yet who wanted to enter medicine. The Scottish universities were technically Presbyterian institutions, but none of them insisted on religious tests for their students in the eighteenth century.

After a striking co-operative beginning (which we shall deal with in a moment), however, the anatomical demonstrations in Edinburgh came to be carried out for a few years by a single member of the Incorporation of Surgeon–Apothecaries, one Robert Eliot. For in 1705, Eliot obtained an annual salary of 15 pounds sterling from Edinburgh Town Council, together with the informal title of 'Professor of Anatomy'. He was joined by another teacher, Adam Drummond, from 1708. The incorporation gave the two of them the use of their theatre 'to teach the said science of Anatomy … both for the publick and private courses and declair they will give him all due encouragement and will allow none of their prentices or servants to be taught by any person who is not a member of their society'.[90] When Eliot died in 1717, the incorporation agreed that John McGill should join Drummond in the teaching.

89 Cunningham, 'Sydenham versus Newton', 1981; Cunningham, 'The medical professions', 1982; Cunningham, 'Sir Robert Sibbald', 1978.

90 Minutes of the Royal College of Surgeons of Edinburgh, 5 August 1708.

At this point this anatomical teaching under the aegis of a surgical institution, together with a town council, was metamorphosed into a major instance of university anatomical teaching. For Drummond and McGill continued teaching together until 1729, when they represented to the incorporation in that

> … the state of their health and business were such that they could not duly attend the said Professorships but that they and the haill Calling being perswaded of the sufficiency of Alexander Monro, one of their number, did therefore unanimously recommend him to the Provost and Town of Edinburgh to be Professor of Anatomy within the said City.[91]

The appointment of Alexander Monro *Primus* as the Professor of Anatomy, first in the City and then in the University, has traditionally been told as a simple and gratifying tale of an ambitious father (John Monro), a gifted son (Alexander), and a dynamic Provost of the Town Council (George Drummond), together planning one of the elements that Edinburgh needed to help restore its fortunes and importance after the Union: a medical school. To construct this story historians (including myself) have somewhat naively followed the account given in the third-person autobiography of Alexander Monro.[92]

Supposedly John Monro, an Edinburgh surgeon, perceiving the need for establishing a medical school in Edinburgh, had his son Alexander educated in anatomy and medicine at all the right places: in London with William Cheselden, in Paris with Monsieur Bourquet for anatomy and Monsieur Grégoire for man-midwifery, and in Leiden with Boerhaave and Albinus, where he learnt to make wax preparations. Returning to Edinburgh in 1719 aged just 22, Alexander so impressed George Drummond, the Provost and John Monro's friend, that the town council was persuaded to appoint him 'Professor of Anatomy within the said City' (as we saw above). With some clever footwork, by 1726 Monro *père* was able to assemble four other former Boerhaave students to teach all the other parts of medicine that Boerhaave personally taught in Leiden, and then to have all five of them be given the title of Professor within the town's college or university.

However, it has recently been quite reasonably argued that this pretty tale of foresight, co-operation and progress in the mutual interests of the development of medicine and of Scotland actually had a much messier origin in the murky world of patronage and in the management of Scotland and Edinburgh by aristocrats in the interests of the British government.[93] The Town Council of Edinburgh, it appears, was hardly an independent body when it came to making professorships in their town's college. Anatomy, like other new subjects taught in the universities of Edinburgh and Glasgow, was a matter of political patronage as much as it was

---

91 Minutes of the Royal College of Surgeons of Edinburgh, 27 January 1720.

92 This autobiography is the basis of the account in Wright-St.Clair, *Doctors Monro*, and others.

93 Emerson, 'The founding of the Edinburgh medical school', 2004.

a matter of the improvement of surgical education. At all events, Monro *Primus* was appointed 'professor of Anatomy in this Citie and Colledge' for £15 per year. World-class anatomical teaching and research, as delivered by the Monro dynasty, was for the rest of the century to be one of the greatest strengths of Edinburgh University and the major attraction that it offered to students – Alexander Monro *Secundus* claims to have had around 200 students in the 1760s, 300 in the 1770s, and a peak of 436 in 1783.[94]

4. Ways of teaching anatomy

Perhaps surprisingly, there was more than one way to teach anatomy, even if one had the constraint of only one body that would inevitably be undergoing decomposition as one worked on it over several days or longer. Different sequences of dissection, diverse methods of demonstration, were adopted by various anatomists. As some at least of these anatomists were private teachers, the reputation of their special way of doing things could directly affect the size of their audience and hence their income. So the choice of sequence and dissection was to our anatomists more than just a matter of personal preference. In general one could say that each had their own concept of how anatomy could best be learnt, and the sequence that should be adopted to best assist the student in remembering what he had been taught. I shall look here at a sample of anatomists and anatomical teaching, taking examples over a hundred years from France, Scotland, the Netherlands, Italy and England, in order to show something of the variety of approaches that were adopted, both towards dissection and towards teaching.[95] Many of these varied courses were running concurrently and over long periods – if they don't die young then anatomists seem to live long – and students who were prepared to travel would thus have had a wide variety of approaches from which to choose.

Paris, 1673–80: Dionis

We begin in the Jardin royal in Paris in the 1670s, where Dionis was the public teacher to whomever wanted to come and learn for free. Here, surprisingly, the arrangement for dissection and demonstration was formally the very traditional one, following fourteenth- and fifteenth-century practice, whereby a doctor,

94 Fraser, 'The building of Old College', 1989, p. 19.

95 Many other examples could have been chosen. For instance William Hewson, at the end of *Experimental inquiries*, 3 vols, 1771–77, has proposals for a course of anatomical and surgical lectures in the 1770s in London. See also Collignon's 1756 (Latin) schema of his course at Cambridge in his *Compendium anatomico-medicum*. Or Valentin's proposed course in Berlin, outlined in his *Amphitheatrum zootomicum … accedit methodus secandi cadav. humana*, 1720.

in this case Cressé, read a discourse,[96] whereas a surgeon, in this case Dionis, performed the dissection and demonstration of the parts. This was an extremely outdated practice almost everywhere else by this date, and had been generally replaced by the teaching innovations introduced by Andreas Vesalius in Padua in the 1540s, whereby the anatomist both lectured and dissected (or at least directed the dissection). Here in the Jardin it meant that the doctor gave a discourse on the parts and their structure and uses, while at a different time and place the surgeon, Dionis, gave an up-to-date anatomical treatment of the parts, including all the discoveries that had been made in the course of the seventeenth century, from the lacteals to the microscopic capillaries in the lungs, in order to demonstrate the circulation of the blood, but all of it presented in a most old-fashioned, indeed medieval, sequence.[97]

With respect to the anatomy, Dionis says to his audience: 'suiting my Discourses to the New Discoveries, I shall endeavour to shew you a Humane Body according to our knowledge of it at present'.[98] He says that the sequence of his book follows the teaching sequence he followed in the Jardin:

> In this Treatise, the Osteology leads the Van; by reason that *that* Part commences our Exercises in the Royal Garden, and that the knowledge of the Bones ought to usher in that of all the other Parts. I divide my Osteology into eight Demonstrations, viz. Two of the Bones in General, Two of the Bones of the Head, Two of those of the Trunk, and Two of those of the Limbs.
>
> In the next place, I carry on our course in Ten Anatomical Demonstrations, viz. Four for the *Abdomen* or *lower-belly*; Two for the Parts of the *Thorax* or *Breast*; Two for those of the *Head*; and Two for the *Limbs*.[99]

The anatomy course began each winter, using 'the first body that offers'.[100] After his demonstration of the bones from a skeleton, Dionis follows 'the order of necessity', and begins with the abdomen, because it encloses the parts most subject to corruption.

---

96 For which see above, Chapter 1.

97 The nature of the anatomical and surgical teaching of Dionis is evident from the books on anatomy and surgery that he published a few years later – Dionis, *The anatomy of humane bodies improv'd*, 1703, and *Cours d'operations de chirurgie*, 1708 – and their popularity is confirmed by the number of editions that came out in quick succession. The anatomical book received seven editions in the original French (1691, 1695, 1698, 1701, 1705, 1706 and 1729). It was early translated into Latin (Geneva, 1696), and also into English (1703, from the third French edition of 1698). The surgical book received eight French editions from 1708 to 1782, and an English edition in 1710.

98 Dionis, *The anatomy of humane bodies improv'd*, 1703, p. 90.

99 Ibid., The Author's Preface.

100 Dionis, *A course of surgical operations*, 1710, p. A2v.

Dionis's account of anatomy is all based on the circulation of the blood, which he treats 'as a standing principle, both for confirming the sentiments of modern Authors, and confuting the Errors of the Ancients'.[101] It seems that, for Dionis, demonstrating the anatomy of the human body following the principle of the circulation meant reinterpreting the extremely traditional three-venters sequence as being the venters where different stages of the blood's formation and its circulation are effected. So in the lower venter or abdomen (1) the victuals are transformed into chyle (in his view the true substance of the blood), and where the blood is prepared and purified (i.e. the gall bladder, liver, kidneys, etc.) for chylification, and all these organs 'contribute to the perfection of the blood and purge its mass of all that can annoy it' (p. 160). The male organs of generation (customarily treated in the first venter) strain out 'the most refined, volatile and penetrating parts of the blood' (p. 169), for the purpose of preparing the semen. Then the middle venter, thorax or breast (2) is portrayed as where the heart is, 'which enlivens and keeps up the body, by raising the blood to a due degree of perfection, and distributing it over the body' (p. 248), 'the heart being the principle which puts in motion all the springs of the machine, and gives rise to all secretions' (p. 276). Here in the second venter the lung 'not only renews the mass of blood, which, without that continual agitation, would pall and corrupt, but likewise subtilizes it by purging it of its excrements, and qualifies it to nourish all the parts of the body' (p. 278). Then finally the third venter or brain (3) 'is of use to be the principal organ of the functions of the soul, and to make a secretion of the animal spirits [themselves created from the blood] in conjunction with the nervous juice, which it distributes to all the parts of the body by means of the nerves' (p. 325); the senses are shown to work via the animal spirits in the nerves. The ninth demonstration is on the upper limbs, and this is where he speaks generally about the muscles; the tenth and last is about the lower limbs.

This is a subtle achievement: the medieval sequence of anatomising in the order of the corruption of the parts is presented effortlessly as anatomising in the order of the blood and its circulation! Up-to-date Cartesian explanations are offered of the hidden workings of the organs. As with the medieval, pre-Vesalian, sequence, Dionis's approach deals with the muscles last and in lesser detail than the internal organs, even though a large proportion of his audience must have been intending to be surgeons.

Edinburgh, 1704: The Incorporation of Surgeon–Apothecaries

When the Incorporation of Surgeon–Apothecaries of Edinburgh began its anatomical teaching in its new anatomical theatre (completed in 1697), the members shared the teaching among themselves in a strikingly collaborative fashion. One of these collaborative dissections was carried out in November 1702 on a male executed

101 Dionis, *The anatomy of humane bodies improv'd*, 1703, The Author's Preface.

for incest, with an eight-day dissection performed on his body.[102] In April 1704 the incorporation carried out another such public dissection and this time recorded the event in detail in their minutes. Ten out of around thirty surgeon members took active roles. The body on this occasion was female.

> The Deacon produced in the presence of the Calling a list of the operators in the course of Anatomy publicly performed in the Chirurgion Apothecaries' theatre 22 April 1704. Which being read, the Calling gave the Operators their thanks and ordained the same to be insert in their books. Whereof the tenor follows, viz.
>
> The first day: James Hamilton, A Discourse on Anatomy in general, with a dissection and demonstration of the common teguments and muscles of the Abdomen.
>
> The 2nd day: John Mirrie, The umbilicus, omentum, peritoneum, stomach, pancreas, intestines, *vasa lactea*, mesentery, *receptaculum chyli*, and *ductus thoracitis*.
>
> The 3rd day: Mr Alexander Nisbet. The liver, *vesica fellis*, with their vessels; spleen, kidneys, *glandulae renales*, ureters and bladder.
>
> The 4th day: George Dundas. The organs of generation in a woman, with a discourse of hernias.
>
> The 5th day: Robert Swintoun. The containing and contained parts of the thorax, with the circulation of the blood and respiration.
>
> The 6th day: Henry Hamilton. The hair, teguments, *dura* and *pia mater*; cerebrum, cerebellum, *medulla oblongata*, and nerves within the head.
>
> The 7th day: Robert Eliot. The five external senses, with a demonstration of the several [sense] organs.
>
> The 8th day: John Jossy. The muscles of the neck and arm, with a discourse of muscular motion.
>
> The 9th day: Walter Potter. The muscles of the back, thigh and leg. The epilogue and conclusion by Doctor Archibald Pitcairn.[103]

Although this anatomical demonstration was performed by and for a body of surgeons, it was as thorough as if it were for physicians. That is to say, the surgeons did not limit their attention to the external parts of the body and the bones, which constituted the range of their normal practice, but explored the innermost parts too, the parts that the physicians still reserved to themselves as the core of their own practice of medicine. Indeed, the surgeon–apothecaries seem to have forgotten about the bones altogether, despite their own practice of setting fractures and so on. This thoroughness was necessary to fulfil the conditions of the 1694 grant from

---

[102] Creswell, *The Royal College of Surgeons of Edinburgh*, 1926, pp. 193–4.

[103] Manuscript minutes of the Royal College of Surgeons of Edinburgh, 18 May 1704.

the town council to show 'as much as can be shown upon one body', with 'ten free labouring days' allowed them to carry it out.[104]

Leiden, 1713–19: Rau

Jacobus Rau (1668–1719) was Professor of Anatomy at Leiden University from 1713 until his death six years later. He had trained as a surgeon, although he gained most of his anatomical knowledge at Paris. In Amsterdam the Magistrates first allowed him to use their anatomical theatre for public teaching. He and the contentious Amsterdam anatomist Ruysch had a considerable controversy over whether there is a true septum in the scrotum of the human male.

But in 1713 Rau was called to the Chair of Anatomy and Surgery at Leiden, where his inaugural lecture was *On the method of teaching and learning anatomy*, and in this he outlined what method the future doctor or surgeon ought to follow. We may take this oration as the idealisation of what he had learnt in Paris, what he had been teaching in Amsterdam, and what he would have taught in the few years that he was professor.

Method, Rau claimed, was essential, for unless we follow the method appropriate to each art, then we understand nothing correctly or in order. Three things are therefore required to learn anatomy. First is the frequent reading of the writings of the ancient and modern anatomists (the moderns being those writing since Harvey), and a proper understanding of what they are saying. Second, live teaching by a teacher (that is to say, from a person, not a book). And third, an introduction to the actual dissection of the bodies of humans and animals, and repeated practice.

'In rightly learning this art, the history of the bones should come first, for it has rightly been called the foundation of the whole of Anatomy. This can partly be done using dried bones; partly from the fresh bones of persons recently deceased'.[105] Adult, child and intermediate bones should be compared in order to see the effects of growth. Here again one should not just use dried specimens, but also fresh bodies.

Rau himself made an extensive anatomical collection that he would have used in such teaching, an *Index* of which was later published by Albinus.[106] It appears that Rau had two major repositories and one minor one. The first major one was completely devoted to skeletons and bones. Male and female adult skeletons were complemented by those of fetuses and children, and contrasted with those of a monkey and a cat. There were over thirty skulls of many human races – an Englishman, a Spaniard, a Frenchman, an Ethiopian – and of women old and young. In the second repository there were more human skeletons – males, females, children, new-born babies – all 'ossa dissoluta', that is to say, with the bones

104 Creswell, *The Royal College of Surgeons of Edinburgh*, 1926, p. 193.

105 Rau, *Oratio inauguralis*, 1713, p. 31.

106 Albinus, *Index Supellectilis Anatomicae*, 1725.

separated. There were also glass jars showing the origin, increase and successive changes of the bones of the trunk and joints.

Then, Rau said in his *Oration*, the student should come to the account ('history') of the nerves. Next one moves on to the history of the blood vessels in the whole body, since the heart is the centre and principal. Finally one should come to the viscera – in which Rau includes the lungs 'and the other viscera which contribute to respiration; and the movements need to be observed of chewing, swallowing, chylification, blood-making, secretion, propagation in both sexes; the astonishing structures of the organs of the voice, of speech, touch, smell, taste, sight, hearing; the life of the fetus in the womb without respiration, the amazing changes of its vessels after birth. All these things should be demonstrated for right learning in human cadavers' (p. 33). Finally, the lacteals should be demonstrated in the live animal.

Here again, Rau's anatomical collection contained many of the necessary parts for teaching this section of the course. For his minor repository consisted of 164 glass jars containing the soft parts of man and brutes 'for demonstrating their structure', including skin, eyes, the tongue and teeth, the spinal marrow, intestines, kidneys, penis, testicles, the secundines of the pregnant womb, and immature fetuses.

Only after he has mastered anatomy should the student attend the lectures of the two professors of the theory and practice of medicine, because only now is he equipped to do so, according to Rau. For this route through anatomical study 'teaches the cause and difference of diseases: which ones are different lesions of the vessels, which are diseases of the brain, and which of the nerves; it distinguishes what are the bad conditions of the lung, which are the problems of the stomach, the many affects and gripings of the intestines, the terrible pains of the kidneys and bladder, and of the other parts, the innumerable diseases of the liver and spleen'. Anatomy commends herself as the mother, fount and origin, without which Medicine could not exist (p. 38).

This of course is an ideal, rather than a course-list, and it certainly corresponds to Rau's own teaching collection. What is especially interesting is that this programme of Rau's would ideally have used anatomy to simultaneously teach physiology (all those movements of chewing, and so on), as well as pathology (all those locations and lesions of the inner parts). If carried out in practice, it would have required masses of corpses – and Albinus records that Rau did indeed 'dissect innumerable corpses of all ages'[107] – for it does not follow the traditional order of dissection at all. I imagine that omission of any mention of the muscles was an oversight, for they are certainly systematically covered in Rau's 'Anatomical course, consisting of osteology, myology [muscles], splanchnology [the inner organs, the viscera], adenology [glands], angeiology [the vessels] and neurology' that was published

107 Ibid., Introduction.

after his death, and that reveals Rau to have been very systematic in his naming and classifying of the parts.[108]

Padua, 1716–71: Morgagni

Until about 1760 Padua was still one of the great places for medical students to go in order to study medicine and anatomy. It had been attracting young men from across Europe since at least the early sixteenth century, and it had had a formidable series of anatomists, each appointed by the committee of three Riformatori, who ran the university on behalf of the Senate of Venice. In the course of the seventeenth century the university came to be rivalled for medical teaching by northern universities, above all Leiden, and in the eighteenth century it came to share in the gentle decline of Italian leadership in learning and culture generally.

From 1716 to 1771 the Professor of Anatomy at Padua was Giovanni Battista Morgagni (1682–1771). There are many accounts of Morgagni's life and achievements written by Morgagni himself, sometimes in the first person, sometimes in the third. To trace the nature of his university anatomical teaching I use at length here two documents by Morgagni, which are so very revealing of the man, of the man in the theatre, and of the man at his anatomical teaching over the years, that they hardly need comment.

The first is a passage from one of Morgagni's several autobiographies, and deals in his typically garrulous and immodest way with the triumphs of his first day teaching publicly in the Paduan theatre.

> From the day he was elected as Anatomicus [15 September 1715] to the 21st January 1716 – and this date was determined by the start of the anatomical course – in that period he had a very limited time, especially for someone who had to concern himself with training his own prosector with many preparations and examinations, and he attended the lectures of all the professors and responded by voice or letter to the congratulations of his friends. But Morgagni busied himself so much so that both at home and at the hospital he managed to instruct Giovanni Battista Volpi in all the necessary anatomical preparations, and he himself would put his hand to dissections first, dissections never known about before in Padua, and so he was able to start on the day fixed. But remembering some of his predecessors who would delay things with the excuse, or sometimes the true reason, of a shortage of the right corpse, he was determined that that shortage should not force him, neither at the time nor later on, to delay the teaching. This is the reason why he established that it would no longer be necessary to start the dissection from the lower abdomen. So on that very day he spoke about the anatomical principles in a theatre that was as crowded as it could be, with an audience of every social class, and he received so much

[108] Published as an Appendix (pp. 34–114) in the *Amphitheatrum zootomicum* of Valentinus, 1720.

applause both during and at the end of his dissertation. He made twenty of these dissertations – nobody had ever heard so many – and without neglecting any part of anatomy. He always talked to an audience that was always more numerous than those of his predecessors, even during terrible seasons, and even when he only had at his disposal corpses in the worst possible state. He dealt with the eye in the presence of Praetor Andrea della Leggia and the Prefetto of Padua, Cavalier Giovanni Pesaro, and in the presence of a very dense audience made up of people of every social class, including very many ladies. He was always speaking impromptu, something that became obvious to every one, for example … on those occasions when the dissection of the corpse offered some altogether unexpected elements for examination, and especially on that day when, almost on the threshold of the theatre, when he was going to be talking about the anatomy of women, he seemed to hear some yobs who were jeering in a very vulgar way, just because the piece to be examined was the womb of a cow, and the topic to be dealt with was the generation of the egg. As the time came to discuss the issue and the presentation of the womb of the cow, Morgagni casually used a sentence from Cicero, taken from the *Rabiribo*: 'Believe me, the Illustrious Senate would never have given me this chair had they believed that I would give up following the steps of my ancient predecessors and not defend the truth because of the jeering of four idiots'.[109] And from that day on it never happened again, and actually, in contrast with what used to happen, the young students never had a more serious behaviour in the theatre and never maintained such a profound silence. And more than one student would try and write down the words of the teacher, but this is less surprising than the fact that hardly ever did our Morgagni find an empty classroom, because apart from the result of re-establishing a simplicity and calmness to the scholastic atmosphere, which had been too rhetorical and not productive enough (but which however pleased the young because sometimes it was obscene and scurrilous), and apart from having raised the lecture from just a few words before to an hour and not rarely an hour-and-a-half, and apart from having eliminated all the philosophical digressions that had routinely been abused and were often false, in order to go back to the pure anatomical account, apart from all these I say, he was the first to teach the discoveries that had followed the circulation of the blood, drawing out all the practical and theoretical consequences that were diametrically opposed to those which, due to the lazy attitude of his predecessors, had enjoyed the greatest honour among the university teachers of Padua. His exposition always remained modern, excusing the ancients sometimes, but sometimes interpreting them so that he could show that Galen and the old professors of Padua did not have teaching concepts different from the modern ones, and the result was he never displeased anyone, not even those most faithful to the ancients.[110]

---

[109] Cicero, *Pro Rabirio*, 7.18: Numquam, mihi credite, populus Romanus hic qui silet consulem me si clamore perturbatum iri arbitraretur.

[110] Morgagni, '*Le Autobiografie*', 1964, pp. 45–6.

From this first account we can see that Morgagni was the complete showman in his anatomical presentations; that he was completely at home speaking in public, able to control his audience; that he was up to date in what he taught; and that he was a master of his anatomical topics. He also ensured a regular supply of corpses, which meant that he was no longer restricted to dissecting in the traditional three-venter sequence, but could take on any topic in which he was interested.

The second of the documents I have chosen in which Morgagni talks about his anatomical teaching dates from towards the end of Morgagni's life. One Antonio Larber, now having risen to the position of Town Physician of Bassano del Grappa near Vicenza, asked him to recall how he had gone about teaching medicine and anatomy in the early years of the century, that is, sixty years earlier. Morgagni's letter of reply, which appeared in print as a pamphlet, is very revealing of the way in which Padua for so long managed to be both the most go-ahead university in Europe in its medical and anatomical teaching, while still having to conform to what by the 1700s was possibly the most antiquated set of teaching regulations of any university in Europe. This letter of Morgagni, though long ago published, is virtually unknown and well worth considering at length. Like the other medical teachers at Padua, Morgagni had to follow a specified sequence of teaching that extended over several years, but his letter reveals how many different ways and sequences a resourceful person could find for teaching anatomy, especially when short of time or bodies.

> How I wish I could satisfy your desire, since it is not enough for you, most dear Antonio, to have heard assiduously the things I taught either in the Hospital or in the Theatre in previous years when you were here, but you want also to know those things that I used to teach before you were born, firstly about the medical art, and secondly about anatomy. But indeed how can I, since, accustomed to speak, as you know, extempore, I have preserved scarcely the main heads of the Exercises in writing? You say, 'Let me know both the way and order you used to use'. If you are content with this, here are a few words.
>
> You are aware which books had to be interpreted here, according to the custom of our ancestors, by the teachers of the Theory of Medicine: the first Fen of the first book of the *Canon* of Avicenna, the *Ars Medica* of Galen, the *Aphorisms* of Hippocrates. I considered that the first two were given to me not so much for their interpretation as for taking up those of their arguments that ought to be treated and expounded by anyone zealous for truth and for the benefit of his auditors. Therefore when I had expounded the thought of Avicenna and Galen briefly and as lucidly as I could, I went on immediately to this, so that I might inquire at greater length about the things that they had once written, and about the things that either the one or the other would have thought 'if by chance they had lived in this our age' after so many new discoveries and experiments had been made. Thus opportunity was given me of putting forward the things that later writers taught differently and why they did so, and of frankly and modestly indicating, by comparing many thoughts, which views I thought to be more true.

From Avicenna's book therefore I drew up arguments so that I might expound one by one the heads of Physiology; from Galen's book, arguments so that I might expound the other parts of the Medical Institutes …

Meanwhile, while I taught my auditors these things in the School and at home, if I had any leisure I either went on to compare my (still unpublished) anatomical observations with those in the writings of the Ancients and Moderns, or, not having forgotten the 'Idea of the Medical Institutes' (a work I had equally promised), I collected those things that were able to help me in writing the books of the Institutes at length.

This went on until, after beginning the fourth year of teaching Medicine, I was ordered to teach Anatomy, and was given barely three months, and this not for preparing myself but for other things and especially for diligently preparing the Dissector, through assiduous practice and by example, so that he could carry out many things that were not usually shown here, or at least not in this way. Therefore the Preface that you once read at my house, was not so much a Preface as a most sincere apology! The day immediately after I had delivered the Preface, the demonstrations of the Elements of Anatomy followed, for the reasons which you see indicated in my *Altera Adversaria*.[111] Thereafter I always retained this arrangement, so that every year I might show and explain those Elements, if not in the Theatre then in my home or in the Hospital, before I taught the anatomy of the whole human body. After the bones had been briefly demonstrated I added to these things the doctrine of the joints and a demonstration of the veins and nerves as they are carried through the neck, thorax and venter, each in their places; for I knew how important it was at the same time to demonstrate the wholes before the pieces are dealt with after they have been cut off and divided into parts, with their own viscera (or, if not divided up, glued to a board and dried), in which case neither the seat of diseases, nor the site, nor the condition, may be apparent to the onlookers. But you have yourself seen these things, and I have certainly not omitted them to this day; I show other things in the Hospital too, since I was afraid that I might not be able to show some things adequately in the Theatre.

You will have been able to notice also that for the most part I observed the Veslingian order in teaching anatomy.[112] However it sometime happened in previous years that, on account of a lack of suitable cadavers, I was obliged to follow a different order so as not to lose time, but in such a way that everything might be seen connected and related to each other. Therefore I sometimes started from the bones, sometimes from the fetus, but never from the head. In

[111] Viz. Morgagni's 'Animadversio I: Anatomes tradendae methodus, ejusque Elementa', 1717.

[112] That is, following the method of J. Vesling (1598–1649), Professor at Padua. The order follows the three 'venters': lower, middle, upper, with modifications. See Vesling, *Syntagma anatomicum*, 1641, translated as Vesling, *The anatomy of the body of man*, 1653.

the second year, when I realised that the time allowed me for teaching would not suffice for my usual number of Exercises, I wanted to find out whether I might be able to include everything in a shorter order. Thus, beginning from the skull and spinal cord, I was led through via the nerves both to the individual organs of the senses and to the instruments of motion, not only the external but also the internal ones – those that serve for speaking, for making the blood circulate, those for digesting and distributing the food, those for expelling excrement, and those for propagating the human species. I don't need to say much about this. I reached the individual parts and thus the particulars of the whole body through this new (and I suggest neither inelegant nor prolix) method, a method requiring a lot of work and not a little concentration. Therefore, having used this once, I never wished to try that other method that goes from the heart via the arteries to the individual parts. But when finally neither the shortness of time nor the lack of bodies any longer prevented me from following the order that I thought was most easy for my auditors and more convenient for me as well, I did not depart from that order that I have said was used here more usually, that is, the order according to Vesling.

It did not, however, please me every year to undertake the same distribution of things in the Exercises of the same argument, nor to deduce either the same animadversions from the *historia* of the same parts, nor the same advices either theoretical or practical. But as far as I was able I always clothed the *historia* with some new dress: and I immediately increased that *historia* or illustrated it. Not my observations only, but also those of other people, were a great help to me so that I might discharge this. For I was accustomed to propose to my auditors observations taken from recently published books, with the authors clearly pointed out and praised, even though I might not always be able to approve or accept the observations, either because, repeated, they did not respond, or because I had not yet had enough time for repeating them. It never caused me regret to withhold assent, since it was possible either to confirm them more securely in following years or to abandon them not only through my experiments but also through those of others repeated without success.

From all these things you will know how it was that, since the same I had 47 lectures every year about the same human body and in the same theatre, the theatre was always full, including, as you know, those people returning to hear (again), of whom several were present repeatedly for many years. You will know equally why it was never disagreeable to me but rather a source of joy that several of them, in different years, tried to write down every word of the teacher – not, as did the more prudent, only try to note for themselves the things and the thought. For even if their industry had this result, that, since they very frequently erred not from the words alone but also from the sense, their writings, whichever were given to me for the approbation they sought, received this frank judgement from me, which in the words of Virgil, I thought should be attached to their writings by my hand:

I am not so ugly; I have frequently seen myself on the shore
When the sea remained still under the winds.

Although, I say, they did not succeed as they wished, it was possible to show by comparing everything that had been written in one year with the writings of other years, how differently I held my exercises in each year.

In order this might occur, additional causes frequently happened that partly called for it: if for instance the viscera, the muscles, or the vessels had either a different seat from normal, or had different shapes or number. I never came across these things but I immediately asked whether it happened that I or others saw this thing, whether frequently, and in which bodies; and if by chance something was absent, [I asked] whence it might be supplied in living animals.

Nor did I do otherwise if the parts showed themselves changed not from birth but by some disease, especially by a rarer one. Indeed you heard these things partly when you were here, and partly you will find them in my books. Other things probably remain in my papers, such as those things to be added at the beginning or end of those exercises at which it pleased the renowned Men to be present who were in charge of the city on behalf of the August City of Venice. Since those things, as far as I remember, were hardly anything but praises or gestures of thanks, they pertain very little to that which you ask from me – I ask myself whether [you do so] curiously or lovingly? – viz. which way I held in teaching. In what manner these little writings of mine are preserved, you will be able, coming here sometime, to see (if you wish) in the great heap of writings which you have often seen at my house, among which are whatever I have written hurriedly for immediate circumstances [that] I have never re-read nor will I again re-read. 'For the eightieth year warns me', as it once warned M. Varro [viz. Terence], 'that I should collect far different baggage before I depart from this life'. While I pray that this life may be longer to you than to me, at the same time I pray that it might be happier. Farewell.

I write this at Padua, March 1st 1761.

We can see here that Morgagni preferred to teach the 'Elements' of anatomy systematically (that is, via anatomical systems), which meant first demonstrating the bones, then the course and ramifications of the veins and nerves, dealing with whole sections of the body (such as the chest or lower belly) before cutting them up. As he says, this is similar to the Veslingian approach, which began with the bones, followed by the muscles, and then by tracing the nerves, veins and arteries throughout the body. It was all about showing the *sites and relations* of the parts (and hence, in pathology, the seats of diseases). The customary way of doing this, Morgagni relates, was to follow the movement of the blood around the body from the heart through the arteries. But having once tried following the nerves – and thus tracing the route from the nerve-centre, the brain, to the senses and to the instruments of motion – Morgagni never again followed the heart–arteries route. Both of these routes, of course, show the functioning of the body *as a whole* and its operations, using methods 'neither inelegant nor prolix'. In addition Morgagni strove constantly to be up to date, mentioning all the latest discoveries and theories, whether or not he had yet tested them to his own satisfaction. Finally we can take his word that his annual course of forty-seven demonstrations on the human body was always full and never the same twice!

Paris, c. 1731–58: Winslow

The anatomising sequence adopted by Winslow, a later lecturer in the Jardin royal, offers a complete contrast with that which had been offered there by Dionis in the 1670s. Although he was not formally appointed as lecturer in anatomy in the Jardin royal until 1743, Jacques Bénigne Winslow (Jakob Winsløv) (1669–1760) had in fact already been teaching there for at least twelve years. For Duverney, while still being the titular holder of the post, had had him perform the duties of it on his behalf, 'his bad state of health not permitting him to undergo that fatigue himself'.[113] Winslow had studied anatomy in Holland under Ruysch, then in Paris under Duverney. Haller admired him as an anatomist, but when he heard him lecturing in French in the Jardin in October 1727, substituting for Duverney, he was not impressed: 'He does not pronounce R and Sch, he did not vary his tone, he speaks soft and low, so that few understand him'.[114]

Winslow was a Danish Lutheran when he first came to Paris; indeed he was the son of a Lutheran pastor. Today the most celebrated thing about Winslow is that he was converted to Catholicism by the preaching and teaching of Bishop Bossuet, receiving the conversion name of 'Bénigne' to add to his Frenchified first name in 1699. Winslow wrote an autobiography that is almost completely taken up with the story of his conversion, ignoring his anatomical life almost entirely.[115] It was Bossuet who persuaded Winslow to stay in Paris and follow medicine and anatomy, rather than theology. In all this, Winslow was treading in the footsteps of his great-uncle Nicolas Steno, an earlier Dane who had gone to Paris forty years before as anatomist and who had converted to Catholicism, similarly under the influence of Bossuet.[116]

However, even more than following this family pattern in religious conversion, Winslow also took Great-Uncle Steno as his model in anatomy (Chapter 1 above). Winslow acknowledged 'with sincere gratitude, that the late M. Steno's *Discourse on the anatomy of the brain* [1668], was the sole original source, and general rule of my conduct in all that I have done in anatomy … [It] contains a great many excellent advices how to shun errors and discover truth, not only in relation to the

---

113 Winslow, *An anatomical exposition of the structure of the human body*, 1733, p. ix.

114 Haller, *Tagebuch*, 1942, p. 30.

115 Winslow, *L'autobiographie*, 1912. See also Anon., *Eloge historique de M. Winslow*, n.d.

116 On whom see Ole Peter Grell, 'Between anatomy and religion', 2007.

structure and uses of the parts, but also in relation to the way of dissecting, and of making anatomical figures'.[117] Winslow's book of his lectures was published in French in 1732 and translated into English in 1733, and he reprinted Steno's discourse within it.

The senses in which Winslow is following Steno are in Steno's advocacy of clarity in reporting; of confessing ignorance rather than feigning knowledge; of recognising that 'the senses alone are capable of informing us whether our ideas be conformable to nature'; of pursuing many and varied methods of investigation ('anatomy ought to be confined by no rules, every new dissection requiring a new method'); and finally, of dedicating his life full time to anatomical research, since this is the only way to complete and true knowledge of anatomy. Here, for instance, is how Winslow opens his work:

> I intitle this work, *An anatomical exposition of the structure of the human body*, because my design is simply to relate that structure, as I have found it by dissections often repeated, and in different manners; and because I confine myself intirely to the human body, I have not enlarged very much on the uses of the parts; I mention those alone, which appear to me to be well founded on the known structure of the body; and sometimes I say nothing of them at all, as knowing nothing certain about them. In that case, I frankly acknowledge my ignorance, in order to excite others who, perhaps, may be more lucky than I have been ....[118]

Winslow's translator into English, G. Douglas (brother of the anatomist James Douglas) describes the volume as 'consisting of two parts; an enumeration of phaenomena or matters of fact, and conclusions or inferences drawn from thence concerning the use or office of the several parts of the body' (p. v). At all stages Winslow has arranged everything for the better comprehension of the beginner.

While Winslow has Steno as his programmatic model, as his practical model for anatomising he adopts Vesalius – in complete contrast to Dionis. That is to say, he arranges his book (and presumably his teaching) on Andreas Vesalius's 1543 book, *Seven books on the fabric of the human body*:

> I have followed the same general order which is observed by Vesalius in his great work *De corporis humani fabrica*; beginning by the bones, and from thence going on to the muscles, arteries, veins, nerves, abdomen, thorax, and head, together with the organs of sensation; and I more willingly pitched upon this method, because I formerly designed to have published a *Vesalius renovatus*.[119]

---

[117] Winslow, *An anatomical exposition of the structure of the human body*, 1733, p. x. On Steno's discourse see above, Chapter 1.

[118] Ibid., p. v.

[119] Ibid. On Winslow's intention to publish a *Vesalius renovatus*, see Chapter 1 above ('Making history').

So Winslow in the early eighteenth century is turning back to an anatomist of the early sixteenth century. But it has recently been shown that Vesalius himself was modelling his work on that of Galen in the second century AD.[120] So Winslow is actually basing himself on Galen! This is how progress is often made: by looking backwards.

This sequence of anatomising is quite different from the medieval one. For instead of starting his dissection with the three venters, in the order of corruption, Winslow instead finishes with them. He thus begins with the bones (including the fresh bones), but then goes on to the muscles, arteries, veins and nerves – that is to say, the great systems of the body – before concluding with the three venters. Of course he deals with the arteries and veins in the light of the circulation of the blood (knowledge not available to either Galen or Vesalius). Everything is explained clearly and tersely, and the presentation of the anatomical facts is kept as distinct from any physiological speculations as in any work I know.

Edinburgh, 1726–40s: Monro Primus

When he started teaching anatomy to medical and surgical students in Edinburgh in 1726, Monro *Primus* did not have an adequate supply of bodies, and as there was as yet no hospital to supply the corpses of paupers who had died without relatives he depended on the grant of executed criminals from the town council.

In the dissection guide he wrote for his son Donald in 1747, 'A treatise on the anatomical encheireses or manual part of anatomy', Monro recorded his practice of teaching anatomy that he had employed and developed over the previous twenty years:

> I suppose myself writing for a publick teacher who has not human subjects always at command but must demonstrate all the parts on two bodies with the assistance of his collection of preparations, which you know is generally my case. [121]

Somehow Monro managed to use these two bodies to teach classes of up to one hundred and fifty students at a time. The dissection guide reveals that Monro (and perhaps his assistants) appear to have performed all the demonstrating themselves – that is to say, the students did not perform dissection for themselves. Nevertheless, with the aid of many wet and dried preparations of human and animal parts, together with experiments carried out in front of the audience on human and

---

[120] Cunningham, *The anatomical renaissance*, 1997, chapter 4.

[121] My account here depends completely on the excellent account of Lawrence, 'Alexander Monro *Primus* and the Edinburgh manner of anatomy', 1988, who is using the manuscript original that is in the University of Otaga Medical School Library, in Dunedin, New Zealand, and that I have not seen; the quotation is from page 198 of Lawrence's article.

animal corpses and on live animals, Monro was able to provide a comprehensive anatomical tour of all parts of the human body. His apparatus for demonstrating features of the body included several microscopes and an air-pump.

After demonstrating the skeleton, following the sequence of his own published book on the subject, Monro turned to the first corpse, on which he would demonstrate the muscles and the inner organs of the trunk of the body. Here Monro frequently used a pipe to blow air into the abdominal wall to distinguish the different membranes around the muscles. The inner organs and their relationship were demonstrated, and a live animal was used to show the peristaltic movement of the stomach and guts.

When he came to the lungs Monro performed experiments on the corpse to demonstrate the intake and expulsion of air. Then there were seven 'experiments to be made on living brutes' concerning the lungs, which the medical historian Christopher Lawrence has summarised thus (p. 206):

1. Laying bare the pleura, observing respiration, and then immersing the creature in water and puncturing the pleura.
2. Making openings into the thorax and, by inserting cannulae if necessary, observing the motion of the lungs and diaphragm.
3. Pricking the lungs through the cannulae to observe the effusion of air or blood.
4. Repeating the above but keeping the lungs constantly distended by means of bellows in the animal's trachea.
5. Inflating an animal's lungs through the trachea but with the whole thoracic contents displayed.
6. Blowing air through the red-hot barrel of a musket into a 'glass' containing a living animal, a thermometer, and a glass tube in mercury.
7. Enclosing an animal in an air pump with a thermometer and a tube immersed in mercury. Observing the animal as it died.
8. Finally with his first corpse Monro demonstrated the brain and sense organs.

The second corpse was dedicated to the showing of the blood vessels – following them out from the heart – and the nerves.

Three elements appear to have been used constantly by Monro: dissection in front of the students; showing preparations made beforehand to demonstrate particular structures or pathways; and animal experiments watched by the students. It was, as Lawrence has described it, 'a theatrical display' (p. 209).

Alexander Monro *Secundus* appears to have followed his father's sequence, according to his announcement in his introductory lecture for 1770–71:

> After having gone through the History of anatomy and recommended the best authors, we shall then proceed to the study of the Bones … After the Bones we shall treat of those substances which serve to unite the Bones together,

> to regulate their motions, and keep them soft, smooth and pliable; after this we shall discourse of the muscles which are small threads all united with one another, and which receive their motion from the fluids. Then we shall treat of the different vessels of the human body, of the arteries and nervous system and of the lymphatic and absorbent veins; after this we shall treat of physiology, its connection with anatomy, of the dissection of brute animals, of the morbid states of the body, their difference from the natural state, and shall conclude our course with the operations of surgery.[122]

London, 1746–83: William Hunter

William Hunter, who established private anatomical teaching in London on a new basis, had experienced dissection in Paris over the winter of 1743/44. The teacher was Antoine Ferrein, teaching at the Collège royal. Hunter later recalled:

> In the course of my own studies, I attended, as diligently as the generality of students do, one of the most reputable courses of Anatomy in Europe. There I learned a good deal by my ears; but almost nothing by my eyes; and therefore, hardly anything to the purpose. The defect was, that the professor was obliged to demonstrate all the parts of the body, except the bones, nerves and vessels, upon one dead body. There was a foetus for the nerves and blood-vessels; and the operations of surgery were explained, to very little purpose indeed, upon a dog.[123]

When he returned to London, Hunter founded a private anatomy school in 1746. He advertised that he would give lectures on anatomy, but, in addition, 'Gentlemen may have the opportunity of learning the art of dissection during the whole winter season in the same manner as at Paris'. What Hunter meant by this was that students would be able to undertake the dissection of human corpses themselves, and in effect have a whole human body to dissect. This was the practice not of the academic or medical professional institutions of Paris, but of the hospitals and, more particularly, of private teachers who bought their corpses from the grave-robbers.[124]

A clear and convenient view of William Hunter's course was presented in print in 1782 – that is, reflecting up to thirty-six years of teaching on his part – in an article in the new monthly journal *The European Magazine and London Review: Containing the Literature, History, Politics, Arts, Manners and Amusements of the*

---

[122] As quoted from student notes by Wright-St. Clair, *Doctors Monro*, 1964, p. 73.

[123] William Hunter, *Two introductory lectures*, 1784, p. 88. Here I follow Gelfand ('The "Paris manner" of dissection', 1972), who places this experience in Paris, while other historians have taken it to be a veiled criticism of Monro *Primus*, whose classes Hunter attended in Edinburgh.

[124] Gelfand, 'The "Paris manner" of dissection', 1972.

Age, published by the Philological Society of London. It was discovered there by the indefatigable Hunter scholar, George Peachey, though he did not discuss its context.[125] The editors of the magazine intended to attract readers by a whole array of interesting articles: 'Another most material article, which we have in design … is, An Account of the Lectures of the celebrated Professors in this Metropolis, and in the Universities of Europe'.[126] 'What we have in view', the editors wrote, 'is simply to collect and bring together the various schemes and plans of study now offered to the world in the several branches of polite and useful science … We now begin the subject with an Account of a Course of Anatomical Lectures which holds the first place in this kingdom, as a medical study …'.[127] So the account and its accompanying anecdotes seem to have been written by a journalist, in consultation with Hunter, rather than just by William Hunter himself.

By 1782 the course was very long, perhaps the longest available in Europe: something like two hundred lectures, given one daily for two hours in the afternoon, from the beginning of October to the end of May.

1. The first three lectures give the history, the uses, and the best method of conducting the study of anatomy.
2. Then the nature and the uses of the similar and constituent parts of the body are explained, to convey a general knowledge of the body, and of its operations, in this order, blood, arteries, veins, lymphatics or absorbents, glands, nerves, muscles, bones and their appendages. After which the particular anatomy is taken up in the following order, viz.
3. The bones.
4. The muscles in a fresh subject.
5. The male organs in the same.
6. The different fresh joints in the same.
7. The viscera, and female organs, in a fresh subject.
8. The organs of sense and integuments in a fresh subject.
9. The brain and its nerves in a fresh subject.
10. The diseases of bones.
11. The diseases of the viscera.
12. Chirurgical diseases more particularly, and the operations of surgery explained and performed upon a fresh subject.
13. The anatomy and physiology of the gravid uterus and foetus.
14. The diseases peculiar to the [female] sex.
15. Of pregnancy and parturition.
16. Of the disorders and management of women in child-bed, and of children. The whole comprehends a full course of anatomy, physiology, surgery,

125 Peachey, *A memoir of William and John Hunter*, 1924, pp. 128–30.

126 'Introduction' to *The European Magazine*, vol. 2, 1782.

127 *The European Magazine*, vol. 2, 1782, p. 166; Gelfand, 'The "Paris manner" of dissection', 1972.

> and mid-wifery; illustrated by a greater number of elegant and curious anatomical preparations than ever were brought together or used in any school of anatomy.

Could one describe this as the best available anatomy course for surgeons in eighteenth-century Europe? Everything in it depended on the exertions of just one man and his one or two assistants, but that man was driven by ambition to succeed, and a desire for wealth, that the class brought him. 'Some people', William pondered, 'thought even my former courses too long. Why? They had been used to see a course of Anatomy finished in thirty or forty lectures; and therefore imagined that, when it took up near four months, it must be unnecessarily minute or prolix. But, let them reflect how imperfect such courses were; let them recollect, that they never saw the human brain and nerves, nor the human lymphatic system, nor the human gravid uterus and its contents: let them recollect, that there were few parts of the body, which they understood thoroughly, after attending even repeated courses of that kind; and then they will see that there was great room for improvement; and own that an opportunity of attending a more compleat course of anatomical lectures, must be a national advantage'.[128]

The most striking thing about this course is the number of 'fresh subjects' that were available. Hunter is not obliged to make the best of one body, nor is he restricted in any way as to which sequence of dissection to follow by the availability (or otherwise) of corpses. At least five human corpses are used in the course, with at least one of them being a woman. But it wasn't just a matter of many corpses for demonstration; within William's school,

> There is likewise a school for practical anatomy kept open and much frequented through the whole winter. Here students see everything that is going on, both dissections and the arts of examining diseases and making preparations; and with their own hands dissect as many bodies, make for themselves as many preparations, and perform as many operations of surgery as they please'.[129]

This constant supply of fresh subjects was for years managed by William's brother John, who did the necessary dealings with the resurrection men.

Two other things are notable about the facilities in William's anatomy school. The first is that he was extremely well supplied with preparations of parts healthy and diseased, and these were all made on the premises, completely under William's control and designed for teaching particular points. It was a most systematic collection, not random or arbitrary.

---

[128] William Hunter, *Two introductory lectures*, p. 100.

[129] Peachey, *A memoir of William and John Hunter*, 1924, p. 130, from *The European Magazine*.

> Besides attending lectures, and dissecting dead bodies, there is here an opportunity of learning Anatomy to the best advantage, by attending the dissecting-room. One winter's attention there, will certainly make a diligent student a good Anatomist. He will, in that time, see the preparatory dissection for every lecture; which will make the lecture itself much more intelligible, and fix it deeper in the mind; he will see all the *principal parts* dissected and demonstrated over and over again; whatever he finds he does not clearly understand, there is such a number of bodies dissected in succession, that he will, at any time, have an opportunity of attending to that particular object, and of getting it explained to him; he will see all the operations of surgery performed and explained, again and again; and he will see the practice of all the arts of making preparations. It is an important piece of education; and as it is not to be had in other places, I recommend it earnestly; and especially to those students who are to be in London, one winter only.[130]

The other thing that made William Hunter's course so outstanding and exemplary was that a significant part of what he taught was of his own discovery.[131] Who else could boast such important discoveries for the education of the eighteenth-century surgeon making his way as the anatomy of the human gravid uterus, and the functions of the absorbent vessels?

The syllabus of William's course had claimed that it was 'a full course of anatomy, physiology, surgery and midwifery'. This was certainly the case for anatomy, surgery and midwifery. But in his lectures William actually disclaimed his – or anyone else's – capacity to provide a coherent account of the physiology of the body:

> In treating of the physiology, it is very difficult to say, what plan we should follow; for, every method which has been yet proposed, is attended with manifest inconvenience … The animal œconomy, indeed, is so extensive a field, and takes so much light from philosophy, and other arts, that in most universities, it has been found necessary to appoint two professors, one for Anatomy, and the other for Physiology. Undoubtedly the man who makes it his business, to investigate every thing relating to the structure of the human body, must be, *caeteris paribus*, the fittest person to explain its operations; and there cannot be a more proper occasion, than when the parts are before us. And, yet in some sense, it is just the reverse: for, every good Anatomist, who has a cool head, and keeps a guard over his imagination, knows, that many of the received hypotheses in Physiology, are built on very loose foundations, and liable to weighty objections; or, demonstrably repugnant to what we already know of the structure of our body. So that if the Anatomist is weak enough, to think that he can, or that he

130 William Hunter, *Two introductory lectures*, p. 109.

131 His brother John did not agree with William here about who had discovered what. See Chapter 5 below.

> must explain, all the operations of the machine, he will certainly make partial dissections, mislead himself and others, and put a thousand little arts in practice, to make the structure correspond with the imaginary use.[132]

Hunter reminded his students, 'There never was a man, perhaps, more followed and admired in Physiology, than Boerhaave. I remember the veneration he was held in; and now, in the space of forty years, his Physiology is – it shocks me to think in what a light it appears'.[133]

William Hunter was, if you like, the ultimate anatomist's anatomist: the facts are before our eyes and we have every facility and opportunity to check and check again. But any speculation beyond what we can see is going to take us into the realms of fantasy and absurdity, so physiology should be approached very gingerly. I have stressed this aspect of William's work because it provides the greatest possible contrast with the anatomical work of his own brother, John Hunter. For that was directed precisely at physiology and pathology – though not in a way other physiologists would have recognised.

London, 1773–93: John Hunter

So I want to conclude this exploration of different approaches to investigating and teaching anatomy by different anatomists, with John Hunter. John attracted much notice, favourable and hostile, in his lifetime, and he has had several biographers, unlike William or almost any other of our anatomists. Yet what his anatomising was about – and hence what his anatomical museum was about – has, I believe, been lost under all this attention. For historians have mostly been concerned with redeeming John's reputation against that of William, and of giving him credit for the disputed discoveries. Quite separately from that issue, the Royal College of Surgeons of London, who were given John's collection after his death, have celebrated him as the originator of 'scientific surgery', though since everyone seems to assume that that expression is self-explanatory, no one explores how (or indeed whether) John had actually created this. No one seems in modern times to have recognised the nature and the distinctiveness of John's anatomical research and teaching, which was quite different from that of any other anatomist in the long eighteenth century.[134]

The *European Magazine* is quite a good place to start, for in the next issue after the one that showcased William Hunter, an equivalent piece appeared about John, and with the co-operation of John himself.[135] Running from October to

132 William Hunter, *Two introductory lectures*, pp. 93–4.

133 Ibid., p. 98.

134 The recent biography of John Hunter by Wendy Moore, *The knife-man*, is excellent on John as a surgeon; she does not deal with him as an anatomist *per se*.

135 *The European Magazine*, vol. 2, 1782, pp. 246–9. John's own account in his lectures closely resembles this one: see John Hunter, 'Lectures on the principles of surgery',

April, every other evening from seven- to eight o'clock in the evening, the course consisted of almost a hundred lectures. The audience consisted of young surgeon apprentices and others. But the lectures were not directly on surgery, nor directly on the anatomical knowledge one needed to wield the surgical knife safely. Rather, they were on *principles* of surgery (as understood by John Hunter), viz. 'To shew the actions of the body and its parts when in the diseased state, with the actions and effects of nature to recovery – and the necessary and proper assistance to be given by the Surgeon' (p. 246). The principles of diseases, said John, 'are to the Surgeon what the first principles of the mathematics are to the practical geometrician, without the knowledge of which a man can neither be a philosopher nor a Surgeon' (p. 246).

This is a quite different anatomical programme and project to any we have hitherto seen among our anatomists. John Hunter was indeed a practical surgeon, being called out to patients at all hours of the day and night, and practical surgery provided him with the bulk of his income. But the content of his teaching came rather from his *research*: 'His doctrines are drawn from personal observations made in the course of an indefatigable life, with the constancy of a most adventurous mind. His ideas, his modes of reasoning, as well as his arrangement of diseases are new; and he therefore has received little aid from books or from other professors', said the *European Magazine* (p. 246), uncritically accepting John's point of view. The lack of any formal education in anatomy, apart from what he had learnt from his brother or acquired for himself in the dissecting room, may in part account for why John had failed to become properly socialised and indoctrinated as an eighteenth-century anatomist: he had no master he was prepared to follow, no doctrines pre-prepared by someone else to assimilate and make his own. This does not of course make John Hunter a man 'ahead of his time', for no one can be ahead of their time, and his concerns, vocabulary and notions are all thoroughly of the eighteenth century. But they are different concerns to those of his fellow anatomists, who for the most part were unable to follow his thinking.

This is how John's course was outlined in the *European Magazine* for 1782:

> The course begins with the Physiology or Natural History of the Animal …
> The Physiology of Diseases follows … [which] brings him to –
> The Consideration of Diseases in general. The Diseases of accident being the most obvious and simple, in preventing the natural operations of the Animal, begin the Surgical part of the Lectures, in which he explains the most simple modes of restoration. Then follow the diseases whose mode of action is similar to those arising from accident.

So John Hunter begins by expounding how the animal – not just man, but the whole range of animals – functions in health. While this may seem speculative (like so many other physiologies, as we shall see in Chapter 3), it is in John's case

---

1786.

in fact founded on very extensive anatomical experiment. Then John surveys how diseases happen in general in the animal. Disease, it has now been established (he claims), prevents the natural operations of the animal. If we understand how diseases operate, we shall appreciate that 'in the animal body the power of restoration to a state of health, arises out of the Animal from its own mechanism and action' (p. 246). What are these natural routes back to health in the animal? The main ones are (1) inflammation – 'it is one of the first principles in most diseases' (p. 247), (2) suppuration (as appears in compound fractures, gun-shot wounds, abscesses), (3) granulations (the formation of grain-like prominences on sores while healing), and (4) cicatrisation (the forming of a scar in healing). So, left to itself, the animal (including man) usually recovers from diseases of accident and other similar diseases, by one or more of these routes. But –

> … if the Animal was in all cases equal to the task there would be no occasion for the Surgeon; but it is necessary in many cases to assist nature by the introduction of artificial powers – to increase the [natural] living powers of the Animal when they are inadequate – to retard them when violent – or to change them when wrong. (p. 246)

The kinds of accidents or disease to which the surgeon is called therefore require (according to John) not just the traditional treatments of the surgeon, such as bandaging, poultices, bleeding, amputation, but a thorough understanding of what nature's principles of healing are, and how they need to be assisted or modified. It is because it is built on the *principles* of nature, that such an understanding was, in John's eyes, a *scientific* understanding. All of John's anatomical research and preparations were aimed in this direction. As the *European Magazine* wrote: 'This Course of Lectures is illustrated by a collection of diseases … he [John Hunter] knows the particular history of the greater part of the diseases, which he has preserved – the patients were under his observation in the Hospitals – he has minuted the progress – and accounted for the various appearances – and effects of each disease, with a fidelity which now renders his Collection a most instructive school for the Student' (p. 247).

All of which meant that a student attending John Hunter's lectures would be presented with a quite new way of thinking about surgery and the surgeon's role, and of the anatomical structures where diseases or accidents occur. From what I can see, only a few acolytes who had extensive exposure to John's way of thinking could fully appreciate the novelty and importance of this approach, and see the relationship between John's anatomical preparations and his views first on physiology, and second on disease and its cure. The great teaching and research museum that John Hunter established in his own London teaching complex in the 1780s was later described by one of his devoted pupils, John Abernethy, as 'the principal record of his [John Hunter's] deeds and opinions':[136]

[136] Abernethy, *Physiological lectures*, 1817, p. 2.

> His object in forming the Museum was simple [Abernethy wrote]; it was to display the anatomical facts which were the ground-work of his physiological opinions; to exhibit the visible means, by which the vital processes were carried on, in all the varieties of living beings, that came within his cognizance … I repeat … that the preparations in the Museum are records of facts. Mr Hunter would have given his money even to his own ruin, to have procured the evidence of any fact warranting an important physiological conclusion ….[137]

The anatomical preparations gave the *facts*: the physiological conclusions were based on these facts – as were the surgical principles about disease and its natural cure.

I imagine that for the ordinary surgical student, by contrast with a devotee such as Abernethy, John Hunter was simply seen as giving an anatomy course with special reference to surgical diseases, and nothing much more.

John Hunter was indeed a maverick anatomist and surgeon, thinking deeply about anatomy, physiology, surgery and cure, more deeply perhaps than any of his contemporaries, and coming to quite distinct conclusions about the role of anatomy for surgery, and hence teaching it in a unique way.

Part 2: Women

To this point what we have seen is a male activity, conducted by men for the education, edification and entertainment of men. Yet then, as now, at least half of the population was female. Women were patients of surgeons and physicians as much as men were. What was the role of women in all this anatomising? Was there a female presence in the anatomy theatre or dissecting room?

1. The first role of women was *to be present, dead, as subjects of dissection*. Where the primary source of bodies was the gallows then males constituted the overwhelming majority of bodies dissected. But where hospitals for the poor were the primary source, there was no distinction: an anatomist had available as many decrepit or diseased bodies of the one sex as of the other. However, strict limitations applied in some institutions when a woman was being dissected. William Hunter, for instance, only admitted fee-paying students at such times to his private anatomy teaching:

> … if a student should wish to introduce a friend to any particular lecture, it will give us pleasure to oblige him; provided he will only take the trouble of presenting his friend, just before the lecture begins. The lectures, however, upon the organs of generation, and gravid uterus, are to be excepted. No visitor can

[137] Ibid., pp. 9–11; Jacyna, 'Images of John Hunter', 1983.

be introduced when we are upon these subjects. The reasons for such exceptions must be obvious.[138]

2. The second role for women was *to be absent from the audience* in the anatomical theatre. There were no female medical or surgical students in eighteenth-century European professional institutions, as there were also no female students at university. The one, sole, female university professor, Laura Bassi at Bologna, was the exception who proved the rule,[139] and she had almost the status of monstrosity.

Fig. 2.9

Sometimes, however, the presence of women in the anatomical theatre is remarked on. For they could sometimes actually be inside the anatomical theatre, but *invisible*. Back in the 1590s the young Thomas Platter had sometimes noticed at Montpellier that 'there are women among the spectators, and they hide their faces behind their masks, especially when the subject is a woman'.[140] Similarly, over a hundred and fifty years later, the Barcelona anatomical theatre of the surgeons,

[138] William Hunter, *Two introductory lectures*, 1784, p. 113.

[139] On Bassi see, for instance, Paula Findlen, 'The scientist's body: The nature of a woman philosopher in Enlightenment Italy', 2003.

[140] Thomas Platter, *Journal of a younger brother*, 1963, pp. 36–7.

built in 1761, had grilled balconies (Fig. 2.9) for those people, including women, who did not wish to be seen in the theatre.

But on a big occasion, such as a professor's inaugural show lecture, women sometimes seem to have been both present and visible. Morgagni recalled that at his first anatomical demonstration in Padua in 1716 there was 'un pubblico fittissimo composto di persone di ogni classe sociale, comprese numerosissime dame', a very full audience composed of people of every social class, including very many ladies.[141]

3. The third role for women with respect to anatomy was *to be repulsed.* It seems indeed that it was the anatomists themselves who assumed that women found anatomical dissections repulsive, which would be in accord with eighteenth-century notions of gentility and of the sensitivity of the female sex of the higher classes.

We can see this assumption about the supposed sensibilities of women with respect to anatomy in a very striking manner from a tale that involved a quite exceptional woman – the Queen of France – who apparently did not share this supposed female weakness. In June 1681 one of the bedchamber women to Madame la Dauphine, a lady of twenty-five years of age and six months pregnant, died in great agony. In the hope of saving the child a surgeon performed a caesarean operation on her after she died, but no child could be easily found. The surgeon and anatomist Pierre Dionis recounts the story: 'The whole court was filled with the noise of this accident. The Queen and the Dauphiness commanded me to open her body, and trace the cause of so sudden a death. Accordingly I went, and in the presence of Monsieur Daquin the King's first physician, and Monsieur Fagon first physician to the Queen, laid open the corpse in the usual manner'. What they found was extraordinary. The womb appeared to have two fundus-es or bottoms, in one of which was a mole or false conception, and in the supernumerary fundus was a child that had been dead at least a fortnight. The growth of the child in this position had led to the womb being ruptured, and the child was actually found lying on the intestines of the mother, whose whole abdomen was full of blood. Establishing the anatomical structure of this odd womb took Dionis a long time:

> But in regard the time commonly allotted for the dissection of a dead corpse, is so short, that one cannot trace all the common particulars; I chose to cut out these two bodies [the mole and the fetus], with the neck of the womb, the testicles [i.e. ovaries], the tube, the ligaments, and part of the spermatic vessels; and wrapping them up in a napkin, sent them to my house, to be viewed at more leisure ... At night I made a nicer dissection of the womb at my own house; but took care not to gash it too much, meaning to keep it as entire as I could, in order to have a draught [i.e. a drawing] of it. The next day the Queen ordered me to bring it to her. She was then at the Dauphiness's apartment, and examined it curiously for

[141] Morgagni, 'Le autobiografie', 1964, p. 45.

> a long time. Monsieur Daquin and Monsieur Fagon acquainted Her Majesty and the court ladies with their sentiments upon the matter. The same day the Queen sent for it again, being then in her closet with only one lady. She reviewed it once more; and I answered all the questions she put to me. Her Majesty had not the same aversion to anatomical demonstrations that other women have: for I have frequently had the honour to show her the dissection of several parts of animals.[142]

4. There were also a few women who were active in anatomical matters, such as Anna Morandi, the *wax modeller*, who will be discussed in Chapter 4. Another female anatomical modeller (though not in wax) was a Frenchwoman, Mademoiselle Marie-Catherine Biheron (1719–95), whose work we know about particularly from a flyer for 'Artificial Anatomy' that she published in Paris in 1761.[143] This was something quite remarkable, and it would be fascinating to know how the artificial anatomy was constructed. For the flyer promised that the public could see an artificial anatomy performed on a body truncated at the extremities (i.e. without head, arms or legs), together with the development of the viscera contained in the three venters. This would be superior to the artificial anatomy recently demonstrated in Paris by M. Desnoües (so Mademoiselle Biheron claimed), because it would present a body covered with 'a true skin' that imitates the exterior and general covering of bodies, and that allows the artificial body to be transported easily and without damage. Moreover, Mademoiselle Biheron claimed to have copied the natural membranes in such a way as to fool the eyes of the spectators, which appears most remarkable in the peritoneum, the omentum and the pleura. Moreover, 'the hollow and membranous vessels such as the stomach and the intestines have been made artificially with the consistence, suppleness and lightness of the natural viscera. One can blow up the stomach, and also the lungs'. Every part, and the relation of the parts to each other, is exactly done, 'to the point that Nature is copied with the greatest precision, and an astonishing truthfulness'. The artificial anatomy was on show from the 13 May 1761, and thereafter was to be on view every day except feast days and Sundays, from eleven in the morning till one, and from four in the afternoon until six, at Mademoiselle Biheron's house on the Vielle Estrapade, at the corner of the rue des Poules.

Mademoiselle Biheron was the daughter of an apothecary, and was initially trained in botanical illustration at the Jardin du Roi.[144] Apparently wishing to cater for a desire by certain women of fashion to study anatomy in eighteenth-century

---

142 Dionis, *The anatomy of humane bodies improv'd*, 1703, pp. 196–200. See also the account in Dionis, *A course of surgical operations*, 1710, p. 89.

143 Biheron, *Anatomie artificielle*, 1761; the flyer has only four pages. Diderot was at one time her pupil; see Diderot, *Éléments de physiologie*, 1964, Introduction, p. ix. Huard, 'L'enseignement medico-chirurgical', 1964, p. 182, gives her dates as 1719–1815.

144 The following information on the life and work of Mademoiselle Biheron is taken from Poirer, *Histoire des femmes de science en France*, 2002.

Paris, she began to study anatomy. In order to do so, she apparently had to bribe people to steal bodies, which she hid in her room; by the time she had dissected them for her ladies, the parts were often in a state of putrefaction. It is this that gave her the idea of making artificial anatomies, using silk, wool, thread, feathers, and waxed coloured paper. An excited observer of the finished article was heard to exclaim that it was perfect: it lacked nothing but the smell! The surgeon Sauveur Morand, who had been charged in 1759 with assembling a cabinet of anatomy for Elizabeth, the Empress of Russia, showed some of Mademoiselle Biheron's models to the Académie des Sciences and the Académie de Chirugerie in Paris, where they were highly praised.[145] But forbidden, as a woman, to take pupils, Marie-Catherine moved to London, where she is said to have been employed by William Hunter and William Hewson: it is claimed that some of Hunter's illustrations for *The Anatomy of the Human Uterus* were borrowed from her. She is said to have turned down royal and imperial invitations to become anatomical demonstrator in Stockholm, Upsala or St Petersburg. With the coming of the Revolution, it was proposed that Mademoiselle Biheron's collection be bought for the new École de Santé in Paris, but the new professors of anatomy declared that only preparations from humans and animals should be used. Her models were deemed unsuitable for public instruction, and seem to have completely disappeared.[146]

5. Possibly there was a woman who was actually a *practising anatomist*. But even if this were so, she held fast to her anonymity in this respect, even though she was quite well known for other literary achievements. The woman in question is Marie-Geneviève-Charlotte Thiroux d'Arconville (1720–1805). The evidence for her being an anatomist is suggestive but not conclusive.[147] Supposedly she drew, or oversaw the drawing of, illustrations of the skeleton, and/or made the French translation of the text, which were actually published under the name of the male anatomist Jean-Joseph Sue *père* (1718–92) in his *Treatise on Osteology, translated from the English of Monsieur Monro* [*Primus*] of 1759. Unfortunately for her posthumous reputation, the name of d'Arconville is not on the title-page of

---

[145] According to Portal, *Histoire de l'anatomie*, 6 vols, 1770–73, vol. 5, p. 2, 'on trouve la description et l'éloge des pieces d'Anatomie que Mademoiselle Bihéron fait avec tant d'art, et qu'elle a fourni pour cet arsenal', in Sauveur Morand's *Catalogue des pieces d'Anatomie, instruments, machines qui composent l'arsénal de Chirurgie à Petersbourg*, 1759; I have not been able to see this.

[146] Haller, *Bibliotheca anatomica*, 2 vols, 1774, 1777, vol. 2, p. 665, mentions that 'Biheron virgo, anatomes perita, fetum cum utero imitata est' ('Biheron, a woman skilled in anatomy, has imitated a fetus with a uterus'), noted in the *Histoire et Mémoires de l'Académie royale des sciences de Paris*, for 1770. I have not seen this.

[147] It is put forward in Schiebinger, 'Skeletons in the closet', 1986, esp. nn. 52 and 53. Unfortunately the referencing is not good enough there for one to follow up the claim and see on what evidence it is built.

the book, nor are the illustrations anywhere ascribed to her.[148] And even if she had actually been both translator and artist, neither role necessitates her having been a practical anatomist.

6. Possibly again, in the mid-eighteenth century, more specifically in the period 1730 to 1790, a new role of women was to have *a separate anatomy*, in the sense of being perceived to have a separate anatomical structure, distinct from that of the male. The case has been put by Londa Schiebinger that the skeleton of the female began to be portrayed separately from that of men, and as having certain distinguishing features, such as a smaller head and a larger pelvis. Schiebinger claims that this distinct female skeleton appeared suddenly and was linked with 'the eighteenth-century movement to define and redefine sex differences in every part of the human body', a movement whose aim was 'to buttress arguments against women's participation in the public spheres of government and commerce, science and scholarship'.[149] Thus, as typical products of their own society, the male anatomists began to 'see' hitherto unremarked characteristics and features in the female skeleton that supported this movement, and to portray them in their illustrations. Whether or not this argument holds (and I am not personally persuaded by it) it is striking that, in Schiebinger's own account, the person most instrumental in making such portrayals of female skeletons was the supposed one *female* anatomist who (she claims) was Marie-Geneviève-Charlotte Thiroux d'Arconville! So this also remains inconclusive for now.

7. Finally with respect to the roles of women in connection with anatomy, there were the very big issues in the long eighteenth century of conception, pregnancy, and the delivery of babies whether by women (midwives) or by men (man-midwives), in all of which women were naturally the persons mostly concerned. Anatomy was found to have much to do with all these issues. They will be dealt with in the course of the next chapter.

[148] Monro, *Traité d'ostéologie*, 1759. Sue says that the translation was undertaken, at his wish, by 'un de mes Elèves avec toute l'exactitude possible' (Préface, p. xxiii). Of the illustrations, he writes 'Tous les Desseins ont été faits sous mes yeux', though without naming the artist; the artist's name on the engravings (where given) is 'Tharsis'.

[149] Schiebinger, 'Skeletons in the closet', 1986, pp. 42, 43.

Chapter 3
Experimental Anatomy and its Sub-Disciplines

1. Anatomy, an experimental discipline

Anatomy, to the modern way of thinking, is not a particularly experimental discipline, that is to say a discipline to which the use of experiment to create new knowledge about the functioning of humans and other animals is absolutely central. I suspect that most of us would offer that palm to different modern disciplines that study the phenomena of animal life, in particular to *experimental physiology*, rather than to anatomy. Yet anatomy had always been a discipline built around experiment, and until the end of the long eighteenth century anatomists in their dozens resorted to experiment with great keenness. To see some of the ways in which anatomy was a fundamentally experimental discipline, I propose looking at a few specimen experiments conducted by different anatomists in the period.

These are experiments to discover what, whether, or how something was the case, or not – that is to say, to discover whether *x* was or was not a fact about the body. A wide range of experimental approaches will be seen to have been adopted, as suited to each situation, and in this respect these experiments follow Steno's call for varied experimentation (Chapter 1 above). All these experiments, all these vivisections, dissections and self-experiments, were done with a purpose. And although some of the more important discoveries in anatomy may have become visible while something else was actually being sought, nevertheless the experiments conducted over the years were in general neither random nor arbitrary. It will be noted that these anatomical experiments do not ask 'why?' questions; as we shall see, that was the domain of physiology. Rather, they ask 'what?', 'how?' and 'whether?' questions most of the time.

Experiment 1

In 1651 Jean Pecquet published his account of how, in the course of an anatomical experiment in about 1642, when he was a young French medical student, he had by chance discovered the receptacle of the chyle, the chyle being the white liquid formed through the digestive process from the ingested food and contained within the 'milky veins' that carry it away from the intestines, and its receptacle (or cistern), being the sac where it is stored before it enters the bloodstream. The 'milky veins' had been familiar to anatomists since their discovery by Gaspar Aselli in 1622.

But where they went, or that their contents reached the bloodstream, was all still unknown. Pecquet recounts some of his anatomical experiments:

> After I had some years ago, by Cutting up of Dead Bodies, acquired a dumb (I may say) and cold Knowledge, I resolved to squeeze forth true knowledge from the Harmony yea of twenty Living Creatures. And because these differ from the former almost only in Motion, which hath its chief seat in the Heart, It was my purpose, having unclothed and cut out the Heart, to contemplate it more manifestly.
>
> Therefore having cloven asunder the Thorax of a Great Hound, I begun my view of the contained parts without delay; I plucked out the Heart, having cut asunder those Vessels wherewith it was tied to the rest of the body; The abundance of blood, which immediately flowed, did at present stop my prying sight; that being spent, I did wonder to see flowing in the Pipe of the *Vena cava* at its connection to the right Ventricle a milky liquor, casting it self out by intermission.[1]

In subsequent animals Pecquet was able at last to find the 'swelling lake of Chyle. And then my eyes, allowing the touch its part, by pressing with my forefinger I oft tried on each part the escape of the enclosed liquor: There, which is wonderful! the passage did render itself flat to the compressing finger; The softness coming by jumps did argue a bladder not of small capacity to hold Chyle lurking under the Center of the Mesentery … So at last was laid open that most desired storehouse of the Chyle, and that *Receptacle* searched out by so much labour'.[2]

Pecquet the anatomist has experimentally intervened, to cut out the living hearts of animals and to trace, touch and squeeze the bladder of chyle.

Experiment 2

Sometime in the 1660s Regnerus de Graaf, a physician of Delft in the Netherlands and an active anatomist, took a male dog, introduced it to a bitch on heat, and when the dog had an erection he then tied a cord around the root of its penis. In the midst of the copulation that ensued the anatomist pulled the cord extremely tight and fastened it, thus keeping the penis in an erect state. He then immediately killed the dog and dissected its penis, whereupon he found that only bright red arterial blood flowed from it and it became flaccid when the flow of blood stopped. This heartless anatomical experiment was part of a long series of anatomical investigations and experiments on the male and female human organs of generation that de Graaf was engaged on. These particular observations and experiments were intended to determine what it is that distends the penis in erection: is it the muscles, the so-called pair of 'erector muscles', as all anatomists had hitherto maintained, or

1 Pecquet, *New anatomical experiments*, 1653, pp. 7–8.

2 Ibid., pp. 26–7.

is it perhaps the animal spirits flowing into the penis via the nerves, or is it the blood brought there by the arteries? De Graaf's observations showed that the muscles cannot have this erecting role, 'since the action of a muscle is to contract and this is the opposite of extending', and the penis would be depressed instead of being erected (which would be, literally, pointless). His experiments showed that the extension of the penis depends exclusively on the blood. 'First, we saw that in cadavers, when water is injected with a syringe through the arteries into the corpora nervosa, the penis extends just as far as it might in living men'. So any fluid arriving with force in the human penis through the arteries extends it. The second experiment, the one with the dog, showed that what the erect penis contains is simply blood, not spirits from the nerves, for blood is all that flows from the dissected erect penis and when the blood has all flowed out the penis becomes flaccid. Observation then showed that the function of the muscles, rather than erecting the penis, was to impede the flow of blood back through the veins and to compress the corpora nervosa, thus maintaining the erection.[3]

De Graaf has here experimented by injecting liquors into dead humans, and cut open the equivalent part in a living dog, to work out what is involved in a particular event – blood or spirits – and to determine the function of particular muscles.

Experiment 3

Pierre Dionis set up our next anatomical experiments at the Jardin royal in Paris, in order to determine whether there is any water present in the pericardium (the membrane surrounding the heart) of the healthy human, or any serum present in the ventricles of the brain of the healthy human, or whether these were pathological phenomena only – that is to say, phenomena of the diseased state, not of the natural state. 'Anatomy gives us to know', Dionis writes,

> … that the whole Surface of the *Dura mater* is crammed with Glands and Lymphatic Vessels; as well as the Membrane of the Lungs, the *Pleura*, the Lungs, the *Pericardium*, and the Great Vessels.
>
> In the same case, the Cavity of the *Abdomen* is all over moist, by reason that the Glands of the Mesentery, the Sweet-bread and the *Peritoneum* are swelled, and so occasion the Separation of the *Lympha*. And for a further confirmation upon this Head, 'tis observable, that in Dissecting a Person that is just Beheaded, we find no Liquor either in the *Pericardium*, or in the Ventricles of the Brain, or in any other Cavity. In like manner, when we Open a Dog as soon as he is tied

[3] Jocelyn and Setchell, eds, *Renier de Graaf on the human reproductive organs*, 1972, pp. 49–53. These modern translators and editors of de Graaf point out (n. 170) that 'the flow of blood through the penis is in fact greater when it is erect than when it is flaccid. The increased amount of blood in the erect penis must therefore be due to dilation of the arteries rather than obstruction of the veins. The action of the muscles does in some way aid erection'.

> to the table, without suffering him to languish, we find not a drop of Water in the *Pericardium*.

From these observations and trials, Dionis can write, 'All these experiments are Demonstrative Evidence, That the Waters of the *Pericardium* and Ventricles of the Brain, is always a consequence either of a disease or of a certain sort of death'.[4] Anatomical experiment has determined the question.

Experiment 4

In 1767 the anatomist John Hunter broke his Achilles tendon by over-vigorous dancing. He invented a quick cure for himself, by compressing the muscles and raising the heel. He was immediately curious about what was happening in the process of cure. His biographer records: 'He divided the tendo achillis of several dogs, by introducing a couching needle through the skin at some distance from it, and with the edge [of the needle] cut through the tendon; in this way the orifice in the skin healed up, and made it similar to a broken tendon. The dogs were killed at different periods to show the progress of union, which was exactly similar to that of a fractured bone when there is no wound in the skin'.[5]

The anatomist intervenes to artificially wound the animal while alive, in order to see the natural stages of recovery in a series of dogs killed at intervals.

Experiment 5

Albrecht von Haller's most famous work was his *Dissertation on the sensible and irritable parts of animals* of 1755. In recounting the experiments on sensibility that he carried out, Haller himself described his disciplinary role in doing so as that of an anatomist:

> I took living animals of different kinds, and different ages, and laying bare that part which I wanted to examine, I waited till the animal ceased to struggle or complain; after which I irritated the part, by blowing, heat, spirit of wine, the scalpel, *lapis infinalis*, oil of vitriol, and butter of antimony … The repeated events of those experiments I marked down faithfully, whatever I found them to be. For what is it to me, in fact, on which side nature decides the question? nay, would it not be very foolish to hazard the reputation of a faithful and accurate observer [i.e. himself] for an imaginary fact, which the simplest experiment would prove to be false to any other anatomist who should choose to repeat it?[6]

---

4 Dionis, *The anatomy of humane bodies improv'd*, 1703, p. 261.

5 John Hunter, *A treatise on the blood*, 1794, p. xx (the 'Life').

6 Haller, *A dissertation on the sensible and irritable parts of animals*, 1755, pp. 659–60.

This project of Haller's was an instance of anatomical investigation and anatomical experimentation because it was concerned with 'a new division of the parts of the human body', based on a distinction between those parts of the body which are susceptible to irritation and sensibility, and those that are not.[7] This was a standard role of anatomy: identifying and dividing the constituent parts of the body according to type. The classification is new in this case, being built on the reaction of parts to artificial stimulus, but *as a classification* it was anatomical. The properties of the parts of the body – even the vital, reactive, properties of the parts – was the business of anatomy, not of physiology, because establishing the properties of the parts was to define them, to establish their identity. Bruno Latour has recently argued that natural objects gain their scientific identity from what happens to them in the laboratory: they 'are completely defined by the list of answers in laboratory trials'.[8] In introducing sensibility and irritability, Haller was engaged in an attempt to add a new pair of trials to the task of giving definition – identity – to the anatomical parts.

In addition to distinguishing between different component parts of the body according to their possession (or not) of this property of sensibility, one of the things that these experiments revealed was the location, the *anatomical seat*, of certain pains, since some parts, such as the tendons, were found to be neither irritable nor sensible (sensitive), hence they could not be the seat of pain. From this experimental finding Haller deduced that the sensitivity of the parts had its source in the nerves: where no nerves are distributed there is no sensitivity to any kind of damage.

This experimental finding – about anatomical *seats* and their relation to nerve *distribution* – is clearly anatomical in nature.

But what about irritability, the other property being investigated here by anatomical experiment by Haller? This, Haller concluded, does not depend on the nerves. For his experiments on irritability showed 'that there is nothing irritable in the animal body but the muscular fibre, and that the faculty of endeavouring to shorten itself when we touch it is proper to this fibre'.[9] He ascribed this faculty to the animal gluten, which, together with the earthy particles, makes up the muscles: 'Experiments have taught us the existence of this property, and doubtless it is owing to a physical cause which depends upon the arrangement of the ultimate particles, though the experiments that we can make are too gross to investigate them'.[10]

In other words, this property is a property of the *matter* of which the basic anatomical parts are made up and of the arrangement of that matter, and as a property it is typical of and exclusive to the muscles: it is one of the things that makes muscle what it is, and hence a concern of the anatomist.

[7] Ibid., p. 657.

[8] Latour, *Science in action*, 1987, pp. 84–94, see p. 88.

[9] Haller, *A dissertation on the sensible and irritable parts of animals*, 1755, p. 690.

[10] Ibid., p. 692.

Experiment 6

In 1780 Alexander Monro *Secundus*, Professor of Anatomy at Edinburgh University, had been investigating the organ of hearing in fishes. He then made some experiments on hearing in water, to see how well sound travelled underwater to the hearing organs of fish. In the process of these experiments the honourable professor obliged himself to jump fully clothed into water while furiously ringing two bells, and probably still wearing his wig!

> I employed two bells, the sound of which I was used to; one of them a small tea-table bell; the other much larger and thicker, so that the sound of it could be very well heard at the distance of a quarter of a mile.
>
> When these were plunged under water and rung, I observed that the sound of them was very sensibly graver; but still the ringing tremor of both was very distinguishable. On performing an accurate experiment, the tea-table bell was found to sound in air the highest G of a common harpsichord; but in water it sounded a fifth false lower, or it sounded the C sharp under the G.
>
> I next plunged my head under the water while I rang the bell in the air, and heard the sound of it distinctly. As the tone of the bell is louder and more acute in the air than in the water, I need scarcely observe that its sound is better heard when the head of the person making the experiment is under the water and the bell is above, it, than when the bell is rung underwater while the head is above it.
>
> I next plunged my whole body with the bells, holding their handles in my hands, under the water; and then rung them, and was surprised with the loudness and distinctness of their sounds, and could readily distinguish their different tones.
>
> In like manner, when, plunged under the water, I struck two stones held in my hands against each other, I was surprised with the shock communicated to the ears.[11]

While these bell-ringing experiments might also qualify as do-it-yourself physics experiments, they are aimed at anatomical conclusions, viz. about the hearing organ in fishes and how it works.

These varied experiments are typical of the hundreds of thousands of experiments undertaken by anatomists in this period. They are all *anatomical* experiments. What makes them anatomical is the fact that they are all aimed at acquiring new knowledge (or testing old assumptions) about the structure, connections, properties and uses of the parts that make up the animal and human body, and about the course of vessels and their contained fluids, the contents of sacs,

[11] Monro *Secundus*, *The structure and physiology of fishes*, 1785, p. 52. For earlier attempts to investigate the hearing of fishes, see Klein, *Historiae piscium*, 5 vols, 1740–49; there were five *missus* or 'courses' in this project, which was completed over nine years.

the function of particular muscles, the presence or absence of particular things in health and disease, the natural sequence of cure of a damaged part, the nature and differences of the component parts, and the stimulus available to particular organs in nature. The experiments are all very much 'hands on', not thought experiments, and many of them are performed on live animals. And, as Haller remarked, they needed to be frequently repeated:

> There is in all experiments a law, for the neglect of which the greatest men sometimes suffer punishment. No experiment, no dissection, should ever be carried out just once. For the truth will not become known except from the constant outcome of repeated trials. There are so many irrelevant circumstances that get themselves mixed up in experiments: they go away on repetition, because they are alien: and the pure remain, which thus always turn out in a similar way, because they flow from the nature of the thing. But nature is also variable, and only by repetition does the sense and will of nature get known. This law, in earlier times little recognised, was first introduced into anatomy by Morgagni.[12]

But my reader is probably wondering why I am not calling these *physiological* experiments, because that is what they look like to the modern eye: are they not instances of *experimental physiology*, especially those experiments conducted on live animals? Historians living in the last two hundred years have instinctively tended to describe them like this. For instance, in his *Lectures on the History of Physiology* Sir Michael Foster simply assumed that all the *anatomical* experiments he describes in the early modern period were *experimental physiology* ones, simply because they were experiments on the live animal![13] This might be an understandable mistake for a professional experimental physiologist like Foster to have made when trying to write some discipline history. But nevertheless it was a mistake: he had mistaken the disciplinary identity of these experiments, since they were not physiological experiments but anatomical ones. In the first place, as we have been seeing, *anatomy* was the experimental discipline par excellence in the investigation of the phenomena of life. In the second, there was no such enterprise or discipline or activity as *experimental physiology*. It did not yet exist. It was created only in the years just after 1800. And in the third place, throughout the long eighteenth century, and for centuries before, there was already a discipline of *physiology*, usually built on anatomy, but in itself in no way experimental. To this we now turn. So the issue for us here is not of distinguishing anatomical experiments from physiological experiments in the long eighteenth century, but rather of appreciating the experimental nature of old anatomy, and the fact that physiology was not yet experimental at all.

---

12 Haller, *Elementa physiologiae corporis humanae*, 8 vols, 1757–66, vol. 1, p. v.

13 Foster, *Lectures on the history of physiology*, 1901. Foster has been followed in this by all modern historians of the field.

2. Physiology: A theoretical science and sub-discipline of anatomy

Our first sub-discipline of the experimental discipline of anatomy therefore is *physiology*, which we can call 'old physiology' in contrast to the new discipline of *experimental physiology* that was only to be created at the beginning of the nineteenth century.[14] One striking contrast between these two distinct disciplines of physiology, old and new, is that in the long eighteenth century anatomists were always claiming that old physiology should be built on anatomy, not on undisciplined and ignorant speculation or on chemistry or some other non-anatomical discipline. In the early nineteenth century, by contrast, proponents of the new discipline of experimental physiology such as Claude Bernard were to claim that the discipline of anatomy was *merely* preliminary to the new physiology. 'Anatomy is indeed only the first step in physiology', wrote Bernard; 'Anatomy in itself is a sterile science'.[15] And as a simpler science than physiology, anatomy should now be subordinate to physiology, Bernard was to claim, instead of dominating it as it had traditionally done. So the relation of old physiology to old anatomy was exactly the reverse of what we would nowadays expect from the relation of experimental physiology to modern anatomy. Thus, strange as it may seem in the context of our customary classification of the disciplines, in moving now from anatomy to physiology in their eighteenth-century disciplinary senses, we move from an *experimental* to a *non-experimental* discipline. But in doing so we also move, in eighteenth-century terms, from a discipline that was both an art and a science, to one that was purely a science. Sciences, like philosophy as a whole in those days, were theoretical disciplines and asked 'why?' questions.

The limit of what was studied in anatomy had traditionally been what was visible to the naked eye. Anything beyond the realms of sight, that was the domain of the physiologist. The microscope, for its adherents, extended the range of what one could see. Thus, for anatomists, observations and experiments made down the microscope, of whatever form, also constituted anatomy. Again, conjecture *beyond* what was immediately visible down the microscope was left to the physiologist. For a few decades at the end of the seventeenth century, the microscope was a very successful anatomical instrument, and it revealed a great deal in the hands of skilled instrumentalists such as Marcello Malpighi.[16] However, the microscope was a controversial instrument, and its findings were subject to challenge. John Hunter, for instance, thought the microscope actively misleading. For most of the

[14] This case has been argued at greater length in Cunningham, 'The pen and the sword', parts 1 and 2, 2001 and 2003.

[15] For the former view see virtually any eighteenth-century anatomist; for the latter, see Bernard, *An introduction to the study of experimental medicine*, 1865, part II, chapter 2, 'Experimental considerations peculiar to living beings', section IV, 'Normal anatomy in its relations with vivisection'. I use the excellent English translation by Henry Copley Greene of 1927, published in 1948; see pp. 105–12.

[16] Bennett, 'Malpighi and the microscope', 1998.

eighteenth century we can see the microscope as a useful adjunct to the anatomist's other investigative tools; but it does not return to being a prime investigative tool into the phenomena of life until the early nineteenth century.

Our anatomists in the long eighteenth century wrote a great deal about physiology, both that of humans and that of animals. Albrecht von Haller, for instance, first issued a popular *First lines of physiology* in 1747, and also put the work of the best part of thirty-six years of his life into a huge eight-volume *Elements of the physiology of the human body* (1757–66). Haller in the *First lines* described physiology as 'animated anatomy'. The contemporary German translation rendered the Latin or Greek term *physiologia* as 'Geschäfte des körperlichen Lebens', 'The activity of bodily life', which expresses it admirably.[17] Withdrawing into his study, the physiologist drew together all the anatomical and other evidence he had acquired from experiment, observation and reading, and *reasoned* his way to understanding how it all functioned together in life. In particular this meant that the physiologist was most interested in the ultimate structure of the body and its parts – beyond what the microscope could reveal – and in the grand functions of the body, ones in which many different parts are involved, such as respiration or digestion. At all stages in his physiological explanations the investigator was going beyond the evidence of the senses. He took the anatomical *facts* and on them built his physiological *speculations*. The more firmly the speculations were built on anatomical facts, the better, more trustworthy and more scientific they would be (the anatomist believed), and we find anatomists repeatedly condemning physiologists whose speculations were not built directly on detailed anatomical knowledge. As Govard Bidloo, Professor of Anatomy at Leiden University, put it in a public dissertation: 'The greatest part of physiology, in medical matters, is the knowledge of the human body and its œconomy; this is acquired from anatomy, which consists in the true perception and demonstration of the structure, motion and use of the separate specific parts making up the whole body, and also of the whole body'.[18]

Other anatomists in their complementary but separate role as physiologists defined the role of the physiologist in similar ways. The anatomist Joseph Lieutaud, for instance, in the 1740s wrote an *Elements of physiology, assembled according to the more familiar and latest experiments of the physicists, and the more accurate observations of the anatomists*. In this he wrote: 'Physiology explores the uses of the parts which Anatomy exhibits', so therefore one should study anatomy first. And 'Physiology, the first part of the Institutes of medicine, is nothing else than Physics accommodated to the Animal Œconomy'. 'The business of respiration appears abstruse and its uses obscure, unless gravity, the mode of expansion and the spring of the air are known. An accurate description of the eye is little help in

[17] *Albrecht von Hallers Erster Umriss der Geschäfte des körpoerlichen Lebens*, 1770 (not seen).

[18] Bidloo, *Dissertationes anatomico-physiologicae quatuor*, 1708, Dissertatio prima, sections v and vi.

understanding vision, if the laws of the reflection and refraction of the rays are not referred to the velocity of light'.[19]

As can be seen, physiology usually depended crucially not only on anatomy but on whatever the current understanding of physics was. Thus in the mid-seventeenth century Descartes and others built physiology on the Cartesian physical principles of matter-in-motion. Many eighteenth-century physiologists lay out a Newtonian or quasi-Newtonian view of matter and motion, especially of forces and pressure, and explain the traditional subject matters of physiology by them. From Boerhaave to Haller this is the case, and it is striking how many different explanations, and reaching how many different conclusions, could be built on Newtonian physical principles, depending on what aspects of Newton's works one privileged. Because physics kept changing, so physiologies kept changing too.

When William Hunter's pupil and assistant in his investigations, William Cruickshank, wrote to show Hunter's priority in claiming that the lymphatic vessels were the absorbent system of the body, he did so as an anatomist, not as a physiologist, and his book was entitled *The anatomy of the absorbing vessels of the human body* (1786). Cruickshank's arguments, like those of William Hunter himself, are all from anatomical experiments and procedures (especially the injection of mercury), from the place of origin and direction of flow of the vessels, from the properties of the vessels, their valves, to their distribution through the body, and so on. The *use or function* of the lymphatics as a system was the physiological aspect of the discovery. When remarking on the discovery in his own lectures, William Hunter had said that, 'if we mistake not, in a proper time it will be allowed to be the greatest discovery, both in physiology and in pathology, that anatomy has suggested since the discovery of the circulation!'[20] Anatomy *suggests to* physiology: Hunter's expression specifies concisely the relation of anatomical practical investigation to physiological theoretical conclusion.

One final important distinction between anatomy and physiology was the discussion of the fluids; as the Parisian Professor of Anatomy, Winslow, put it: 'The history of the fluid parts comes into a system of anatomy, only occasionally, and by the bye; because it properly belongs to what is called Physiology or the Animal Œconomy'.[21]

---

19 Lieutaud, *Elementa physiologiae*, 1749 (1746), pp. 2, a4v.

20 Cruickshank, *The anatomy of the absorbing vessels*, 1786, p. 4; William Hunter, *Two introductory lectures*, 1784, pp. 58–9.

21 Winslow, *An anatomical exposition of the structure of the human body*, 2 vols, 1733, vol. 2, p. 101. Winslow's translator, George Douglas, elaborated this position: 'the foundation of a complete physiology, is the description of the fluid as well as the solid parts; that is, the nature, properties, motions, etc. of the chyle, blood, and all the liquors secreted from the blood, are to be inquired into and illustrated by hydrostatical, chymical and mechanical experiments, before the animal oeconomy can be explained' – Preface, p. xiv.

In his lectures William Hunter was scathing about the pretensions of physiologists who did not know enough anatomy:

> In treating of the physiology, it is very difficult to say, what plan we should follow; for, every method which has been yet proposed, is attended with manifest inconvenience … Undoubtedly the man who makes it his business, to investigate every thing relating to the structure of the human body, must be, *caeteris paribus*, the fittest person to explain its operations; and there cannot be a more proper occasion, than when the parts are before us. And, yet in some sense, it is just the reverse: for, every good Anatomist, who has a cool head, and keeps a guard over his imagination, knows, that many of the received hypotheses in Physiology, are built on very loose foundations, and liable to weighty objections; or, demonstrably repugnant to what we already know of the structure of our body. So that if the Anatomist is weak enough, to think that he can, or that he must explain, all the operations of the machine, he will certainly make partial dissections, mislead himself and others, and put a thousand little arts in practice, to make the structure correspond with the imaginary use.
>
> … I must therefore expect, that you will not hereafter be surprised, when you find me avowing great ignorance, in many of the most considerable questions relating to animal operations; such as, sensation, motion, respiration, digestion, generation, &c. In my opinion all these subjects are much less understood, than most people think them. Our vanity deceives us, and persuades us that we have got the whole, as soon as we have acquired a smattering of natural knowledge. Hence it is, that the different sects of Physiologists, have endeavored to explain animal functions upon such different principles. Hence, for example, to account for digestion, some have made the stomach a mill; some would have it to be a stewing-pot; and some a wort-trough [for fermentation of malt for beer]: yet all the while, one would have thought that it must have been very evident, that the stomach was neither a mill, nor a stewing-pot, nor a wort-trough, nor anything but a stomach.
>
> This species of philosophy, has prevailed in many parts of Physiology; and makes up a great part of what has been commonly taught, as sound and useful physiological learning. One thing, of peculiar properties and power, has been explained by another, of different properties and powers, as absurdly, as if colours had been explained by sounds. Animal functions, generally speaking, are of a peculiar nature; and like nothing which is to be found in the works of art, or wherever there is not animation and life: and, had Physiologists spent that time in making accurate observations upon animals themselves, which has been thrown away upon mechanical and chemical visions, by this time we might have understood animal principles and processes, better than we now do.[22]

[22] William Hunter, *Two introductory lectures*, 1784, pp. 93–6.

The many different eighteenth-century physiologies that were either dreamt up out of thin air or carefully built on anatomical facts, have fascinated historians, and most of what they have written on eighteenth-century anatomists concerns their physiological systems; I refer my reader to their accounts, since my concern here is primarily with anatomy and the making of anatomical facts, and not with physiological theories that might (or might not) have been built on these facts.

The distinction between the *anatomist*, the man who-observes-and-does-experiments and the *physiologist*, the man who-works-it-all-out-in-his-head, is very nicely brought out in a journal report of 1748, written for the uninitiated, of the then-current dispute between an anatomist, Professor Albrecht von Haller, Professor of Anatomy at Göttingen University, and a physiologist, Professor Georg Erhard Hamberger, Professor of Medicine at Jena University (1697–1755). It is all amusingly written, and while proclaiming total impartiality it totally favours the approach of the anatomist who-observes-and-does-experiments, for he alone reaches facts.[23]

Twenty years ago, the anonymous journalist writes, Professor Hamberger published a thesis on respiration, maintaining that the internal intercostal muscles, being closer to the vertebrae by their lower attachment than by their higher attachment, must pull the ribs down when they function. It was actually an old opinion, that of Galen, but now supported by a geometric demonstration. Professor Hamberger also claimed that there was a certain amount of air between the pleura and the ribs, whose rarefaction and elasticity served the alternating dilatation and compression of the lung. He claimed that he had seen the mediastinum of a living dog emerge like a vessel full of air, through a hole made in the chest, thus showing that the lung was not contiguous with it.

Three years ago, the researches of Haller brought him to the question of respiration. Naturally (according to our journalist) he followed the line of his teacher, Boerhaave, maintaining that there is no air between the pleura and the ribs, and that the internal intercostal muscles raise the ribs, just as the external ones do. This was the opinion of all the modern anatomists, such as Winslow and Albinus.

Professor Hamberger's physiology was based on the system of 'adhesion', with the larger minute particles of matter attracting and attaching themselves to the smaller. By these laws of physics he had explained most of physiology: animal secretion, transpiration, heat, muscular movement, etc. According to the journalist, he had built his explanations on descriptions in books rather than dissections. 'This physiology could not prevail at the same time as that of Herr Haller, whose principles were totally opposed to it. The Professor of Göttingen based his views on anatomy itself: he has learnt how to bring together, in the new Theatre of Göttingen, founded under his direction, so many laws favourable to

[23] 'Mémoire sur une controverse au sujet de la respiration', 1748. I translate and paraphrase.

Anatomy, that he is able to check as many times as necessary all the anatomical descriptions that physiology needs' (pp. 415–16).

A paper war now began, with Haller first being accused of not understanding mathematics. Haller had to respond. He took the time for the necessary experiments and dissections. Then he replied. Dissections show there is no air between the pleura and the ribs. Experiments show that any wound that does admit air forces the lung to retract. The mediastinum that Hamberger saw emerge in a dog like a vessel of air, is not a true mediastinum, but a membrane peculiar to brutes, which humans lack. Haller exposed the pleura in a live animal, and saw the lung stuck tight to it, and how a wound in the pleura led immediately to the retraction of the lung, leading in turn to the death of the animal.

When it came to the intercostal muscles, 'the Savant of Jena' had made a little model of wood and string, which Haller reproduced. But Haller now applied the same mechanism to the thorax of a man: he had prepared a corpse whose ligaments were conserved in their natural suppleness; he attached twenty-eight pullies (in place of the muscles) and as many strings, which he worked via a weight placed beyond the vertebrae. In this way he represented the internal intercostal muscles, with the strings showing the direction. The result was quite different from the model: the weight constantly raised the ribs (not lowered them), all together and each separately. Hamberger had assumed that the ribs are all equally mobile. 'But Nature had not made the ribs like this' (p. 420). They are not equally mobile, but the first rib is very strongly fixed, with the others progressively more mobile. Haller measured the length of the ribs and the obliquity of the intercostal muscles. He found precisely which weight was necessary to move which ribs. He produced numbers and figures.

What was Hamberger's response? He wrote seven 'programmes' against Haller, without paying any attention at all to Haller's experiments! Our journalist noted that in the 'programmes' of Jena there was not an experiment, a description, a new test, indeed nothing added to what Hamberger had written twenty years before.

It cost many more animals their lives, which Haller sacrificed to the controversy. Since the Savant of Jena continued to insist that there really was air between the lung and the pleura, Haller plunged live animals into water, then opened the pleura, he called witnesses to see if a single bubble came out of the cavity of the chest, he did everything possible to make the air emerge if it was there, but nothing appeared. He repeated and varied the experiment on live animals and on corpses: not a bubble appeared, and so the dispute appeared to be resolved. To solve the question of the use of the internal intercostal muscles Professor Haller proposed a decisive experiment. He exposed the muscles in a live animal, and then watched what happened in respiration. These muscles would act either in inspiration or in expiration. If the latter occurred, then Hamberger was right. So Haller made this experiment in front of many witnesses, he repeated it in different species of animal, he varied it in many ways, and 'Nature declared itself constantly on his side' (p. 426): the internal intercostal muscles work vigorously in inspiration, they flatten and lengthen in expiration. Haller pleaded with Hamberger to see his error on a human corpse, or on a dog, or even on a printed plate from Albinus's anatomical atlas. But all in vain. However, our journalist concludes, from this error of the physiologist, that two principal phenomena of respiration, long argued about, have now been finally settled by an anatomist.

What can we see from this dispute? We can see that the ideal anatomist of the eighteenth century builds his opinions on facts acquired from anatomy and anatomical experiment, well planned, precisely carried out, and repeated. We can see that the physiologist, if he is a Haller, builds his theories on anatomical facts, but if he is the other kind of physiologist, a Hamberger, then he builds his castles in the air on assumptions, false physics, misapplied mathematics and observational errors, and no kind or quantity of anatomical experimental facts will persuade him. But more than this: the fundamental thing we can see from this dispute is that the anatomist performs experiments, and the physiologist doesn't.

This relationship between anatomy and physiology can be seen very clearly in an engraving that Haller himself placed at the front of the second volume of the *Elements of the physiology of the human body* (Fig. 3.1). It has special significance because it is the only illustration in all eight volumes, apart from two on the heart and lungs in volume 1. The muses and their attendant *putti* represent the anatomist and the physiologist and their assistants. It is a scene with two parts. On the left is anatomical investigation. On the right is physiological contemplation. On the left is the manual *art* of anatomy. On the right is the mental *science* of physiology, taking its place with geometry and astronomy, other sciences that reach higher, theoretical, truths. The anatomist cuts; the physiologist reflects. The anatomist is active, knife in hand. The physiologist contemplates and writes, in the conventional philosopher's pose with cheek on hand. The anatomist deals with means; the physiologist with ends. The anatomising is about *what* and *how*; the physiologising is about *why*. The anatomist deals in *findings and experiments*, while the physiologist deals with *causes*, something not accessible to the anatomist. It makes no difference that in reality it may have been the same person who conducted both the anatomy and the physiology, for this does not change the fact that the operations belong to different disciplines, and are different in kind and in status, and probably also in location and time. Oh that the allegorical figures in this picture were wearing hats, for then we would see that they would be different hats!

Obviously this practice and this usage changed either some time towards the end, or perhaps at or after, of the long eighteenth century, when the physiologist became for the first time an experimenter and a creator of facts about the body. With respect to word usage, the earliest use that I know of the term 'physiological researches' occurs in a small work published right at the end of the eighteenth century: in 1797 in Paris, by the anatomist Jean-Joseph Sue the younger (son of the anatomist Jean-Joseph Sue who had published 'The human-cutter', *Anthropotomie*, in 1765). In the wake of two signal events – the discoveries by Galvani of the effect of electrical current on living- and dead animal tissue, and the invention of the guillotine for humane decapitation of the enemies of the French Revolution – the Institut national had determined 'not only to repeat the experiments of Galvani, but also to try new experiments and to penetrate the mysteries of the remarkable movements which organic fibre presents in this kind of electricity'.[24] Sue's own experiments involved the decapitation of chickens, turkeys, rabbits, and larger animals, and observing how long, and in what manner, 'animality' survived in the cut-off head and, in the case of frogs, how long it survived in the four quarters into which he had sectioned it. The term 'physiological researches' is appropriate since Sue was investigating life and sensation – 'the phenomena of the movement and sensibilities of animals' – and he was living in a period when 'the general laws of the organisation of animals' were for the first time being actively pursued (see Chapter 6, below).

Fig. 3.1

Similarly, the earliest use that I know of the expression 'experimental physiology' occurs only in the first year of the next century, in a work published in Paris in 1800, Antoine Portal's *Memoirs on the nature and treatment of several illnesses*, where all is confusion.[25] Portal was now Professor of Anatomy at the Muséum d'histoire naturelle. It seems probable that the publisher introduced the term for what Portal would have called 'anatomical experiments' at the time that

---

24 Sue, *Recherches physiologiques*, 1797 (An VI), p. 2; the other quotations are from pp. 3 and iv respectively.

25 It has recently been claimed that the expression was used with respect to Haller in 1748 in 'Mémoire sur une controverse au sujet de la respiration', which I have used extensively above. It is not in fact used there; cf. Steinke, *Irritating experiments*, 2005, p. 109. Even a 1720 reference to a work by Christian Vater of Wittenberg as *Physiologia experimentalis* (1701), turns out to be a work called *Physiologia experimentis et demonstrationibus mechanicis, chimicis et anatomicis illustrata et confirmata* – that is, *Physiology illustrated and confirmed by mechanical, chemical and anatomical experiments*, which is something quite different; for the reference see Heister, *Oratio de incrementis anatomiae in hoc seculo XVIII*, 1720, p. 66. Heister himself said that in Vater's book 'experimenta anatomica egregia quam plurima describuntur', i.e. the experiments are anatomical ones.

he made them back in the 1770s. On the title-page a 'Précis des expériences sur les animaux vivans, d'un cours de physiologie pathologique' is announced. The publisher comments that

> … at the end of this collection I have felt I ought to reprint the précis of the experiments on living animals that Citizen Portal made in his course of physiology at the Collège de France, which one of his pupils published in 1771; this way of teaching is not only more precise, but also the one that instructs best: you don't forget what you have clearly seen.

When one reaches the précis itself it is entitled 'Lettre de M. Collomb, étudiant en Médecin à Lyon; sur un Cours de physiologie expérimentale, fait cette année 1771, au Collége [*sic*] de France, par M. Portal'.[26]

Portal had been professor at the Collège royal from 1769. Whatever this course had been when Portal gave it in 1771, by 1800 in the hands of its new editor it has become both 'experimental physiology' and 'physiological pathology'! Interestingly, no such 1771 publication is listed by Portal himself, nor can it be found in the great libraries of the world today – at least not by me. The experiments themselves are mostly the Hallerian anatomical ones on irritability, tested on dogs and cats. One deals with blood transfusion to demonstrate blood circulation to students, while another, using a new-born puppy, shows (against the view of another anatomist, d'Hérisant) that the lungs do not have their own movement. Yet others show that irritating the diaphragmatic nerves increases the movement of the diaphragm. Nowhere within the text does Portal refer to his discipline as physiology. Indeed, with respect to the experiment just mentioned, he is here recorded as having said, 'Many other *anatomists* have repeated this experiment'![27] So it looks as though this first usage of the term 'experimental physiology' was introduced by Portal's later editor.

Thus right to the very end of the eighteenth century the role distinction continues: the anatomist performs experiments; the physiologist theorises.

However, with the creation of the new Schools of Health in France by the revolutionaries in 1795, a different amalgamation of anatomy and physiology was adopted in an attempt to improve teaching. In the Montpellier school, for instance, the new first course was called both 'Physiological anatomy' and 'Anatomical physiology' –

> It means that we shall adopt a plan that is not simply anatomic, but that is accommodated equally both to the demonstration of the parts that Anatomy describes [*décrit*] and to examining the functions that Physiology explains [*explique*]. In this plan we shall divide the human body into different systems of

---

[26] Portal, *Mémoires sur la nature et le traitment de plusieurs maladies*, 4 vols in 2, 1800, 1808, vol. 1, p. vii, vol. 2, pp. 239–318.

[27] Ibid., vol. 2, pp. 298, 303–4, 306 (my emphasis).

> organs, to which systems we shall relate [*correspondre*] the functions attributed to them. We shall commence with the description of each system and of its parts, which shall be constantly accompanied with the explanation necessary to make known the order of the functions that belongs to it. This division of anatomico-physiological systems will provide the division of our course ….[28]

The systems in question were to be those of the bones, the muscles, the nerves, the vessels (that is, the veins and the arteries), the viscera, the glands and lymphatics, and the sexual system. So in this brave new approach the 'systems' at issue are still anatomical rather than physiological (the viscera, for instance, rather than respiration), and anatomy continues to *describe* while physiology purportedly *explains*. Physiology has not yet been made into an experimental science, nor has anatomy lost its pre-eminence.

3. Generation and man-midwifery: Sub-disciplines of anatomy

Our second sub-discipline of anatomy is *generation*. The modern scientific discipline of *embryology* was created in the early nineteenth century, and it is one of the new disciplines that came to replace generation, a traditional sub-discipline of anatomy, that had flourished during the long eighteenth century and had been pursued on and off for centuries before that, indeed since the time of Aristotle. The term 'embryology' was coined by the creators of the new discipline to distinguish it from the old discipline. While students of generation did study the embryo – though usually under the title of the 'fetus' – they also studied much more.[29] Moreover, generation was not limited to the study of humans – nor even of animals, since the most diligent anatomists such as William Harvey investigated it even with respect to insects. Indeed, due to its Aristotelian origins, the discipline of generation took as its domain all aspects of 'coming-to-be' of the new animal or human (or insect) or even plants: the total process of the generation of one perfect animal or plant from another. For *generation* was concerned with the formation of the seed, with the process of conception, with the development of the fetus, with gestation and growth, and finally with birth: all the processes from start to finish that go to make up the perfected, complete, animal. Our modern discipline of 'embryology' by contrast studies just a sub-set of these topics, and according to quite different mental understandings of what is going on. Similarly, just as embryology has displaced part of the old study of generation, so similarly 'reproduction' has replaced another part of it. The Dutch physician and anatomist

---

28 *Programmes des cours d'enseignement dans l'École de Santé de Montpellier*, An. III [1795], p. 5.

29 In modern medical parlance both terms are used about humans: an *embryo* is the developing child up until the end of the third month; a *fetus* is the developing child for the remaining six months before birth.

Renier de Graaf investigated in the 1660s the organs of generation in humans. His books published in 1672 were therefore called *A treatise on the organs of men serving for generation*, and *A new treatise on the organs of women serving for generation*. His modern translators have mistakenly rendered these titles as *On the human reproductive organs*, thus unconsciously transforming the work into a modern idiom and into a different scientific discipline – their own.[30] Reproduction, sexuality and embryology are all terms to be avoided when we speak of the old anatomical discipline of *generation*, not just to show respect to actors' categories, but because the use of them in connection with the old anatomical discipline of *generation* will muddle our thinking, and we shall find ourselves tempted to write histories of disciplines that did not yet exist. For these are all concepts of nineteenth-century origin, to meet nineteenth-century needs, and to develop nineteenth-century research.

In 1651 there appeared in Latin the greatest work on generation to date: that is, William Harvey's last work, *Anatomical exercitations, concerning the generation of living creatures*. Within two years there appeared an English version that some scholars (myself included) believe probably represents Harvey's original text.[31] Sir George Ent had gone to see the aged Harvey and found him, Democritus-like, happy in the investigation of Nature's works as (in part) a way of distancing himself from the current political troubles. For Harvey's party, the royalist cause, was at a low ebb, with the king executed and no end in sight to the republican government. Anatomy was Harvey's chosen study in such circumstances: 'It hath ever been the delight of my Genius', Harvey said to Ent, 'to make strict Inspection into Animals themselves: And I have constantly been of opinion, that from thence we might acquire not only the knowledge of those less considerable Secrets of Nature; but even a certain Adumbration of that Supreme Essence, the Creator'. Ent managed to persuade Harvey to let him see his manuscript, his *Anatomical exercitations, touching the generation of animals*, and, after much protest, to take it away with him and publish it to the world. In a charming use of the idiom of the anatomical sub-discipline of generation, Ent wrote that, for himself, 'in this great

---

30 Jocelyn and Setchell, eds, *Renier de Graaf on the human reproductive organs*, 1972. The *Oxford English Dictionary*, which records historical usage, has only one instance of 'reproduction' before the nineteenth century that might conceivably be taken in this sense, William Smellie (not the man-midwife and anatomist) translating Buffon's *Natural history* in 1785. To this we can add the use of the term in the 1769 English translation of Lazarus Spallanzani's *Prodromo di un opera da imprimersi sopra le riproduzioni animali* (1768). But in both cases it means the re-growth of amputated parts in certain animals, and the constant renewal of certain parts during life such as skin and nails. It does not mean any aspect of the generation of the complete animal. But when speaking of the process of fertilisation in the animal (and in particular about the 'germ' that pre-exists in the female before fecundation or impregnation by the male), Spallanzani and his translators properly describe such researches as part of *generation*.

31 White, 'The 1653 English edition of *De motu cordis*', 1999.

Business I have performed no more then the meer office of a Midwife: producing into the light this noble Issue of His Brain, in all its parts and lineaments perfect and consummate, as it is now presented to your Veiw [*sic*]; but staying long in the Birth, and fearing, perhaps, some injurious Blast of Envy or Detraction'.[32] This perfectly encapsulates the role of the anatomist studying generation: from the *issue* of one perfect creature (that is, the discharge of some special fluid or matter from the body, viz. the semen), he studied the generation of all the parts and lineaments perfect and consummate of a new creature, and its birth or completion, until it appeared in the light.

Only he who studies things with his own eyes and experiments, Harvey argued, will understand natural phenomena such as generation. We should, he says, in these matters be like those two great masters of anatomy and generation: Aristotle and Fabricius ab Aquapendente (who had been Harvey's own teacher of anatomy at Padua).[33] Harvey argued in the Preface that the best method of studying generation in animals (including man) was to start from the perfect creature and then work backwards, until we have discovered 'out of what *first matter*, by what *efficient* [cause], and what *procession* the *plastick power* hath its original; and then also what progress Nature makes in this work' (b5v). What was a man in his mother's womb, before he was an embryo? Was he three bubbles, some rude and indigested lump, a conception or coagulation of mixed seed, or what? In the Preface, Harvey outlined the plan he had adopted:

> We therefore (according to the Method proposed) will explain, first in an *Egg*, and afterwards in other *Conceptions* of several creatures, what is constituted *first*, and what *last*, in a most miraculous order, and with a most inimitable prudence and wisdom, by the great God of nature; and at length we will reveal what we have found out concerning the *first matter* out of which, and the *first efficient* [cause] by which, the *fetus* is made, as also of the *order and Œconomy* of *Generation*: that thence we may attain to some infallible knowledge of each faculty of the formative and vegetative *Soul*, by the effects of it; and of the nature of the *Soul* it self, by the *parts*, or *organs of the body*, and their functions.[34]

This, then, was the role of the anatomist as student of generation.

Ex ovo omnia, Harvey announced on the frontispiece of the Latin edition of the book – 'Every creature comes from an egg', a sentiment sometimes expressed as *Omne vivum ex ovo*. The main novel claims that Harvey makes in the book have been summarised as follows: 'that all living things arise from ova or living primordia (in his, not in our, sense); that they do not arise from any formed material

32 Harvey, *Anatomical exercitations, concerning the generation of living creatures*, 1653, A7v.

33 Cunningham, 'Fabricius and the "Aristotle project"', 1985.

34 Harvey, *Anatomical exercitations, concerning the generation of living creatures*, 1653, b6r.

contained in such ova nor out of material contributed by the male and female during sexual intercourse; and that fertilization is due to an influence, to nothing material'.[35] Although he discusses the development of the fetus to some extent and also birth, Harvey's main interest in generation was in its initial stages. Harvey's primary experimental animal was the domestic chicken and its egg, and in this he was following both Aristotle and Fabricius. But Harvey extended his experiments to as wide a range of animals as he could, including the royal deer.

Harvey's own solutions to these questions about generation were not widely followed. In the first place Harvey was far too Aristotelian in his commitments to serve as a model for the next generation of investigators, who mostly inclined to mechanistic and atomistic approaches in their physiological explanations. Moreover, the new 'preformation-with-*emboîtment*' theory came to be preferred by most people studying generation; that is, the view that the development of the fetus consists entirely in an expansion of a tiny perfect being that was there all along, together with the view that, encapsulated within the first parent of every species, all future generations were present. Some people maintained that the preformation was in the ovum of the female ('ovists', such as Jan Swammerdam), while others maintained that it was in the semen of the male ('animalculists', such as Antoni van Leeuwenhoek). Both forms of this approach were supported by research with the microscope, which became very popular from the 1660s, some of the research being undertaken by people who would have regarded themselves as anatomists, such as Malpighi, but most by people who would have regarded themselves rather as natural historians. But for a considerable time it appears that generation was more a theoretical study than a practical anatomical one. In other words, this kind of generation was discussed in works of physiology, rather than of anatomy.[36]

What one finds on generation in anatomy lectures and books, however, is usually an account of the parts of both men and women serving generation (i.e. their genitals), their actions and their uses. Pierre Dionis, for instance, teaching in the Jardin royal in Paris in the 1670s, devoted a whole demonstration to the genital organs of man and woman. But when he published his lectures he included an extended dissertation on generation more generally as well. And it is here that he put to the anatomical test the theories of Swammerdam and others. 'I am very sensible that an Anatomist ought to examine all the events of Nature, and to conduct his reason by what appears to his senses', Dionis writes,[37] and he resolved 'not go along with the moderns where their arguments are not grounded upon the structure of the parts that are visible to my own eye; and backed by certain experiments'.[38]

---

35 Meyer, *An analysis of the De generatione animalium of William Harvey*, 1936, pp. 1–2.

36 On all these issues see the excellent work of Gasking, *Investigations into generation 1651–1828*, 1967, and Cole, *Early theories of sexual generation*, 1930.

37 Dionis, *The anatomy of humane bodies improv'd*, 1703, p. 195.

38 Ibid., p. 210.

He relates several recent unusual cases where dissection of the womb of women dying pregnant had revealed peculiar circumstances, all of which taken together persuaded him that 'the system of the *ovistae*', that is, those who claimed that humans begin as eggs, 'gives the most probable account of generation' (p. 210). The structure of the parts, and the mechanics of copulation, persuaded him that the so-called vesicles are more probably eggs, and the so-called female testicles are ovaries, that is, egg-making glands. Each egg falls one at a time down the tube (now known as the Fallopian tube) to the womb, where it encounters the male semen, which somehow mysteriously fertilises it. In his researches Dionis was not able to see the mammalian egg, something not possible before the early nineteenth century. Instead he worked from the visible anatomical structures and the balance of probabilities to decide between the different theories. The ovist theory was new and not original with Dionis: he had simply found a particular version of it the most likely, given the anatomical structures. He had changed his mind from the first edition of his *Anatomy* to the third; reflection and dissection had persuaded him of the ovist position. Echoing Harvey, he wrote: 'Upon the whole, we conclude, That no generation is performed without eggs: *Omnia ex ovo*';[39] every creature comes from an egg.

Although so much eighteenth-century discussion of generation was theoretical (that is, physiological) rather than practical (that is, anatomical), one of the greatest anatomists and physiologists, Albrecht von Haller, returned to a brief but intensive anatomical research programme on generation, which formed the factual basis of his theoretical accounts of generation in his celebrated *Elements of physiology*. According to his own account, Haller found himself for a while unable to undertake the dissection of human bodies, nor could he gain access to large animals. So for three years he took up experiments that he could make on his own. He turned to microscopic studies on the blood and vessels of live animals, and 'to the discoveries that the egg of the hen might furnish'.[40] In experimenting on the egg of the hen Haller was following the oldest tradition from Aristotle. The results appeared in two volumes of experiments on generation, *On the formation of the heart in the chick*, first published in Latin in 1757. The significance of Haller looking at the heart was that the heart is the first organ to become visible, whether it came into being from pre-existing materials of a different kind, or was simply enlarged from something that already existed. Haller hoped that these experiments would produce anatomical facts that would enable one to choose between these two theories. Haller changed his views on generation twice in his working life, from (1) believing in animalculist preformation, following his teacher Boerhaave, to (2) being an epigenicist (believing that 'the parts of animals are successively generated from a fluid in accordance with definite laws'), to (3) believing in ovist preformation. This particular series of experiments was crucial to Haller's making that second switch. Historians have argued about whether the chicken came

[39] Ibid., p. 247.

[40] Haller, *Sur la formation du coeur dans le poulet*, 2 vols, 1758, vol. 1, p. 3.

before the egg here; that is to say, they have suggested that these experiments were undertaken by Haller to provide himself with evidence for his already-changed opinion, rather than being the evidential ground for his making that mental switch.[41] At all events, for our present purposes what is significant is that, for these researches on generation, undertaken by Haller as an anatomist, he spent three years on his experiments, opening the eggs of twenty-six different clutches of different birds at intervals from the seventh to the five-hundred and twenty-eighth hour (i.e. the twenty-second day) after laying, looking at the development of the heart and other parts.[42] Although the conclusions he came to from his experiments and observations were ones related to issues topical in his day, nevertheless this kind of experimentation was the traditional domain of the anatomist as student of generation.

In the long eighteenth century, the generation of humans – or certain aspects of it – became more important for the medical or surgical practitioner than ever before. The anatomists showed a renewed interest in it too, as a topic that medical students wanted and needed to learn, and thus a topic on which the anatomists needed to undertake fresh research and provide instruction. For this was the period when medical men first began to be routinely involved in normal human birth.[43] Hitherto attendance at birth had not been seen as a medical event at all. Rather, as in many traditional societies, it was attended by women with whom the future mother felt secure: her midwives. As childbirth is predominantly a safe and straightforward event, the ignorance of all matters anatomical and generational by the midwives was not usually of great consequence. Birth, for the most part, looks after itself. However, it had come to be the case that when a child was obstructed in its exit to the world, or when the labour was protracted over several days, or when the infant died before being born, then in such cases midwives or other bystanders would summon a surgeon. Thus when the surgeon arrived at a birth event, it meant that something had gone horribly wrong, and the surgeon's role was usually to try to save the life of the mother by extracting the dead fetus with his specialised instruments, especially the ghastly hook. As a Scotsman preaching to midwives put it: 'then the chirurgeon comes at length with his instruments to finish the tragedy, and puts both out of pain, and makes a separation, that the infant

41 Roe, *Matter, life and generation*, 1981, chapter 2. The quotation from Haller is on p. 24, translated by Adelmann. See also Monti, ed., *Albrecht von Haller: Commentarius de formatione cordis in ovo incubato*, 2000.

42 Haller, *Sur la formation du coeur dans le poulet*, 2 vols, 1758, vol. 1, tables, pp. 14–22. This work was first presented as a memoir to the Royal Society of Sciences of Göttingen in 1757, in Latin.

43 For information on the origin of the man-midwife I rely on the excellent account by Wilson, *The making of man-midwifery*, 1995. See also his 'William Hunter and the varieties of man-midwifery', 1985.

may not be buried in its mother's womb: for thus it happens, that where the man comes, one or both must necessarily die'.[44]

In a few places in Europe some medical oversight of midwives was given to medical men, and this was especially so in the Netherlands, where the Amsterdam model was widely copied. From 1688 the examination of midwives there was made compulsory, and their competence and knowledge was tested by the Collegium Obstetricum and the town's lector (reader) in anatomy. From 1674 for over half a century this role was filled by the anatomist Frederick Ruysch. To improve the anatomical knowledge of midwives Ruysch suggested to the town council that every three months a 'recently dead body be brought back in the quiet of the evening from the hospital, and afterwards brought back to the place where it had come from' for demonstrating to midwives and their pupils.[45] In Leiden the comparable municipal lecturer and examiner for midwives was Conelis Trioen, who served in this capacity from 1710 until his death in 1746.[46]

By the mid- to late seventeenth century a few men – and even some midwives – had developed ways of 'turning' the fetus when it was obstructed, and bringing it out by the feet rather than the head. Similarly some new instruments were devised that would assist in bringing out a live (rather than a dead) child, though the design and use of these 'forceps' was long kept a trade secret. Both these developments, though not widely known or practised, helped make a space for the medical man in the birthing room.

Knowledge and experience of normal birth could be obtained in certain places, most notably of course in Paris. The description of his time there in the late 1710s by a young Scotsman is quite striking.[47] James Houstoun had studied to be a surgeon at Glasgow, taken classes at Edinburgh, and then gone to Leiden, where he was taught by Boerhaave. He graduated as a physician there in 1714. He went to Paris where, torn as ever between his pleasure in idleness and his desire to gain a fortune, for a few months he enjoyed Parisian debauchery. Then he remembered his scheme of becoming man-midwife to the rich:

> I began to think of what I came to France for, my principal design having been to get into the Salle des Accouchements de l'Hôtel Dieu. This is an institution founded by Louis XIV, and perhaps one of the best in the world, calculated as well consistent with policy as charity. In this large and convenient hospital, Hôtel-Dieu, there are two very large halls set a-part for receiving women with child.

---

44 McMath, *The expert mid-wife*, 1694, Preface to the Reader, not paginated.

45 Marland, 'The "burgerlijke" midwife', 1993, p. 196.

46 Camper, *Itinera in Angliam*, 1939, p. xiii.

47 Houstoun, *Dr Houstoun's memoirs of his own lifetime*, 1747; the quotations are from pp. 64–6 and 73–4.

> The manner of which is thus, all women in that condition of whatever nation or religion, within a month or six weeks of her time of delivery, are there received, maintained, and well accommodated with all necessaries, at the King's charge, till they are brought to-bed, and entirely recovered from child-bed.

Houstoun saw at least one English noblewoman come there to conceal her condition, and the institution was used by both poor and rich. The babies could be taken away by the mother, or entered into a foundling hospital for 10,000 children, also maintained at royal charge. Houstoun saw this as Louis XIV's way of repopulating his army at small cost. This double institution, says Houstoun, 'is not only a nursery for children, but a seminary for educating and instructing midwives, who must all be strictly examined before they profess midwifery, which also saves 10,000 more children from being murdered by the hands of unskilful midwives'.

Admission to the hospital for a medical man was no way automatic, and Houstoun had to seek patronage in royal circles to get the King's order:

> I at last succeeded Mr. Clayton in the Salle des Accouchements de l'Hôtel-Dieu, where I met with more laborious and fatiguing business than any I ever had in the course of my life; every twenty-four hours, *communibus diebus*, we brought about ten children into the world, and as there were only four midwives and myself to go thro' all, night and day, we were always employed; and I continued in this hospital four months, without setting my foot out of it four times. This is the only place, for thorough instruction in this business; I brought near three hundred women to-bed in the time, and every one of the four midwives in proportion: so that we had cases of all sorts; we assisted one another, and had a Mistress Midwife who directed the whole; and, in very extraordinary cases, we called in the Master-Surgeon of the hospital's assistance. This hospital is served, in the meanest, servile work, even to cleaning the close-stools, by religious nuns, of the very best families in France, and none but such can wait there, which makes them better and cleaner than any other.

Houstoun's experience of normal and abnormal birth was exceptional for his time, but in the end he never put it to use as a practising man-midwife.

Gradually in the course of the eighteenth century the medical man, both surgeon and physician, made his way into the birthing room, the room temporarily set aside in the home for this purpose. Consequently, his anatomy teachers tried to ensure that he came equipped with the necessary anatomical knowledge. Anatomists had occasionally been able to portray the fetus in the womb, where they were able to obtain the body of a pregnant woman who had died before giving birth, as Bidloo for instance had done in his 1685 book, portraying a seven-month fetus. Nevertheless such knowledge of gestation and normal birth was relatively difficult for the anatomist to obtain initially, since he needed a supply of human fetuses at different stages of development, and of women who had died at different stages of pregnancy, on which to make his dissections and observations. However, where there was a will there was a way, and anatomists found that it was indeed possible to produce a detailed and accurate anatomical account of the gestation and birth of humans by systematic investigation, once such knowledge was a commodity marketable to young medics who were expecting to add this new role of man-midwife to the traditional roles of the medical man. There was much anatomical activity in this respect in the 1740s and '50s.

The 'man-midwife' did not make the same impact everywhere: for instance, in Italy this new occupation seems not to have appeared at all. The phenomenon of the 'man-midwife' is most evident in England in the eighteenth century, so we will expect to find English medical men, and more particularly the Scottish practitioners who dominated anatomy and medicine in London, to be involved in researching, writing and teaching new practical and anatomical skills in the realm of gestation and childbirth.

As an example of an ambitious surgeon and anatomist keen to make his mark in the field of man-midwifery we can take John Burton (1710–71), a Cambridge- and Leiden-educated doctor who had been involved in the establishment of York Infirmary in 1740. As he remarked in the Preface to his book, *An essay towards a complete new system of midwifry* [*sic*], *theoretical and practical* of 1751, he was able to advocate methods of practice that were entirely new and 'founded upon Reason and Experience, which is the surest Foundation in the Practice of all Branches of *Physic*. For, as Nature discloses herself in an obscure Manner, we must strictly observe her Operations, by which we shall see the Facts; and then a thorough Knowledge of *Philosophy* and *Anatomy* will enable us, by such Guides, to penetrate into her secret Principles'.[48] The first part of the work consisted of 'an *Anatomical* and *Physiological* Description of such Parts, as are necessary to give a thorough Knowledge of the Bones of the *Pelvis*, and their Structure; the true Fabric and Situation of the *Womb* before it becomes *pregnant*; and the various Alterations it undergoes from the Beginning, and during that State, till its Reduction to its former Condition after *Delivery*'.[49] Burton had been able to acquire this knowledge from a quite extensive practice of man-midwifery, many cases from which are recorded in the book. But he was able to display it in copper-plates because he had been fortunate enough to come across a woman near York who died, undelivered, at her full term. Her body was brought surreptitiously to York, where it was dissected and sketches and then engravings were made of it in various stages of dissection at Burton's request by the young artist George Stubbs (whom we shall meet again in Chapter 5). It has been suggested (and later actions by Stubbs make it likely) that Stubbs based one or two of these (Fig. 3.2) on images in Albinus's *Seven tables of the uterus of a pregnant woman* of 1748.[50] Poor Burton became immortalised, for political and personal reasons, by Laurence

48 Burton, *An essay towards a complete new system of midwifry*, 1751, p. xiii.

49 Ibid., p. xv.

50 Fountain, 'Stubbs … as an anatomist', 1968, p. 641.

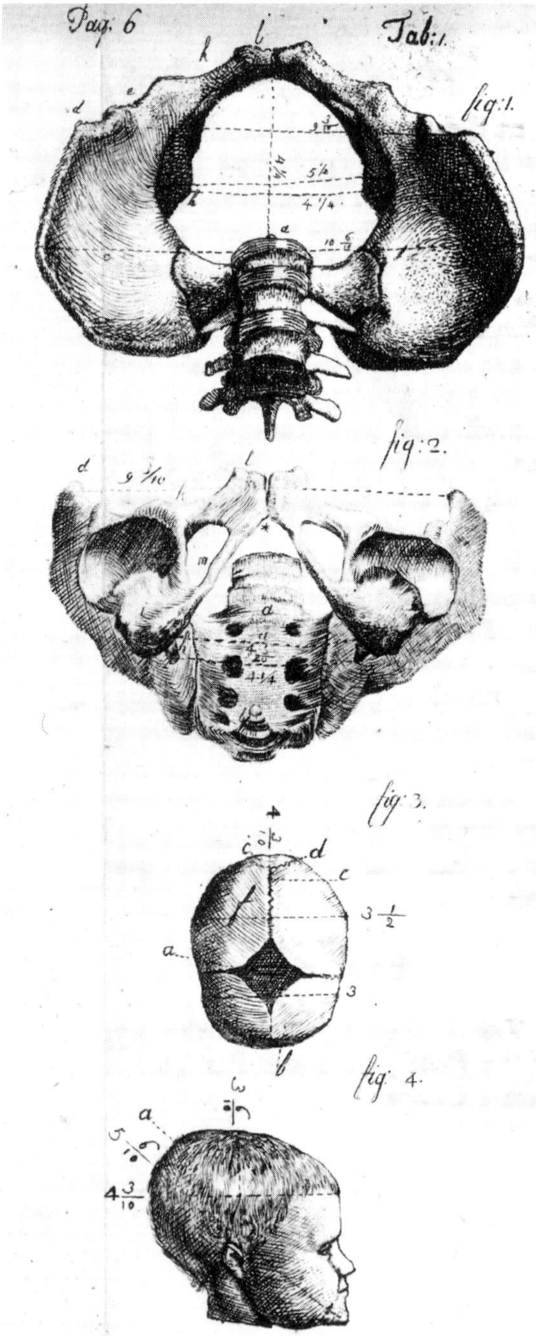

Fig. 3.2

Sterne as the ignorant and maladroit man-midwife 'Dr Slop' in *Tristram Shandy*, which appeared in parts from 1759.

In the Netherlands Bernhard Siegfried Albinus had already produced *Seven tables of the uterus of a pregnant woman who died in childbirth* in 1748, as a continuation of his 1747 *Tables of the skeleton and muscles of the human body* (Fig. 3.3), though as there is no text to these images (except for legends) it is not clear from them whether Albinus saw these tables as having a special role.

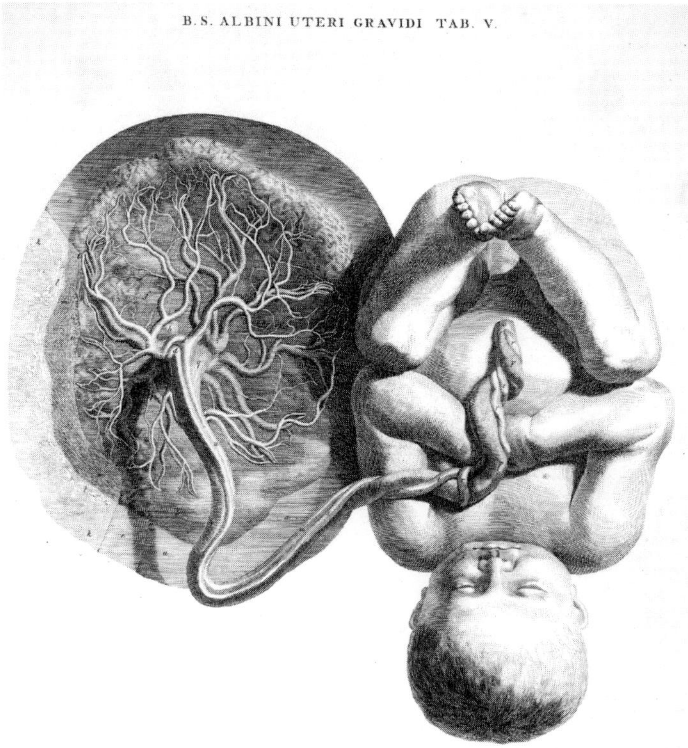

Fig. 3.3

Similarly in Edinburgh both Dr Donald Monro and his younger brother Alexander Monro (later called *Secundus*), at the urging of their father, each dissected a woman, one who had died at five months' pregnant and the other at six months into her pregnancy, and produced illustrations of the gravid uterus in 1754 based on those of Albinus and Bidloo for the sake of comparison. Dr Donald Monro, who had himself attended courses of midwifery, presumably those of William Smellie (see below), had particular instructions from his father to look for the connections that supposedly existed between the vessels of the womb and those of the placenta (secundines). The dissection took three days, during which

Professor Alexander Monro *Primus* demonstrated to his pupils the parts in their natural situation. Donald was busy dissecting and injecting four different coloured fluids into the blood vessels. All four injected materials were extravasated (that is, they flowed out of the blood vessels) between the placenta and the uterus: 'not one single vessel of the placenta was filled with any of the injection that came from the uterus; nor was a vessel of the womb filled with that which came from the secundines … I must be of opinion [Donald wrote] that there is no anastomosis [i.e. no connection from vessel to vessel] between the uterine and secundary vessels'.[51]

Burton, using his one female body, had managed to get into print first with his text and pictures and thus to pip at the post the greatest teacher of man-midwifery in Britain at the time, the Scot William Smellie (1697–1763). Later Burton was to attack Smellie, so it may well be that there was a race on to publish first, at least in Burton's eyes. Smellie's early education in medicine had begun as an apprentice apothecary in his home town of Lanark. Then he served as a surgical mate on a naval vessel for two years. Returning to Lanark he set up in practice, which included his being called as a surgeon to cases of obstructed birth for a number of years, before he went briefly to Paris in the hope of learning something of the use of the forceps in midwifery from the Grégoires, father and son.[52] In his early years in London Smellie attended the lectures on normal anatomy and pathology of Frank Nicholls. He started teaching the theory and practice of midwifery, to both women and men, at his own London house in 1741 in a two-week course. As part of the course he took his students to attend poor women in their confinements; the students paid to attend, the women were attended for free. By 1750 Smellie could claim that his pupils had seen over fifteen hundred normal and abnormal deliveries in his presence.

In the course of the 1750s Smellie put into print the results of his researches into human generation insofar as they related to the practice of midwifery. They are the product of above thirty years of practice, with well over five hundred cases related. First came *A treatise on the theory and practice of midwifery* in 1752, followed by *A collection of cases and observations in midwifery* in 1754. Finally, also in 1754, he produced his anatomical atlas on midwifery, which was a companion to the first two publications: *A sett of anatomical tables, with explanations, and an abridgment, of the practice of midwifery*.[53] The illustrations are life-size. He had intended to publish it in both Latin and English, but in the event it appeared only in English.[54] In his Preface he says that, 'in a long course of teaching and practice in Midwifry [*sic*], I hope I may without vanity say, that I have done something

---

51 Donald Monro, 'The dissection of a woman with child', 1754, pp. 411, 419.

52 Butterton, 'The education … of William Smellie', 1986.

53 Smellie, *A sett of anatomical tables*, 1754. At least ten editions in English were published in the next fifty years, with translations also into Latin and German. See Russell, *British anatomy 1525–1800*, 1987, nos. 753–766.

54 Johnstone, *William Smellie the Master of British midwifery*, 1952, p. 85.

towards reducing that Art, into a more simple and mechanical method than has hitherto been done'. Thirty-six plates deal with the anatomy of the human female pelvis; the anatomy of the uterus at different stages of gestation; the descent and curves followed by the fetus and the different types of presentation of the fetus at term; and the interventions with hand and/or instruments to assist obstructed or difficult birth. Most of these figures, Smellie wrote, 'were taken from Subjects prepared on purpose, to shew every thing that might conduce to the improvement of the young Practitioner, avoiding however the extreme Minutiae, and what else seemed foreign to the present design; the situation of the parts, and their respective dimensions being more particularly attended to, than a minute investigation of their structure'. Some of the subjects (i.e. the corpses) Smellie had acquired and prepared himself, while some had been prepared by others, such as the dissection of a uterus at term with twins, by a Mr Mackenzie (Fig. 3.4). Most of the Tables had been drawn by Jan van Rymsdyk, and a few by the important Dutch anatomist Peter Camper, who was a student of Smellie and a regular visitor.

Smellie has here used the terms 'mechanical', 'mechanism', and 'machine' in the discussion and practice of midwifery. He spoke of his aims as being directed 'towards reducing that Art, into a more simple and mechanical method' than any hitherto. His use of these terms demonstrates how fundamentally his understanding of childbirth was built on anatomy. Everything may be said to have been based on his insight that, during the course of labour and birth, the head of the fetus turns (rotates) – indeed *must* turn if it is to emerge successfully – since the passage through the pelvis of the mother that the head of the fetus must follow is so narrow. 'Hence', as a relatively modern obstetrician has summed up the core of all Smellie's advice to the prospective midwife, 'the importance of bringing the head through with its widest diameter lying in the widest diameter of the pelvis at all stages of its descent'.[55] All the assistance that the midwife might give had to have this as its aim, whether using the hand or the forceps. Thus, as is clear in the *Anatomical tables*, Smellie as midwife–anatomist is concerned with the measurements of the pelvis of the mother and of the head of the fetus (taken from skeletons) – and then with all the dangers arising from the many varieties of distorted pelvises; and he is concerned with the stretching and other changes in the uterus and vessels that occur in the course of pregnancy and at the onset of labour, all of which is revealed by the study of the anatomy of these parts. The skills of the man-midwife are to be established on the anatomical knowledge of the mechanics of the processes that childbirth involves.

Smellie improved the design of the forceps 'mechanically' too; that is to say, in a manner that accords with and assists the natural anatomical mechanics of childbirth. Smellie called the forceps 'artificial Hands' and illustrated the use of them to help the fetus along, when necessary for the safety of either the mother or the child (Table 16), 'in imitation of the natural progression by the Labour-pains'

[55] Ibid., p. 46.

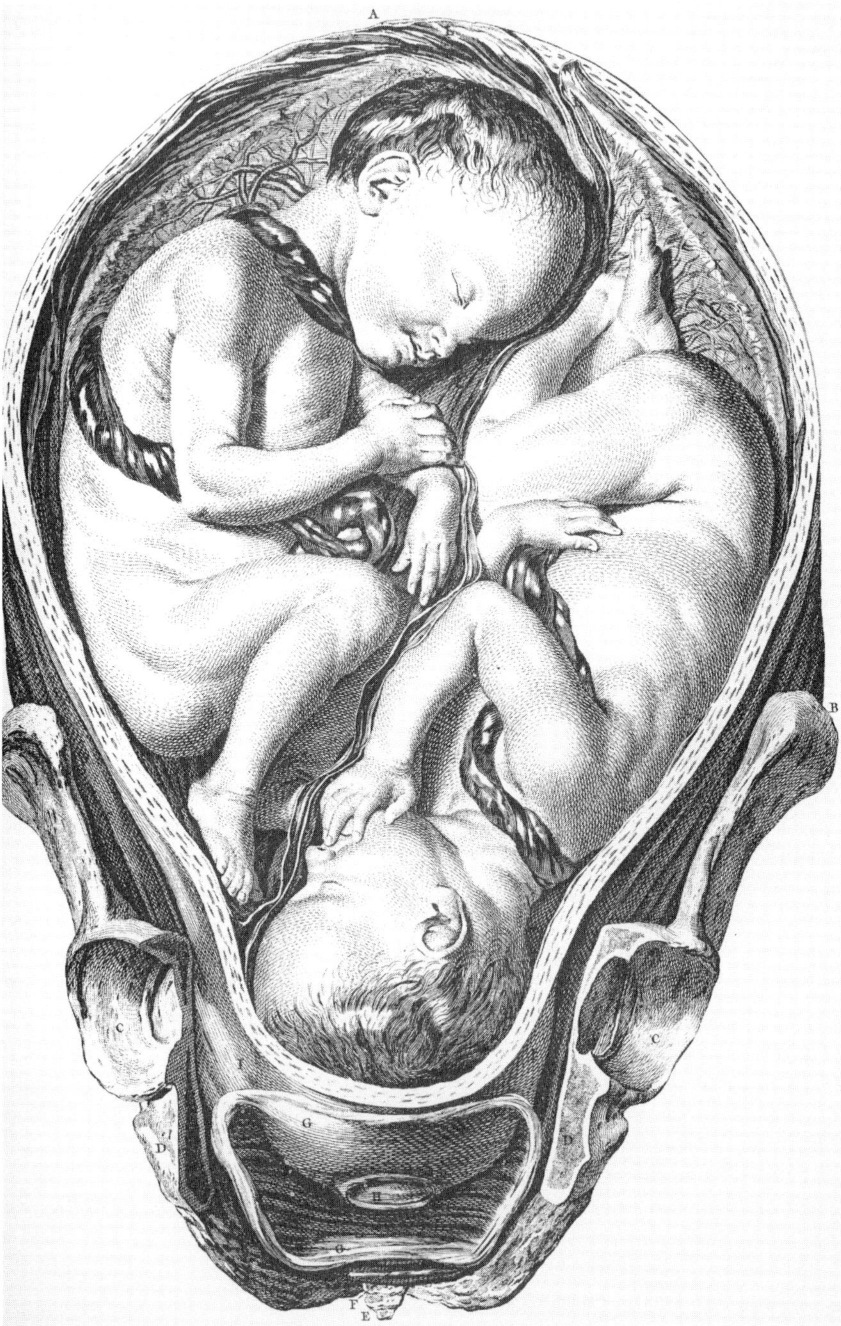

Fig. 3.4

(Fig. 3.5, Table 17). However, the forceps should be a last resort, to be used only when the operator's manipulations fail or are impossible.

But for Smellie there was a further aspect to learning 'The Mechanism of Parturition' (as he entitled the most important section of his 1751 *Treatise*). This uses the term 'machine' in a way more familiar to us today, and was again built directly on anatomical knowledge. For the one positive thing Smellie had learnt from the younger Grégoire in Paris was to use in his teaching a dummy of the pregnant lower female body together with a manikin or phantom of the fetus. Grégoire's dummy was a piece of basket-work covering the bones of a real pelvis, which was hardly satisfactory. But from seeing it Smellie (as he later wrote) 'considered that there was a possibility of forming machines which should so exactly imitate real women and children as to exhibit to the learner all the difficulties that happen in midwifery; and such I actually contrived and made by dint of uncommon labour and application'.[56] One account of his success in this says that these models

> … are composed of real Bones, mounted and covered with artificial Ligaments, Muscles and Cuticle, to give them the true Motion, Shape and Beauty of natural Bodies, and the Contents of the Abdomen are imitated with great Exactness. Besides his large Machines (which are three in number) he has finished six artificial Children with the same minute Proportion in all their Parts; so that with the apparatus he can perform and demonstrate all the different kinds of Delivery with more Deliberation, Perspicuity and Fulness than can be expected on real Subjects.[57]

What one can conclude from Smellie's achievements is that knowledge of the mechanics of natural and obstructed childbirth is essential for the birth attendant, whether male or female, and that such knowledge can *only* be discovered by anatomical observations and be taught by anatomical instruction. Putting anatomy at the heart of his teaching meant that Smellie could help his students avoid the ghastly scenes that he had come across himself as a young practitioner in midwifery still ignorant of the anatomy and mechanics of childbirth, in which the arm or even the whole body of an obstructed child would be sawn off in desperate (and unsuccessful) attempts to help the mother. Anatomical knowledge makes better midwives.

In the same way, the most celebrated work on understanding how gestation is carried out was also an anatomical one, produced by the anatomist William Hunter. He published his *Anatomia uteri humani gravidi tabulis illustrata* (*The anatomy of the human gravid uterus exhibited in figures*) in 1774, built on investigations he had been conducting since 1751.

---

[56] Quoted in ibid., pp. 18–19.

[57] Quoted in ibid., p. 25, without reference. Three other descriptions of the machines are given there.

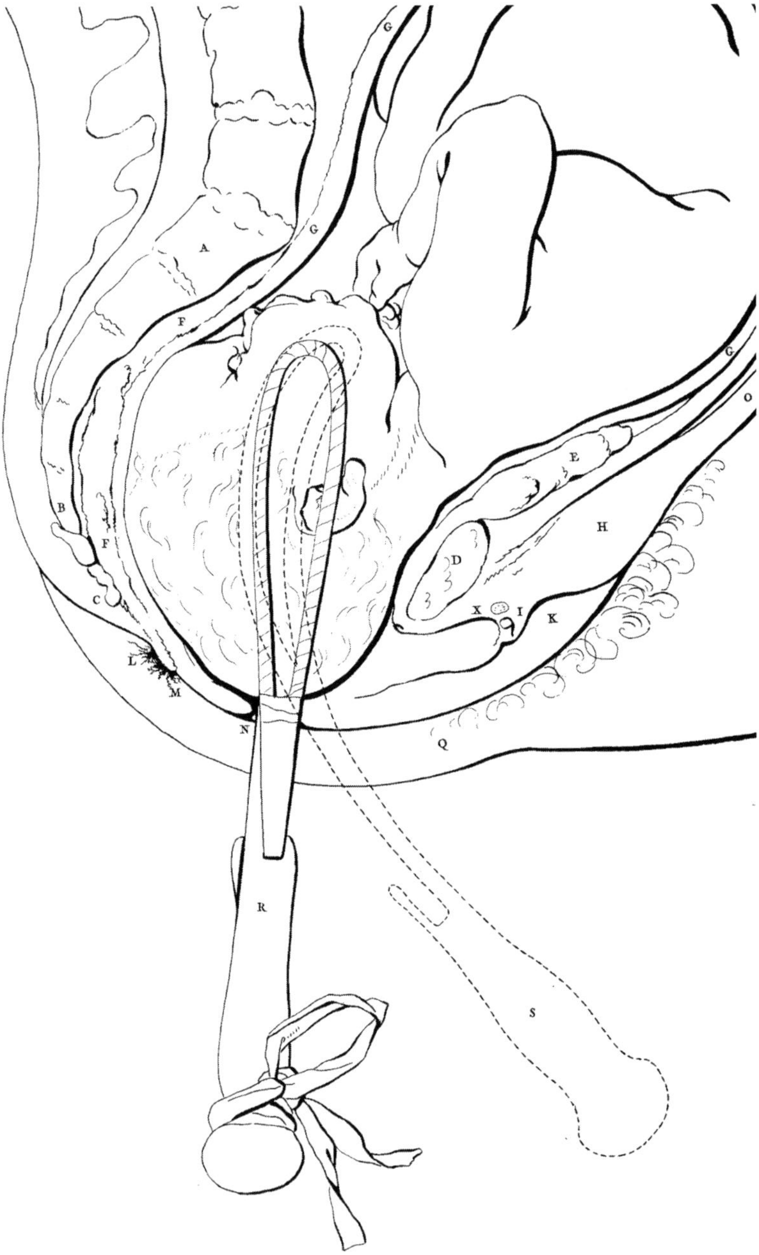

Fig. 3.5

He had been Smellie's living-in pupil for a year, and the two men had a continuing lively relationship. Among his other roles William Hunter was elected one of the surgeon men-midwives at the Middlesex Hospital in 1748, and then at the British Lying-in Hospital in 1749. While there, according to his brother John, William 'endeavour'd to improve Midwifery. He allowed the women to deliver themselves, or at least give nature leave to exert her own powers. He even allowed the uterus to dispose of its Placenta; all of which gave a new turn to midwifery; and instruments which had been too much used [esp. the forceps] were by Dr. H. almost laid aside'.[58] Hunter has been called 'the man who entered the womb and resolved the anatomy of the placenta'.[59] On the title-page of his book he calls himself 'Physician Extraordinary to the Queen, Professor of Anatomy in the Royal Academy, and Fellow of the Royal and Antiquarian Societies', and he dedicated the work to the King.

Strikingly, the book is in both Latin and English – as Smellie had intended for his own book – on facing pages; Hunter's adoption of this dual-language device marks the period when Latin was giving way to English as the scholarly language for English-speakers, and it also enabled William to make claims to an international audience for his work, even though his primary audience for it was a local one. Hunter had the letter-press printed in Birmingham by John Baskerville, the innovative printer, font-maker and paper-maker, in order to ensure that the superior quality paper protected the prints. The drawings were made (yet again) by the indefatigable Jan Rymsdyk, though engraved by a variety of engravers.[60] In the first edition there were thirty-four plates, each often illustrating more than one object.

Hunter indicates that he had been investigating the gravid uterus and the development of the fetus for some time in brutes before he was able to use human subjects. He describes how he had come across suitable bodies of pregnant women and fetuses at different stages of development, the first being in 1751 when he obtained the body of a woman who died at full term without having given birth. 'The body was procured before any sensible putrefaction had begun: the season of the year was favourable to dissection; the injection of the blood-vessels proved successful; a very able painter, in this way, was found; every part was examined in the most public manner, and the truth was thereby well authenticated' (Preface). The first ten plates that were prepared were all from this first body, in which the arteries and veins had been injected with wax of different colours.

Having got this far Hunter decided to publish a book on the human gravid uterus. But as other bodies became available, so the project expanded. After he had procured a third body, he wrote that he 'foresaw that, in the course of some years, by diligence he might procure in this great city [of London], so many opportunities of studying the gravid *uterus*, as to be enabled to make up a tolerable system; and

---

58 Brock, ed., *William Hunter ... A memoir*, 1983, pp. 6–8.

59 Wilson, *The making of man-midwifery*, 1995, p. 175.

60 Thornton, *Jan van Rymsdyk*, 1982.

to exhibit, by figures, all the principal changes that happen in the nine months of utero-gestation'. *The anatomy of the human gravid uterus* was the result.

The difference between Hunter's book and the *Anatomical tables* of his teacher Smellie was that where Smellie was concerned centrally with the mechanics of childbirth, and how to assist it, Hunter by contrast is concerned with the development of the uterus and of the fetus within it over time, and barely at all (at least in this book) with the delivery of the child as a technical matter that might require the intervention of the midwife or man-midwife. Fig. 3.6, for instance, shows 'the child in the womb, in its natural situation'. Thus these two great anatomical books served different purposes and were complementary.

These investigations had resulted in new knowledge, and William Hunter claimed it for himself:

> The gravid uterus is a subject likewise, which has afforded me opportunities of making considerable improvements; particularly one very important discovery; viz. that the internal membrane of the *uterus*, which I have named *deciduas*, constitutes the exterior part of the *secundines*, or after-birth; and separates from the rest of the *uterus* every time that a woman either bears a child or suffers a miscarriage. This discovery includes another, to wit, that the placenta is partly made up of an excrescence or efflorescence from the *uterus* itself.
>
> These discoveries are of the utmost consequence, both in the physiological question about the connection between the mother and child; and likewise in explaining the phenomena of births and abortions, as well as in regulating our practice.[61]

While the relation of anatomy to midwifery is most evident in England, there were practitioners and researchers elsewhere, both in Britain and on the continent, who were developing comparable studies, and chairs were established in a number of medical faculties whose holders had the duty to teach obstetrics or midwifery. One particular instance is the new Georg-August-Universität founded at Göttingen in 1737 by the Elector of Hanover (who was also George II of England), where the incomparable Albrecht von Haller held sway in the medical faculty. Here an 'extraordinary' professorship of anatomy and obstetrics (that is, a professorship out of the ordinary line) was created in 1751, at the initiative of Haller, for the young Johann Georg Roederer (1726–63).[62] In 1753 Roederer published his lectures as *Elements of the obstetric art for the use of students* in Latin. Having the lectures in this printed form, rather than having to constantly take notes, the students would be better able to follow the explanations of the teacher. The work was composed (Roederer wrote),

---

61 William Hunter, *Two introductory lectures*, 1784, p. 61.

62 On Roederer, see Schlumbohm, '"The pregnant women are here for the sake of the teaching institution"', 2001. On the childbirth institute at Göttingen, see Bueltzingsloewen, 'Die Entstehung des klinischen Unterrichts an den deutschen Universitäten', 2004.

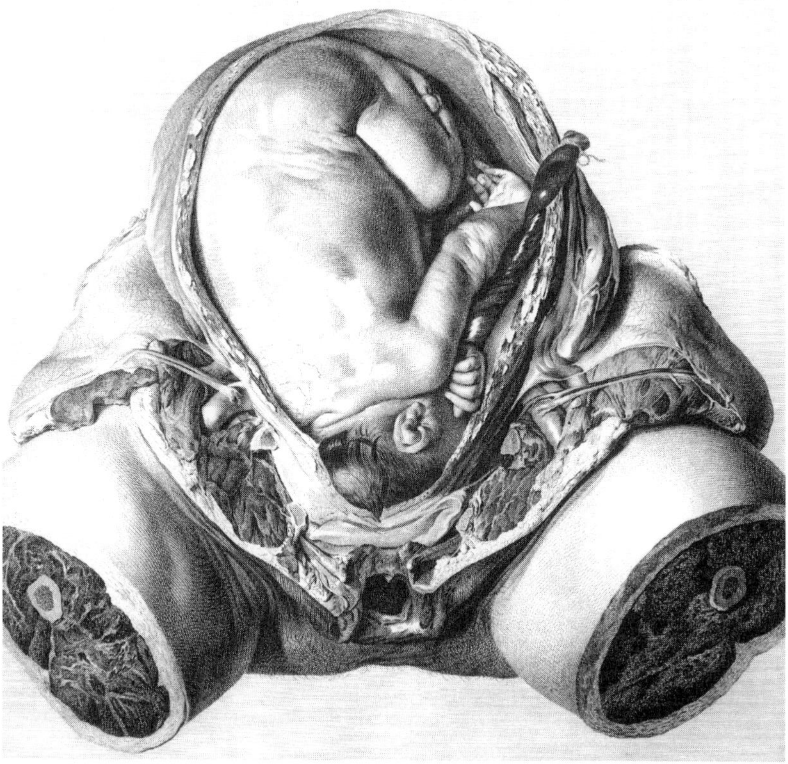

Fig. 3.6

> … according to my own observations, according to the teachings (to which a long experience guaranteed the validity) of the celebrated Fried whom I had as my master, and according to the best writings on this subject that we have. I have had recourse to Anatomy, particularly to the Anatomy of living creatures, which is the basis of all Medicine, and without which one cannot know the shape of the pelvis, of the uterus, and of the parts of generation. I assert nothing that is not supported on certain and incontestable principles. The Hospitals that I have long frequented have been a great help to me, but the drawback is that one cannot always make the artificial observations that one needs, and which one terms 'experiments', and that one has to wait patiently for the phenomena that nature deploys … As for myself, I now have the advantage of being able to contribute my own observations, and of being able to demonstrate to my Disciples, on the subject itself [i.e. on the pregnant mother] the truths that my observations enabled me to discover.[63]

[63] Roederer, *Elementa artis obstetriciae*, 1753, Preface. I translate from the French edition: Roederer, *Élémens de l'art des accouchements*, 1765, which was more accessible

Roederer, like Smellie, used many episodes of his own practice in his teaching and, again like Smellie, he included cases with bad outcomes (the death of both the mother and child) and cases with good outcomes, for 'the obstetric art is still in its infancy'.[64] Again like Smellie, Roederer begins his book with anatomy: the anatomy of the female pelvis, both in its natural state and in a distorted state.

> The pelvis of women being naturally shaped as I have just described, facilitates the exit of the fetus at the time of childbirth: for the diameter of its 'upper opening' is hardly five inches, and the conjugate axis or small diameter around 4 inches; given a well-proportioned head of a fetus, not exceeding these dimensions when it is compressed by the violence of the labour pains, this opening will be large enough to give a route for the fetus.[65]

The sentiments here could be Smellie himself talking, at least thus far. And the reason for this is that Roederer had indeed studied under Smellie in London, as well as under William Hunter, and had also studied in Paris with André Levret (Accoucheur to the Dauphine and author of *L'art des accouchements*, 1747) and Grégoire, and even in Leiden with Albinus. His book is nowadays sometimes regarded as the beginning of the scientific era in childbirth in Germany. However, apart from Dr Fried of Strasbourg, he acknowledges no debts to his teachers, except to compare their instruments briefly. He went on to produce an anatomical atlas *Images of the human uterus explained with observations* (1759), with seven tables, from dissections of women dead at or before term.[66] Fig. 3.7 shows 'the external surface of the gravid uterus with its blood vessels', while Fig. 3.8 illustrates a malformed female pelvis. Roederer was given a small three-room hospital to deliver poor women, in which about twenty-five women were delivered per year. From 1785 his successor as professor, Johann Heinrich Fischer, had a grand new purpose-built obstetric hospital, which is still standing, though now devoted to music (Fig. 3.9).[67]

---

to me. Fried is Johann Jakob Fried, who taught Roederer in the medical faculty at the University of Strasbourg.

64 Roederer, *Élémens de l'art des accouchements*, 1765, p. 443.

65 Ibid., pp. 4–5.

66 Roederer, *Icones uteri humani*, 1759.

67 On the later Göttingen story and instrument collection, see Tröhler, 'Tracing emotions, concepts and realities in history', 1993.

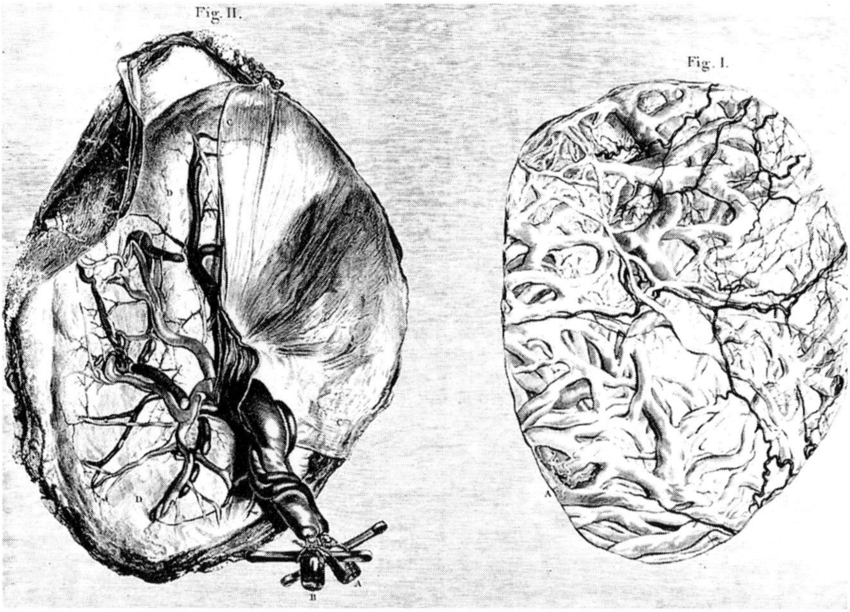

Fi. 3.7

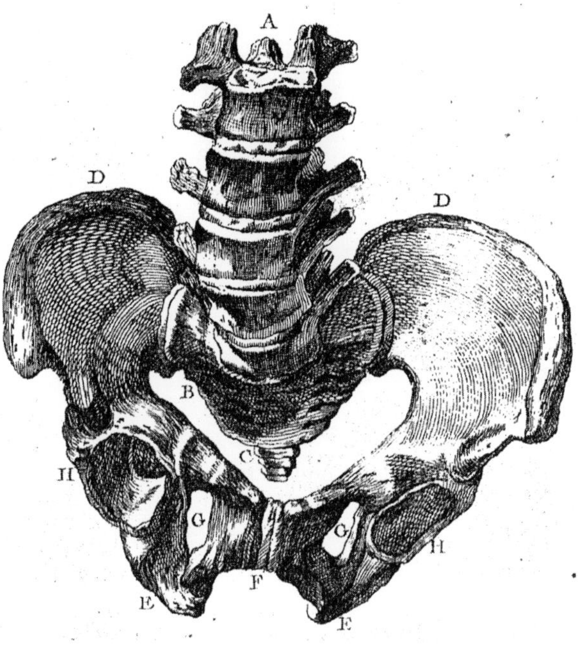

Fig. 3.8

Fig. 3.9

4. Comparative anatomy: A sub-discipline of anatomy

Comparative anatomy became an area of great interest and novel investigation in the long eighteenth century, and was of enormous popular interest as well as to anatomists. This is our third sub-discipline of anatomy. The various meanings of 'comparative anatomy' and the complex relation of comparative anatomy to the discipline of anatomy requires the subject to have a chapter to itself; it will be dealt with at length in Chapter 5 below.

5. Pathology: A sub-discipline of anatomy

When we turn now to pathology as our fourth sub-discipline of anatomy, we shall find that not only did all the most eminent practitioners devote much attention to pathology in the long eighteenth century, but that in doing so they changed the nature of it from a largely non-anatomical to an exclusively anatomical study.

Why anatomy in pathology?

Pathology is the discipline in which the nature and incidence of disease is studied in the body in order to determine cause of death in individuals and with the hope or expectation that thereby treatment or cure can be improved in the long term. As a discipline, pathology, like physiology, was usually taken to be a relative of the art and science of anatomy, and its role as a sub-discipline of anatomy became much more important in the course of the eighteenth century than it ever had been before. For an anatomist, pathology is pursued and developed by dissecting and inspecting the bodies of people who have died of disease (rather than from natural causes).

There is no *necessary* relation between pathology and anatomy. Whether there is or not depends on two things. First, it depends on one's belief about how the body functions naturally – that is, on one's physiological viewpoint. If you think that physiological understanding should be built on anatomical knowledge, then an anatomy-based pathology will be relevant to you. It was in this sense that John Hunter described pathology as 'the physiology of disease' in 1786. Similarly, at the end of the eighteenth century the Dutch anatomist Peter Camper described pathology as 'practical physiology', which is the counterpart I suppose to Haller's description of physiology as 'animated anatomy'.[68] However, if you think that the body functions exclusively in chemical ways (for instance), then no pertinent knowledge is likely to be gained from anatomical investigation about the cause and cure of disease. This was the position of the mid- to late seventeenth-century Helmontians and Paracelsians of England.

Second, whether or not there is a necessary relation between pathology and anatomy depends on one's understanding of the relation of cause and effect. For instance, the most celebrated practical physician of seventeenth-century England, Thomas Sydenham, thought there was no meaningful link between, on the one hand, the symptoms and the course of the disease and, on the other, any changes evident to the senses in the inside of the body after death. Indeed he thought such 'pathological' anatomy was actively misleading. Sydenham's objection to anatomising the healthy dead body and to cutting up the body after death from illness was based on his view of the nature and accessibility of cause: nothing can be established about the cause of disease, he believed, from anatomy. For Sydenham, disease is not the effect of some state of, or damage to, the internal organs, but a condition of the body as a whole as it fights against some (invisible) morbific matter.[69] 'A disease … is nothing else than the effort of Nature endeavouring with might and main the expulsion of the morbific matter for the health of the sick person'.[70] It is not some imperfection of an internal organ. So even though Sydenham can look very much like a modern to the historian today, given his passionate concern with compiling generalised case-histories of diseases in order to work out better treatments for them, he did not hold a remotely modern view of the importance of pathology as pursued by anatomists.

---

68 John Hunter, 'Lectures on the principles of surgery', 1786–87.

69 Cunningham, 'Thomas Sydenham', 1989.

70 Greenhill, *Thomas Sydenham … Opera omnia*, 1854, p. 26.

Thus the crucial point would seem to be that if you believe that the experience of illness comes about from – that is, that it is *caused by* – or is an incidental but constant *consequence of* deleterious changes within the body, and especially of some disablement of one or more organs, then the pathological-anatomy view makes sense. If you think illness has some other identity, and/or some other form of cause, then this view may not make sense. If you think that causes, by their very nature, are inaccessible to our senses or our reason – that hidden causes must inevitably remain hidden, because of the limits of the capacity of human understanding – then again this approach makes no sense. So the conditions under which the pathological-anatomy view is considered to be valuable and is actively developed are not 'eternal' but time- and place-specific. The eighteenth century was one such time for many anatomists. Here we shall be exploring why such anatomists vigorously pursued this approach now – and what particular line or lines they pursued individually within this general approach.

Most anatomists of the seventeenth and eighteenth century took it for granted that pathological investigation was part of their disciplinary role, following the ancient tradition represented by Galen in his book *On affected parts* (*De locis affectis*). As with the relation of physiology to anatomy, pathology also had a theoretical side that made it more of a *scientia*, but the anatomist *qua* anatomist rather pursued the tangible, visible, sensory side of disease as evident in the body during life and after death.

But until the early eighteenth century, if pathology was taught or investigated by anatomists, it was not done systematically. In normal anatomical teaching what was demonstrated was a healthy and usually young body, because often it was the body of a criminal young man. Any aberrations from the normal with respect to the inner organs might be pointed out by the anatomist, but had no further significance than as single observations. There was no systematic teaching offered by anatomists on the diseased body as such. One thing they sometimes did, as they performed the anatomy of a healthy person, was to discuss which diseases were incident to which parts. At the College of Physicians of London a series of lectures had been established in 1632 by Theodore Gulston, 'to be read from time to time by one of the four youngest Doctors of the said College upon two, three or more diseases as the Seniors of the College should appoint'. However, the Seniors soon decided that the lectures should be on regions of the body and the diseases affecting them, rather than on diseases as such.[71]

Some anatomical lectures on the stomach, given at the College of Physicians around 1660 by Francis Glisson, Professor of Physic at Cambridge University, and possibly part of the Gulstonian series, show a typical discussion of pathology within anatomy. Glisson is at one point speaking of the figure or shape of the stomach or ventricle, and particularly of the two orifices, the one to receive the

[71] Clark, *History of the Royal College of Physicians*, 2 vols, 1964, 1966, vol. 1, p. 251; and Royal College of Physicians MS. 28 (Sir William Browne's copy of Extracts).

food and the other to put out the excrements. One kind of disease possible here is a tumour, Glisson says:

> … if there happen a tumour in either of these orifices, whether it be humoral or flatuous, it will fill up the passage so as nothing can pass easily up or down. And, as it falls [out] in the *angina interna et latente* [internal and obscure suffocating pain], the tumour fills the passage of the trachea arteria, and so stops the breath, so here a tumour in the very entrance of the ventricle fills and stops, or at least straitens, the passage of the meat and drink into the ventricle, and though it pass down by the throat and gula well enough, yet when it comes to the orifice of the stomach, it recoils into the mouth, and sometimes with such force as it comes out at the nose. And not only humoral tumours will cause this recoiling and difficulty in swallowing, but flatuous tumours also. I know some [people] at this time that for many years have had great difficulty in swallowing, and sometimes their drink returns and is forced out at their nostrils; I suppose this comes by some tumour flatulent betwixt the coats of the ventricle, the rather because this symptom especially troubles them when they are empty and when otherwise they are troubled with windiness in the ventricle. The same may be said of tumours in the pylorus, but they are rare diseases; yet if any tumour chance to happen there, it must be a hindrance both to the chyle and excrements in their descent, and no doubt sometimes cause them to be rendered by vomit.[72]

Glisson's pathological anatomy here, then, consists in discussing all the possible unnatural alterations that each part of each organ is, or may be, subject to, with some of the examples being taken from his own practice as a physician. But the body that Glisson would have been dissecting as he lectured would have been a healthy one.

In this light it can be seen that the most celebrated member of the London College of Physicians in the seventeenth century, William Harvey, had been very unusual in his views, at one point writing about a plan of his to

> … undertake to relate from the many dissections of sick bodies, and the most grievous and wonderful diseases of dead persons, in what manner and how the inward parts of them are changed in place, bigness, condition, figure, substance, and other sensible accidents, from their natural form and appearance, which all Anatomists commonly described, and how diversely and wonderfully they are affected.[73]

Harvey was here comparing his views to those of Jean Riolan, the Dean of the Paris Medical Faculty, who had claimed in his *Anthropographia et osteologia*

[72] Cunningham, *English manuscripts of Francis Glisson (2)*, 1998, p. 31.

[73] *De circulatione exercitatio prima* to Jean Riolan (1649), as translated in Harvey, *The anatomical exercises*, 1653, p. 122.

(Paris, 1626) that if anyone knew how to dissect human bodies with skill and diligently inquired into the individual parts, he would easily understand the causes and seats of hidden diseases, and be able to prescribe appropriately. Harvey, then, is calling for the anatomist to dissect *sick* corpses, not just healthy ones.

Pathology in the library

Pathology of a certain kind became popular among anatomists from at least the 1660s. For the most part, however, it was conducted in the library rather than in hospitals or anatomical theatres. One of the first volumes to be published recording this kind of pathology was by Théophile Bonet (1620–89), physician to the Duc de Longueville of Neufchatel. Bonet was obsessed with practical medicine, and 'his concern with studying the course of diseases and their causes made him so good at prognostics and so successful in his cures, that no reputation was ever more solidly based than his was'. After forty years of practice and collecting material Bonet eventually 'retired to his study with his papers' and spent ten or twelve years turning them into books, as an eighteenth-century chronicler of medical men reported.[74] One of the first of these was his *Forerunner of a practical anatomy, or on the hidden causes of diseases, revealed from the dissection of cadavers; the first part of the first book, on illnesses of the head evident from opening it.* He sent an advance copy to Thomas Bartholin (son of Caspar), Professor of Anatomy at Copenhagen, in 1674.

Bartholin was a bit stunned, because he had already spent twenty years preparing a similar work, and now he had been pipped at the post by Bonet! He had first conceived the idea when on his ten-year peregrination around Europe to pursue his medical education. Bodies were freely available from the hospitals of Rome and Paris, he recalled, and even when he returned to Copenhagen he could readily acquire them, but then the supply dried up. Prompted by the sight of Bonet's book, Bartholin quickly issued an 'Advice' about building a practical anatomy from diseased bodies, rather than producing one himself. He spoke about the desirability of constructing a practical anatomy, and he paid much attention to preparations to protect the anatomist against the dangers from the odours of decaying bodies. At the end he wrote: 'I have expounded the reasons of my plan and the means by which I wanted to struggle through to a Practical Anatomy, equipped and supplied by so many of my own experiments of many years, and by those of other people from a virtually universal literary Encyclopedia'.[75] As a return, Bonet dedicated the *Forerunner*, when it appeared in 1675, to Thomas Bartholin 'prince of anatomists'.

Meanwhile Bonet continued hard at work in his study with his scissors and paste. In 1679 he published the work that the forerunner had preceded. It is the

[74] Eloy, *Dictionnaire historique de la médecine ancienne et moderne*, 4 vols, 1778, vol. 1, p. 402.

[75] Bartholin, *De anatome practica*, 1674; quotation from p. 39.

work that, as we shall see in due course, inadvertently prompted into action the man who was to be the greatest anatomical pathologist of the eighteenth century.[76] The book in question had the unappealing title of *The grave-yard; or, Practical anatomy from cadavers dead from disease. Setting out histories and observations of virtually all illnesses of the human body, and revealing their hidden causes.*[77] The frontispiece of the book shows Bonet himself, and the title-page shows him at a post-mortem. Anatomists had been frequently portrayed in dissection scenes as *reading* a book by some authority, whether it be Galen, Mondinus, or someone else, while looking at a body or some anatomical preparation. Here, by contrast, the anatomist is portrayed, or portrays himself, as having put down his knife and taken up his pen, in the act of *writing* a book (Fig. 3.10).

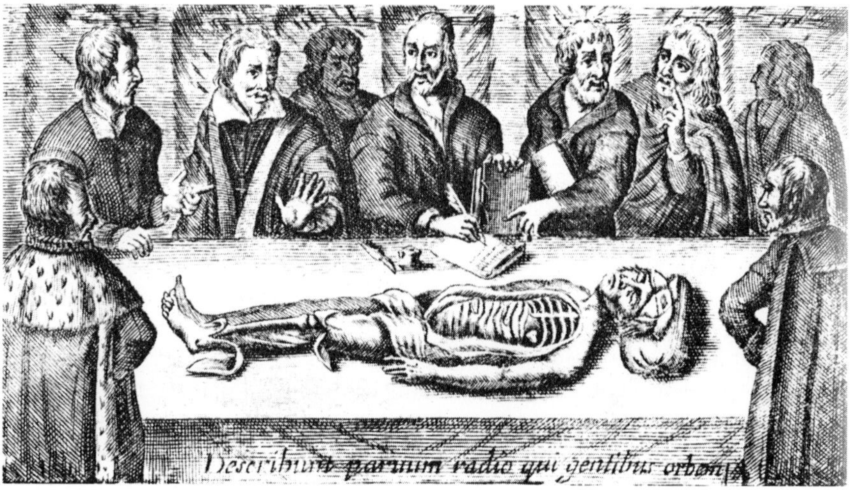

Fig. 3.10

Books in those days sometimes (like today) got advance reviews from eminent and famous people to quote in order to encourage sales. Bonet had lined up many eminent academic physicians to say what a good idea his *Grave-yard* was and how well he had fulfilled his objectives. One of them was a certain Johann Conrad Peyer of Basle, who wrote:

> Your work of Anatomico-Medical histories is so valuable both for its novelty and its utility that it cannot but be sought after by all those whose business is medicine. Nothing more suitable could be desired for the perfecting of our art; for by such pathologic accounts [*historiae*] there is greater capacity for diagnosing

[76] The following section on Bonet through to Morgagni is based on Cunningham, 'Pathology and the case-history', 1996.

[77] Bonet, *Sepulchretum sive anatomia practica*, 1679.

> the parts affected and the causes of diseases than by the bare precepts and dubious rules of so many systems that, in not deploying examples, are like a dish without salt. In natural philosophy today nothing is considered to be true that cannot be perceived by the senses or demonstrated by experiment. If medicine were taught by this method, much of the certainty and solidity in it would be evident, and its detractors would no longer ridicule it as conjectural and more uncertain than the wheel of fortune. I always come back from the dissection of some body or from the reading of anatomical accounts wiser and more observant, so I think those people who consider dissection worthless to be completely blind.

What was Bonet's *Grave-yard* actually like? Did it really take nothing to be true 'that cannot be perceived by the senses or demonstrated by experiment'? Was it based on personal experience of anatomical dissection, and if so was it Bonet's own experience? In 1682 he was to publish an enormous work on practical medicine, the *Mercurius Compitalitius, or Medico-Practical index*, assembled from his very wide reading in medical writers, and arranged by alphabetical order of disease names so that, as he said, one could find the disease immediately (supposing one knew its name in the first instance, of course). In this, each disease is dealt with in a complete way, and all the alternatives that different writers had suggested for cure were listed without discrimination. A few of the instances were taken, he shyly says, from his own practice – 'The observations that have no author's name affixt to them, crept hither out of my own diary'[78] – but the vast bulk of them were from other writers.

In the *Grave-yard* Bonet had used this same method of building virtually everything on his reading of other medical writers. What Bonet had done in this instance was to cull from as many medical writers as he could find – which was very many – every account of every illness he could find where there had been some post-mortem investigation and some written record of it. To call the accounts he used 'reports' would make his examples sound far more systematic than they were. It is said that he gathered over three thousand such records.[79] He took 'Examples' of each of these post-mortems, and printed them at length. He then added a Scholium to each such illness, in which he included whatever other medical writers had written about the cause and the cure of this illness, and glossed anything of particular interest. This produced a very large work, running to two (and then three) volumes. The arrangement is different from the alphabetical sequence of *Mercurius Compitalitius*, for in the *Grave-yard* he deals with disease from head to toe, which was the traditional (indeed, the Galenic) sequence for discussing the diseases of the parts. Each disease is dealt with simultaneously by description and by cause, in a series of 'Observations'.

[78] Bonet, *Mercurius Compitalitius*, 1684, Author's Preface.

[79] Foster, *A short history of clinical pathology*, 1961, p. 5.

Bonet's technique, which was that of showing by example, can best be shown by example. I choose one of the very shortest ones, from very early in the first book, which is, of course, on the head.

> OBSERVATION III
>
> Pain of the head from overfull-ness of the meninges; a mass of fluid in the ventricles
>
> An example [*Exemplum*]
>
> 'Here we have a 35 year old tailor with a warm and bilious temperament, *easily enraged*, living quite temperately; in the year 1670 he fell into a grave, compressive and searing illness of the bone of the forehead … the pain was without fever, the *pulse* had a hardness *similar to melancholics*, otherwise healthy in the whole body. Thenceforth sleeplessness, deliriums at intervals, sleep for 3 to 4 hours at a time, so profound that if someone lacerated his whole body he scarcely woke him; when he finally awoke he felt no remission of the pain; he yielded to his fate after a space of 14 days without a convulsion. I had predicted to those present, using language suitable for common understanding, that a cold abscess was present.
>
> I open the head in the dead man; see: the vessels in the meninges are turgid … by exploring the inside, in the front ventricles there are eight ounces of yellow fluid, the choroid plexus somewhat white, otherwise nothing remarkable.' F. Illmerus, physician in Wartenberg, in his *Observations on miscellaneous curiosities for the year 1670.*

This, as can be seen, is a real case-history and post-mortem: the case of an individual patient, and one that really happened. In this instance Bonet gives just the one case-history, but with other illnesses he sometimes gives several. Then follows the 'Scholium'. In this instance it is taken from the same source as the Example itself, but usually Bonet scours other writers for material for the Scholia.

> I judge that either a lymphatic vessel or some capillary blood vessel has broken, on account of the excess anger in which the ebullient humours, propelled in too great quantity, have distended the smallest vessels, and by affecting a small area have broken through the vessels; whence the small transudations of fluid, by filling, distending and pressing on the particular vessels and cavities of the brain, have caused these symptoms. For cure I bled from the foot, arm and hand, applied vesicatories, leeches, cupping-glasses to the sides of the head, gave purgatories, &c. I gave suporifics, but all in vain. If anyone is interested and reads Hippocrates *De morbis* [*On diseases*], he will find several pains in the head, but will find no text that can be applied as a whole, for his patients either

> die between 2 and 4 days, or a fever occurs, which in our case was not observed. But if anyone wants to work by analogy [from the Hippocratic cases] I would not object.

Bonet then glosses phrases he had put into italics, and other interesting points:

> *Easily enraged*] In the next section Illmerus makes the generalisation that he found the vessels of the meninges turgid with blood most often in convulsives and irate people; he especially remarks on the ire of this tailor.
>
> *Pulse similar to melancholics*] On the pulse of melancholics, see Ercole Sassonia, *On pulses*, book 3, c. 38 & 100.

It can be seen from this instance that Bonet's concern is not with using the instances he gives to draw wider conclusions about the relation of anatomical structure to particular diseases, but simply with facilitating the work of the practical physician. This is what happened in this case, he says: does it match the case with which you, the practical physician, are confronted? If so, here is the outcome of this case, and here are the means of cure that were tried. The whole work certainly conforms at one level to what Peyer said about it: 'In natural philosophy today nothing is considered to be true that cannot be perceived by the senses or demonstrated by experiment', for it is built on the experience (*experimentum*) and the observations of reliable people – though not of Bonet himself; it depends on anatomising people who have died of disease; and it is devoted to very practical ends. But, on the other hand, it is simply a compilation, and no grand conclusions are drawn from it. It is typical of the general approach to pathology before the eighteenth century, for it deals in individual cases, working from occasional post-mortems, and does not deal in a systematic way with what is common in the anatomical circumstances of many people with the same disease.

This tradition of compilations of pathological cases was nevertheless still frequently followed by eighteenth-century anatomists and, to be frank, constitutes the greatest quantity of pathological writing, even after anatomists had been shown a better way by some of their number (as we shall see below). The great Haller himself, for instance, produced seven large Latin volumes of *Disputations contributing towards the history and cure of diseases* (Lausanne, 1757–60), in which he reprinted some two hundred dissertations and university theses from the 1660s to his own day, from many of the medical faculties of Europe. All these came from his massive personal collection, and he selected them because to him they showed the Sydenhamian or Hippocratic spirit that he was looking for. 'I want a picture of the disease, created at the bedsides of the sufferers, so drawn by a faithful pen that he who sees the picture cannot fail to recognise the disease', Haller wrote in introducing the series; 'I want cures that combine these two virtues: viz. that they are not commonplace nor taken rashly from too few examples, but confirmed by long experience'.[80] The volumes follow the antique sequence of listing diseases from head to toe. Some of the dissertations are very up to date, such as the cures by electricity, or the accounts of scurvy, leprosy or rickets, or that of spina bifida, here presented at Vienna by Ignatius Gild in 1754 (Fig. 3.11). Many of these dissertations were built on anatomical dissection. But for the most part – and this is a consequence of how and why they were created – these pathological dissertations deal with single instances of morbid conditions, or with the very rarest of cases, or with extremely limited trials of particular medicaments. Thus, despite the several thousands of pages in these volumes, we do not have here the systematic approach necessary for building a consequential pathology.

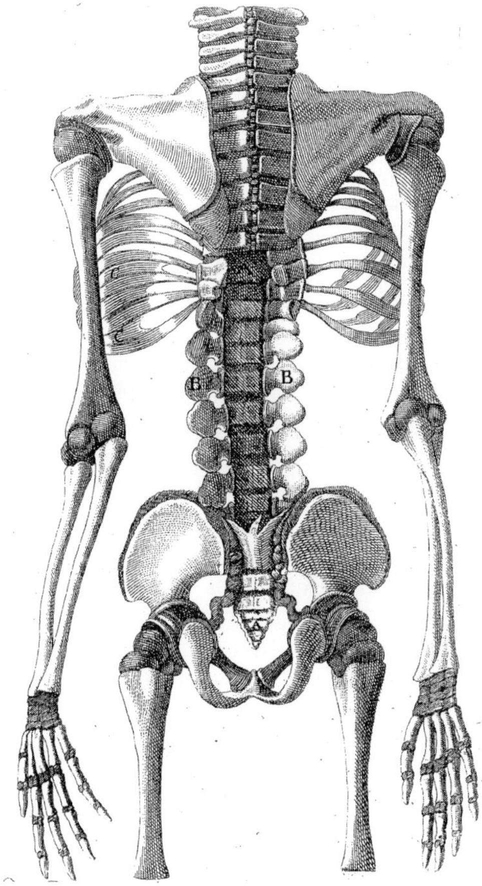

Fig. 3.11

[80] Haller, *Disputationes ad morborum historiam et curationem facientes*, 7 vols, 1757–60, vol. 1, p. i.

This limitation of pathological investigation to single cases can also be seen in the case of the highly ambitious young French anatomist Antoine Portal, who edited in 1767 two large volumes of the work of Joseph Lieutaud: *An anatomico-medical history, consisting of very numerous divinations [extispicia] of human cadavers, by which the genuine seat of diseases comes into view, and their causes are revealed and their effects are visible.*[81] This too owed a great deal to Bonet's *Grave-yard*. It began with the internal lesions of the abdomen, moving on in book two to internal lesions of the chest, in book three to lesions of the brain, and in the fourth to external wounds. In other words it followed the order of dissection, rather than the head-to-toe tradition. The text consists of about three thousand short case-histories or 'Observations' – every single one of which is taken from the writings of other medical men! Despite Portal's protestations in his Proemium about the virtues of making observation primary in the investigation of disease, it is not his own or Lieutaud's own observation, and the book was compiled in libraries and not at the bedside or in the post-mortem room.

Pathology in the hospital and anatomy theatre

The long tradition of making post-mortems on important persons to see whether they had been poisoned, or on interesting cases because of their rarity, must now be distinguished from a new series of approaches to pathology in the eighteenth century: the *systematic* investigation of *many* corpses over *many* years, with *comparison* of the results with each other. The advantage of *systematic* investigation in general is that only systematic knowledge builds into general knowledge. And the advantage of *systematising* one's information is that only if the information obtained is arranged in some meaningful way will it enable one to elicit consistent relations of *cause and effect*. So one-off post-mortems are not systematic, and long lists of corpses and their inner lesions produce no knowledge about cause and effect, and are never more than lists, as the example of Bonet and his *Grave-yard*, and even of the great Haller, show.

To move from the occasional post-mortem to a systematic view of anatomical pathology requires and required three things. The first was regular and frequent access for anatomists to numerous corpses dead of disease. But this is a necessary condition, not a sufficient one. For access to the bodies of the poor who died in the hundreds of hospitals of medieval and early modern Europe had been notionally available to anatomists for centuries – had they wanted it. So the second requirement was a desire on the part of some anatomists to build a systematic pathology based on anatomy. And this in itself assumes the third necessary thing: a view that most if not all diseases are anatomically based, that diseases have what people in the eighteenth century called 'anatomical *seats*', and that disease was in some way a product of some defect or failure in the *structure* of organs and

[81] Lieutaud, *Historia anatomico-medica*, 1767. The 'divinations' is a classical reference to the use of dissections of animals to foretell the future.

other parts, which would in turn affect the *connections and flow* of liquids to and from organs and other parts. (The view that disease is primarily a phenomenon not of organs but of their constituents, and particularly of tissues and cells, is a nineteenth-century one).

It will readily be appreciated that this view of diseases as having 'seats' is a quintessentially anatomic concept, depending as it does on all those things that the anatomist was skilled in doing: recognising structures, connections and flows. But even here there was something special, something new. For Galenists had long realised (for instance) that the liver has a distinctive structure that can be analysed in detail by the anatomist. They had also recognised that the liver might be subject to lesions. But for them the liver performed its functions not simply by virtue of its structure (though that was relevant), but mainly through its 'faculty'. In the case of the liver this was its 'blood-making faculty', something whose workings lay in the very special material of which the liver was made, and which could not be reduced to anatomical structure, nor simply be made subject to anatomical analysis. Thus a defect in the structure of the organ did not in itself mean a defect in its action, and similarly a defect in its action did not necessarily entail a defect in its physical structure.

Moreover, in the Galenic and Hippocratic medical traditions, which persisted in use until at least the mid-seventeenth century, it would be true to say that pathological interpretations had tended to concentrate on morbid conditions of the *fluids* – the blood and those highly active but anatomically elusive 'humours' and 'spirits' – rather than on the *solids*, the tangible and non-flowing material that made up the organs of the body.

That extra something then, which for the anatomists made pathology *purely* an anatomical matter, seems to have been the near-complete success of the mechanistic view of the body promoted in the seventeenth century first by René Descartes in his *Discourse on method* (1637) and his posthumous *On man* (1662), and later by followers of Isaac Newton and others, and for anatomists most particularly by Marcello Malpighi, the originator and co-ordinator of the Italian approach to pathology (on which we shall hear more below).[82] It was all put rather well in 1693 by Malpighi himself:

> Since the limits of anatomy are now so well established and mechanical philosophy has been linked to medicine, and since the seats of disease, especially chronic disease, have become known by the frequent dissection of cadavers, the animal œconomy, both in health and disease, stands revealed to such an extent that it is now possible for medicine to be restored and strengthened both a priori and a posteriori. Practice can therefore be enriched by collecting accounts of

[82] On which see the excellent thesis: Brown, *The mechanical philosophy and the 'Animal œconomy'*, 1981.

> post-mortem examinations, by the mechanical examination of fluids, and by the addition of remedies.[83]

Thus, as a modern historian of pathology has written, 'only with the advent of frequent and systematic dissection of the human body performed for the purposes of localizing disease can we find what counts as a meaningful pathological tradition. This phenomenon … was primarily an outgrowth of eighteenth- and nineteenth-century modernization … in the second half of the eighteenth century, in many places – certainly, and to varying degrees, France, Germany, the Baltics, Russia, Italy, Britain and America – surgeons and physicians were already keenly focused on morbid anatomy as a way of correlating post-mortem findings with ante-mortem diagnosis'.[84]

By 1700 such mechanistic views of the working of the body were already very widespread if not universally held among medical men. There were some celebrated exceptions – men who believed that the phenomena of life could only be explained in terms of life, not mechanism, such as the rival Pietist medical teachers at the University of Halle, Georg Ernst Stahl (1659–1734) and Friedrich Hoffmann (1660–1742). Their respective physiological systems, though radically different from each other, had no use for anatomical pathology. The one believed the soul explained everything, the other that Cartesian mechanism indicated that any pathological problems were at a level below the visible.[85]

If we were to look for some turning-point after which the majority of anatomists began to see anatomical pathology of one kind or another as part of their disciplinary roles, then we probably need to place it with Hermann Boerhaave, the most important medical teacher in the whole of Europe in the early decades of the eighteenth century. Boerhaave managed to take on board and to teach an anatomy and physiology that embraced both the older pathology of fluids, together with the Malpighian and the Newton-influenced pathology of solids of more recent days.[86]

In medical thinking about pathology in the eighteenth century there was another pressure that made the seats of diseases important. This was the passion for fashioning *nosologies*. A nosology is any classification of diseases, usually presented in tabulated form and drawn up by an individual or organisation, established on the basis of some medical principles or other, with the diseases arranged usually according to some theory about cause, and where that is not possible or desirable, by some other supposedly meaningful classification, such as

---

83 Malpighi to Jean Antoine Chouet, Geneva publisher, 1693; Schullian, ed., *The Baglivi correspondence*, 1974, pp. 114–15 (translation by Schullian, modified).

84 Maulitz, *Morbid appearances*, 1987, pp. 170, 172.

85 French, 'Sickness and the soul', 1990; Geyer-Kordesch, 'Georg Ernst Stahl's radical Pietist medicine', 1990.

86 Cunningham, 'Medicine to calm the mind', 1990; Knoeff, *Herman Boerhaave (1668–1738): Calvinist Chemist and Physician*, 2002; Lindeboom, *Herman Boerhaave*, 1968; Kegel-Brinkgreve and Luyendijk-Elshout, eds, *Boerhaave's orations*, 1983.

anatomical location in the body. The term 'nosology' was popularised in 1763 by Boissier de Sauvages for his tables of diseases,[87] and the term was rapidly adopted by other medical teachers in the century. It is still in use, though the expression 'classification of disease' is more common today. Nosologies were not usually very stable. If someone came along who had a different view of the range of possible diseases, of the relative importance of different diseases, of the natural relations of diseases, or of the causal relationships of different diseases, then he could simply draw up his own nosology if he felt strongly enough about it. So we find many different such nosologies created in the latter eighteenth century, and anatomical location is important in most of them.[88]

What things should the intending pathological anatomist take systematic note of in order to make his histories worthwhile and valuable to other people in the long term? Johann Conrad Peyer, whom we met just now as a 'puffer' of Bonet's *Grave-yard*, had, just before doing that, drawn up and published a suitable checklist himself in his *Method of anatomico-medical histories*, first published in 1678 in Paris, and later to be taken very seriously by the greatest pathological anatomist of the age.[89] Peyer's little work was dedicated to Joseph Duverney, Anatomist in the Royal Academy of Sciences in Paris. 'It was not long ago', Peyer wrote to Duverney, recalling a recent visit to Paris, 'that we had a great talk about anatomy as we walked in the Jardin royal. I was astonished and I then lamented that in a town so well provided with hospitals there were so many cadavers everyday with no-one dissecting them in order that the cause of death might be discovered' (p. 61). What was needed was a series of anatomical–medical histories. Peyer said to his reader: 'An anatomical–medical history is a narrative expressed in speech or writing of the noteworthy things concerning health in the human body during life, and of the things occurring and to be observed after death', while a 'method', said Peyer, 'is an art of appropriately gathering together such histories, and of arranging them in order' (pp. 67, 68).

> For this is the only way to a true pathology, without which there is only error and the absurdities of opinions … What, I ask, is more evident than that the knowledge of diseases is to be acquired from dead bodies, just as natural philosophy gets the knowledge of things from nature itself, not just from books? (p. 64–5)

Basically what he recommends is that the individual should be observed in the healthy state: what is typical of him or her, what is their diet, etc. Then an account should be given of the onset, progress and outcome of the disease. And if death

---

87 This was a development of his *Nouvelles classes des maladies*, first published in 1731. See Martin, 'Sauvages's nosology', 1990.

88 DeLacy, 'Nosology, mortality, and disease theory', 1999. For an example, containing several different nosologies, see Cullen, *Synopsis nosologiae methodicae*, 1775.

89 Peyer, 'Methodus historiarum anatomico-medicarum', 1682 [1678].

results, then dissection is necessary. 'It is a good thing to have bodies dead of every kind of disease to dissect and describe, and to have lots of them. For this is the only way to gain solid knowledge of diseases, better than six hundred verbose "systems" of the art' (p. 75). Finally an exegesis or comment on the case should be made at leisure.

To illustrate all this Peyer gives an extensive case from his own experience of a patient dying from dropsy arising from the heart. I now abbreviate greatly. First comes the case-history. A young man in Basle, Nicolaus Schweiger, a boy of good but very poor parents, who therefore knew only very vile food and hard labour, always had difficulty in breathing. Then at the age of 15 different parts of his body began to swell up: first his scrotum, then his legs and feet; then his left arm swelled and went down; and his umbilicus and whole abdomen puffed up. His pulse was weak, his breathing difficult. He urinated with great pain and difficulty, since his penis was both swollen up and also twisted by a tight frenum (the frenum, or 'bridle', being the ligament that attaches the inside of the foreskin to the glans). Thus far the description of the illness.

At last the poor boy died in agony on 14 May 1674. Now comes the dissection. 'At night the body was dissected by Henrick Glaser, then professor of anatomy, with me and others present and helping' (p. 83). Removing the coverings of the abdomen, they found serum everywhere, like gelatin. All the internal organs of the lower belly seemed in good order. On removing the sternum the lungs were found to be adhering to the diaphragm and pericardium. 'The heart was pallid and flaccid and distended with much clotted blood, and with twin polyps' (p. 84). There was no time to examine the brain.

Finally, the scholium or exegesis. It is clear, writes Peyer, that the heart – the most important organ of the body – of the young man had struggled since birth, and his breathing had been obstructed since he was a boy. Cases where the lungs were adherent or the heart pallid are sought in earlier writers. Bartholin, Willis, Harvey, Lower are consulted and their observations compared to the case at hand. Our own experience is also valuable, Peyer says, and he discusses other recent dissections of Basle people where similar or comparable phenomena appeared.

One positive function that the scholium has – though Peyer does not discuss this – is to keep the facts elicited in the dissection separate from the reasoning built on them. Nor does Peyer make any play with the fact that the crucial underlying assumption here is that the lesion discovered in the body at postmortem is the *cause*, not the *effect*, of the illness.

Another lone pathologist of a similar kind is Stephen Blancard, who published his little *Anatomia practica rationalis*, or *The practical rational anatomical inspection of rare corpses dead from diseases*, in Amsterdam in 1688. Here the frontispiece indicates that the dissections of people dead from disease were carried out in the hospital itself (see Fig. 2.7 in Chapter 2 above). Blancard gives a hundred short cases, usually including a very brief history of the disease, an equally brief 'anatomical inspection', and a slightly longer etiology of the disease, that is, an account of its cause.

We saw above how Albrecht von Haller engaged in scissors-and-paste pathology. But he had also earlier been heavily involved, at least for a four-year period from 1749–53, with hospital- and anatomical theatre-based pathology. His observations here were in his own eyes typically Sydenhamian in that they 'faithfully related what I myself saw, in as plain and concise a manner as possible'.[90] Dozens of bodies were brought to his anatomical theatre at Göttingen (see Fig. 2.2, Chapter 2 above), both criminals who had been hanged and were to be publicly dissected, and suicides, and a wide range of other corpses on which he was able to perform both private dissections and post-mortems. We saw something of his generous supply of corpses in Chapter 2. These included aborted fetuses; still-born babies; babies who had died within hours, days or weeks of birth; mothers who had died in or shortly after childbirth (on at least one occasion he was cutting up a woman within half-an-hour of her death in childbirth); children; and men and women of all ages and walks of life, from noble families to beggars found dead in the street and poor people who had died in the local hospital.

With his researcher's eye Haller was always looking for what could be learnt from pathological findings towards improving medical practice. Yet even with these large numbers of bodies available to him, Haller could only talk of single cases or of one or two histories of a particular disease. To illustrate the nature of Haller's pathological observations and their limitations I choose a very sad case for which there were actually two histories, and my reasons for choosing it will become obvious.

> Observation 14: A Peripneumony from a transudation of blood. History II.
>
> Since [the case in History I] I have met with a still more melancholy instance of the same disease, in a strong manly boy of seven years of age, who was slightly indisposed for about eight days with frequent bleedings at the nose. His face was swelled and pale, and towards night he used to complain of a chillness without any fever. In other respects he was healthy enough. At length, on a Sunday after supper, he was seized with a difficulty of breathing, and next morning he was not able to remain any time out of bed. He felt an uncommon heat, and the same difficulty of breathing as before. I advised bleeding immediately; but another physician, who was called in, apprehending it to be an intermitting autumnal fever, or that the disease might be owing to worms, gave his opinion for deferring it. A clyster was given him, and he vomited twice spontaneously. But on Tuesday the case discovered itself too plainly, and what he spit was round, thick, and peripneumonic, with a mixture of blood. Upon opening a vein, the blood appeared fizy and pleuritic; he became light-headed, and would hardly take any medicines. After the bleeding was twice repeated, without giving any relief, he had oxymel of squills given him; but the anxiety increased upon him,

90 Haller, *Disputationes ad morborum historiam et curationem facientes*, 7 vols, 1757–60, vol. 1, p. i.

> he grew gradually cold, weak, and unable to expectorate, and died at the end of two days after the disease had declared itself. – Farewell, my dear child, and expect thy father to follow thee! – In order to be convinced of the cause of this cruel disease, I had the body opened by persons well versed in dissections, who told me that the left lobe of the lungs was full of thick blood, and sank [rather than floated] when put in water; that the left ventricle of the heart was empty, and the right, together with its auricle, distended with blood; that a polypus adhered to the mouth of the Aorta; and that there were evident signs of an inflammation of the Pleura. Here also appeared that cruel effusion of blood all over the cellular substance of the lungs, as appeared in the preceding case; and hence we see that in children, true mortal Peripneumonies are also produced.[91]

The former history appears to have been that of Haller's first son, but the poor father, in losing both his children to the same disease, is nevertheless able to come to a conclusion about the incidence of this particular disease in children.

In addition to making observations on one-off cases, Haller clearly had some systematic method too, for he speaks of certain observations as being 'taken from my diary on the common diseases of the lungs, hoping that they will throw some light on the aetiological controversies relating to the pleurisy and peripneumony' (p. 22). It may be that he had diaries on the common diseases of other parts of the body. But nevertheless even with his own extensive medical and anatomical practice and his unremitting energy, he was simply not seeing enough cases over a long enough period to draw major conclusions about either the incidence or the anatomical cause of diseases.

There are several other eighteenth-century anatomists who sought single-handedly to build the basics of a pathological discipline from their own experience, as for instance John Hunter in London or the Monros in Edinburgh. In Amsterdam the anatomist and supervisor of midwives Frederick Ruysch in 1690 published *A hundred anatomico-surgical observations*, hoping thereby to contribute his experience of practice and dissection to the understanding of disease, showing 'the things that I have been able to observe through twenty-three years (in which in this vast and populous town I have not only practised medicine, but I have also carried out very frequent dissections of bodies both privately in our hospital, and publicly in the Anatomical Theatre)'.[92] But his hundred observations, of prolapsed wombs, ulcerous tongues, and worms in the intestines, were and remained single cases. Nothing could be built on them.

---

[91] Haller, *Pathological observations*, 1756, pp. 26–8. This is a translation of his *Opuscula pathologica*, 1754. These cases were again published in his *Opera minora*, 1763, vol. 3, with others.

[92] Ruysch, *Observationum anatomico-chirurgicarum centuria*, 1691, Lectori benevolo.

But there was also one group of anatomists who did try to develop pathology as a collaborative activity. They were a group of outstanding anatomists in Italy, inspired by the example and work of Marcello Malpighi.

A number of Malpighi's personal records of post-mortem dissections that he performed on patients survive in manuscript, running from 1666 to 1693, mostly of people dying in Bologna.[93] Some of them are of named members of the church hierarchy or nobles. Others are of poor people who died in one or other of the Bologna hospitals. The age at death varied from six to over sixty. Here is an instance:

> On the 5th of December 1684 there died signor Giovanni Battista Gornia, my dearest friend, Professor Ordinarius at Pisa, and personal physician to the Most Serene Grand-duke of Tuscany. He used to labour under a melancholic hypochondria, with frequent suppression of urine, vomiting, cold at his extremities and around the bladder. He was taken with an acute fever, after excessive urination, lasting a month, and a wasting-away of his whole body. Pain in the head was very great, the pulse urgent and tense, a burning of the eyes, thirst, unable to sleep, bloody urine, sharp and migrating pain throughout the whole body. After venesection everything calmed down. Towards the fourth day blood poured from his nose, until the eighth day, when all the symptoms became less and the pulse weak and infrequent. Gradually there was difficulty in breathing with very small pulse and the internal senses were affected. Thus he died on the twelfth day.
>
> In the dissection the left kidney was found flaccid and eroded, with blood clots in the pelvis. The pancreas hard. The stomach contracted like an intestine. Urinary bladder brawny internally, its neck calloused and hard. In the right ventricle of the heart a polyp the size of a thumb, with its tail prolonged towards the lung. In the left lobe of the lung there was seen a substance hard and with the colour of liver. When this part was cut a transparent serum flowed out.

Among the significant points for us to note here are firstly, that Malpighi is interested in the general health of the person, and especially in the illness and circumstances of death. His other observations all concern variations from the norm, as discovered in normal anatomy. Equally, however, we need to note that Malpighi makes no suggestion about cause and effect with respect to the patient. It is possible that Malpighi intended to assemble many such autopsies and compare and contrast them systematically, but he did not do so. Finally, it is striking that Malpighi, the master of the microscope, does not use the microscope in his pathological anatomies.

---

93 Some of the Latin originals have been transcribed from a Bologna manuscript in Weber, *Aspetti poco noti della storia dell'anatomia patologica tra '600 e '700*, 1997; some of them have been translated into Italian in Malpighi, 'Storie anatomico-mediche', 1967. The example is taken from p. 78 (Latin), p. 438 (Italian) – the translation is mine.

However, it is clear that the followers of Malpighi did indeed build on his modest example what amounts to a long-term and systematic programme in pathology. There were three great individual stars of this approach – Giovanni Maria Lancisi (1654–1720), Antonio Maria Valsalva (1666–1723) and Giambattista Morgagni (1682–1771) – but there were many other active pathologically inclined anatomists in the 'team', perhaps a dozen or so at any one time in the great Italian cities, who sometimes appear as side-actors in the writings of these three and of others. They were seeking to link diagnosis of a case to the origin (or cause) of the illness, and eventually to treatment. They could open up the bodies of people dead of a particular lesion of the inner organs 'no longer guided by a questionable mental image of it but by our actual senses, the anatomical knife leading the way'.[94]

Several things were crucial to and distinctive of the work of this whole school of anatomo-pathologists. The first was that they conducted or were present at *many dissections* of bodies dead of disease, both the poor in the hospitals and the rich in their homes. Some of them were attending one or more such dissections every week. Second, they worked together over *many years*, unhurriedly investigating the many freely available corpses. Third, they *kept notes*. Not brief notes, but very extensive ones; not for a while, but over the course of many years. Eventually their records of the hundreds of dissections they had conducted together went back over more than half a century. Fourth, they *worked together*, always dissecting with one or more colleagues whenever possible, inviting each other along to what promised to be specially interesting cases, sharing the work of dissection, and sharing their thoughts about the cause and course of the diseases. One of the advantages of this way of working together was that they shared a trust in each other's findings, and a common (if tacit) standard of factuality about what they found. All the elements for a systematic anatomy-based pathology are there and being deployed.

We can take Lancisi as our first example of such a 'Malpighian' pathological anatomist. Lancisi was a Roman who had acquired his education there first with the Jesuits and then at the papal university, the Sapienza. He was educated in geometry as well as in medicine. Even before he obtained his doctorate he was a regular visitor to the hospitals of Rome where he examined 'continually and with the utmost attention the cadavers dissected',[95] inspired, he claimed, by the example of Eustachius. When he became assistant physician in the San Spirito Hospital in Rome in 1676 he 'constantly attended at the bedsides of the patients, carefully noted their signs and symptoms, used his skill to try to explore the causes of their diseases, and was present to observe the issue [i.e. postmortem]. Since this was all written down with great exactness, he completed many volumes of case histories of diseases and for the rest of his life kept improving them and thus gathered together a great acquisition for the benefit of medicine'.[96] He followed

94 Lancisi, *De aneurysmatibus*, 1952, p. 99.

95 Life of Lancisi by Assalto, as translated in Lancisi, *De aneurysmatibus*, 1952, p. xiii.

96 Ibid., pp. xiii–xv.

this self-education with five years of reading the medical classics and taking notes. From 1684 for thirteen years Lancisi was Professor of Anatomy at the Sapienza. Thereafter he became physician to three popes.

The subject that Lancisi chose early as his, as it were, 'speciality' was the causes and characteristics of aneurysms, especially as a cause of death. An aneurysm is a ballooning of an artery at some point, for whatever reason, sometimes called a tumour of an artery. The artery balloons as a response to the pressure of the blood in it because at that point it is not as strong or thick as usual. In severe cases the aneurysm will burst, and in the case of the aorta – the blood vessel into which all the arterial blood is pumped out of the left ventricle of the heart at the beginning of its passage through the body – this can lead to immediate death. It is only because Lancisi and others were now dissecting so many dead poor people from the hospitals, and the occasional rich former patient, that they began to see the significant incidence of aneurysm of the aorta as cause of death. The question that arises for the anatomical pathologist from such observations is: how and why do the arteries come to suffer from this condition? The answers that Lancisi eventually gave were all equally anatomical: they were first, that anatomical inspection reveals that the walls of the arteries consist of four layers of fibres, and second, that in cases of aneurysm these fibres can be seen to have worn away or been dilated either by some blow from outside, or by the pressure of the blood coming from the heart (compression, spasm and erosion). Lancisi lectured on this theme as a professor, and completed his book shortly before he died in 1720, *On the motion of the heart and on aneurysms* (published in 1728). The value to him of his early study of geometry is shown in the way in which he applies geometrical reasoning to assessing the force of the blood and the necessary effects of that force on the blood vessels, especially the aorta. Lancisi was especially interested in developing ways to diagnose and treat aneurysms for the benefit of 'the surgeon, who carries the torch of anatomy in his hand'.[97]

Much of Lancisi's book is devoted to case-histories, 'since instruction illustrated by examples sinks more deeply into the mind than that which relies solely on reasoning'.[98] One such example is that of a 37-year-old man whose condition was brought on by the stooped posture that comes from constant letter-writing, together with most immoderate drinking. The man refused to mend his ways when he suffered from a dyspnoea (breathlessness) and palpation of the heart. Mental worry at last led him to consult his physician Antonio Pacchioni, who, Lancisi writes,

> … had attended with me other patients afflicted with this kind of illness. He observed that the man was breathing with difficulty but could not sit up, for in an upright posture he became much worse, but was now lying prone and more often on the right side, with violent palpitation of the jugular veins and carotid

[97] Lancisi, *De aneurysmatibus*, 1952, pp. 102–3.

[98] Ibid., p. 73.

> arteries, also the pulse was rapid and irregular. Pacchioni at once suspected that there was increase in the bulk of the heart and that it had sunk down. So he consulted me immediately and told me what had happened, and I encouraged him to hope that, when the man was dead, the opinion that he was turning over in his mind [i.e. that the heart was hypertrophied] would be found to be the actual fact. At last, on January 2, 1717, the patient died, and on the following day the cadaver was dissected. There were present the distinguished physicians Antonio Pacchioni and Francesco Soldati, while I for my part examined the viscera with especial care.[99]

The dissection, conducted by the surgeon Sebastiano Tamanti, showed that the patient did indeed have an enlarged heart, 'larger than any known ox heart' (p. 223), which was impeding his respiration, and a polyp in the coronary vein. Lancisi draws from all this the conclusion that such illnesses are brought on by rich food, a sedentary life and mental distress, together with a defect of structure in the coronary vessels.

What is typical about this event among such keen anatomical pathologists is the way in which they work together. Not only do two physicians share the diagnosis and prognosis, but there are four people present and participating at the dissection of the cadaver. One of them, Lancisi, notes their findings in great detail, and later in life compares them with all the similar cases of which he has records. It would seem that these pathologists rarely had a problem in gaining consent to dissect the cadavers of their former patients, for Lancisi records one exceptional event where 'an autopsy of the cadaver was omitted because of the old-womanish scruples of the relatives and the lack of this is the reason why I am obliged to speculate like a philosopher without direct evidence'.[100]

The most important pathologist of this new kind was Giambattista Morgagni (1682–1771), not only for his pathological work, but because he effectively also wrote the book of the great project. Morgagni called himself the 'grandson of Malpighi'. Morgagni was a holder – the culminating genius, some have said – of the anatomical chair at Padua, whose previous incumbents had included Vesalius and Fabricius.[101] He was Professor of Anatomy from 1715 until his death fifty-six years later. Over the course of sixty years he dissected, either for demonstration or for post-mortem purposes, at least seven hundred human corpses, mostly from the public hospitals of Padua and Bologna, which gave him an experience of normal and abnormal human anatomy that was probably unrivalled. This is, on average, about one per month throughout every one of those sixty years. 'The more opportunities hospitals are offering us to observe unusual diseases and even more the usual ones', Morgagni wrote, 'the oftener I feel sorry for the physicians of antiquity who were deprived of this advantage', as hospitals had not yet been

[99] Ibid., pp. 221–3.

[100] Ibid., p. 285.

[101] See Long, *A history of pathology*, 1965, p. 67.

founded.[102] Morgagni's book-of-the-project (as it were) was *On the seats and causes of diseases*. It is a very rich book that still awaits thorough study, either with respect to Morgagni and his career, or as a window into eighteenth-century Italian medicine and disease.[103]

Morgagni was a student at Bologna University and graduated in philosophy and medicine in 1701. At the university and in the intellectual circles around it Morgagni found a set of heroes whose influence on him was immense. At the age of 17, in 1699, while still a student at the university, Morgagni was admitted to the Accademia degli Inquieti. The Inquieti was one of many groups, mostly short-lived, that gathered together to pursue the 'new philosophy'. In Bologna it followed in the line begun with the Coro Anatomico (1650–?60), in which Malpighi had first become attracted to the study of anatomy. The Inquieti was founded in 1690 by Eustachio Manfredi.[104] In the Inquieti Morgagni's immediate hero was his teacher Antonio Maria Valsalva (1666–1723), who had himself been a pupil of Malpighi.

In one of his autobiographical fragments written in the third person, Morgagni recalled of himself:

> He continued in Bologna until the beginning of 1707, applying himself to Mathematics, to modern Philosophy, to Botany, to Medicine, and especially to Anatomy, to which most useful science he always had a particular inclination from his childhood. Thus while, under the direction of many pupils of the famous Malpighi, he studied Medicine and attended all three Hospitals of Bologna, dissecting in these such an incredible number of cadavers, and assisted the late distinguished Signor Valsalva every year at the Public Anatomical Theatre; to the great labour of dissecting and of observing he in this way joined another no less labour, that of deliberating on and reading virtually all the modern and ancient [medical] Authors.[105]

It was his membership of the Inquieti that was responsible for Morgagni's first becoming a published author, and on his beloved subject of anatomy, by giving him a forum for, and encouraging him to publish, the work that first showed the world that he was a great anatomist, his six sets of *Adversaria anatomica*, published between 1706 and 1719.

---

[102] Morgagni, *The seats and causes of* diseases, 3 vols, 1769, vol. 3, p. xxxi, letter to J. F. Schreiber, translated by Paul Klemperer.

[103] The best assessment, at least in English, still seems to be Jarcho, 'Giovanni Battista Morgagni', 1948, which is now sixty years old. Valuable also is Nicolson, 'Giovanni Battista Morgagni and eighteenth-century physical examination', 1992.

[104] On the Inquieti see Rosen, 'The Academy of Sciences of the Institute of Bologna, 1690–1804', 1971; Cavazza, *Settecento inquieto*, 1990.

[105] Morgagni, *Autobiografia*, 1956, p. 1, my translation.

Then, in 1700, into this situation of emulation and concern with medicine and anatomy, there came a new edition of the *Grave-yard* of Bonet.[106] How did Morgagni respond to this work? According to his later account, when Morgagni first saw it he was full of hope about it, especially when he saw that the title-page said that it was equipped 'with the necessary indexes'.[107] But his enthusiasm for it was short-lived. For he found it omitted much it should have included, it repeated much that it did include, it confused natural with diseased appearances, and the scholia were useless. Worst of all, the indexes singularly failed to assist the user:

> … from this defect the great and primary advantage of the work is totally cut off; as this advantage could only have arisen from having a great number of similar symptoms at hand, so that you might readily compare them with many morbid appearances; which were either similar, or not similar, to each other; and by this means to be able at once to conceive, which of those symptoms are most frequently, most rarely, or never, joined with any particular species of internal morbid constitution.[108]

Nevertheless, despite these many defects in plan and execution, Bonet's *Grave-yard* obviously seemed such a good idea to him that Morgagni the young man forthwith resolved to do the task properly. He was about eighteen years old.

> I remember, likewise [he later recalled], that as young men are generally presumptuous enough to entertain thoughts of the most difficult and laborious undertakings, I did not even then despair, but if I should have sufficient leisure in future time, I should not only be able to supply the deficiencies I have pointed out in the *Sepulchretum* [*Grave-yard*], and others besides these, but also that I should be able to reform the indexes; and I even thought of a plan whereby this might be done, and communicated my plan to that respectable society, which is now called the Academy of Sciences [i.e. the Inquieti].[109]

Thus was born the plan that was to come to fruition some sixty-one years later in *De Sedibus*, his book *On the seats and causes of diseases investigated through anatomy*. Throughout this whole period Morgagni continued to be obsessed by the *Grave-yard*, and after sixty years was still forever criticising it for errors, confusions, repetitions, omissions, and the like, while still taking it as his model. In one way one could say that *Seats and causes* is both a commentary on as well as a replacement for the *Grave-yard*. It is in these ways that *Seats and causes* is a product of Morgagni's early years under the tutelage in anatomy of Valsalva, and of his association with and membership of the Inquieti.

---

106 Bonet, *Sepulchretum*, 1700.

107 Morgagni, *The seats and causes of diseases*, 3 vols, 1769, vol. 1, p. xviii.

108 Ibid., vol. 1, pp. xviii–xix.

109 Ibid., vol. 1, p. xix.

For almost all the sixty years between making this resolution to rewrite the *Grave-yard* properly, and actually fulfilling it, Morgagni was a professional anatomist. In the first few years he was often in the company of Valsalva in Bologna, assisting with the preparations for the public anatomies, and present at the many post-mortems Valsalva conducted on the patients from the three public hospitals in Bologna: first, the tiny medieval St Mary of Life, originally founded by an order of flagellant monks ('the Company of Devoted Beaten Ones') in the late 1200s, connected to a church dedicated to St Vitus; second, the enormous fifteenth-century hospital of St Mary of Death; and, third, the Hospital for Incurables (i.e. patients with the pox). Some of the dead patients that Morgagni dissected were dissected in the hospitals where they had died; others, and especially those dying in the cold winter months, would be taken into the anatomical theatre and dissected there in front of the students. In Bologna the location of the hospital of St Mary of Death was particularly convenient, because the new university buildings (the Palazzo Archiginnasio) were built immediately adjacent to it in the early seventeenth century. In the first photograph (Fig. 3.12) the edge of the cathedral is on the right, the hospital on the left, and beyond it the university.[110] An archway passage at the first floor level, seen here from the rear in Fig. 3.13, meant that dead bodies could easily and without fuss be moved from the hospital to the anatomical theatre in the university, which was at the same level.

Morgagni was also involved with other anatomists from the earliest years. For instance, he assisted Lancisi in the publication in 1714 of the *Anatomical plates* of Eustachius, from the engraved plates preserved in the papal collections (see Chapter 1 above). He attended post-mortems performed by other physicians, such as Giovanni Domenico Santorini, who was demonstrator of anatomy at Venice from 1706 to 1712. He also performed post-mortems with other young physicians. For example, he recalled a dissection at Bologna in 1704 or 1705:

> At that dissection were present the celebrated Jacob Bartholomew Beccario, now professor at Bologna, and president of the Academy of Sciences there; and that very experienced physician Heraclito Manfredi, who assisted in the dissection, and who then discharged the office of assistant physician to the hospital of St Mary de Morte, with great diligence and reputation. And I would have you understand, that these two excellent men were not only witnesses, but companions of my labours, whenever I say that histories were observed in that hospital: for it was our custom, at that time, to see the patients together, and to examine in conjunction into the latent disorders after death, both from a desire of improving our studies, and from a mutual regard to each other.[111]

---

110 Since 1881 the hospital of St Mary of Death has been the Archaeological Museum of Bologna.

111 Morgagni, *The seats and causes of diseases*, 3 vols, 1769, vol. 1, pp. 83–4.

Fig. 3.12

Similarly Morgagni could later recall sharing thirty-eight years of anatomising with Nicolas Mediavia, 'public demonstrator, and most deservedly doctor of the sacred college of physicians and philosophers … for it is so long since I began to exercise him in anatomical business; and so great is his assiduity, that he has not only been present at almost all the dissections I have performed at Padua, but also frequently, by dissecting and demonstrating with me, has given me very useful, and very laudable assistance'.[112] In 1711 Morgagni left Bologna permanently on being offered his first job at the University of Padua. In Padua he was successively the holder of the second chair of theoretical medicine, and then of the first chair of anatomy from 1715 until his death.

Morgagni gives so much detail about the pathological dissections he conducted that one could probably reconstruct the entire sequence of his dissections at Venice, Bologna and Padua. But what is most important and striking here is that Morgagni regularly performed public anatomical demonstrations – traditionally performed on the bodies of healthy criminals – on *bodies dead of disease*, even including *lues venera*, the pox. This was to turn the anatomical theatre into a pathological anatomy theatre.

Interestingly, for Morgagni, pathology is not a particularly visual subject, unlike anatomy. There is not a single illustration in the whole of the *Seats and causes of diseases*, and even the portrait frontispiece presents Morgagni as an academic, with a pen and a book as his emblems, with no evident reference either to his anatomising, his pathology or even his medical practice (Fig. 3.14).

Fig. 3.13

One immediately striking thing about Morgagni's work with respect to disease-histories is that he had a totally different view about confidentiality of patients' identities than obtains today. Indeed he had no view at all about confidentiality as such. Rather, he had a view about status. Thus if a patient or post-mortem corpse was eminent, famous, rich, well connected, or a personal friend, then Morgagni is sure to mention

112 Ibid., vol. 1, p. 106.

Fig. 3.14

his or her name and fame and the history of their medical complaints, no matter how personal or intimate. But if a patient was poor, of low class, or a common patient at one of the charity hospitals, Morgagni treated them as if they had no name or personal identity of their own. Only if the poor patient was particularly memorable to him, as he looked at them with his anatomist's eye, did he bother to find out and make note of their name, as happened for instance with a poor woman with thirteen ribs, something of peculiar interest for an anatomist. Obviously this behaviour on his part reveals something about social class and social mores in eighteenth-century Italy. It certainly reveals something about the recentness of the concept of patient confidentiality on the part of the medical profession. Possibly it reveals something about Morgagni's authenticating what he is saying by reference to the eminence of the gentlemen, aristocrats and churchmen he treated and dissected.

Morgagni actually came to write the great volume, the *On the seats and causes of diseases investigated by anatomy* quite late, in the years after 1740. He says that 'a friend' urged him on to it. This friend, a 'young gentleman of strict morals and an excellent disposition' (vol. 1, p. xx), used to engage him in discussion of medicine and of Valsalva and Albertini; when Morgagni told him his intention of revising the *Grave-yard*, this young man urged him on, suggesting that he should write in as familiar a manner as possible. Morgagni started, he says, by writing some sample letters to this friend. Hence the form of the book. This long-suffering young man may, of course, never have existed. For the epistolary format was one very popular at this time, for novels as well as more serious works, and it was very common for authors to modestly claim that they were only publishing because they were urged on by their admirers. Moreover, other of Morgagni's major works had also appeared as Epistles. The real advantage of this format was that it allowed Morgagni to add 'what I had said in conversation, or medical conferences, or any thing of that kind, though never so minute' (vol. 1, p. xxi), and at whatever length he felt appropriate (though the letters do, nevertheless, have quite a firm and consistent structure). It also indicates that he may have expected a relatively wide readership for these letters, for although they are in the choicest Latin, they are indeed most interestingly written. The care with which Morgagni cross-referenced these letters, and provided them with four kinds of copious index (1. of the subjects of the letters; 2. of diseases and their symptoms; 3. preternatural appearances in dead bodies; 4. names and passages most worthy of notice), indicates, however, that he expected them to have a lasting value beyond any appeal they might have as 'letters'. But how he envisaged his medical readers using them, except via the indexes, is something of a mystery, for the letters are dense with learning and commentary – accounts rather than distillations of a lifetime's anatomical and pathological learning. The work is patterned on that of Bonet in dealing with diseases from head to toe, and in dealing with pretty much the same list of diseases.

The work consists of seventy long letters. To see Morgagni's approach at work, and to see what he means by 'the seats and causes of diseases investigated by anatomy' we can take as our example, as we did with Bonet, apoplexy or 'stroke' (which is today's preferred term, even though it was initially short for 'stroke of God'), the sudden-falling-down disease that usually leads to paralysis of one or both sides of the body.[113] This is sometimes fatal of itself, and sometimes it accompanies other diseases, as result, trigger or side-effect. It so happens that Malpighi, Valsalva, and Morgagni's friend Ramazzini all died of apoplexy, so it makes a particularly pertinent example for us. Morgagni regards apoplexy as caused in general by an excess of fluid in the brain, and especially in its lateral ventricles (the side cavities within the brain mass). The fluid was either blood (sanguineous apoplexy) or serum (serous apoplexy).[114]

[113] On the history of stroke in general, see Fields and Lemak, *A history of stroke*, 1989; Fields and Lemak say that it was Morgagni who 'first presents the evidence that the pathological lesion in apoplexy is on the side opposite the paralysis' (p. 11). Morgagni duly attributes priority for this observation to his teacher, Valsalva.

[114] On serous apoplexy see Jarcho, 'Some lost, obsolete, or discontinued diseases', 1980.

What kind of causal relation does Morgagni deal in, connecting the evidence of the case-history to the explanation provided by the pathological account? It was said of Morgagni that 'he was no metaphysician'.[115] He worked, of course, within the context of four major kinds of 'cause' in medicine, a tradition of very great age by his time, and which was to continue into the nineteenth century. These 'causes' were the predisposing, the exciting, the proximate and the immediate. The predisposing causes involved the patient, his age, condition, general health ('habit'), and so on. The 'exciting' causes usually came from without, and are what bring on the disease. But the 'proximate' cause is the one (or ones) that actually brings the disease about. Morgagni was concerned with and discusses all these kinds of cause, and one needs to bear this in mind when he might otherwise seem to us to be rambling on about the patient and the circumstances of his death. But it was the 'proximate' cause that was of course of greatest interest to him, since this is the one that anatomy could reveal: in apoplexy this was the flooding of blood or serum into the ventricles of the brain, hindering or preventing or bringing to an end its usual behaviour. As a commentator on Morgagni remarked, 'the proximate cause of diseases is still involved in so much obscurity, that it is discovered only in those disorders which are seated in a single organ, and in some particular part of the structure of that organ'.[116] This was the kind of cause that Morgagni was most concerned with, but it was not the only kind, and Morgagni in his discussion takes the proximate cause as revealed by anatomy and works back from it, seeking to discover what exciting and predisposing causes had brought it on.

Morgagni drew his cases from several sources. First in importance and in his presentation are the cases that his revered teacher Valsalva had observed; Morgagni has all Valsalva's medical papers to hand. Then come those that Morgagni himself had observed, either with Valsalva or on his own. Then there are the cases from the practice of trusted friends and colleagues. And finally there are the cases that he finds in books in the course of his enormous reading. Thus he is adopting precisely the opposite of the criteria used by Bonet, who went first to cases reported in books, and only hesitatingly added cases from his own experience. As his case-histories in *Seats and causes* show, Morgagni himself came across instances of apoplectic (and other) diseases either because the patients were brought into the hospitals shortly before their death, after which he could dissect them, or because he was performing post-mortems on the Great and the Good in their homes. So he knew the immediate happenings before the death (and had often observed these himself) and the circumstances of the death itself, and usually something about the very long-term medical history and complaints of the deceased, often gleaned from relatives, bystanders and friends. To these he added, of course, the findings from dissection, in which he notes every appearance and condition in the cadaver that differs from the normal, not just those that might be suspected as underlying the disease or death. In the case of apoplexy, for instance, he usually has a complete

[115] Hamilton, *Observations on the seats and causes of diseases*, 1795, p. xxi.

[116] Ibid., p. xxiii.

post-mortem account of each corpse on the basis of which to make correlations of all the organic lesions found, with the precise pathological findings about the apoplexy proper in that particular case. In this way he could investigate the relation of the proximate cause to the exciting and predisposing causes in the case of each individual. So one could say that he produces a full pathological history of each corpse, though he does not usually have a full medical history of the patient; indeed he is found sometimes expressing impatience when, for instance, the relatives of a peasant did not notice which side of the patient's body was first paralysed in a case of apoplexy brought into hospital. The pathological history he produces deals with organs, and is acquired by the use of all the five senses, recording what is visible, palpable, audible, smell-able and taste-able (though Morgagni could not face with the equanimity of Valsalva the tasting of all the different fluids found in dissection),[117] and in producing the history he uses certain tests such as burning or heating unusual things, such as concretions or peculiar fluids, found in the bodies being dissected. But Morgagni does not seem to deploy the microscope.

The (immediate) cause of apoplexy 'and that which contains the disease', Morgagni says, is a sudden diminution of the internal motions performed in the brain. This can be brought on by a number of causes (proximate causes); these are many and various. Some fall under the senses; 'I shall spare no pains to describe clearly and distinctly those which are generally manifest and appear within the skull'. There are also other causes, not evident to the senses, and Morgagni promises 'to form a conjecture about these causes, where the case shall sometimes compel us'. Those proximate causes that are evident to the senses, and that can be perceived in the skull on dissection, lie in the blood or the serum: that is, these proximate causes produce the sanguineous and the serous apoplexies respectively.

Morgagni works through Valsalva's cases (Letter 2), then his own cases (Letter 3), then cases from other writers (final section of Letter 3). He details the age, sex and (usually) the occupation of the sufferer, and their medical history insofar as it is available to him. If they are famous or of the ruling classes, he mentions their name, fame and nobility. He often notes the date and place of the occurrence of the attack of apoplexy and of the dissection. As he is dealing with apoplexy he mentions whether the person was addicted to wine or tobacco, indulgence in both of which he obviously took for granted made people vulnerable to apoplexy (and in this he was following a long tradition). He remarks many times on the 'maxim' of Valsalva that in cases of apoplexy one finds the brain affected on the opposite side of the body to that on which paralysis occurs. He deals with cases that show concreted blood in the lateral ventricles, in the substance of the brain, in and around the cerebellum. He discusses at length cases where cavities were formed in the substance of the brain by, he reasons, explosive aneurisms of the delicate blood vessels. He discusses whether certain abnormalities found within the dissected skulls of people dead of apoplexy, such as bony growths or the hardening of the

[117] Morgagni, *The seats and causes of diseases*, 3 vols, 1769, vol. 1, letter 4, para. 27.

blood vessels, might in some way be bringing on apoplexies, and if so by what kind of mechanism. He discusses the season of the year, and the relation of this to the occurrence of the illness. In particular he discusses the weather of May 1729 in Padua, a time 'when numbers were swept away' by apoplexy 'to the great consternation of everybody'. Some people at that time died of blood extravasated into the ventricles, others of 'blood obstructed and collected together in the vessels of the brain'.[118] Morgagni made this distinction on the basis of the dissections he then performed. What caused this great incidence of death from apoplexy, he concluded, was a very cold autumn, winter and spring being followed very suddenly by very hot weather: the blood expanded more quickly than the blood vessels could expand, 'so that the blood, for this reason, expanding itself suddenly, the same thing happened as if the vessels were suddenly distended with a double quantity thereof' (vol. 1, p. 62). It followed from this that people with a tendency to apoplexy should avoid such weather, or take great care of themselves at such seasons. Morgagni also has various things to say about therapy and the care of the apoplectic patient: whether it is safe and wise to let blood, give emetics, etc.

It is clear, at least here in the case of apoplexy, that no particular problems arise for Morgagni in recognising clinical cases of apoplexy: the manner of the fit, or of the death, and the medical history of the patient make almost all such diagnoses unproblematic. Moreover the post-mortem findings also indicate in a generally unproblematic way where the 'seat' of the disease is: that is, it is located in the brain, and within that it can be localised where the main abnormalities or lesions are found. It must also be said that Morgagni is very keen to work out whether there is any possible causal relation between such anatomical 'causes' and abnormalities and lesions in other parts of the body. For instance, did the fact that in a particular corpse the left vertebral artery took its origin from the curve of the aorta, instead of from the subclavian, mean that the blood came into his brain with more force than in other people, and thus bring on the apoplexy?[119]

Finally we may look at Morgagni on the serous apoplexy (Letter IV). There had been since ancient times differences of opinion as to whether serous apoplexy ever actually occurred: was there ever really an unusual amount of watery fluid in the brain in cases of apoplexy? And if there was, was it consequence rather than cause? Morgagni is well aware of this debate (which continued into the nineteenth century, when serous apoplexy was discontinued as a possible disease).[120] The principal case of serous apoplexy that Morgagni brings forward, though admittedly an 'imperfect kind of apoplexy', was that of Valerio Zani, who was one of a number of close friends and patients of Valsalva and Morgagni whom they had the pleasure and honour of dissecting after death. Zani's father had died of apoplexy, and his uncle of bladder stone. Zani had both. He was fat, sedentary, and given to

118 Ibid., vol. 1, letter 3, paras. 11–13 and 26–7.

119 Morgagni, *The seats and causes of diseases*, 3 vols, 1769, vol. 1, letter 3, paras. 20, 23.

120 Jarcho, 'Some lost, obsolete, or discontinued diseases', 1980.

good living. His teeth all rotted away from a discharge of salt saliva that continued for many years. At the age of 63 he had a pain in the head, dullness of the senses, and intermittent temporary loss of motion on the right side. In the autumn his legs swelled, and the right one discharged much serum; he became very sleepy and speechless. Then he died. He had had no great fit, or stertorous respiration, but the other symptoms pointed to apoplexy. 'Having sawed through the skull, the dura mater appeared corrugated; but under the pia mater, within the sulci or furrows of the brain, so limpid a serum stagnated, that this membrane, before it was cut through resembled a transparent glass' (vol. 1, p. 65). There was more serum in the left and right ventricles. Was the serum from the leg of the same nature as the apparent serum found in the brain?, Valsalva asked, for that would indicate that this was a true case of serous apoplexy. Valsalva used the experimental test: he tried to coagulate the serum from different parts of the brain (and presumably, as usual, he tasted it too) and that from the leg. The serum from the leg coagulated like white of egg, but that found in the brain did not coagulate; he mixed it with spirit of salt ammoniac and spirit of vitriol, to see if any changes occurred (they didn't). You see, Morgagni says to his hypothetical correspondent, 'how long a series of experiments is required, and how cautiously and ingeniously collected' before we should dismiss the existence of serous apoplexy.[121]

After his treatment of apoplexy, Morgagni worked with similar care and a similar amplitude through the other diseases of the head, such as epilepsy; and on to diseases of the thorax, such as dropsy of the thorax and spitting of blood; and then to diseases of the belly, such as vomiting, suppression of urine and hernia; and lastly to surgical and general diseases, such as fevers, fractures, gout and, finally, to a special obsession of his: the poison with which Cleopatra killed herself.

We can see that Morgagni's work is experimental; even the conjectures are based on data accessible to the senses. The account of causation is organ-based (though not using the microscope) and ultimately mechanical, in a very loose sense. The work as a whole also does what Morgagni complained that Bonet's *Grave-yard* did not do: it provided

> … a great number of similar symptoms at hand, so that you might readily compare them with many morbid appearances; which [are] either similar, or not similar, to each other; and by this means to be able at once to conceive, which of those symptoms are most frequently, most rarely, or never, join'd with any particular species of internal morbid constitution.[122]

---

121 Morgagni, *The seats and causes of diseases*, 3 vols, 1769, vol. 1, letter 4, para. 3.

122 Ibid., vol. 1, pp. xviii–xix.

Pathology elsewhere

Morgagni was in his time, and has continued to be, the most famous and single-minded seeker of the seats and causes of diseases via anatomy in the eighteenth century. But the long-term fate of his project was not good.[123] It was of course a project conceived sixty years – a whole lifetime! – before it appeared, and thus probably destined to be always out of date. Nevertheless, while *Seats and causes* received several reprintings over the next ninety years or so and was laboriously translated into modern languages, by the early 1790s it had already begun to look outdated. One Edinburgh physician, James Hamilton, Jr., thought that Morgagni needed to be rescued from himself, since the 'facts' in Morgagni's work were in danger of being lost under the verbiage that no one any longer read. Hamilton boldly volunteered himself to edit, rearrange and reduce the work.

> So extensively has medical knowledge been diffused since the publication of Morgagni's writings [Hamilton wrote], although no more than a period of about thirty years has elapsed, that they are now deemed chiefly valuable on account of the facts which they contain. As, therefore, readers in general find they have to wade through a prodigious quantity of uninteresting matter, before they can arrive at what they regard as useful; these writings are at present seldom looked into, being only occasionally consulted as a dictionary.[124]

So Hamilton set out to systematise *Seats and causes*. He rearranged the diseases according to the system of Macbride, 'dividing diseases into Universal, Local, Sexual, and Infantile' (p. xiii); he left out all the information about the patients – that is, he omitted all the crucial information that Morgagni complained Bonet had omitted! – and improved the 'uncouth' style of Morgagni, that style which had been so admired for its classical Latinity. Unfortunately, or perhaps fortunately, Hamilton only completed one small volume of this travesty of Morgagni's intentions.

The Italian hospital approach to pathology, as typified by Morgagni, had some followers elsewhere who paid something more than lip-service to it. For instance, Antoine Portal, whom we met before as a young man writing scissors-and-paste pathology, in 1803 when he was Professor of Anatomy at the Muséum d'histoire naturelle (what had been the Jardin royal until the French Revolution) published a work that shows considerable advance on the compilation volumes of Lieutaud that he had edited thirty-five years before. This was his final *Course of medical anatomy, or the elements of the anatomy of man, with some physiological and*

---

[123] An attempt was made to follow it up at the Royal College of Physicians of London by William Heberden, Benjamin Alexander (who translated the work into English), and a few others, which lasted for a while. See Vaughan, *'Improvements in the art of healing'*, 2004.

[124] Hamilton, *Observations on the seats and causes of diseases*, 1795, p. xxi.

pathological remarks, and the results of observation on the seat and nature of diseases, through the opening of bodies. Anatomy, he writes, not only teaches us the structure of the human body, but it also throws light on the causes and seats of disease, 'without which Medicine will be nothing but a blind empiricism' (vol. 1, Preface, p. i). The editor of one of his works said of Portal that from 1768 his doctrine 'is that of Valsalva, Morgagni, Senac, Lieutaud, etc., which M. Portal has adopted and spread throughout his books and his lessons, according to the confirmative results of his own numerous observations'.[125] In writing his *Course of medical anatomy* Portal claimed that Morgagni, with all his merits, had not produced a book usable by the practical physician. Hence his present one. 'My lectures have always had as their object the parallel between the natural state and the diseased state, which is so instructive that the knowledge of the one facilitates that of the other, and one knows each of these two parts better if one knows them both'.[126] This method had proved so popular in his public lessons in the Jardin royal and the Collège de France (previously the Collège royal), Portal claimed, that for nearly forty years 'the major part of the physicians and surgeons of France, and a great number of foreigners, had frequented them', and taken up the same approach in their own teaching.[127]

In practice what Portal actually offers in this five-volume work is a relatively limited account (without illustrations) of the structures of the body 'in the order generally adopted by the anatomists' (vol. 1, p. xv) (this time going from the bones to the muscles, to the blood vessels including the heart and lymphatics, to the brain and spinal cord, and finally to the lungs and internal viscera). Linked to, but not integrated with, these descriptions are *précis* of the diseased conditions to which the different parts are prone, many of which come from observations he had himself made on bodies dead of disease.

Despite all this eighteenth-century interest in pathology, our pathologists have not so far shown any interest in illustrating their works, even though anatomy books were usually considered to be better for teaching if they were illustrated. The first pathological work to be illustrated was Matthew Baillie's book *The morbid anatomy of some of the most important parts of the human body*, which first appeared in London in 1793, and whose illustrations appeared somewhat later, from 1799 to 1802, as *A series of engravings, accompanied with explanations, which are intended to illustrate the morbid anatomy of some of the most important parts of the human body*. Baillie was a nephew of both William and John Hunter. On William's advice he had first acquired an Oxford medical degree, so that he could become a physician and, in time, a Fellow of the Royal College of Physicians of London, thus bypassing the laborious surgical route to becoming a physician that William himself had had to take. So, at least for Britain, Matthew Baillie

---

125 Portal, *Mémoires sur la nature et le traitment de plusieurs maladies*, 4 vols, 1800, 1808, vol. 4, p. vii.

126 Portal, *Cours d'anatomie médicale*, 5 vols, 1803, vol. 1, p. xii.

127 Ibid., vol. 1, pp. ix–x.

was in a most unusual position: someone who was professionally a physician, but with his own anatomical school (inherited from William), free access to the two greatest anatomical collections in Britain (those of William and John), and a very extensive knowledge of anatomy as a teacher of the subject. It seems to have been this unusual position that first put Baillie to thinking about developing pathological anatomy. For, according to the division of labour that the professional bodies tried to enforce in eighteenth-century London, the physician dealt with illnesses occurring within the body, and hence was preoccupied with diagnosis, *symptoms* and prognosis, while surgeons supposedly limited their interventions to broken bones and illnesses of the surface of the body, yet were given a thorough education in *internal anatomy*. Here, with Baillie, was someone with intimate anatomical knowledge of the healthy condition of the inner organs of humans, and with another side to his professional life, which consisted of diagnosing from symptoms, and attempting to cure, illnesses of the inside of the body. He was thus probably as aware of the relation of symptoms to inner, pathological, changes as any other British practitioner of his day. 'It appeared to me,' he wrote in his autobiography, 'that Medicine and Surgery were so connected together, that by knowing something of the latter, I should become more satisfactorily acquainted with the former … During my situation as Physician [to St George's Hospital in London] I was not only as attentive as I could be to the Cases of my Patients, but embraced every opportunity of examining the Morbid appearances after death'.[128] He had access to William Hunter's collection of morbid preparations, and his role as a hospital physician gave him frequent opportunities to examine diseases in dead bodies. He wrote that he was led to writing his book 'because I had observed with much attention the morbid appearances in the numerous Bodies which were examined at St George's Hospital, and because I thought that the Books hitherto published upon this Subject were too diffuse, and the descriptions of the diseas'd appearances were often indistinct and very inaccurate'.[129]

Baillie studied 'the changes of structure arising from morbid actions in some of the most important parts of the body', and this is what his book is about.[130] What is most striking here is that he holds a reversed view of the causal relationship between disease and organic lesion that we have seen with Morgagni. For Baillie, the pathological changes in the internal organs are the *result* – not the cause – of the disease, which itself is constituted by somewhat mysterious 'morbid actions' operating in the body. Symptoms (at least the more important ones) are therefore the signs of 'morbid actions' that are actively affecting inner structures. By studying pathology intently, it should be possible, in time, to see how the symptoms and the morbid appearances go regularly together, and hence improve diagnosis:

128 Crainz, ed., *The life and works of Matthew Baillie*, 1995, p. 24.

129 Ibid., p. 41.

130 Baillie, *The morbid anatomy of some of the most important parts of the human body*, 1797 [1793], Preface to the first edition, p. i.

> A person who previously had attended very accurately to symptoms [i.e. someone trained as a physician], but was unacquainted with the disease, when he comes to examine the body after death, and finds some of the appearances that are described in this Treatise, will acquire a knowledge of the whole disease. He will be able to guide himself on such knowledge in similar cases, and also to inform others. It may, perhaps, too, lead him to a proper method of treatment.[131]

So 'morbid anatomy' meant something very particular to Baillie: 'the changes of structure produced by disease'.[132]

Unusually, Baillie presents his 'morbid anatomy' not in the form of cases, as Morgagni had done, but as generalisations. This allowed him to be extremely terse and precise, at least as compared to other pathologists. As with the original *Morbid anatomy*, the set of engravings was limited to the major organs and the significant changes of structure that diseases cause in them. The drawings were made by the young William Clift, the last pupil and amanuensis of John Hunter. The engravings were published in sets, twice a year, which made the publishing process less daunting, and the ten sets dealt successively with the heart, the lungs, the stomach, the intestines, the liver, the spleen and kidneys, the urinary bladder, the organs of generation in both sexes, and the brain.

A couple of brief instances will show Baillie's working. First we can take the final diseased appearance that he dealt with: the morbid condition of the brain caused by apoplexy (Fig. 3.15). Here the 'morbid action' is the rupture of a blood vessel in the brain and the flowing out from it of blood into the substance of the brain or into the ventricles. The first figure shows the diseased appearance in a person who had survived a fit of apoplexy some months, and the extravasated blood seems to have been absorbed and a serous fluid substituted in its place. So the changes of structure produced by disease in this case are these: 'The substance of the brain immediately surrounding this serous fluid, becomes tough and firm, so as to resemble a good deal in its appearance a membrane'.[133]

However, a second example will show that Baillie was not always consistent in his method. This example is the one he is most celebrated for, the description of the tubercules in the lung and their development over time, in the disease of phthisis pulmonalis (known also at the time as consumption, and later to be known as tuberculosis) (Fig. 3.16).

---

[131] Ibid., Preface to the first edition, p. v.

[132] Ibid., Preface to the first edition, p. vi.

[133] Ibid., p. 227. Fig. 1. represents the greater part of the half of the cerebrum upon the right side, its anterior extremity being cut off. The flattened side of the hemisphere is presented to view, viz. that which lies in contact with the falciform process of the dura mater. B is a large mass of extravasated and coagulated blood.

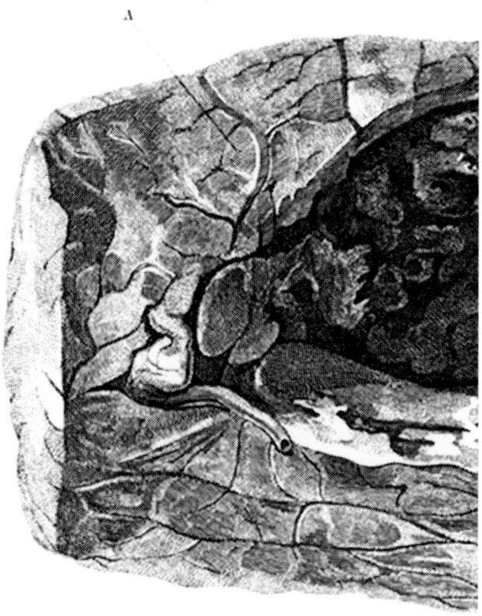

Fig. 3.15

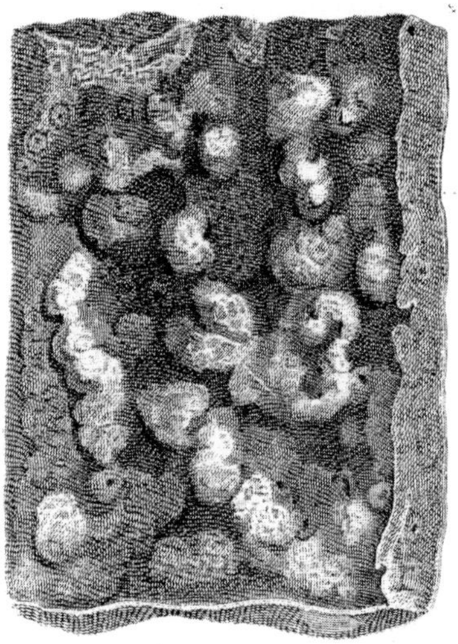

Fig. 3.16

> There is no morbid appearance so common in the lung as that of tubercules. These consist of rounded firm white bodies, interspersed through their substance. They are, I believe, formed in the cellular structure, which connects the air cells of the lungs together … They are at first very small, being not larger than the heads of small pins, and in this case are frequently accumulated in small clusters … The most ordinary size of tubercules is about that of a garden pea, but they are subject in this respect to much variety. They adhere pretty closely to the substance of the lungs, have no particular covering or capsule, and have little or no vascularity. When cut into, they are found to consist of a white smooth substance, having great firmness, and often contain in part a thick curdly pus … In cutting into the substance of the lungs, a number of abscesses is sometimes found, from pretty large tubercules having advanced to a state of suppuration. In the interstices between these tubercules, the lungs are frequently of a harder, firmer texture, with the cells in a great measure obliterated … When tubercules are converted into abscesses, pthisis pulmonalis is produced, one of the most destructive diseases in this island …. (pp. 66–9)

No 'morbid action' action is identified here, bringing on the disease. Moreover, the morbid changes to the structure of the lung (the turning of the tubercules into abscesses) are for Baillie this time the origin of the disease, rather than its product.

But despite all his pioneering pathological work, as Baillie himself is reputed to have said to his family: 'I know better perhaps than any other man, from my knowledge of anatomy, how to discover a disease, but when I have done so, I don't know better how to cure it'.[134]

Finally, we need to note that many of the anatomical museums created by anatomists throughout Europe had their pathological specimens, mostly of monstrous or unusual parts of humans or other animals. One museum, possibly the largest of all, that of Leiden University, had the collections of some of the most important anatomists of the eighteenth century: those of Rau and the brothers Albinus (Bernhard Siegfried and Frederic Bernhard). The curators persuaded the Professor of Anatomy at Leiden, Eduard Sandifort, to prepare an illustrated catalogue of the most useful preparations. Two enormous volumes eventually appeared in 1793. Sandifort was a great believer in the value of pathology to anatomy, and turned these volumes in that direction. In his inaugural oration as Professor of Anatomy and Surgery at Leiden in 1772, he had spoken on the theme 'Careful inspection of cadavers, the best aid to practical medicine', taking respiration as his example and pointing out that hospitals offered the greatest opportunity of furthering such a programme.[135] So Sandifort used the diseased, deformed, damaged and monstrous bones and skulls in the collections in the anatomical theatre as the basis for a pathological anatomy. The 127 plates of the second volume of this series are dedicated to this purpose.[136]

---

134 Wardrop, *The works of Matthew Baillie*, 1825, p. xl.

135 Sandifort, *Oratio de circumspecto cadaverum examine optimo practicae medicinae adminiculo*, 1772.

136 Sandifort, *Museum anatomicum Academiae Lugduno-Batavae*, 2 vols, 1793.

Chapter 4
Human Bodies: Getting, Keeping, Picturing, Publishing, Arguing

In Chapter 2 we saw something of how careers were made and audiences attracted to the discipline of anatomy. Here I want to look at the human corpses that the anatomist used, and at issues of acquisition, preservation, representation and dissemination. In addition I look at some specimen controversies that occurred during the making of new facts in anatomy, to see something of the heat and light that anatomical investigation could produce.

1. Acquisition: Getting bodies

In early eighteenth-century London up to eight times a year, at Tyburn, now the site of Marble Arch, there were public hangings of criminals from the infamous gallows known as 'Tyburn tree'.[1] They were very public events – stands were erected for the crowds; thousands attended; it was a great day out for families – and the military often had to be present to maintain order. For after the criminals had been hanged, there was very often a fight for their corpses, pulling their bodies down from the gallows. The novelist Samuel Richardson witnessed such a riot in 1740:

> I have this day been satisfying a curiosity, I believe natural to most people, by seeing an execution at Tyburn … That I might the better view the prisoners, and escape the pressure of the mob, which is prodigious … I mounted my horse, and accompanied the melancholy parade from Newgate to the fatal tree …
>
> As soon as the poor creatures were half-dead, I was much surprised before such a number of peace-officers, to see the populace fall to hauling and pulling the carcasses with so much earnestness, as to occasion several warm encounters, and broken heads. These, I was told, were the friends of the persons executed … and some persons sent by private surgeons to obtain bodies for dissection. The contests between these were fierce and bloody, and frightful to look at … [Afterwards] the face of everyone spoke a kind of mirth, as if the spectacle they had beheld had afforded pleasure instead of pain, which I am wholly unable to account for.[2]

---

1 Linebaugh, 'The Tyburn riot against the surgeons', 1975.

2 Richardson, *Familiar letters*, 1928, pp. 217–19.

In Georgian London both the College of Physicians and the Incorporation of Surgeons had the right to a number of bodies annually to dissect in their halls, in order to teach anatomy to their junior members. The surgeons had since 1540 had the right to 'have and take, without contradiction, four persons condemned adjudged and put to death for felony by the due order of the King's laws of this realm, for anatomies, without any further suit or labour to be made to the King's highness ... to make incision of the same dead bodies ... for their further and better knowledge, instruction, insight, learning, and experience in the said science or faculty of surgery'.[3] The physicians, similarly, had the right enshrined in a grant under the great seal to four bodies a year of those executed for theft, homicide or any other felony to anatomise.[4] These rights were exercised irregularly and (it seems) without controversy.

It was only from the late seventeenth century that these bodies were demanded more regularly, and it is no surprise that the criminals and their families therefore came to fear the dissection of their bodies, even more than their hanging, and wanted a family burial for the corpses. So the fights between the two parties got very violent.

In 1728 the Company of Surgeons complained to the Court of Aldermen of Middlesex:

> First, The Masters or Governors of the said Company do take the liberty to represent that the night before the execution several sand carts bricklayers carts and other carts and wagons are planted round the gallows on pretence of receiving and giving the spectators a place on the day of execution and for which they pay twopence, threepence and sixpence apiece. From this practice the gallows is so blocked up that your petitioners beadles cannot bring away the body without danger of their lives ...
>
> That it is become a late practice to sell the bodys of such malefactors to private surgeons and others, which always stirs up a contest about the body to be delivered to your petitioners beadles and is a means whereby the knowledge of anatomy will be confined to very few hands.
>
> That the common executioner himself is likewise ... engaged in such unwarrantable sales whereas he or such other persons have no property in the bodys of such malefactors.[5]

So there were third parties involved too: the bodies were being sought by 'private surgeons', that is to say, anatomists who wanted the bodies to dissect privately for paying students. In practice it seems as though the hangman delivered the bodies to whoever paid him most. So the surgeons' company were frequently powerless to implement their rights to their four bodies a year.

---

3 As cited in Peachey, *A memoir of William and John Hunter*, 1924, p. 2.

4 Ibid., p. 4.

5 Quoted in Flint South, *Memorials of the craft of surgery*, 1886, pp. 236–7.

This situation was made worse when it was turned into a moral issue by the so-called 'Murder Act' passed by Parliament in 1752 in order to 'strike a greater terror into the minds of mankind, and thereby at least lessen the number of murders'.[6] In 1759 the 'Master of Anatomy' who was conducting the public dissection of one Richard Lamb at the Incorporation of Surgeons of London, made an opening- and closing speech drawing attention to this moral role of anatomy.

> It being well known, in how great horror dissection was held by almost all mankind, more especially the lower class, as the most harden'd villains, tho' they braved death, still shuddered at the thoughts of being made an Otomy (as they called it) [Parliament] very wisely ordained, that every such malefactor (not hung in chains), should be delivered to the surgeons for dissection …
>
> Curiosity more than improvement, has, I am persuaded, drawn the greater part of this audience together … Happy would it be, if this publick occasion, this sight of death, may prove a monitor to every individual here, and by them be repeated to their acquaintance (especially those who are prone to wrath) always to have an eye on this table, whenever they find themselves urged by the passions of malice and revenge … Let therefore this Anatomical Table in the Surgeons Theatre, be a preacher to all this audience; and should their passions run high, and the voice of reason and religion be forgotten, may this dread table present itself to their view, and restrain the arm raised to deprive a fellow creature of life ….

The surgeon, Mr Tate, repeated the moral lesson at the end of the dissection:

> … I think few who now look upon that miserable mangled object before us, can ever forget it. It is for this purpose our doors are opened to the publick, that all may see the exemplary punishment of a murderer ….

Struggles for the corpses of hanged men, between their families on the one side and the surgeons' men on the other, continued until 1783, when the hangings were transferred to Newgate Prison: they were still public, but most opportunities for disorder had been removed.

Normally, a dead body would be interred by its close relatives, and once interred it would be left to rest until the Final Resurrection or, more prosaically, until more room was required in the cemetery, as was the case in the Cemetery of the Holy Innocents in Paris, where the bones of the dead were regularly dug up and exhibited to make room for fresh corpses. No one had any interest in delaying the normal interment process or in intervening in it – except the anatomist.

---

6 All these quotations are from the Old Bailey Session Papers, as printed in Forbes, 'To be dissected and anatomised', 1981, pp. 490–92.

The anatomist's desires were peculiar.[7] As he himself was aware, everyone shared a sense of the transgressive nature of the act of dissecting a dead human being in public. People didn't like it – not for themselves or their friends – even though as viewers they were absolutely fascinated by it. It *felt* like a crime for an anatomist to dissect a fresh corpse. And yet, as Diderot put it, 'Cadavers are the only books from which one can properly study Anatomy'.[8] Hence, from the very beginning of public anatomising in the medieval and early modern period, the anatomists had sought 'legitimate' bodies. That is to say, they appealed to the state or local authorities for the right to the bodies of those poor who had died without relatives, and the bodies of the malefactors that those authorities put to death judicially. Thus in Italy we find that the Bolognese anatomists got their bodies from the hospitals of St Mary of Life and St Mary of Death. This latter hospital was immediately next to the medical school, and the physicians there such as Morgagni and his colleagues could even 'reserve' interesting cases in the hospital for pathological investigation after they died (see Chapter 3 above). In Paris the anatomists got their dead bodies from the public hospitals such as the Hôtel Dieu, or the Salpêtrière, a hospital for poor women.

Things were easier when there was an all-powerful ruler, such as the Emperor. In Vienna, where the university was under the direct control of the Emperor, the anatomists initially got their bodies from the hangman; and from the 1670s sometimes from hospitals. Empress Maria Theresa issued an 'anatomical act' in 1742 that specifically authorised the Viennese Professor of Anatomy to take bodies that had been executed. But seven years later she issued a new order:

> My physician and protomedicatus Herr Gerhard von Swieten has been told by many that anatomical and surgical demonstrations are rarely organised because, especially during the winter, there are not enough executed bodies (this year the number was the smallest, only three). Therefore for the use of the public and for the benefit of the studies of medicine and surgery, not to hinder but to continue with the aforementioned demonstrations without the slightest further obstacle, Her Highness has graciously deigned to recommend that for want of the executed criminals the bodies necessary for the anatomical and surgical demonstrations should also be taken from the local hospitals and given on request to the professors of anatomy. Also, for the surgeon Molinari, a facility should be arranged in the St Marx hospital [a lying-in hospital] to hold demonstrations for the local midwives, and he should be allowed to use the bodies of women who die after delivery ....[9]

---

7 An exceptional analysis of these issues for early nineteenth-century Britain has been made by MacDonald, *Human remains*, 2006, Introduction and Chapter 1.

8 Diderot, *Encyclopédie*, 17 vols, 1751–65, vol. 2, 1751, p. 511, 'Cadavre'.

9 Codex Austriacus, p. 475 (1749). I am indebted to Dr Tatiana Buklijas for this information and the translation.

When the Empress's son, Joseph II, abolished capital punishment in 1787, the unclaimed bodies from the hospitals were made available to the anatomists.

Haller's experience in Göttingen was similar, in that the controller of the university could effectively do as Haller wished, and he gave him first refusal on all local bodies. Haller recorded:

> I arranged that the bodies of criminals from everywhere should be passed on to me for dissection, so that even clandestine births should be brought to me; and great was the supply of human bodies, especially those of children. And this facility allowed me to investigate the anatomy of the arteries with great application, which at that age can be shown better than the muscles.[10]

Indeed, so ample was the supply of bodies for Haller that he could even offer them to his students:

> When I saw numerous and diligent students coming to the new University, and who sought to gain the highest honours in the Georgia Augusta, if there was any of the highest intelligence, I used to persuade him to take up some difficult part of anatomy or a physiological question, in investigating which he might take up two winters [anatomy sessions]. There were enough cadavers for them to use, and I used to help the studies of these young men, as much as I could, I made many experiments on various animals myself ....[11]

The greatest problems about the supply of bodies for dissection occurred in Britain, especially in London and Edinburgh where there was the greatest density of anatomical teachers. The Edinburgh anatomists acquired their bodies from the town council, which was the absolute authority within its town walls. A local physician, Archibald Pitcairne, who had just returned from a spell as Professor of Medicine at Leiden University, applied to the town council in 1694 for the right to have for dissection the bodies of those who died in the workhouse. He wrote to a friend in London about his scheme:

> We offer to wait on those poor for nothing, bury them after dissection on our own charges (which now the town must do on theirs). And yet, Sir, there is vast opposition made to it by the chief-surgeons, who neither will eat hay, nor suffer the oxen to eat it. I do propose, if it be granted, to make better improvements in anatomy than have been made at Leyden these thirty years, for I think most or all anatomists have neglected or not known what was most useful for a physician.[12]

---

[10] Haller, *Bibliotheca anatomica*, 2 vols, 1774, 1777, vol. 2, p. 296.

[11] Ibid., vol. 2, pp. 196–7.

[12] British Library, MS Sloane 3198, f. 15, letter from Pitcairne to Dr Gray, 23 September 1694.

Indeed the surgeon–apothecaries were so dismayed that they felt compelled to apply to the town council the very next week for a similar right to bodies for dissection. As we saw in Chapter 1, the town council made it a condition of its grant of bodies that the incorporation build its own anatomical theatre.

However, once the teaching of Alexander Monro *Primus* began in earnest, first in the Incorporation of Surgeon–Apothecaries and then in the University at Edinburgh, there seem to have been many students who wanted to engage in extracurricular anatomising themselves, and there were several incidents of grave-robbing with its resulting public rioting in Edinburgh in the course of the eighteenth century. Both Monro himself, as Professor of Anatomy, and the Incorporation of Surgeon–Apothecaries tried to dissociate themselves from this. Trouble had begun in 1711, when sepulchres in the Greyfriars churchyard were violated; the incorporation strenuously denied any involvement.[13] The next bad year was 1725. In February the Corporation tried to prevent Monro from having dead bodies delivered to their hall in his absence, for fear that they may have been acquired illegitimately, and that this would redound to the discredit of the Corporation:

> The Calling ordered that upon the Professor of Anatomy's lodging a human body in their Hall or Theatre, he forthwith by a letter acquaint the Deacon, Treasurer or in case of their absence the Eldest Deacon in town for the time thereof and that it was regularly procured and obtain their allowance for the same. But in case the Professor do not in the terms of this order … obtain their allowance as said is, the Professor is hereby required immediately in obedience to their order to remove and carry off the subject and shall be answerable to the Calling for any consequences that may arise from his bringing in or carrying off the said subject.[14]

In April Monro wrote a letter to the local paper, the *Caledonian Mercury*, declaring his 'just abhorrence of that vile, abominable, and most hideous crime of stealing human bodies out of their graves, and which must directly tend to the ruin of my profession' because it would give it a bad name. He offered a reward of 3 pounds sterling for information leading to the conviction of the culprits.[15] The issue of 'the raising of the dead' recurred in 1739 and 1742.[16] On the latter occasion a dead body had been found in a warehouse, and a 'tumult and disturbance' had resulted. The Corporation threatened their own apprentices and members with fines and

---

13 Minute printed in Creswell, *The Royal College of Surgeons of Edinburgh*, 1926, p. 196.

14 Manuscript minutes of the Royal College of Surgeons of Edinburgh, 11 February 1725.

15 *Caledonian Mercury*, April 1725.

16 Manuscript minutes of the Royal College of Surgeons of Edinburgh, 31 March 1739, 12 and 13 March 1742.

expulsion if they were found dissecting or inspecting bodies raised from the dead. To reassure the Edinburgh public they decided to have their decisions printed and posted up everywhere, ordering the document 'to be printed without loss of time, and to be dispersed through the City and suburbs by affixing them on the Church-doors, City-gates, Coffee-houses or other public places, and to be published in the Edinburgh newspapers, in order to show the World the sentiments of this Society against such a monstrous Crime'.

Public rioting over the supply of recently dead corpses continued in Edinburgh well into the nineteenth century, even though resurrectionism had turned into a regular trade. The legal situation continued to present dilemmas to those responsible for administering the law. Medical science had to be defended, but 'the feelings of most men, particularly among the lower orders, are strongly repulsive to such a practice' and therefore had to be taken into account, in the opinion of the Lord Advocate of Scotland. He had to advise in a case as late as 1821, in which the resurrectionist George Lawrie had been caught supplying the class of Alexander Monro *Tertius*. For the Lord Advocate the issue revolved around the fact that the trade was engaged in for gain:

> In Lawrie's case, not only was the thing gone about in the most slovenly manner possible, but this Mr Lawrie assumed the trade of a Resurrection-man *for hire*. He agreed to furnish the Anatomy Class at Edinburgh with certain subjects at the price of six guineas each. To me this appears a very ugly feature in the case. If this person had been a Lecturer, or a Medical Student, or had been in any respect to derive benefit in the way of science from the dissection of such bodies, his case would have assumed a different aspect, but where a professional man of this description chooses for the sake of *pecuniary gain* to take upon himself an occupation which few of the lowest order are found willing to engage in, I do think that he ought to be held as doing so under all the risks attending such a calling.[17]

Circumspection was requisite on the part of the anatomists in ensuring their supply of bodies. Brazen behaviour on the part of anatomist or supplier would lead to rioting and punishment for the supplier.[18]

But yet, as we saw in Chapter 2, it was in practice possible to teach a great deal from a single course and just one or two bodies, as was the case with Monro *Primus* at Edinburgh. The fact that he had to economise on bodies might have

---

[17] Letter of Sir W. Rae, Lord Advocate, June 1821. Edinburgh University Manuscripts Gen 1429/18/22.

[18] See the series of pamphlets and broadsheets for and against, in Edinburgh University Manuscripts P.46 15–23, beginning with John Barclay's *The Medical School of Edinburgh*, 1819, and including 'The Humble Petition of the Worms to the Magistrates; setting forth the lamentable Condition to which they are reduced, in consequence of the Diabolical Practice of Raising the Dead'.

concentrated his mind in this respect. But at all events, grave-robbing was not essential to his method.

In London, however, practical anatomical teaching could not have been possible without a regular trade of grave-robbers. While there had been private anatomical teachers in London from the beginning of the eighteenth century – the historian Peachey has counted twenty-four practising at some time between 1700 and 1744[19] – they had not been well supplied with fresh human corpses. When William Hunter took a course with Frank Nicholls, there were only two human corpses. James Douglas had had to use dogs because he had so few human corpses.[20]

The turning-point came in 1746. As William Hunter recalled,

> In the earlier part of my life, I found such advantage in putting my own hand to the knife, and in examining the different parts of the body, at my leisure, and in my own way; that I was convinced, I might be of considerable service to the public as a teacher, by opening a school for practical Anatomy, which had not been done before in this great city.[21]

But Hunter's school would be unlike all others in London, for he promised that 'Gentlemen may have the opportunity of learning the art of dissection during the whole winter season in the same manner as at Paris'.[22] What this meant was 'providing each of his pupils with one entire body and from time to time inspecting the dissections himself'.[23] In fact, in his first course, William gave between sixty and seventy lectures 'into which he introduced six subjects [i.e. bodies for dissection], viz. one for the muscles and male parts, one for the viscera, brain and spinal marrow with the female parts, one for the blood vessels, one for the nerves, one for the peculiarities of the fetus, and one for the operations in surgery, in which he shewed the art of embalming'.[24] But in subsequent years, and particularly once his brother John had arrived and began to manage the trade in bodies, there were many more available for the students.

This demand greatly enlarged the trade of 'resurrection men': enough human bodies – *fresh* bodies – had to be supplied, legally or illegally! It was dark, dirty and dangerous work in the cemeteries at night – if they were discovered, resurrection

---

[19] Peachey, *A memoir of William and John Hunter*, 1924, p. 8. This work contains all the information one could want on anatomical teaching in England from 1700 to 1746, culled from newspapers, minutes, books and prospectuses.

[20] See Douglas's *Myographiae Comparatae Specimen*, 1750.

[21] William Hunter, *Two introductory lectures*, 1784, pp. 108–9.

[22] *London Evening Post*, 16 September 1746, as quoted in Gelfand, 'The "Paris manner" of dissecting', 1972, p. 99.

[23] Letter to *Westminster Journal*, 19 December 1746, from 'Publicus'; discovered by Peachey, *A memoir of William and John Hunter*, 1924, pp. 97–8.

[24] John Hunter's account, from Brock, *William Hunter … A memoir*, 1983, p. 5.

men could be mobbed – but the job paid well. A good team could raise eight bodies a night to sell to the anatomists. This 'Paris method' set a new standard in London, and William was also an excellent teacher. But he warned his students –

> In a country where liberty disposes the people to licentiousness and outrage, and where anatomists are not legally supplied with dead bodies … it is to be hoped that you will be upon your guard and, out of doors, speak with caution of what may be passing here, especially with respect to dead bodies.[25]

The whole business of course culminated, both with respect to Edinburgh and to Britain as a whole, in the case of William Burke and William Hare. In Edinburgh these two resurrectionists were caught in 1832 trying to cut out the middle-man. For they murdered their victims to order, thus doing away with all the inconveniencies of having to dig them up at dead of night. They were tried; Hare turned against his friend, and Burke was condemned to be hanged and dissected.

2. Preservation: Preserving bodies better and longer

Preparations dry and wet

Human and animal bodies were difficult to keep in good condition for dissection, since they began to decay from the moment that they died, and there are numerous tales of how revolting the dissection room smelled after a few days, especially if it was poorly ventilated. Anatomising took place in the colder months of the year in order to slow the decay, but as the anatomists themselves had to be kept warm there was often an open fire burning. 'Fresh subjects', as the anatomists referred to them, were therefore always in demand.

But the practising and teaching anatomist did not work just from fresh whole human or animal bodies. He supplemented these with preparations of particular parts, and it came to be possible to preserve these for very long periods of time. The correct term for these objects is 'preparations', a term that reeks (as it should do) of the 'preparation room' where they had been 'prepared' specifically for use, primarily in teaching. The term 'specimen' was not used of such things until the early nineteenth century, and invokes a scenario in which objects are collected as perfect or exemplary instances, often with an eye to making a collection complete. So we should not use the term 'specimen' here, as it takes us down all the wrong roads for understanding what these preparations, collections and museums were for in the eyes of the anatomists who created them.[26]

The preparations could be typical parts in perfect condition and dissected out, or ones hard to see in the course of a dissection, or ones of special interest as being

---

25 William Hunter, *Two introductory lectures*, 1784, pp. 108–9.

26 I owe this point to a public talk by Dr Samuel J. M. M. Alberti.

unusual or remarkable for some feature or sign of disease. Their primary function was for teaching and demonstration. This is William Hunter's view:

> Preparations serve two purposes chiefly, to wit, the preservation of uncommon things, and the preservation of such things as required considerable labour to anatomize them, so as to shew their structure distinctly. Of the first sort are, the pregnant uterus, diseases, parts of singular conformation, &c. Of the second class are, preparations of the ear, the eye, and, in general, such as shew the very fine and delicate parts of the body, which we call the minutiæ of Anatomy … Preparations should not be used as substitutes for a body; but supplementally, to demonstrate such circumstances clearly, as are intricate, confused, or invisible in the fresh subject ….[27]

William Hunter used such preparations very extensively as practical teaching aids in his lectures:

> When smaller parts of the body are to be demonstrated here, or such as are obscure from their situation, they will be shewn at two or three places [in the room] successively, that every one present may get a distinct view of them … Objects that are still more minute, and most of the *preparations*, must be sent round the company; that every student may examine them in his own hand. To prevent confusion, you will please to observe, that, in the first seat, the preparations are to go round from right to left; in the second bench, from left to right; and so alternately, to the farthest seat of all. To prevent loss of time, when you give a preparation to your neighbour, be so good as to point out the *part*, or *circumstance* which is then to be examined; as I shall do, when it is first handed round ….[28]

William Hunter also recommended that the students should themselves acquire their own anatomical preparations, both to learn the location of the parts as a student, and to refresh the memory when in medical or surgical practice:

> I must likewise earnestly recommend it to every student, to make and collect as many anatomical preparations as he can. He should not only have a skeleton for his own use, but he should have several skulls, divided by different sections, to shew all the interiors of that part. He should have a preparation of all the blood-vessels in their natural situation, and two preparations of the trunk of a child, the one presenting a fore-view, the other a back-view of the whole viscera; and as many preparations of the organs of sense and generation, and of the particular viscera as he can easily procure.[29]

---

27 William Hunter, *Two introductory lectures*, 1784, pp. 89–90.

28 Ibid., pp. 111–12.

29 Ibid., p. 110.

The preparations that Hunter was recommending here were both 'dry' and 'wet'. Dried preparations had been common for centuries, and as long as they are kept from water and insect infestation, dry preparations of skins, skeletons, bones, wings etc. last indefinitely, though they may shrivel or flake and the anatomically interesting feature may thus become difficult to see.

The Rotterdam anatomist Louis de Bils claimed to have perfected a method in the 1650s of both dissecting a human body without spilling any blood, and of preserving bodies and parts of bodies intact, 'embalmed' in some way, which remained perfect for many years. Bils put out a sworn and notarised 'Act', claiming legal protection for his discoveries.[30] Bils claimed to be building an anatomical theatre in a house that had previously been the court of the English Merchants of Rotterdam, and that he had permission from the States General 'to take all the bodies of those that shall be executed by the hand of justice, whether military or civil, as also of those strangers that shall die in the Hospitals, in all parts belonging to the said States General; and to dissect the said carkases in such manner as himself shall think good for the satisfying of his own curiosity and the promotion of knowledge for the common good' (pp. 2–3). As a result, Bils claimed he was able 'to set up a *merely new anatomy of the body of man*, by many dead bodies dissected and embalmed *in an extraordinary manner*, in which bodies shall be shewn all the veins, arteries, sinews and fibres severed from one another, but remaining fast, both where they first arise, and where they end. As also the heart, liver, lungs, and entrals, eyes and brains shall remain manifest in the said bodies' (pp. 3–4). What Bils wanted was for friendly people to give him money towards his scheme to prepare some fifty bodies in this way, and he promised that he would in due time give them the secret of his procedures '*for the dissecting of a whole body without spilling any blood*, and for the *embalming it for whole ages*' (p. 8). Robert Boyle was interested enough in this proposal to have it published in English, together with a letter from an unnamed French philosopher who could say, 'I have seen his four embalmed bodies, which are extremely fine; and divers smaller parts of the body by themselves. One of those is embalmed with all the excrements in the guts, and a little fetus (that he hath) is embalm'd whole' (pp. 14–15). It is not easy at this distance to envision quite what Bils had done. The corpses were not in a fluid, since one enthusiastic witness claimed that Bils gave his followers 'complete liberty to let their sight and their hands wander into all parts of the body without exception'.[31]

Whether Bils could be believed or not – and many people thought him a charlatan – it was Boyle himself who in mid-seventeenth-century England showed the method by which 'wet' anatomical preparations first became possible. These were preparations of the soft parts of human and animal tissues, preserved in spirit (alcohol) in a transparent container. Here too, as long as the liquid is kept topped

30 Bils, *The coppy of a certain large act*, 1659.

31 Wittenbergius, *Declaration*, 1657, p. 3. See on Bils, Luyendijk-Elshout's introduction to Ruysch, *Dilucidatio valvularum*, 1964 reprint.

up, the preparations can survive for years. Boyle presented this method in his 1663 book *On the usefulness of experimental philosophy*, with special reference to how medicine and anatomy could be improved:

> Nor were it amiss that diligent trial were made what use might be made of spirit of wine, for the preservation of a human body. For this liquor being very limpid, and not greasy, leaves a clear prospect of the bodies immersed in it, and though it does not fret them, as brine, and other sharp things, commonly imployed to preserve flesh, are wont to do, yet it hath a notable balsamic faculty, and powerfully resists putrefaction, not only in living bodies ... but also in dead ones. And I remember that I have sometimes preserved in it some very soft parts of a body for many months (and perhaps I might have done it for divers years, had I had opportunity) without finding that the consistence of shape was lost, much less, that they were either putrefied or dried up.[32]

Certain objects in the museum of the Royal Society of London were preserved by this method from the earliest years of the institution. In 1680 Thomas Cox, a physician, gave the Society a male human fetus that had been preserved like this from about 1665: 'The skin hath been kept white and smooth for so long a time, *scil.* above fifteen years, by being included with rectified spirit of Wine in a Cylindrical Glass; to the middle of which the *Foetus* is poised, by means of a Glass Bubble of an Inch diameter, the Neck whereof is fastened to the *Anus* of the *Foetus* by a wire'.[33] Similarly the Society had a linnet, given by Boyle himself, that had been preserved in rectified spirit of wine for seventeen years by 1681.[34]

As can be seen from these descriptions, it was usually necessary to suspend the dissected part in the glass jar, using silk thread or gut, in order to display it most clearly. To limit evaporation the top was initially sealed with a wet pig's bladder, which stuck to the rim as it dried. From the early eighteenth century pitch was used to make the seal with a glass top. A number of new designs for such jars were brought out in the eighteenth century, such as the one by Claude le Cat, Royal Demonstrator in Anatomy and Surgery at Rouen, in 1749, illustrated here in Fig. 4.1. This one was designed to reduce evaporation of the spirit by having a double groove filled with mercury or oil, making a tight seal between the glass lid and the glass vessel; the lid had a double hook underneath, from which to hang the threads suspending the preparation. Le Cat thought his contrivance 'vastly preferable' to one he had seen M. Réamur present to a meeting of the Académie royale des sciences in Paris in 1746.[35] The main problem with preservation in alcohol is that

---

[32] Boyle, *Some considerations touching the usefulnesse of experimental natural philosophy*, 1664, pp. 25–6.

[33] Grew, *Musaeum Regalis Societatis*, 1681, pp. 3–4.

[34] Ibid., p. 58.

[35] Le Cat, 'An account of glasses of a new contrivance', 1749–50.

it tends to bleach out the colour, so preparations red with blood soon turn white. The preparation also sometimes disintegrates in the fluid.

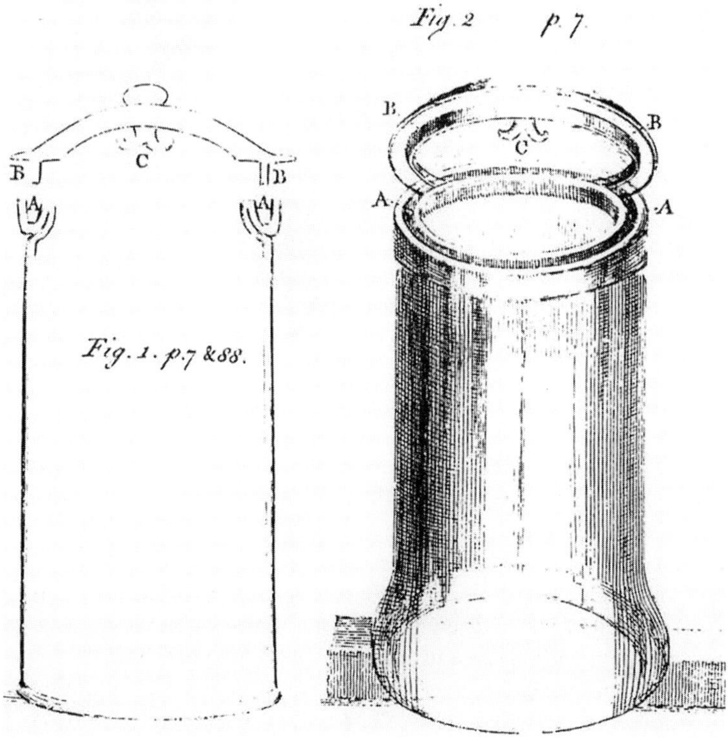

Fig. 4.1

We have seen William Hunter using his preparations in teaching. But there was a further important use for preparations for some anatomists, and that was *research*. John Hunter, William's younger brother and sometime prosector, assembled a vast collection of preparations, both dry and wet, that numbered some seventeen thousand by the time of his death in 1793. We saw something of this unique project of John's in Chapter 2 – his curious surgical/pathological/physiological approach – where we looked at his teaching.

Injections[36]

Injections of air or water (usually coloured) into the vessels or organs of dead bodies by a tube either blown into by the anatomist or attached to the bladder of

[36] In this section I am greatly indebted to Cole, 'History of anatomical injections', 1921, but I have myself seen most of the works he discusses.

some animal, were familiar techniques for anatomists from the sixteenth century, and were used by William Harvey in the 1620s and Francis Glisson in the 1650s. They were used both in research and in anatomical demonstrations. A bladder attached to a tube can be used both to suck excess fluids up, and to push them out, and is in all essentials a primitive syringe. Brass or pewter syringes were already common medical instruments for administering enemas and the like, and were soon adapted for anatomical use.

A more powerful and efficient syringe was invented by the Delft physician Regnier de Graaf. He discussed it most fully in his *Little treatise on the use of the siphon in anatomy* of 1678, saying that he had invented it some fifteen years earlier. It was an adaptation of an instrument, a siphon, he had invented for giving clysters. He wanted something more flexible than a metal pipe. He tried the intestines of a hare, the trachea of a bird and the penis of a bull, but none was satisfactory. Eventually he had to settle for curved bronze tubes and a leather washer. Ideally the whole contraption, shown here in Fig. 4.2, could inject fluids with some force and without leaking too much. The new syringe would help anyone trying 'to investigate the intimate course of Nature of the human body', according to de Graaf. Instead of the anatomy teacher having to dissect out the systems of vessels through their whole course or within a particular organ, with the syringe it would be possible in just one day to demonstrate (for instance) all the arteries and veins of the human body. Different systems of vessels could be filled with water coloured with different tinctures. The siphon would allow one to trace the circulation of the blood through the different parts of the body. And among its other uses, the syringe could produce an erection of the penis in the bodies of dead men – this sort of experiment tied in with de Graaf's long-standing interest in the anatomy of the organs of generation. These Graafian injections are all temporary and for immediate demonstration: the injected fluids gradually make their escape, and no permanent preparation is left.

Transfusion, in its most basic form, consists of the simple injection of a fluid, but from one creature into another, and where the injecting power comes not from a human operator but from the force of the heart of the other animal itself. The transfusion of blood between two living animals of the same or different species (first from dog to dog, later from sheep to man) became something of a public rivalry for a short time between the Royal Society of London and the Académie royale des sciences of Paris in the late 1660s, but was cut short by the death of one of the experimental humans. Initially a series of quills pushed together was used as the connection between one animal and another; later an artery of a horse was used, which was more reliable and flexible.

In the mid-seventeenth century a new method was hit on that served both to display and to preserve preparations: that of *injections of **wax*** and other substances. According to the historian of the topic, F. J. Cole, 'The period of injection lies between 1650 and 1750, after which time interest in injections gradually declined as it was found that the new method was unable to realize the great expectations

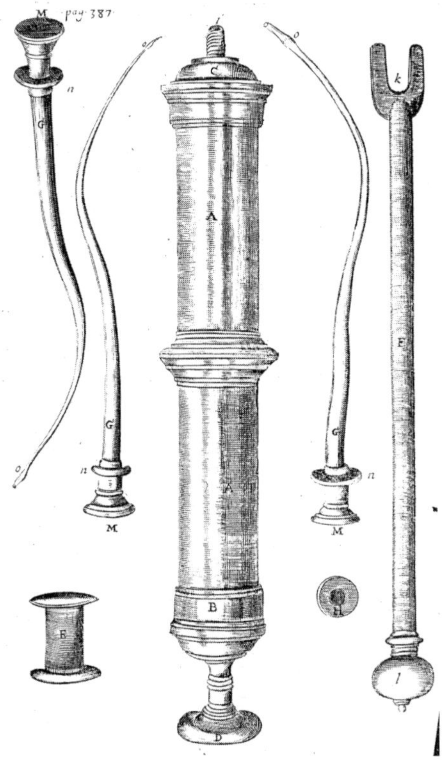

Fig. 4.2

of its professors'.[37] Cole speaks of this period as witnessing a 'frenzy for injections',[38] with William Hunter saying, 'there is no making a good practical anatomist without it'.[39] Jan Swammerdam, working in Amsterdam, showed in 1672 in his *Miracle of Nature, or the structure of the womb in women*, that in order to make a *permanent* preparation from an injection it is necessary to use a warm fluid, which then sets and becomes a solid. A secret wax mixture was used, injected with a copper syringe. After the injected wax had set, the whole preparation was dried and sometimes lacquered to preserve it. By this method Swammerdam was able to prepare a human womb injected with red wax, with its attached vessels, which he sent to the Royal Society of London. By this means he could also make the capillary arteries and veins of the face visible. Charles Jenty, an eighteenth-century London anatomist, recounts what happened next: 'But he [Swammerdam] did not long persist either in the use or improvement of his new invention: for an excess of piety, soon after, spoiled his anatomical turn, and made him look upon such practices as impious. The devout Swammerdam was, no doubt, afraid of rivalling the Almighty in the perfection of his works: but his fears, in this particular, were ill founded'. Swammerdam conveyed the secret to the anatomist Frederick Ruysch of Amsterdam, who performed it even better and (as Jenty joked) 'without any fear of offending God'.[40] Frank Nicholls invented the technique of injecting hard-wax resin into systems of vessels, and then

[37] Cole, 'History of anatomical injections', 1921, p. 287.

[38] Ibid., p. 286.

[39] Cited from a manuscript of Hunter's lectures in his possession by Cole, ibid., p. 334.

[40] Jenty, *A course of anatomico-physiological lectures*, 1757, p. cxxxix.

corroding away the flesh in acid; this results in a delicate wax structure displaying the relations of the vessels.[41]

Of course a consequence of the dedicated use of the technique of injections is that it reveals that just about *all* the parts and organs of the body consist of systems of vessels – that's what injections do, reveal the vascular structure! Certainly the blood vessels (for instance) vary in their incidence and appearance in a way unique to each organ, so the vascular structures created or revealed by injections in the various organs are very different from each other. The injectors could thus ignore the fleshy matter (the parenchyma) around the vessels when trying to explain how the various organs worked. So one might say that the vascular view of the body and its structure, which was the dominant anatomical view throughout the long eighteenth century, was a consequence of the success of the injecting technique, and especially in the hands of Ruysch, who could demonstrate the presence of blood vessels even in the bones. 'The whole body is made up of vessels', as Ruysch claimed in 1696.[42]

Ruysch (according to Jenty) made these injections 'his principal business, and the employment that engrossed most of his time, consisted in carrying anatomy to that noble height of perfection which it had never before arrived at';

> The vessels were so curiously injected, that the remotest parts of their ramifications, which were as slender as the threads of a spider's web, became visible; and, which is still more surprising, sometimes were not so without the assistance of a microscope. What, then, must the nature of that substance be [which Ruysch injected] which is, at once, so fine as to enter the imperceptible cavities of these canals, and, at the same time, is possessed of such a quality, as to indurate itself there?[43]

The extraordinary impact that Ruysch's injection work achieved can be seen in some of his most famous exploits, as recounted by Jenty:

> The entire bodies of children were injected: for the operation was thought very difficult, if not entirely impossible, in adults. Nevertheless, in the year 1666, by order of the States General, he undertook to inject the body of the English [Vice-] Admiral [Sir William] Berkeley, who was killed, on the 11th of June, in the engagement betwixt the Dutch and English fleets. This body, though very much spoiled, before Ruysch put his artful hand to it, was yet sent over to England, as curiously prepared as if it had been the fresh body of an infant ... Every part

[41] According to William Hunter and Cuvier; but Cole ('History of anatomical injections', 1921, p. 326) says that Bidloo, Cowper and Homberg anticipated him by forty years.

[42] Cole, 'History of anatomical injections', 1921, p, 305, with no further reference. I cannot find a work by Ruysch published in 1696.

[43] Jenty, *A course of anatomico-physiological lectures*, 1757, pp. cxxxviii–cxl.

> of the injected matter preserved its consistence, softness, flexibility, and even acquired fresh degrees of beauty with time … All his injected bodies glow with the striking lustre and bloom of youth; they appear like to many living persons fast asleep; and their pliant limbs pronounce them ready to walk. When Ruysch began to produce such surprising preparations, abundance of incredulous people pronounced the facts impossible; but he gently opposed their obstinacy, with these words: 'Come, and see' … Considering the advantages of his secret of injections at that time of day, and the strong curiosity that naturally reigned in Ruysch's breast, we need not be surprised, if he discovered things, that had escaped the notice of all that went before him, such as the bronchial artery, which supplies the lung with nourishment, before unknown to the most minute and accurate anatomists; the periostium of the ossicula auditus, which were formerly looked upon as bare; the ligaments belonging to the articulations of these ossicula. He likewise found, that the cortical substance of the brain was not glandular, as was commonly thought, but consisted of vessels infinitely ramified ….[44]

But more than this. Ruysch did not stop at simple injecting, even of the most challenging subjects. For he had an extremely fanciful mind and constructed tableaux with his preparations, some of them very weird indeed, but clearly fascinating at the time. In 1691 he published his illustrated *Ruyschian anatomical museum, or catalogue of rarities which are preserved in the house of the author*. Here scenes of life in death were presented, with moralising slogans such as 'Being born we die'. The whole series of exhibits presented at one level an enormous *memento mori*.[45] But at another it presented a tour through human anatomy from fetus to old person, sometimes illustrated by animal preparations. Dry preparations, wet preparations, balsamically preserved preparations (one of Ruysch's specialities) and injected preparations were interspersed. Fig. 4.3 here is built on a pile of human bladder stones, either passed while the patient was alive, or extracted by lithotomy, or prised from the dead body, together with kidney stones and other stones coughed up. The enormous central stone was from the bladder of an 80-year-old man, and at the back (letter 'O') 'a stone of prodigious size, like a rock, found between the legs of a corpse in a coffin'.[46] Around this rock stand three skeletons of aborted four-month fetuses. According to Ruysch, the one on the right of the picture 'holds and as it were wipes its eyes with a small chaplet made from the finest membrane endowed with myriads of the reddest arteries, as if weeping, experiencing the miseries of humans' (p. 9). It has the inscription, 'Man, born of woman, living a brief time, is filled with many miseries'.

---

44 Ibid., pp. cxl–cxli.

45 See Hansen, 'Resurrecting death', 1996, passim.

46 Ruysch, *Thesaurus anatomicus primus*, 1737, p. 8.

Fig. 4.3

Peter the Great, Czar of Russia, saw Ruysch's collection in 1698, and on his second visit to Amsterdam in 1717 he bought up the whole collection and had it taken to Saint Petersburg for the use of the Petersburg Academy; much of it can still be seen today in the Kunstcamera. It was catalogued and worked on there by an early appointee to the Saint Petersburg Academy of Sciences, the German Josias Weitbrecht, who later published an account of the ligaments of the human body. Peter obliged Ruysch to supply an account of how he made his preparations, but, as might be expected, this is not particularly revealing of his techniques.[47] Ruysch quickly made another collection of preparations to replace the one Peter had bought.

While wax injection of various kinds was essential to make permanent preparations, *injections of* ***mercury*** were introduced for certain kinds of investigation. While Ruysch had been the first to suspect and then prove that the lymphatics had valves within them,[48] it was Anton Nuck, Professor of Anatomy

47 Cole, 'History of anatomical injections', 1921, pp. 307–8.

48 Ruysch, *Dilucidatio valvularum*, 1665.

at Leiden University, who began to use mercury in the 1680s – 'our Mercury', *nostrum Mercurium* – to explore the glands of the human body, and then the lymphatics.[49] Using mercury, he found that the lymphatics are distributed in all sorts of hitherto unsuspected parts of the body. Fig. 4.4 is his depiction of the lymphatic veins, injected with mercury, around the kidneys of a dog.

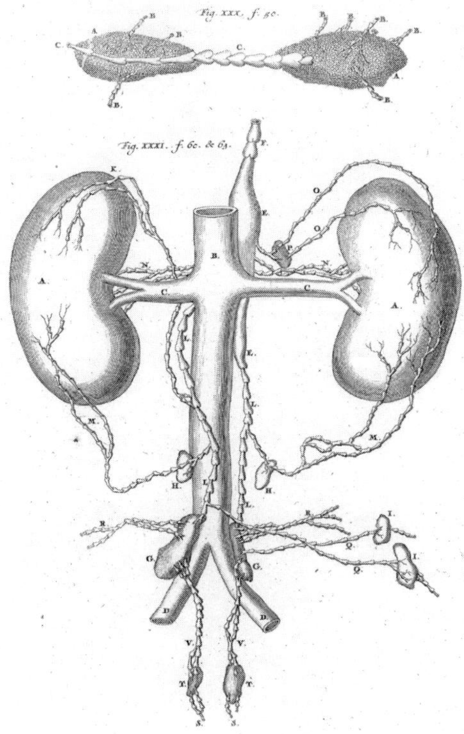

Fig. 4.4

This technique was crucial to the further investigation of the lymphatic system. A century later John Sheldon, a London anatomist and Professor of Anatomy at the Royal Academy of Arts, intended in the 1780s to write a complete account of the lymphatics in all animals, and he began with a historical account of the discovery itself and the techniques necessary for injecting the lymphatics.[50] First he gave due credit to the recent discoverers of the use or function of these vessels: the two 'indefatigable English anatomists of the present age', John Hunter and his pupil the late William Hewson:

---

[49] Nuck, *Adenographia curiosa*, 1691, pp. 12, 13.

[50] Sheldon, *History of the absorbent system*, 1784. In the event only this volume appeared, and was limited to the lacteals in the human.

> The former had observed the lacteals full of white chyle in the crocodile, an amphibious animal of the lizard tribe; and had likewise discovered lymphatic vessels and lymphatic glands upon the necks of birds, particularly upon the neck of the swan, a considerable time before Mr Hewson's discovery of the whole of the absorbent system in those animals. The latter, Mr William Hewson, by innumerable laborious dissections, instituted on living and dead animals, discovered and injected the whole of the absorbent system in birds, amphibians, and fishes. By these important discoveries he proved the universality of the system of absorbents, and by demonstrating their existence in the above-mentioned animals, rendered it most probable that the veins, even in them, do not absorb; since they are provided, as well in man and quadrupeds, with a peculiar system for that purpose.[51]

Sheldon is very detailed on how one goes about the practical side of injecting the lymphatics, in part in order to encourage other anatomists to bring their tacit knowledge and skills into the public arena. It is the best account we have of such work from the eighteenth century. First, one needs to work indoors, lest the preparations dry out too much; to work in winter (the anatomical season); to work from 10am to 2pm; to work with horizontal light from the window; and to work without microscopes, which do not assist one in acquiring the necessary skills for injecting. A table is needed with a depression that will collect spilt mercury and allow one to steep membranous parts. Because of the presence of valves in the lymphatics it is necessary to inject them against the direction of flow, that is, to inject them from the smallest branches towards the larger, which requires great skill since if the vessels are damaged in the operation all the mercury pours out.

All depends on the skill of the operator and the quality of the instruments. Fig. 4.5 shows Sheldon's instruments. A new and delicate kind of tube is required: 'the success of the anatomist depends very much upon the perfection of this instrument'.[52] The operator 'should hold the instrument in his hand like a pencil or pen'. The tube is to be made of glass or brass, with a funnel at one end and a collar of steel cemented to a pipe via a screw-thread. This must be iron or steel since all other materials amalgamate with the mercury. The pipe has an aperture no bigger than the diameter of a wire, and if made of glass it needs to be made by someone skilled in making thermometers or barometers. One should puncture a lymphatic vessel with a very sharp lancet:

> The tube, with the pipe affixed, is previously to be filled by the assistant, with dry pure quicksilver; the anatomist keeping his eye upon the puncture, takes the blow-pipe, and observes by inflating the puncture, whether the vessel be opened; if he has succeeded, he takes the tube from the assistant, still keeping his eye upon the puncture, and holding it just above the collar like a pen, he introduces

51 Ibid., pp. iii–iv.

52 Sheldon, *History of the absorbent system*, 1784, p. 5.

it, by means of the puncture, into the cavity of the vessel … If the quicksilver should pass on with freedom in the vessels, we need only pour more of it into the tube as it descends; but if it be at a stand, we must press on the mercury in the vessel with our fingers. This pressure must be carried on higher and higher, in the course of the lymphatic, till we come near the gland, or glands, into which the vessels terminate … When the vessels are sufficiently filled (which the anatomist determines by the turgescency of them, evident to the sight and touch, and by the resistance made upon the quicksilver in the tube), the assistant passes a thread by means of a needle, if necessary, under the vessel, and ties it above the puncture, before the anatomist withdraws the injecting pipe.[53]

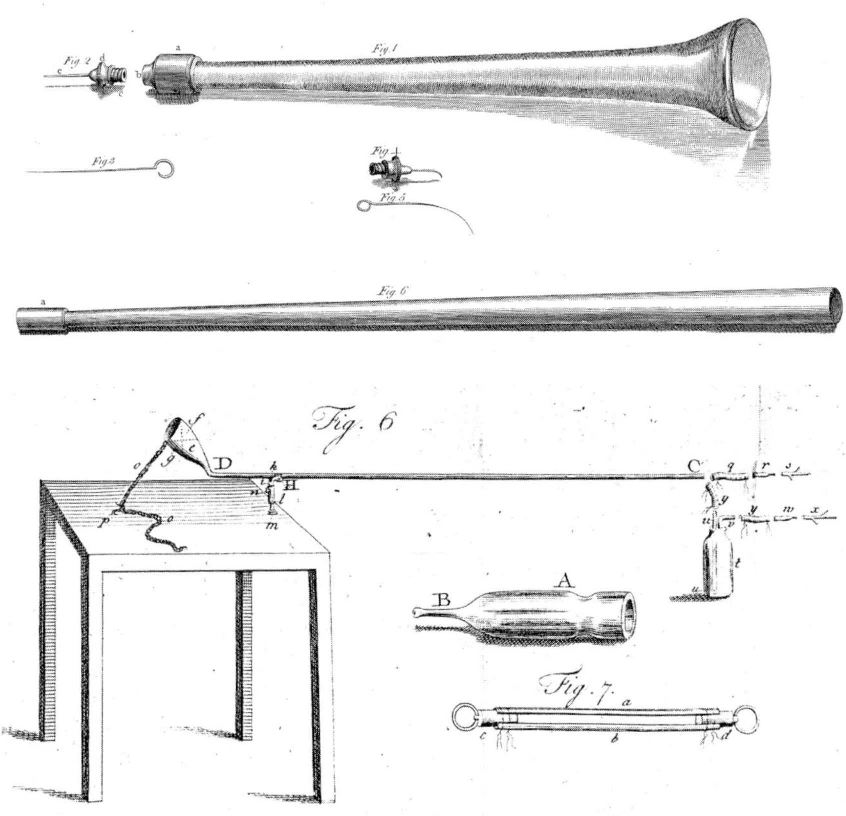

Fig. 4.5

53 Ibid., pp. 10–11.

If one has been successful so far, one can make a dry preparation of the larger vessels thus injected with mercury. First the preparations are dried in air, and then

> … both the vessels and fleshy parts should be varnished over, either with good white spirit, or copal varnish, to preserve them from insects, to render them more beautiful and lasting, and to make the vessels more conspicuous. They should then be inclosed in glass cases, handled as little as possible, and remain always in the horizontal position.[54]

It is also a good idea to make drawings of these preparations, since in time the mercury decomposes or leaks, or the varnish becomes opaque.

One can also make a wet preparation of these vessels. In this case you first feed the experimental animal, and wait while the food in the stomach is turned into chyle as it is absorbed into the lacteals. Then you kill the animal. Next, you should 'take the intestines with the lacteals full of chyle and the mesentery tied off, and plunge into rectified spirit of wine' and preserve the preparation in a bottle. The lacteals have thus been, as it were, 'naturally' injected with chyle. If one wants to replace the chyle with mercury, one can use the putrefaction or maceration method, putting the preparation repeatedly into fresh water until 'the fixed air that is let loose by this process will get into the absorbents [viz. the lymphatics], they will be easily seen, may be opened, and injected with quicksilver'.[55] Such preparations often needed some form of support since, being full of heavy mercury, they otherwise lose their natural shape.

Towards the end of the century, in 1784, when much work had been done on the tracing of the lymphatics, the Paris Académie royale des sciences set up a prize for the best account of the lymphatics, built on experiment. In Siena, Paolo Mascagni, who had been encouraged to investigate them systematically by the Professor of Anatomy there, Pietro Tabarrani, set out to produce a full account of them in humans. He sent off his entry, but it did not reach Paris in time. However, Mascagni persisted, and he was able to publish his major work on the lymphatics in the human body in Siena in 1787. His systematic and long-term work had enabled him to discover lymphatics all over the body, in places unsuspected by previous investigators. His primary investigative tool was his simple injecting instrument, which he had designed and built himself. It consisted of a vertical glass tube to act as a reservoir of mercury – any other material, except steel, would have bonded with the mercury – forming a right angle at the bottom to provide a firm handhold, and ending in a fine pipette of glass, which was inserted in the fragile lymph vessels and through which the mercury was delivered by gravity in a controlled fashion. It is illustrated here in Fig. 4.6.

---

54 Ibid., p. 13.

55 Ibid., pp. 8–9. Alternately, one could inject the vessels with mercury, and then preserve the whole preparation in alcohol, thus preserving the tissue.

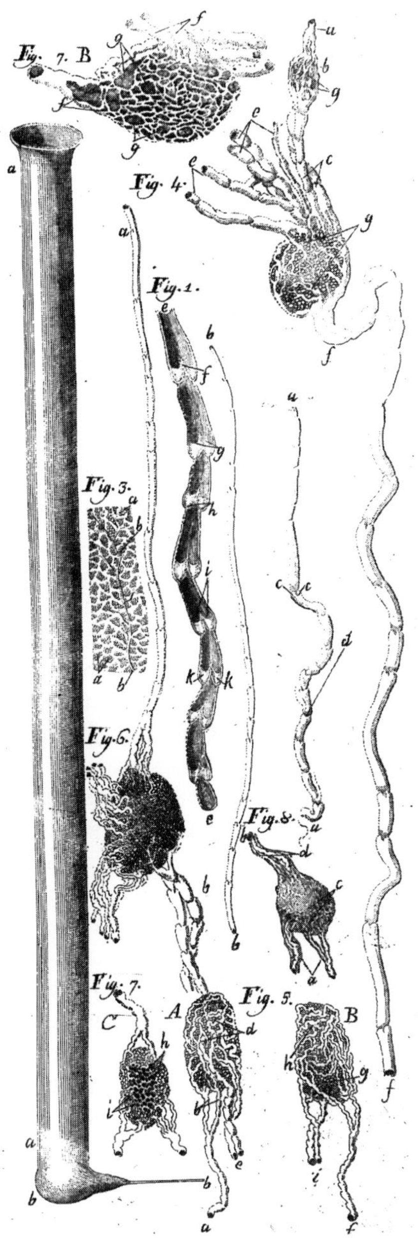

Fig. 4.6

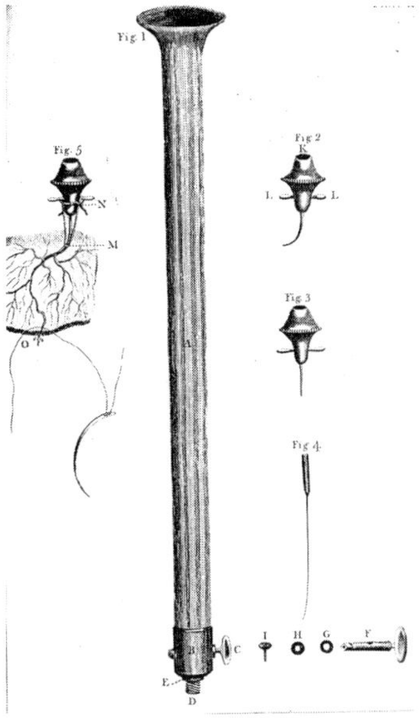

Fig. 4.7

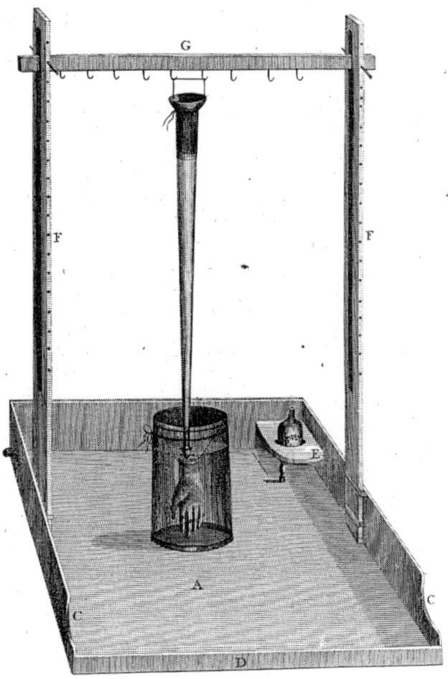

Fig. 4.8

Thomas Pole, a London surgeon, produced *The anatomical instructor* in 1790, to assist students to become 'complete Anatomists' (p. xiv). He is very thorough in expounding the means of injecting mercury. On a wooden tray, illustrated here in Fig. 4.7, that could be filled with water if necessary, a bodily part can be filled by an injecting tube suspended from an adjustable cross-bar. The next illustration (Fig. 4.8) 'is a glass jar containing water, in which is immersed a hand; with the quicksilver injecting pipe fixed in the artery, as in the process of filling the vessels. The hand is suspended by a string from the edge of the jar' (p. 121). The injecting tubes were made of glass fixed in a steel cock, and could be from five- to twenty inches long, depending on the pressure needed to inject the vessels.

As will be clear from the obsession with the lymphatics that eighteenth-century anatomists showed, being able to demonstrate them in this way was a great prize. Not only have the invisible vessels been made permanently visible, but their function has also been discovered.

Wax modelling

Wax in various forms, usually made from animal fat, was a basic material for the anatomical injector, as we have seen. But wax was also important for another new anatomical entrepreneur: the wax modeller, who copied what he or she saw in dissection and preserved it in full-colour three-dimensional models shaped from wax. The basic ingredients are beeswax, Venice turpentine, and some ingredient to give the desired colour. Most of the anatomical wax modellers worked on their own, and kept their formulas and techniques strictly secret. Whatever it was they added to this basic formula has proved to have made these models extremely resistant, and even some that have been left inadvertently on heated radiators for years have not melted.

Wax modelling has been practised by sculptors from ancient Greece well into the nineteenth century. It has many virtues in this respect:

> Perhaps the single most important aspect of any sculpturing material is its plasticity. Rivaling clay, one of man's first plastics, wax can be poured, carved, made malleable with hands, remelted, corrected, painted, and can be worked with the most simple tools or just by fingers and fingernails. Wax does not flake, peel, chip, or warp incurably. It takes centuries to dry out; it is not injured by insects nor affected by moisture. Size is no problem. With an armature support pure wax can be modelled into any size, no matter how large.[56]

The wax sculpture can either be built up, modelling it piece by piece, and assembling the pieces by using a soldering iron or pegs, or it can be cut or carved out of the solid wax block, like marble sculpture.

---

[56] Newman, *Wax as art form*, 1966, pp. 99, 104.

In the case of anatomical wax modelling the initial idea came from the Sicilian abbot Gaetano Giulio Zumbo (or Zummo, 1656–1701), and was named after him, a 'Zumbo' being 'a wax picture which was really an oil painting that had some of its parts modelled in wax relief and then locally colored'.[57] Zumbo had created such scenes in Florence, such as 'On the French disease' and 'The five stages of decomposition', grisly scenes of death and decay presented as moral lessons. The move to anatomical modelling has been well retold recently by Martha Burmeister:

> At the close of the seventeenth century, Zumbo's career took him to Genoa where he formed a partnership with Guillaume Desnoues. Desnoues, a surgeon and professor of anatomy, had been producing models made through wax injection to illustrate his bi-weekly lectures. After preparing a model of a woman and child who had both died during childbirth (which he showed before an audience of 2,000 people open to nobility), he was frustrated to find that the figures still suffered from decomposition. Eager to find a way of preventing decay, Desnoues joined with Zumbo; Desnoues would dissect bodies and Zumbo would recreate their likeness in wax forming an artificial anatomy.[58]

Desnoues fell out with Zumbo as he preferred his wax models without the horror of the tomb, which was Zumbo's speciality, and he continued on his own.

The resulting wax anatomical models soon became well known as Desnoues showed them in Paris in 1711 to the Académie royale des sciences, and then his nephews took the collection to England, touring it from 1719. Later the models became the basis of Rackstrow's Museum of Anatomy and Curiosities, which continued in London from 1746 to 1798.[59]

One of the advantages of anatomical waxes is that, if they are well done, they obviate the need for a constant supply of fresh human corpses for students to dissect. Once the waxes have been modelled on the basis of dissected corpses, and suitably coloured, they function as teaching aids in which all the salient parts are more evident to the eyes than in the dissected body: for instance, if the arteries and veins are coloured differently, their respective routes through the body become self-evident to the eye. It all happened first in Bologna, a city still in the mid-eighteenth century at the forefront of investigations into nature. The Archbishop of Bologna, Prospero Lambertini, had been trying to encourage anatomical studies in

---

57 Ibid., p. 93.

58 Burmeister, *Popular anatomical museums*, 2000, pp. 30–31.

59 On Desnoues's and other anatomical waxworks in London from 1719, see Peachey, *A memoir of William and John Hunter*, 1924, pp. 25–31; on the fate of this and other popular anatomical displays in nineteenth-century England, see Burmeister, *Popular anatomical museums*, 2000. William Cheselden in his *Osteographia* of 1733, table LVI, refers to a human fetus supposedly in its mother's [fallopian tubes] for twenty-six years, 'Communicated to me by Monsieur Arneau owner of the french anatomy waxwork'.

the town.[60] The wax sculptures of a horse and a human kidney by a worker in wax, Ercole Lelli, apparently inspired Lambertini to establish an anatomical museum in the Academy of the Sciences of the Institute of Bologna. When Lambertini became pope himself as Benedict XIV (1740–58), he employed Lelli to sculpt a comprehensive anatomical display in wax:

> … eight life-size sculptures, of which two were a male and a female nude, and six were écorchés of the human musculature; twenty representations of the muscles of the larynx, the pharynx, the eyes, the ears, and the generative system; skeletons one each of a male, a female, and a fetus; six tables representing all the segregate bones of the body; a single sculpture of the cranial and jaw bones detached; six tables of the large and small bones encased in their membranes; four tables of the cartilaginous parts of the body; four tables of sectioned bones; and two tables one each of the inner and outer ear.[61]

This commission made Bologna a great centre for the production of wax anatomical models. Lelli took on Giovanni Manzolini to assist him, and when these two fell out, Manzolini opened his own wax modelling studio. He was assisted in this by his wife, Anna Morandi (1716–74) and together they provided hundreds of models for patrons across Europe, and especially for the School of Obstetrics in Bologna that was run by the Professor of Surgery, Giovanni Antonio Galli. Anna Morandi appears to have rivalled and perhaps even excelled her husband in the quality of her anatomical waxes, and she gave anatomy and dissection lessons to medical students privately. After the death of her husband in 1755, Anna Morandi received direct patronage from the pope, who allowed her an annual stipend, while the Bolognese Senate gave her the title of Chair of Anatomical Modelling at the University of Bologna.[62] The waxes produced by Manzolini and Morandi are still preserved in Bologna.[63] In the striking self-portraits in wax that the couple produced, seen here in Fig. 4.9, Morandi portrays herself as dissecting the human brain, the centre of the senses, while Manzolini shows himself dissecting the heart, the centre of the emotions.

---

60 Messbarger, 'Waxing poetic', 2001, pp. 68–9.

61 Messbarger's paraphrase and translation of the commissioning document; see Messbarger, ibid., p. 70. Presumably some are described as 'tables' because they were mounted on tables.

62 Ibid., pp. 73–6.

63 An illustrated catalogue of the anatomical waxes of Lelli and the Manzolinis is given in Armaroli, *Le cere anataomiche Bolognesi del settecento*, 1981.

Fig 4.9

The Bologna anatomical waxes in turn inspired Giuseppe Galletti, a physician, to try to have an equivalent publicly accessible collection established in Florence, the capital of Tuscany. He approached Felice Fontana, physicist to the court of the Grand Duke of Tuscany (Peter Leopold of Lorraine), to raise the subject with the Grand Duke to improve facilities for medical training. Fontana, however, presented it – successfully – to the Grand Duke as both 'one of the most beautiful components of modern physics', and also as a project that would render the new Grand Duke's name immortal.[64] The project was begun, and although Fontana was not the modeller but the organiser, it became associated with his name, since it was he who decided what models were to be produced and how. The exhibition was treated as part of the new 'Imperial and Royal Museum of Physics and Natural History', which was in the planning and organising stage, and which was intended to 'enlighten the people and make them happy by making them civilized', following physiocratic political principles.[65] So when the museum opened in 1775 there were already six rooms full of anatomical waxes, produced under Fontana's direction. By the end of the century the whole museum complex was popularly called 'La Specola' after its observatory tower. In its day La Specola was regarded as the finest collection of anatomical waxes in Europe open to the public.[66]

---

64 Maerker, 'The anatomical models of La Specola', 2006, pp. 296–7.

65 Felice Fontana, quoted in Maerker, 'Turpentine hides everything', 2007, p. 257.

66 Hilloowala et al., *The anatomical waxes of La Specola*, 1995, pp. 42–3.

Fontana built up and arranged the anatomical wax collection in such a way that a visitor or student could (he thought) educate him- or herself by simply learning through their eyes: no text was needed except the artificial body in its perfect form, through visual study of which the laws of nature underlying its form and function would spontaneously appear to the spectator. 'Everything remains well enclosed on the shelves by majestic and great glass panes, and at one glance everything is seen, everything is known. The symmetry that reigns within and without the shelves, the order and regularity of the pieces is such that it captivates the observer, and at the same time instructs and delights him'.[67] However, this belief in the possibility of the innocent self-education of the visitor was not shared by everyone – not even by Fontana's own assistant. There were plenty of visitors, for the museum was open to the public several hours a day, and more than a hundred visitors a day are recorded.[68]

The wax models here were fashioned after illustrations in famous anatomy books. For instance Albinus's celebrated images of the human figures, with the muscles on show, were faithfully copied into wax.[69] Three-dimensional corpses had been transferred to two dimensions on the page, and were now restored to three dimensions in the wax models! But dissection of human corpses was another resource for Fontana's wax modellers, and an ample supply of corpses was granted to Fontana of adults from the hospital of Santa Maria Nuova and of children from the orphanage, the Istituto degl'Innocenti. The museum's sweeper went round these institutions twice a day with a basket to collect the latest available bodies.[70] Understandably, there was sometimes conflict over these corpses with the teachers and medical students at the hospital.[71] The technical procedures were quite elaborate, with plaster-casts often being taken from the human parts, and the wax model being cast from the plaster-cast. A sculptor always worked under the supervision of an anatomist, 'and this is done because without such supervision not even the most excellent sculptors can copy nature with exactitude'.[72]

Wax anatomical figures from La Specola, made on commission, can be found across Europe, in for instance Montpellier or Pavia. Fontana's collection was seen by Emperor Joseph II of Austria. He asked Fontana to have some figures made for the Military Medical Academy that he had founded in 1786 in Vienna, the Josephinum. Fontana made them, and they were delivered in forty cases by mule.

---

67 Maerker, 'The anatomical models of La Specola', 2006, p. 302, quoting and translating the 1775 Italian guide to the museum.

68 Ibid., p. 304.

69 Some of the correspondences are illustrated side by side in Hilloowala et al., *The anatomical waxes of La Specola*, 1995, pp. 22–37. For a complete catalogue, reproduced in colour, of the waxes, see Düring et al., *Encyclopedia anatomica*, 1999.

70 Hilloowala et al., *The anatomical waxes of La Specola*, 1995, p. 51.

71 Maerker, 'The anatomical models of La Specola', 2006, pp. 301–2.

72 A medical visitor, Dr Des Genettes, writing between 1789 and 1792, quoted in Hilloowala et al., *The anatomical waxes of La Specola*, 1995, p. 56.

Joseph paid 30,000 florins for them. They are still there today. According to a contemporary,

> The preparations are set up in caskets of rosewood with gilt edges and walls of the finest glass. In this way they are protected from dust, without being obscured from the eye of the observer. The caskets may be opened for study. The wax preparations are surrounded by drapery of white silk and pillows of blue satin decorated with silver fringes, and with curtains of green silk which serve partially for decoration and partly for conservation. Over each preparation hang framed colored drawings with labels in Latin and in Italian. The drawers beneath the caskets contain the descriptions of the preparations in German and in Italian.[73]

However, it appears that the students in the Josephinum hardly used the waxes, as enough bodies were available for them to dissect from the Garrison hospital.

3. Picturing bodies better

There were many anatomical atlases produced in the eighteenth century – that is, large (sometimes elephant folio) books illustrating the anatomised human body or its parts, or various animal bodies and their parts. On the negative side we can say about these atlases that they very rarely made money or even met their costs. They were sometimes abandoned midway. There really was not a serious market for them, especially as every stage of the printing processes that they had to use was so very expensive. On the positive side we can say that they were produced to spread the word about anatomy. The marvellous colour prints of Gautier d'Agoty (on whom more below) were produced, he said, 'to facilitate the study of anatomy by all sorts of persons, especially those who are studying Medicine, Surgery, Painting and Sculpture, and who have as their goal the health of the human body, and the greater part of whom lack the necessary faculties and facilities to arrive at precise and sure knowledge, which it is almost impossible to arrive at without a collection of a great number of dissected pieces, all which one finds prepared and demonstrated in this work'.[74] As one might suspect, the different styles of portrayal that each anatomist adopted were of a piece with the fashions and styles of art more generally, and this is an important reason for the variety of imagery in anatomical portrayal.

---

[73] Quoted (and presumably translated) from Jantsch, 'Zur Geschichte der Wiener anatomischen Wachspräparetensammlung', 1951, in Brazier, *A history of neurophysiology*, 1984, p. 145.

[74] Gautier d'Agoty, *Anatomie de la tête*, 1748, 'Avertissement'.

Better portrayal

The problem underlying all illustrations of anatomical dissection is that they are all, to a greater or lesser extent, idealisations. Indeed this is why engravings in the eighteenth century or photographs in the twenty-first century are attempts at solving the same problem: that of bringing into view, and to the viewer's attention, the things that the anatomist wishes to make visible. For anatomising is not only a very messy business because of the nature of its subject material, but distinguishing all the structures that are visible to the eye of the trained anatomist is very difficult for those who are not yet anatomists. Memoirs of doctors, looking back to their own training in the last generation when medical students learned their anatomy on the human corpse – that is, the final decades of the twentieth century – recall how 'difficult it was for them to identify any structures beyond the major vessels and muscles with certainty'. In Glasgow in the late 1970s, a former medical student recalls, 'the lecturer would prepare us [medical students] … drawing on the board what we already knew we'd be unlikely to see, even with the eye of faith, and then we'd all troop through into the laboratory with our white coats, dissection sets and the latest edition of the anatomy textbooks … These were … our "atlasses", regional topologies of the human body'.[75] So illustrations, while presented as objective records of what can be seen in the body, have actually always served as very practical guides to the eye when young anatomists have first looked into the body, because they are idealisations.

Many of the most dedicated anatomists of the eighteenth century were concerned with developing the legacy of accurate and aesthetic illustration of human anatomy, a legacy that they traced back to Vesalius and Eustachius in the sixteenth century and whose works some of them reprinted (as we saw in Chapter 1). William Hunter for instance, expressed the view that accomplished illustrations functioned, as it were, as a universal language, like music:

> The art of engraving supplies us, upon many occasions, with what has been the great *desideratum* of the lovers of science, an universal language. Nay, it conveys clearer ideas of most natural objects, than words can express; makes stronger impressions upon the mind; and to every person conversant with the subject, gives an immediate comprehension of what it represents.
>
> From the time when this art came more generally into use, it has been much more easy both to communicate and to preserve discoveries and improvements; and natural knowledge has been gradually rising, till it is at length become the distinguishing characteristic of the most enlightened age of the world.
>
> Anatomy has, at least, kept pace in improvement with the other branches of natural knowledge. Many of the moderns, through much labour and patience, as well as ingenuity and judgement, have thrown considerable lights upon the structure and operations of the human body; and they have particularly by

[75] Bamforth, 'Splendid cadavers', 2006, pp. 103, 101.

> engravings, made the study of that art, in which humanity is so much interested, both more easy and pleasant. Most of the principal parts of anatomy have, in this manner, been successfully illustrated.[76]

Interestingly, the work in which this paean of praise to anatomical engravings appeared, Hunter's outstanding anatomical atlas of the human gravid uterus, was a work essentially without a text, or rather, where the illustrations themselves serve as the text.

A few anatomists invented or applied new techniques in order to make more perfect visual representations of anatomy, and I shall discuss two of these below.

The Albinus method

It was the intention of Bernhard Siegfried Albinus – the 'great' Albinus – Professor of Anatomy at Leiden University, to produce the most beautiful anatomical tables ever seen, and he eventually believed that he had actually achieved this. His obsessive perfectionism and his overwhelming concern with accuracy in portrayal created many problems and delays for himself in executing his project. The first and greatest of these problems was that he simply could not draw! So he had to guide the pencil of somebody else. But in the second place, he had a very strong mental artistic vision of what he wanted. And what he wanted was to portray the 'ideal' human body, the 'homo perfectus': not the body of any individual, but of the idealised man or woman.[77] Yet all he could use as his raw material were the corpses of men and women with all their natural variations. The ideal needed to be distilled from all these imperfect versions. The project took him many years, during which he was 'pushed on both by a love of the work, and an earnest inclination to improve Anatomy, to try if I could perfect what I had conceived in my mind'.[78]

The artist and engraver that Albinus employed was Jan Wandelaar. Writing in the 1740s, Albinus could say of this artist that, 'for a great many years by-past, he has worked for very few besides myself; and for these last ten (most part of which he has been wholly employed on these tables) almost for me only. And he both drew and ingraved them all by my direction. In the first place I endeavoured to make him understand, as well as possible, what was to be drawn; and I was constantly with him, to direct how every thing was to be done, assisting him in the drawing, and correcting what was drawn. And thus he was instructed, directed, and as entirely ruled by me, as if he was a tool in my hands, and I made the figures myself'.[79] Not only did Wandelaar have the patience of Job, tolerating Albinus

---

76 Hunter, *Anatomia uteri humani gravidi*, 1774, Preface.

77 Punt, *Bernard Siegfried Albinus*, 1983.

78 *The explanation of Albinus's anatomical figures*, 1754, 'An historical account of the work', pp. iii–xxiv, see p. xiv. The Tabulae were first published in Latin in 1747. The historical account was also translated in Brisbane, *The anatomy of painting*, 1769, pp. 5–21.

79 *The explanation of Albinus's anatomical figures*, 1754, pp. xxi–xxii.

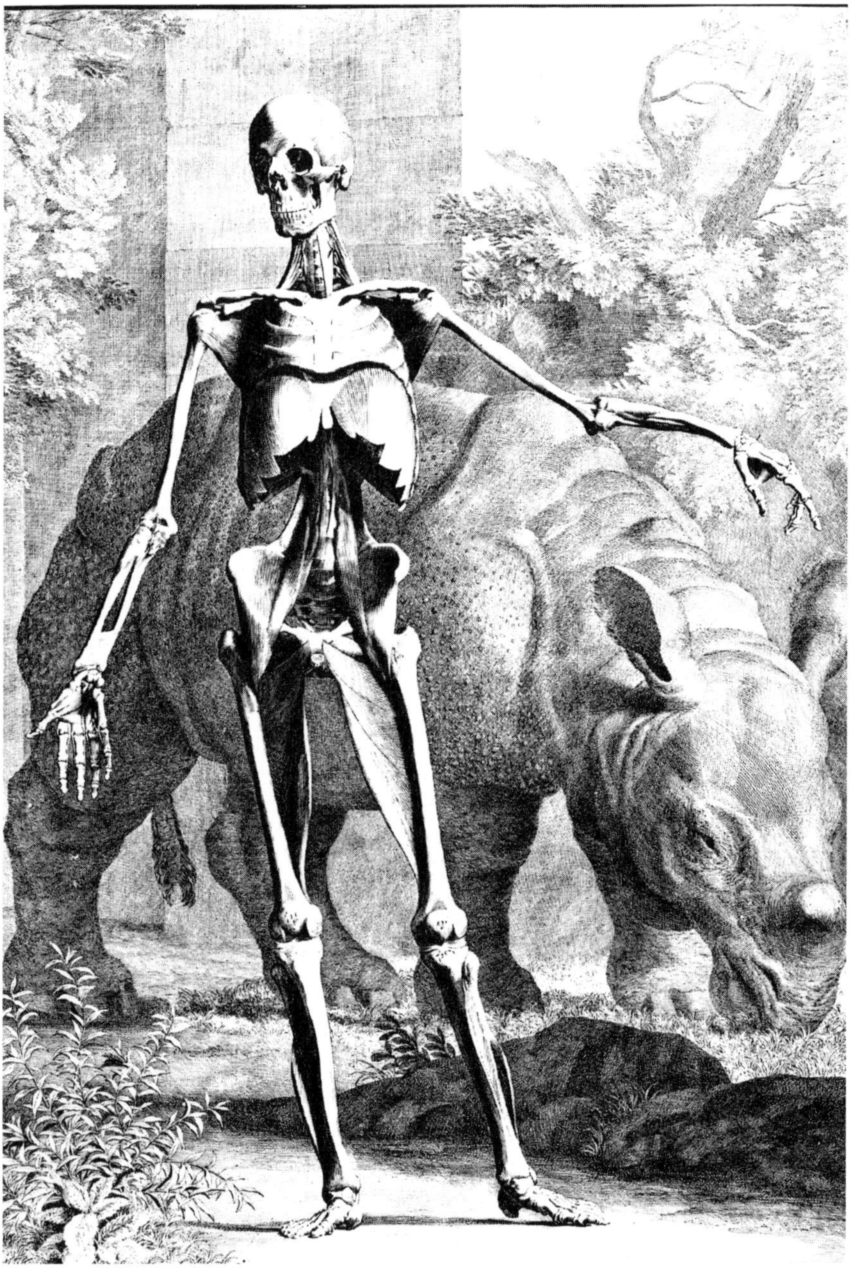

Fig. 4.10

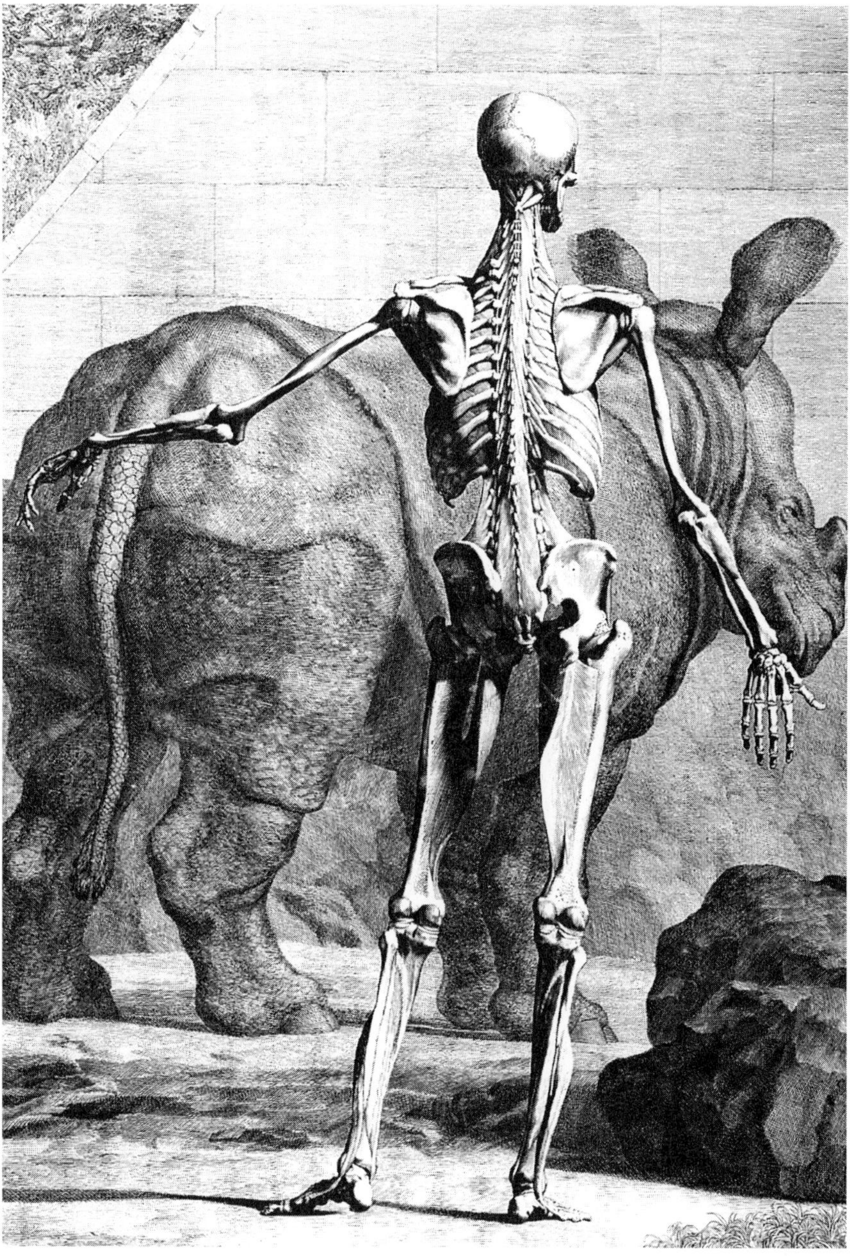

Fig. 4.11

fussing around him the whole time, but he even went out of his way to make the illustrations yet more beautiful: 'To the general figures he [Wandelaar] has added ornaments, not only to fill up the empty spaces of the tables, and make them appear more agreeable; but likewise that by the shading of those ornaments upon the margins of the tables, the light and shade of the figures might be preserved, and heightened' (p. xxii). One such ornament was Clara the so-called 'Dutch' rhinoceros whom Wandelaar drew from the life when she was being exhibited in Amsterdam, here in Figs 4.10 and 4.11, shown in front and rear view, matching the pose of the human. We shall meet Clara again in Chapter 5.

Albinus's account of his procedures is extremely detailed and recounts both his failures and his successes over two decades. He began in the early 1720s by planning the portrayal of the muscles, and (as he wrote) he 'was extremely desirous to have them exactly expressed in the figures, not only singly, but in the order in which they lie with one another'. This scheme failed because when the outer muscles are removed, the proper situation of the inner muscles is not preserved. 'What I wanted was something more than even the best anatomists trouble their heads about', he wrote; 'it being usual for them to make only random figures of the parts, without considering either the orders, dimensions, continuations, or connexions of them with one another'.[80]

So Albinus began his work over again using a new method, and this time decided to start with the skeleton 'as the basis or foundation to build my figures upon'. This schema had great advantages, 'for as architects, having first laid a certain foundation, upon it build the edifice, together with all its parts; so we may consider the muscles connected to the skeleton, as to the foundation; the viscera, and other parts in their proper situations: then the nerves, veins, and arteries, properly disposed, and whatever else is necessary towards the construction of the human fabric. It is by far the best way then for art to imitate this method, which nature has pointed out' (p. v).

Albinus chose the best skeleton he could find for this work:

> I chose it of that age in which the bones are arrived at their full growth and firmness … It was of the male sex, of a middle stature and well proportioned; of the most perfect kind, without any blemish or deformity, either as to the bones themselves, or their connexions with one another … I made choice of one that might discover [i.e. reveal] signs both of strength and agility; the whole of it elegant, and at the same time not too delicate … in short, all the parts of it beautiful and pleasing to the eye. For as I wanted to shew an example of nature, I chose to take it from the best pattern of nature. But even those skeletons, which may be reckoned the best, are different from one another, and I intended only to exhibit one of them; I proposed that that one should be an example for the others. And I cannot help congratulating my good fortune in this respect, for happening

80 Ibid., p. xxii.

> upon a body, which as it promised, so likewise it contained such a skeleton as I wanted.[81]

The concern for beauty and for following the best patterns in Nature is very evident throughout Albinus's whole approach. This project was not intended simply as a first-class technical portrayal of the human body, but a work of art too, reflecting Nature's work of art in 'the book of nature'.

But how to prepare a human skeleton with all the cartilages still intact after the bones had been well cleaned? Towards the end of 1725 Albinus prepared a skeleton carefully, 'but the ligaments being naturally weak, and adhering but loosely to the bones, it required a good deal of skill to manage it, to keep the bones in their proper positions' (p. vi). Albinus then conceived the method that had to be adopted: the skeleton needed to be supported on a tripod, as shown here in Fig. 4.12, re-imagined after Albinus's description. The tripod had to be slightly taller than the skeleton, whose feet must not reach the ground. The legs and lower part of the trunk were laid on the tripod, which was itself placed upon a low table. 'Then I tied a cord to the upper part of the spine, where it is firm and less flexible, and pulling it up streight to the ceiling of the room, I passed it through a ring which I had fixed into the ceiling, and then fastened the end of it to a hook which was drove into the wall' (p. vii). Albinus then fastened other cords to the spine of the skeleton, and fixed them in the walls, and then used more cords to suspend the arms from the ceiling. The skeleton was eventually held by a dozen or more cords 'and partly by pieces of board, paper, pasteboard etc.' (p. viii) in a position that suited Albinus. Having arranged his skeleton Albinus then next checked it against a living human: 'I next looked out for a thin man, of the same size with my skeleton, and making him stand naked in the same position, I compared the skeleton with him, especially the hip-bone, spine, thorax, scapulae, and clavicles; because if these parts were put into proper positions, there would not be any great difficulty in the rest' (p. viii). Once the skeleton was well prepared and drawn, it would be the foundation for drawing the muscles.

Albinus now applied his obsessiveness to the question of how the skeleton was to be drawn. Initially he expected the image to be engraved directly onto a metal plate by an engraver. Here again Albinus made a technical innovation by using two 'quadrates' in order 'to measure the whole skeleton, and mark out the position, size, and figure of every part'.[82] The first, large, quadrate was a frame as large as the skeleton, with cords stretched across vertically and horizontally making a number of equal squares. The engraver's plate was similarly divided into the same number of equal squares. Finally a small frame, divided as before, was placed four feet in front of the larger frame. This can be seen in Fig. 4.13, again re-imagined

---

81 Ibid., pp. xiii–xiv.

82 These quadrates should not be confused with the grids artists sometimes used for drawing perspective, as by Albrecht Dürer; on which see Gombrich, *Art and illusion*, 1972, p. 306.

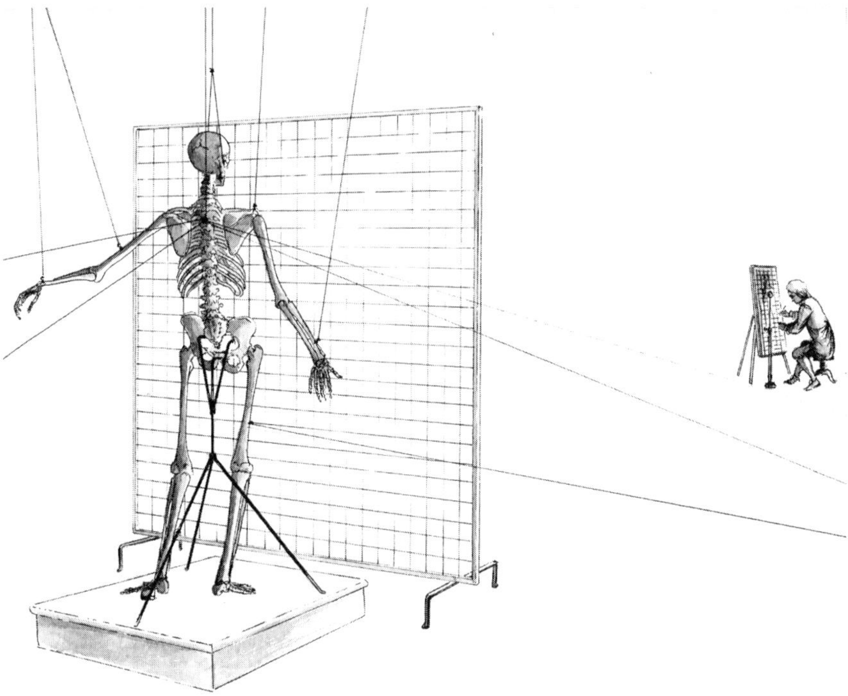

Fig. 4.12

after Albinus's description. The engraver's task was now to portray within each square on the plate what was visible in each square of the frame in front of the skeleton. The engraver could now look through both frames and see where parts of the skeleton intersected ('decussed') with the cords forming the squares in both frames. These he could mark on his plate and then fill in the missing lines. It worked, but even Albinus noticed that 'it occasioned an incredible deal of trouble to the engraver' (pp. ix–x). A front view, rear view and side view of the skeleton were prepared in this way. The whole operation took three months, in which the skeleton had to be kept moist lest it dry out. For some time there was a hard frost that kept the skeleton perfectly preserved for copying. The skeleton 'was hurt likewise by the fire, which we were obliged to have always when the naked man stood; for he neither could nor would stand without it' (p. x). Some years later, in 1733, Albinus returned to these images of the skeleton and had the artist reduce them from their natural size. They were then engraved and printed.

The result of Albinus's use of the tripod and the many cords to position the skeleton, was that he could portray the skeleton in very life-like poses in which the relation of the bones to each other was as in action in life. In the first two pictures (front and back) the skeleton

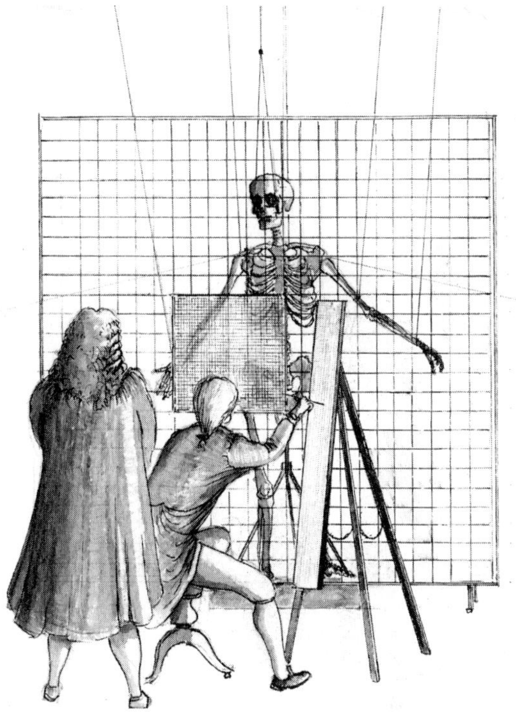

Fig. 4.13

> … stands firm upon the right foot, resting but very gently on the left. The right foot rests upon the heel, and upon the anterior heads of all the bones of the metatarsus, especially of that which belongs to the great toe, the sesamoid bones in the mean time intervening. The toes are bended downwards, as it were to catch hold of the ground; and by this it stands the firmer. The extremity of the foot is turned outwards, as it naturally is in that posture, when the foot stands most firm. The right knee is straight, etc.[83]

After the skeleton, the muscles next needed to be portrayed. Albinus was busy at this for a number of years from 1738. Here, 'the muscles appear connected to the out-line copies of the skeleton, in the same manner as they are to the skeleton in the dead subject' (p. xvii).

Finally, to ensure that the beauty of the figures not be hurt, Albinus came up with the idea of double tables: one of the figure fully finished, with all the shading that was required, and one of the outline figure, on which the key to the names of the parts could be clearly seen. See Figs 4.14 and. 4.15.

---

83 *The explanation of Albinus's anatomical figures*, 1754, p. xi.

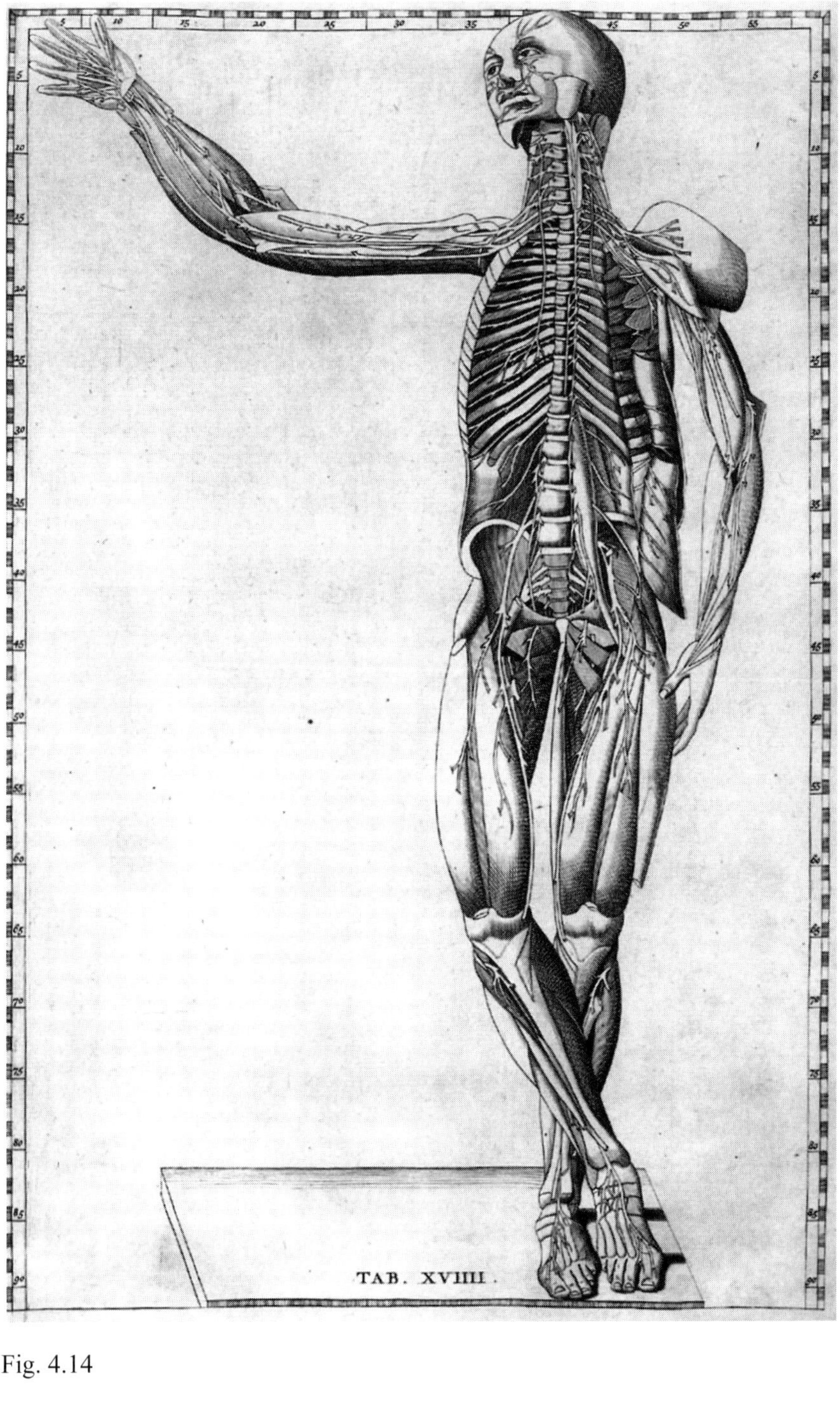

Fig. 4.14

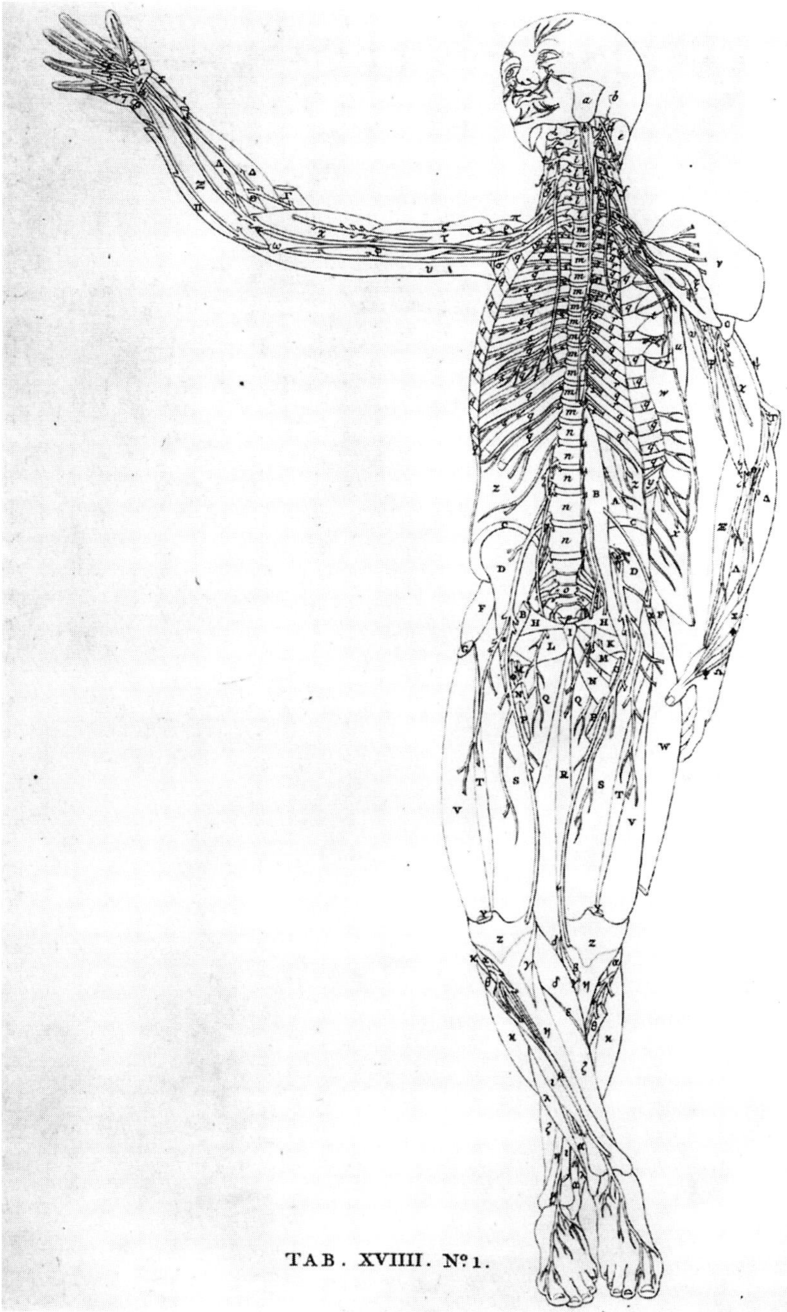

Fig. 4.15

No project of anatomical illustration can have had more time, thought, labour, artistry and money spent upon it – Albinus spent 30,000 guilders on publishing it – than this atlas of the human body.

The Cheselden method (Camera obscura)

Albinus's atlas had rivals, however, both for beauty and completeness. One of them was by William Cheselden, the London surgeon and anatomist, who worked for many years on his great folio volume on the bones, human and animal, the *Osteographia, or The anatomy of the bones*, which was published in 1733. In this work, as the title-page of one of the editions claimed, there were fifty-six plates, and 'every bone in the human body is here delineated as large as the life, and again reduced to lesser scales, in order to shew them united to one another'.[84] He recounts in the Address to the Reader that he and his artistic collaborators had started by copying poses from Vesalius's book of almost two hundred years earlier, starting with the skull. The use that Cheselden makes of Greek statuary models, such as the Venus de Milo and the Belvedere Apollo, are in this spirit of emulating Vesalius, as is a general playfulness in the choice of animal images, with 'the sceleton of a cat frighted at the sceleton of a dog' (Fig. 4.16), and 'a pocky sceleton, in which there is scarce one bone sound' (Fig. 4.17).

But this process, apart from revealing that there were significant disproportions in Vesalius's images, did not give good results. As he wrote to the reader: 'Then we proceeded to others, measuring every part as exactly as we could, but we soon found it impossible to execute the difficult parts of such work in this way; upon which I contrived (what I had long before meditated) a convenient camera obscura to draw in, with which we corrected some of the few designs already made, throwing away others which we had before approved of, and finished the rest with more accuracy and less labour, doing in this way in a few minutes more than could be done, without in many hours, I might say in many days … My engravers, Mr Vandergucht and Mr Shinevoet, not less skilled in drawing than in their own proper art, knew too well the difficulties of representing irregular lines, perspective, and proportion, to despise such assistance, always declaring that it was impossible to do these things so well without'.

Cheselden put a picture of the camera obscura (Fig. 4.18) on the title-page of his book on the bones. In Chapter 8, on 'Comparative skeletons', he explained the structure and use of the camera: 'About six inches within that end where the draftsman sits is fixed the table glass, upon the rough side of which he draws with a black-lead pencil, which he afterwards traces off upon paper; towards the other end, in a sliding frame, is put the object-glass, which being moved backward and forward, makes the picture bigger or less, and the inside of the case is made black to prevent reflections of light'. A pupil of Cheselden's elaborated this explanation a little, to specify that the object-glass was convex, 'through which the rays passing

84 Russell, 'The *Osteographia* of William Cheselden', 1954, p. 39.

Fig. 4.16

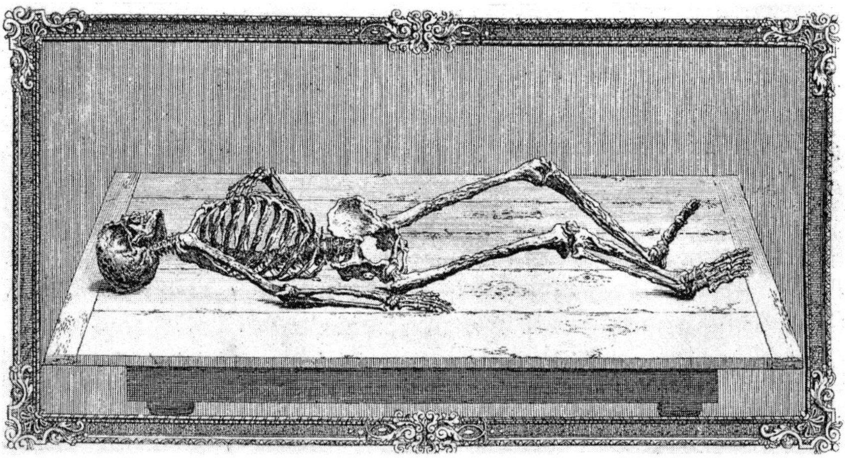

Fig. 4.17

form the object, converge and meet in a focus upon the table-glass placed near the other end, analogous to the crystalline humour and retina in the eye'.[85] The plate of the skull of a woman who had died of the venereal disease (Fig. 4.19) shows, for instance, how the use of the camera obscura could make complicated structures much clearer.

85 Russell, 'The *Osteographia* of William Cheselden', 1954, p. 46.

Fig. 4.18

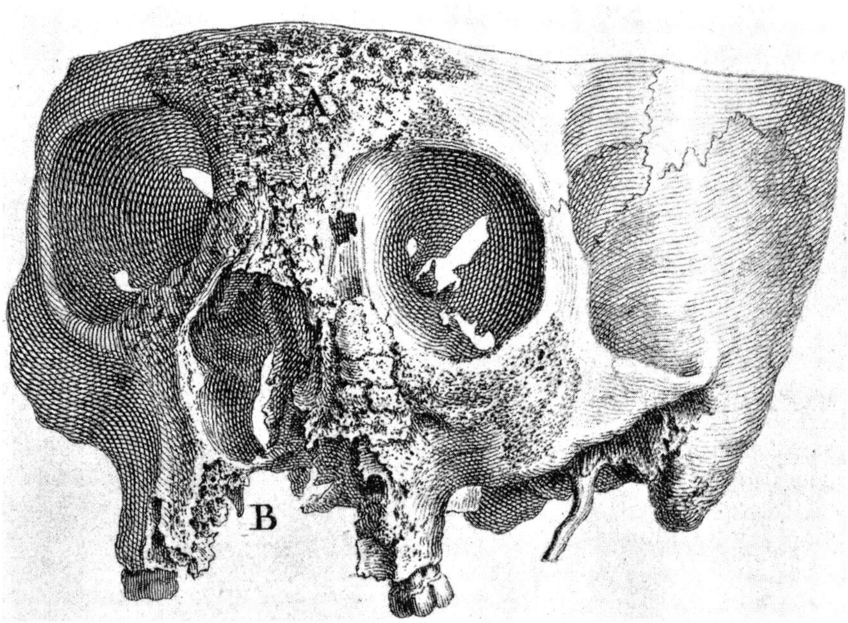

Fig. 4.19

Though Cheselden believed this anatomical atlas was 'the most magnificent work of its kind now extant',[86] it was not a publishing success (unlike his *Anatomy of the human body*, which went into many editions), despite Cheselden's persisting in bringing out further editions at his own charge. One of the problems was that Cheselden had an idea similar to that of Albinus, in having a double set of prints

[86] Ibid., p. 47.

'one before they are lettered, to shew them in their full perfection; and the other with letters, answering to their descriptions'.[87]

The anatomist and the art student

Finally on this theme of 'picturing bodies', mention must be made of a new role for anatomists: demonstrating bodies – live and naked for superficial anatomy, together with the skeleton and the muscles – to artists in order to teach them how to portray the surface anatomy of the human body effectively in paintings, sculptures and engravings. Some artists, such as Leonardo da Vinci and Michelangelo in the sixteenth century, had conducted their own anatomical studies on dying patients and corpses, in order to improve their own skills in portraying the human body. But attending 'life classes', or 'the life school', was not universal for artists and aesthetes until the eighteenth century, and it was in association with this class that anatomical teaching was introduced. This innovation was linked with the creation of schools and academies for drawing, painting and exhibiting.[88] The art academy had originated in Italy, but the modern model for such an institution was the Académie royale de peinture et de sculpture, founded in Paris early in his reign by the infant Louis XIV, in 1648.[89] This Parisian Académie was to be the model for new institutions in Antwerp (1663), Vienna (1697), Berlin (1697), Brussels (1711), Copenhagen (1754), Saint Petersburg (1757), Venice (1756), Amsterdam (1767), Stockholm (1768), and elsewhere.

The assumption underlying this kind of academy was that painting and design in general (*dessein*) can be taught – it is not just a matter of inspiration and practice. Here the senior members, in the role of professors, shared the teaching duties. The most favoured form of painting at the time was that of 'history' painting, in which a story, or a moment from a story, is told, usually from classical or Biblical sources, and often on a grand scale with many figures, mostly male. The 1654 statutes of the Académie reveal that masters were called in from outside to give the students special classes in geometry, mathematics, architecture, perspective and anatomy. By the 1663 statutes these teachers are giving public classes twice a week.[90] But even at the Académie in Paris such anatomical teaching as there was, was given by a succession of surgeons or artists, and professional anatomists, as such, do not seem to have been involved in the teaching of artists for another century – until 1746 – when J.-J. Suë *père* gave classes at the Charité hospital to a mixed class of students of surgery and of the Académie. It was only from 1776 that the course took place at the Académie itself 'with demonstrations on the corpse' (*sur le naturel*).[91]

---

87 Ibid.

88 Pevsner, *Academies of art*, 1940.

89 On its history, see now Comar, ed., *Figures du corps*, 2008. The name of the Académie was changed to L'École des beaux-arts in 1804.

90 Aucoc, *L'Institut de France*, 1889, pp. cxx, cxxx.

91 Huard, 'L'enseignement medico-chirurgical', 1964, p. 204n.

And only in the 1777 statutes do full professors of geometry and anatomy appear, with the anatomy professor obliged to give one course a year and to teach the matters 'de cette science qui conviennent aux arts de peinture et de sculpture'. The anatomist had no great status in the Académie: the order of precedence of all the fifty-four professors and functionaries of the Académie at this date are listed, and the professors of geometry and anatomy come last of all.[92]

The first book on anatomy that came out of the Académie was published (though not written) by the painter François Tortebat in 1667: *An epitome of anatomy suited to the arts of painting and sculpture* (*Abrégé d'anatomie accommodé aux arts de peinture et de sculpture*). The text was by Roger de Piles, and the ten figures were copies of Vesalian muscle-men. This remained in print and in use for a century or more.

It was the idea of Colbert – the king's first minister – that the best students from the Académie should go to Rome to improve their skills and to copy ancient scenes and sculptures and send them back to France, as models and schemes to reproduce to the glory and beautification of France and the French monarchy. The students competed in Paris, and those awarded the 'Grand Prix de Rome' were sent to 'L'Académie de France à Rome', established there in 1666 by the artist Charles Errard the Younger (1606–89). At any one time about a dozen students received grants and their keep, and worked under the supervision of the Rector, Errard. From the beginning Errard drew up a statute that there should be annual anatomy teaching:

> The knowledge of anatomy being of great value for painters and sculptors who wish to become knowledgeable [*scavans*] and who want to understand the different effects that the muscles produce according to their different movements, the Rector of the Academy will have the dissection of a corpse undertaken every winter and will take care also to have it cast, in order for the students to learn the situation of the muscles and the effects of their movements.[93]

Presumably the casting would have been done in plaster. There is a record from the early 1680s of a teacher of mathematics and a teacher of anatomy being paid, the anatomist for teaching for six months from All Saints Day to Easter.[94]

In Rome Errard initiated a great volume on anatomy for artists, 'a work most useful for painters, sculptors, and all other people studious of the noble art of design', as the title-page says. Printed in 1691, it used the dissection skills of Bernardino Genga (d. 1690), 'Royal anatomist' and Professor of Anatomy and Surgery in the San Spirito Hospital in Rome, while the young Lancisi drew up the brief commentary and prepared the index. Errard himself did the illustrations

92 Aucoc, *L'Institut de France*, 1889, pp. cliv, clviii, clix–clx.

93 Article XI of the 1666 statutes, in Alaux, *Académie de France à Rome*, 2 vols, 1933, vol. 1, p. 16.

94 Lapauze, *Histoire de l'Académie de France à Rome*, 2 vols, 1934, vol. 1, p. 65.

of the bones and muscles. Most significantly Errard and Genga also used certain celebrated antique Roman sculptures as their models, including the Laocoön (in the Vatican) 'considered anatomically' to show the muscles.[95] For this purpose the sons and serpents of the original sculpture (Fig. 4.20) have been dispensed with (Fig. 4.21).

In the Netherlands an Academy of Art was founded in Amsterdam in 1767, to which a professor of anatomy was appointed. This was the anatomist Peter Camper, who was an excellent artist as well as dissector. As he recalled, 'Painting, and whatever is relative to the Art, have been my favourite amusements from my earliest years: and as the characteristic differences in men and animals appeared to me the most interesting objects in nature, I was disposed to pay more than ordinary attention to these'.[96]

It was a theme Camper developed over his professional lifetime. The first inaugural thesis he presented at Leiden University, as doctor of philosophy in 1746, was 'An optical dissertation on vision'. Later the same day he also presented a thesis for the degree of doctor of medicine, 'A physiological dissertation on certain parts of the eye'; both dissertations were published.[97] Camper was appointed Professor of Anatomy successively in Franeker in 1749, then in Amsterdam at the Athenaeum Illustre from 1755 to 1761, with the use of the town's famous anatomy theatre, and from 1763 to 1773 in Groningen University. He had a very large supply of different corpses, especially in Amsterdam, on which to develop his theories about artistic representation.

As a young artist Camper had learnt to draw human faces by using ovals and triangles, as taught by Albrecht Dürer and other great masters. But –

> When I gave lectures in the public college [the Athenaeum Illustre] at Amsterdam, as Professor of Anatomy, I found, by comparing bodies of various ages that were brought to me for dissection, that the oval was not calculated for the delineation of the features with any degree of accuracy or expedition. With this idea I sawed several heads, both of men and of animals, perpendicularly through the middle; and I was fully convinced that the ball of the head forming the cavity destined to contain the brains, was in general very uniform; but that the position of the upper and lower jaws was the manifest cause of the most striking differences. The same observation may be extended from quadrupeds down to the finny race …
>
> Having contemplated the inhabitants of various nations with greater attention, I conceived that a striking difference was occasioned, not merely by the position

[95] Genga, *Anatomia*, 1691, title-page. Strangely this book is not mentioned by either of the historians of the Académie at Rome, Lapauze or Alaux.

[96] Camper, *Works*, 1794, p. 1.

[97] Camper, *Dissertatio optica de visu*, 1746; Camper, *Dissertatio physiologica de quibusdam oculi partibus*, 1746. The first of these has been translated into English by ten Doesschate. On Camper's biography, see Visser, *The zoological work of Petrus Camper*, 1985.

Fig. 4.20

Fig. 4.21

> of the inferior maxilla, but by the breadth of the face, and the quadrangular form of this maxilla.[98]

Camper worked along these lines in his drawing and his anatomising for many years, and once he retired he was able to write up his theory. And when he was appointed first Professor of Anatomy at the newly founded Academy of Art in Amsterdam, he gave a series of lectures outlining his views –

> Delighted with my discoveries, as is generally the case, I communicated them to several admirers of the fine arts. They imagined that a considerable service might be rendered to the science of Painting, and particularly to the Academy of Drawing, established at Amsterdam, were I to deliver some public lectures upon the subject, before the members of that society. These I delivered on the first and second days of August, in the year 1770, before a numerous and respectable audience.[99]

Camper's lectures deploy his anatomical experience in favour of improving the skills of the artist. They deal with representations of the human face, of all races,

98 Camper, *Works*, 1794, pp. 7, 8.

99 Ibid., p. 11.

from Antiquity to the present, and with the skulls of people of various races in the present. It is, according to Camper, the differing angle that the 'facial line' makes that basically distinguishes the faces of the different races: the 'facial line' can be found by placing a ruler or other straight line from the forehead to the lips (ignoring the nose). 'An idea suggested itself', Camper recalled, 'that in drawing or painting of the head, the best method would be to imitate the process of Nature; first to form the cranium or skull, then mark the facial line in the direction required, and afterwards arrange the other parts according to given proportions'.[100] His schema is given in Fig. 4.22. Camper's printed lectures are explicit that the 'facial line' is 'natural' – learnt from observing Nature – and he presents it explicitly as a means to help the artist portray nature, and not as any kind of generalisation about a hierarchy of creatures or human races. Unfortunately it was later used by other people as a basis for scientific racism.[101]

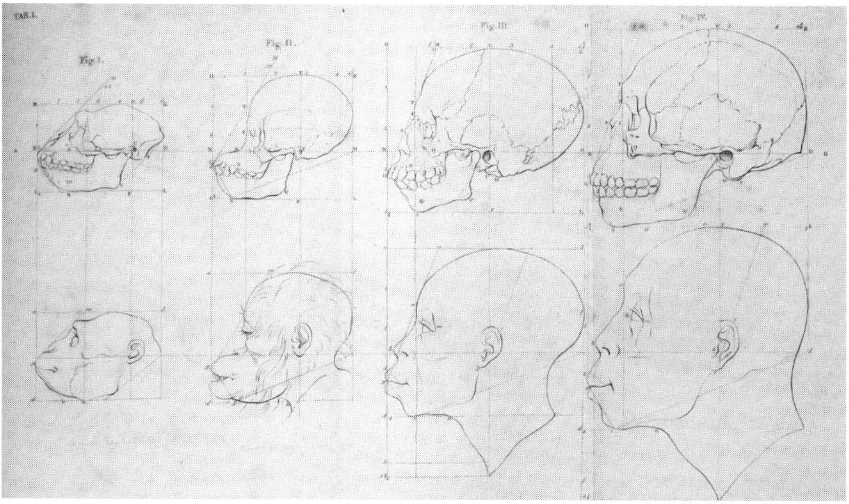

Fig. 4.22

In London there were private schools of design in the early eighteenth century, and Genga and Errard's great style book from Rome was published twice in English in London (1723, 1767) to meet this market.[102] There were also repeated attempts in London to establish an academy with royal support, but these came to fruition at last only in 1768. Anatomy teaching was provided for in the founding

[100] Ibid., p. 114.

[101] Meijer has recently shown how undeserving Camper was of such a reputation, in her *Race and aesthetics*, 1999.

[102] [Genga], Anatomy improv'd and illustrated with regard to the uses thereof in designing, 1723; for the 1767 edition, see Russell, *British anatomy, 1525–1800*, 1987, p. 80.

document of the Royal Academy of Arts: 'There shall be a Professor of Anatomy, who shall read annually six public lectures in the Schools, adapted to the Art of Design; his salary shall be thirty pounds a year; and he shall continue in office during the King's pleasure'.[103] The first professor was William Hunter, who held the post until his death in 1783.[104] This was an excellent choice since Hunter was an art connoisseur in his own right, and had a considerable collection of paintings old and new. 'A general taste for the Arts prevails', Hunter told his students at the Academy; 'the best works of Artists are sought for with eagerness and purchased at almost any expense; and artists themselves are respected by the public, and cherished by a King who delights in everything that can improve and exalt his people'.[105] Hunter had strong – if unsophisticated – views on the relation of Nature to artistic design. His attitude was that, 'as the superiority of Nature over Art seems to shine forth in almost every thing, in the Fine Arts the more precise the imitation of Nature is, that is, the nearer we come to the point of realising [Nature], the more *striking* I should suppose the effect will be'.[106]

The core of Hunter's teaching in the Academy consisted of three annual lectures on the bones and three on the muscles: 'The study of the bones and muscles', he announced, 'will be allowed to be very essential to painters and sculptors, if it be known, and it is really true that every variety which any living creature can bring out, colour only excepted, depends entirely upon the play of the muscles'; 'All these observations put together prove that a very correct knowledge of Anatomy must be of the utmost consequence to every artist who is to make resemblances of the human body. It enables him to observe and distinguish clearly all the variations of form, because it explains their causes'.[107]

Hunter continued to develop his enthusiastic if somewhat naïve views about the desirability and possibility of the artist directly and simply imitating Nature, and he expatiated on them at length in 1774 in the Preface to *The anatomy of the human gravid uterus*. While in the Royal Academy Hunter's theme was the importance of anatomy for painting, but here by contrast he was more concerned with the importance of painting and engraving for anatomy:

> Anatomical figures are made in two very different ways; one is the simple portrait, in which the object is represented exactly as it was seen; the other is a representation of the object under such circumstances as were not actually seen, but conceived in the imagination. *Bidloo* has given us specimens of the first kind; *Eustachius* of the latter.

---

[103] Kemp, ed., *Dr William Hunter at the Royal Academy of Arts*, 1975, p. 13. This includes the text of Hunter's lectures, 1769–72.

[104] On William Hunter as professor, see Darlington, 'The Royal Academy of Arts and its anatomical teachings', 2 vols, 1991, esp. vol. 1, chapters 3 and 5.

[105] Kemp, ed., *Dr William Hunter at the Royal Academy of Arts*, 1975, p. 33.

[106] Ibid., p. 38.

[107] Ibid., pp. 36–7, 45.

> That figure which is a close representation of nature, and which is finished from a view of one subject [that is, of one body], will often be, unavoidably, somewhat indistinct or defective in some parts: the other, being a figure of fancy, made up perhaps from a variety of studies after Nature, may exhibit in one view, what could only be seen in several objects [that is, several bodies]; and it admits of a better arrangement, of abridgement, and of greater precision. The one may have the elegance and harmony of the natural object; the other has commonly the hardness of a geometrical diagram: the one shews the object, or gives perception; the other only describes, or gives an idea of it. A very essential advantage of the first is, that as it represents what was actually seen, it carries the mark of truth, and becomes almost as infallible as the object itself.[108]

What Hunter meant by representing 'what was actually seen, it carries the mark of truth' becomes clearer when we look at the legend for Plate 6 of this work, which 'represents the child in the womb, in its natural situation' (see Fig. 3.6 in Chapter 3, above). Here Hunter writes about the fetus –

> Its body was covered with a white, greasy *mucus*, which is commonly seen on children at their birth. This is represented at the upper part of its back, where it was intersected with lines, from the wrinkles and motion of the child's body. Every part is represented just as it was found; not so much as one joint of a finger having been moved to shew any part more distinctly, or to give a more picturesque effect.

It would seem thus that 'the picturesque' may have been what Hunter was trying to avoid, and which he seemed to think could be avoided by some simple (or simplistic) trust in the artist's or anatomist's capacity to make a direct representation of what lies before his eyes.

In fact the *picturesque*, as a theory and practice of art, was all the rage in England in the late eighteenth century. While all definitions of it appear circular to the modern eye – 'that peculiar kind of beauty, which is agreeable in a picture', as William Gilpin, one of its proponents, defined it[109] – in practice its advocates generally used it of landscapes and took it to mean rough, unpolished: nature in the raw, but framed in a refined way. It was also a theory of landscape gardening, put into practice in great gardens such as that at Stowe, where new landscape 'pictures' appear to the eye suddenly at every turn.[110] So when Hunter had the fetus portrayed without any picturesque effect, he probably meant without any elaboration to make it more presentable or picture-like, and with nothing added and nothing taken away – in his vocabulary, 'just as it was found'. An anatomist

---

108 William Hunter, *The anatomy of the human gravid uterus*, 1774, Preface.

109 Gilpin, *An essay on prints*, 1781, p. xii. On the picturesque in general, see the incomparable Hussey, *The picturesque*, 1967.

110 See Price, *An essay on the picturesque*, 1794.

whose principles of portrayal ran counter to Hunter's views in every respect was Albinus, as we have seen. For Albinus went to enormous lengths to portray the 'ideal' skeleton or muscle-man – not the body 'just as it was found' – and, with his approval his artist, Wandelaar, deliberately added 'picturesque' ornament to his images, such as Clara the rhinoceros.

But despite William Hunter's views on portraying subjects 'just as it was found', the picturesque was actually also put forward by a contemporary of Hunter's as the proper way of approaching *anatomy for artists*, and it is possible that it was against this that Hunter was propounding his simplistic view. John Brisbane, an Edinburgh-educated physician and amateur of painting, had produced his book *The anatomy of painting … giving a short view of picturesque anatomy* in 1769, which reproduced, though on a smaller scale, some of the anatomical illustrations of the great Albinus. 'Painters, statuaries, and engravers', Brisbane claimed, 'should assert their rights, and teach, and write upon this science, in a picturesque manner, suited to their own art'.[111] He was writing, he said, for 'those, who pursue the arts of design, in order to awaken their attention to this part of their profession, and as a specimen to form the taste of youth, early, to a love of elegance in anatomy, and to show them how much shorter, more easy, and agreeable it is, to be introduced to that science by means of figures [viz. illustrations], than by tedious systems, and lectures upon dead bodies alone'.[112]

Brisbane's picturesque anatomy for painters was actually quite circular: a painter should look at and learn the anatomy of the body with the eye of a painter!

> A lover of the arts of design, or indeed any anatomist of true taste, will look upon the human body and all its parts, with the eye of a painter, otherwise, he will see and describe it in an ignorant and rustic manner: this picturesque turn we observe in few modern anatomists, but rather a great ignorance of it … Observing the human body with the eye of a painter, enables us to see it in all its beauty and perfection, and raises in our minds a thousand ideas of the uses and propriety of the several parts, whereof one ignorant of painting will be totally insensible: and in describing the human body upon this plan, we naturally do it in the most clear, short, and agreeable manner, far different from the dull pedantic descriptions and tedious trifling of vulgar anatomists. (pp. v–vi)

Brisbane took Albinus's illustrations as the height of picturesque anatomy, and something more particular about what Brisbane meant can be seen in his description of the muscles, as seen in Albinus's illustrations:

> By these figures we perceive, that the number and strength of the muscles, are every where suited to the several parts they serve, and the joints they move. Thus

111 Brisbane, *The anatomy of painting*, 1769, p. iii.

112 Ibid., pp. v–vi.

> the muscles on the trunk are few, but generally large, broad and flat, serving for walls and coverings as well as moving powers; but on the extremities the muscles are numerous, and mostly oblong, suited to the size, form, and action of these parts, and the many joints may be moved. About the shoulders and hips, the muscles are short, large, thick and strong, giving the idea of that strength necessary for strong and violent action, and to command large members that are so constantly in motion ….[113]

Florence, where Renaissance art had started, was actually among the last major cities to have a modern school for artists. The school was founded by the Grand Duke Pietro Leopoldo, as the Galleria dell'Accademia di Belle Arti in 1784, and he gave some of the rooms of the hospital of Saint Matteo for the use of the Academy. For the crucial role of anatomy teacher there Paolo Mascagni was chosen in 1801. He had been Professor of Anatomy at Siena (1774), then at Pisa, and now at Florence. Mascagni

> … taught viva voce every year all those young people who attended the studies at the Academy, sometimes in the room of the life class, sometimes in the local anatomical theatre. As a very generous lover of the glory of Tuscany, he was not only happy to dictate his lectures, but he also wrote them up and he was keeping the autograph in such a way that he could respond to the plan to have it published.[114]

Like other anatomists in the art schools, what Mascagni showed was

> … the size, the figure, the position, the connections, the articulations, of all the bones, and the entire arrangement [*complesso*] of that layer of muscles which are just below the skin, which in the various attitudes of rest, movement, and expression of the passions, show their presence and give greater or less relief to the surface of the animal body.[115]

After his death Mascagni's magnificent book was published in Italian in Florence in 1816: *Anatomy for the use of students of sculpture and painting*.

The developments of the market for paintings and painters that led to the creation of academies of art in so many cities of Europe, encouraged many anatomists to plan (and sometimes finish and publish) their own guides explicitly meant for artists. In addition to those already discussed here, the historian Ludwig Choulant has studied publications of this type by Carlo Cesi (1697), teaching in Rome; Crisóstomo Martínez from Valencia and working in Paris (first published before

113 Ibid., p. xviii.

114 Preface of the editors to Mascagni, *Anatomia per uso degli studiosi di scutura e pittura*, 1816.

115 Ibid.

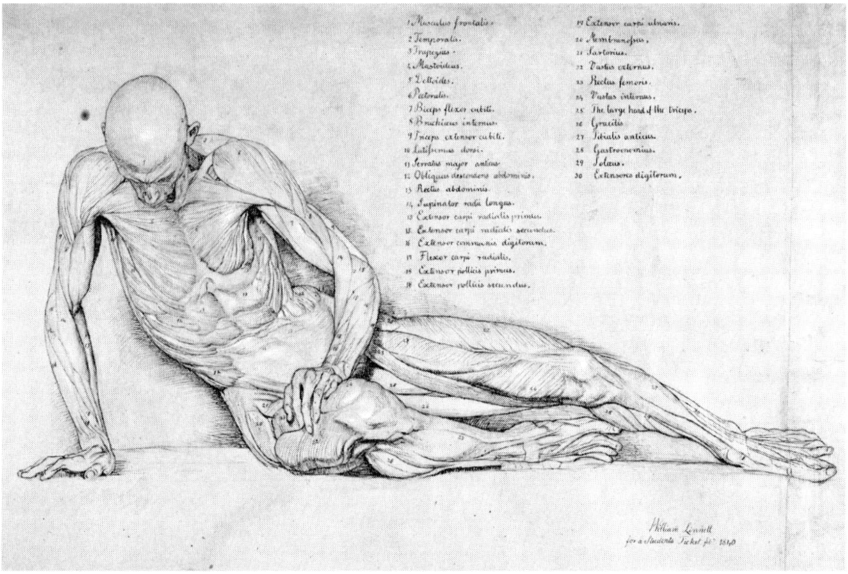

Fig. 4.23

1692); Erco Lelli, the wax modeller of Bologna (before 1766); Lambert Sigisbert Adam, a graduate of L'Académie de France à Rome, teaching later at the Académie in Paris (1773); Cornelis Ploos van Amstel, an artist and engraver of Amsterdam, whose illustrations were used in works for artists produced in Amsterdam by J. Ynterma (1783), and in Zurich by Johann Heinrich Lavater in German (1790), and in French (1797); Jean Joseph Sue, the younger, in Paris (1788); and other anatomists through into the early nineteenth century.[116] Altogether these indicate a strong perceived demand for artistic anatomy by professional anatomists. As we shall see in Chapter 5, this same market for paintings and painters also led the artist George Stubbs to teach himself the hard way the bones and muscles of the horse.

A particular form of anatomical representation was frequently now created for this market of art students: the *écorché* figure, that is, a figure of any size and made out of any suitable material, which showed the superficial muscles of the body with the skin stripped off (*écorcher* meaning 'to skin', and *écorcher vif* meaning 'to flay alive'). The name indicates that it had a French origin, and artists are known to have used them from at least as early as the sixteenth century. In the early Royal Academy such figures were occasionally made from casts of real human bodies. At some date in the 1760s some members of the Academy were invited by John Hunter (William's younger brother) to view some corpses of men hanged at Tyburn: 'When they had fix'd upon one, he was immediately sent to their appartments. All this was done in a few hours after death, and as they had not

[116] Choulant, *History and bibliography of anatomic illustration*, 1920.

become stiff, Dr [William] Hunter conceived he might first be put into an attitude and allowed to stiffen in it, which was done, and when he became stiff we all set to work and by the next morning we had the external muscles all well exposed readying for making a [plaster] mold from him, the cast of which is now in the Royal Academy'.[117] Another such body was put in the attitude of the classical figure known as the Dying Gladiator, and a cast of it still exists at the Academy, known as 'Smugglerius'.[118] It is shown in a contemporary drawing in Fig. 4.23.

4. Better printing

In the long eighteenth century the standard way of printing an illustration was to use engraving on copper. Woodblock cuts had been popular in the sixteenth century and, as Vesalius's book of 1543 on the fabric of the human body shows, could be developed to an astonishing degree of beauty and faithfulness. Woodblock cuts would reappear in the nineteenth century with the coming of the steam-powered newspaper printing press, for the woodblock was very tough and could withstand thousands of impressions. But from about 1550 to about 1800, the engraving held the field.

An engraving is almost always preceded by a drawing or a painting. The two professional skills of drawing and engraving were quite distinct. Exceptionally, as we shall see later, an extraordinary artist like George Stubbs could learn and practise both, as could Le Sieur Gautier d'Agoty; indeed, both of these men also did their own dissections as well. But normally, in order to make anatomical illustrations the anatomist could either be his own artist, or he needed to find an artist whose work he liked or whose pencil he could steer (as was the case with Albinus, as we saw above). All prints usually have the artist's name in one lower corner, who *delinavit* (drew), or *pinxit* (painted) it, and that of the engraver in the other, who *sculpsit*, or *fecit*. Wandelaar sometimes writes 'Wandelaar did it (*fecit*) from the life' in his many engravings for Ruysch. When it comes to the illustrating of natural phenomena in this period, then, it is useful to remember that, by the time a natural subject came into visual print, it was often at least two steps removed from 'nature': first by passing through the medium of the artist, and second by passing through the medium of the engraver. In the case of Albinus, it had passed through three mediums, including that of the anatomist himself. All sorts of errors could be inadvertently introduced into this process, especially given the fact that the engraver had to produce on his copper plate a reversed image of the artist's image. Anatomists demanded much technical detail, and often wanted the many parts minutely numbered or even labelled. On top of this the printer had to be skilful to produce even prints from the engravings, something that became more

[117] Kemp, *Dr William Hunter at the Royal Academy of Arts*, 1975, p. 16; the account is by John Hunter.

[118] Ibid., p. 17.

difficult the larger the plate. The whole process was extremely expensive, since it called on the highest skills of three trades.

It was often the case that a purchaser wanted his anatomical illustrations in colour, and there was a trade of *hand colourists* to cater for this taste, and for whom anatomy was only one of many subjects to be coloured. The colourist could copy one of two 'originals' in applying the colour to the printed sheet: either 'nature' in the form of the original anatomical subject, or a coloured painting of the anatomical subject. Those anatomists whose plan was to produce engravings that could be properly coloured by colourists therefore had to start the proceedings not with a black and white sketch, but with a full-colour painting. This was the case for instance in 1757 with the French anatomist who lived and taught in London in the mid-eighteenth century, Charles Nicholas Jenty, who wished to produce 'a more natural compendious view of the human structure than any which has hitherto been produced' in his 'Essay on the demonstration of the human structure, half as large as Nature, in four tables. From the pictures painted after dissections, for that purpose. Disposed in such a manner, as to represent gradually all the capital parts of the human body in their natural situation, as they appear in dissection when the sanguiferous vessels are injected. Calculated to convey a clearer artificial idea of the animal œconomy than has hitherto appeared in any other anatomical figures'.

> I thought that if this Work were represented in Colours as nearly as possible imitative of Nature [Jenty wrote], it would make a stronger Impression on the Memory; but the Difficulty was, how to meet with a proper Artist in anatomical Painting; such, it is well known, being very scarce, especially in England: And, accordingly, having tried various Artists, I found none imitated Nature better than Mr. Van Riemsdyk; who, as he has been for some Time employed by the Doctors Smellie and Hunter in Pieces of Midwifery, &c. and is well known to be the ablest Person we have in London, in anatomical Performances; so, I will venture to say, he has executed them as well as any Man can, who is not himself an Anatomist and Painter ....[119]

If all is not clear in these figures, writes Jenty, 'I must beg Leave to observe, that it is not my Fault, nor owing to any Want of Endeavours in me to get the Artist to represent them better; being no Artist myself, and having done nothing in the Work, except preparing Nature for the Painter, with as much Care as my Abilities would permit, and disposing the Parts as the Reader sees them'.[120]

Anatomical coloured illustrations represented a particular challenge and opportunity to a handful of printers in the eighteenth century who were developing different kinds of *colour printing*.[121] That is to say, in these processes the different

119 Jenty, *An essay on the demonstration of the human structure*, 1757, pp. 6–7.

120 Ibid., pp. 8–9.

121 Here I rely on Gascoigne, *Milestones in colour printing*, 1997, pp. 8–17. See also Rodari, *Anatomie de la couleur*, 1996, where many of the early pictures are reproduced in

colours are added not by hand using a paintbrush, but as part of the printing process itself, using one or more engraved plates with coloured ink. The first printer to turn his hand to printing coloured anatomical plates was Jakob (Jacob) Christophe Le Blon, who hailed from Frankfurt, but who printed in London. Around 1723 he had a scheme with Nathaniel St André, the King's anatomist, to produce an anatomical book with coloured illustrations. He wrote of it to Robert Walpole, Chancellor of the Exchequer, 'The anatomical tables, which I am now performing under the direction of the King's anatomist, Mr. St André, will confirm the usefulness of the said imprimerie beyond question'.[122] The book seems not to have appeared – probably due to St André being mocked over his gullible involvement in the famous case of Mary Toft, who supposedly had given birth to rabbits. A Dutchman, Jan L'Admiral, adapted the process he had learnt from Le Blon and produced at least five coloured mezzotint plates between 1736 and 1741, all published at Leiden.[123] Two of these were made, on L'Admiral's initiative, to illustrate dissertations of Albinus, the first (1736) on the arteries and veins, with a picture of a portion of the membrana nervosa of the human intestine, and the second (1737) on the seat and cause of the colour of Ethiopians, with a print of skin sections. The other four were published under L'Admiral's own name in 1738, including the dura mater of the head of a human fetus of eight months, shown in one print from the concave and in another from the convex perspective, following preparations by Ruysch.

Finally, in 1741, L'Admiral produced a print of a preparation of the human penis injected with red wax, showing some newly discovered anatomical facts, viz. the lacunae of Morgagni, plus some lacunae not known even to Morgagni. This last print appears to be part of a projected series on the parts of man serving for generation.[124]

The modern historian of colour printing, Bamber Gascoigne, stresses how difficult the technical issues are for printing in colour in the process known as *mezzotint*, in which the plate is first hand-roughened so that it can hold a continuous film of ink, and then the image is achieved by smoothing the plate to different

---

full colour. The first coloured printed images of anatomy were from woodblocks: Gaspare Aselli's *De lactibus sive lacteis venis*, 1627.

[122] Krivatsky, 'Le Blon's anatomical color engravings', 1968, see esp. p. 155–6; Le Blon's letter is undated.

[123] All of these are bound together at the Wellcome Library for the History and Understanding of Medicine, London.

[124] Unfortunately there is no point in reproducing it here in black and white, since the whole point is that it is a colour image. The version of the print in *Préparation anatomique des parties de l'homme servantes à la génération, faite sur les découvertes les plus modernes* (a folded sheet, with the illustration mounted on blue paper), uses the anatomical description by William Cockburn in his book on gonorrhea of 1713; see British Library 1830b2 (2). Russell, *British anatomy*, 1987, items 184–6 claims that this image is by Le Blon and dates from 1721 – and that is was then copied by L'Admiral and published in 1741. In Rodari, *Anatomie de la couleur*, 1996, the two images are presented opposite one another on pp. 70–71.

degrees, reaching white at the smoothest stage. Each of the three plates (red, yellow and blue) has to be separately prepared, each depicting only what is appropriate to the colour it will print; the depth to which the plate is smoothed depends on the depth of colour it has to print; register has to be exact, and the colouring has to be balanced by eye. However challenging the technical process, the results are, to my modern eye, nevertheless captivating. The advantages of the new technology of mezzotint for the anatomists was that it gave a soft line (as in Nature) rather than a hard line, together with realistic gradations of colour that remained constant from print to print, not subject to the whim of the colourist.

When Jenty was preparing his *Essay on the demonstration of the human structure* (published in 1757) he pondered whether to use engraving or mezzotint for the half-life-size illustrations. He decided on mezzotint, 'but, at the same time, unfortunately met with a Man who persuaded me, that he could do the Work in Colours, as follows, by scraping four different Plates for one single Print, in such a Manner, one for a black, one for a blue, one for a yellow, and one for a red; as, by a due Combination of these Colours, and the same Sheet of Paper going over one another, it would represent any Picture whatever, as is well known by the Laws of Optics, to be the Case' (p. 7). But this printer turned out not to be a master of this art, so Jenty wasted hundreds of pounds, and eventually had to settle for printing only in black and white.

But there was one printer who really could achieve what Jenty was promised, and that was the very colourful Frenchman Jacques Fabien Gautier d'Agoty (1711–85), who had already published such a volume on the muscles of the human body in 1746, and whose images are perhaps the most striking of all printed anatomical representations in the whole eighteenth century.

Le Sieur Gautier d'Agoty (as he styled himself) seems to have had a remarkable success as a producer of coloured anatomical atlases, not only finishing the entire course of human anatomy, but also able to put out reprints. Gautier's printing system went one step beyond that of L'Admiral, in that Gautier used four colours, adding black. Initially Gautier had the help of Duverney to make the dissections of the muscles of the head and contribute the anatomical information, and when Duverney died, he had the assistance of Mertrud for a while for the same service for the trunk, internal organs, and the organs of generation of woman. But finally he started undertaking every role: everything was 'dissected by my hand or under my eye, such that I am the Demonstrator, the Painter and the Engraver all in one, which gives me a certain advantage in this sort of project'[125] – and of course he was also the printer. Here is the all-round anatomist who can do everything!

---

125 Gautier d'Agoty, *Anatomie général des viscères*, 1752, final page.

5. 'Warm controversy': Discoveries, credit, and the making of anatomical facts

It was in 1762, in the course of his third public dispute over the making of anatomical facts, that the London anatomist William Hunter spoke, not without humour, of the vital role that controversies between anatomists played:

> It is remarkable, that there is scarce a considerable character in anatomy, that is not connected with some warm controversy. Anatomists have ever been engaged in contention. And indeed, if a man has not such a degree of enthusiasm, and love of the art, as will make him impatient of unreasonable opposition, and of encroachments upon his discoveries and his reputation, he will hardly become considerable in anatomy, or in any other branch of natural knowledge …
>
> It has likewise been observed of anatomists, that they are all liable to the error of being *severe* on each other in their disputes. Perhaps from being in the habit of examining objects with care and precision, they may be more disgusted with rash assertions, and false reasoning. From the habit of guarding against being deceived by appearances, and of finding out truth, they may be more than ordinarily provoked by any attempt to impose upon them; and for any thing that we know, the passive submission of dead bodies, their common subjects, may render them less able to bear contradiction.[126]

Hunter also observed here that 'literary disputes are disagreeable to the greatest part of mankind; and the disputants are, for the most part, condemned by the world'.

Hunter's position here about disputes is that they are an unavoidable result of men passionately pursuing their discipline and trying to make discoveries. We find certain terms constantly being used in such disputes, terms such as property, honour and reputation. Discoveries are intellectual property; they belong to their discoverer. Discoveries build the reputation of an investigator, and reputations need to be defended if challenged. A gentleman's honour is at stake. Discoveries redound to the honour of the discoverer. But on the other side, Hunter claims, disputes are desirable, since 'by engaging the passions of men more warmly, [they] rouze a spirit of emulation, and give a spur to enquiry'. A cartoon by Rowlandson, Fig. 4.24, shows happy days in William Hunter's dissecting room, full of active anatomical researchers seeking to make new discoveries, with Hunter himself portrayed at the highest point and his pupil Mr Hewson engaged in the central dissection. An attempt at identifying each dissector has been made by Cecil Wall, who found that they are all qualified surgeons – apart from Dr Hunter himself, of course – not students, so they are being portrayed as researching rather than learning what is already known.[127]

---

126 William Hunter, *Medical commentaries*, 1777, Supplement, p. 1.

127 Wall, *The history of the Surgeon's Company*, 1937, pp. 85–6.

Fig. 4.24

William Hunter certainly knew what he was talking about with respect to disputes. He was as embroiled in public disputes about anatomical discoveries as any anatomist of the time, and he was equally certainly as proud and defensive of his discoveries in the field as anyone. And it is the case that most of the greatest anatomists of the period were involved in one public dispute or another.

We need to keep in mind these multiple roles of disputes when looking at the paper wars of our eighteenth-century anatomists. Disputes were matters of personal dignity and honour, and this is most evident in the name-calling and intemperate language regularly employed by disputants. They were matters of establishing or defending social status, with disputants constantly claiming that they are lowering themselves to public dispute only because their friends urge them that otherwise their silence will make them look guilty as charged by their opponents, and their social status will be undermined; 'When the characters of men are *staked* in a dispute, it grows too serious and important to be neglected', as Hunter said.[128] The lining-up of friends around each disputant, and the frequent appeal of one of the disputants to the best friends of the other, that they make him listen or withdraw, or see sense, make these disputes very much like duels, with the seconds and other friends insisting that each gentleman defend his honour, lest they be shamed.

128 Hunter, *Medical commentaries*, 1777, Appendix, p. 26.

New discoveries in anatomy were fairly hard to come by in the eighteenth century, as so much had been discovered already.[129] Big discoveries were of course even harder. Perhaps the most extraordinary anatomical discovery was that made by Albrecht von Haller, which was of great consequence and controversy in its time, but only of little importance today: the sensibility and irritability of the individual parts of the body (see Chapter 2 above). Among the other most important discoveries of the period and that involved the participation of several anatomists were the *villi* in the small intestines (1744)[130] – this was related to the subsequent discovery of the function of the lymphatics; the structure of the testicle (an issue closely related to the lymphatic business); the descent of the testicle from the abdomen after birth; and the true structure of the placenta. They are all about structure, and about function deduced from structure. These comprised, as it were, cutting-edge topics in the century, and hence mostly controversial. They primarily involved anatomists in Göttingen, Berlin, London, Edinburgh and Leiden.

Most of the accusations are about theft: the stealing of ideas, facts or discoveries. If part *x* was first discovered by me on date *z*, then it cannot have been first discovered on date *z*+1, as claimed by my opponent. Witnesses are assembled to confirm dates and facts. Sometimes the disputes are about anatomical facts: does a part that anatomist A has discovered actually exist or not? Of course the anatomists in these disputes were generally talking past each other – as in most disputes at most times. Public presentation of one's own position was the strongest course one could take, since there was no neutral tribunal to which one could appeal, anymore than there is in science today. To the modern mind the arguments over facticity – whether the part is actually there or not – are hard to sympathise with, since the anatomists all prided themselves on their use of hands and eyes. But even the keen eyes of different anatomists saw different things in the same materials.

There were many notable anatomical disputes in the long eighteenth century, such as those between Morgagni and Bianchi, between Albinus and Haller, between Albinus and Camper. While some of these disputes were very witty as well as very rude, it can nevertheless be tedious in the extreme to trace every step of such rows, every moment when an ego was bruised, or a trivial claim defended, and it is pointless for us to try to offer retrospective judgements about who was right and who was wrong. So what I shall do is offer a few examples of different *topics* of dispute that actually occurred in anatomy: over charlatanism, plagiarism, facticity and theft. I stress that the disputes I have chosen were strictly *anatomical* ones, regarding claims about facts that could be found in the body, and not physiological – theoretical – ones of the kind that historians have generally been concerned with.

---

[129] See the quotation from John Barclay to this effect in Chapter 6, section 2, below.

[130] Lieberkuhn, *Dissertationes quatuor*, 1782 (1744). See also Lassus, *Essai ou discours*, 1783, pp. 241–3; Lassus is very good on new anatomical discoveries in the period.

These disputes took place in learned institutions, such as the Royal Society, and their journals; in periodicals not associated with institutions, such as the *Medical Commentaries*, edited by Monro *Primus* in Edinburgh, or the *Critical Review* in London, which, under the editorship of William Smollett, was always ready to stir a dispute; in books on larger subjects, as authors criticised each other or defended themselves on minor points; and in the hundreds of pamphlets printed usually at the charge of the disputant himself, distributed as widely as possible by him and also sold in bookshops.

A charlatan case: Ruysch vs. Bils, 1665

Frederick Ruysch loved a good anatomical dispute, and in his long lifetime (he died in his nineties) he engaged in many. Indeed, his career as an anatomist started with a dispute with the Rotterdam anatomist Louis de Bils (1623–69) over the existence (or otherwise) of valves in the lymphatic vessels.[131] Though it had happened while he was still a young man, Ruysch could reproduce much later – with remarkable recall – his public dispute with Bils, in which Bils denied the existence of such valves, and Ruysch repeatedly demonstrated them to him; whereupon Bils claimed he had seen them frequently before! Ruysch announced to his reader, 'I bring to light only those things which I have proved by experience (the one and only master of truth)'. While other anatomists had seen these valves, Ruysch claimed that he was the first to explain the technique of making them visible and the first to print illustrations of them. Ruysch related this story in his *Clarification of the valves in the lymphatic vessels and the lacteals*, in 1665. Ruysch's reward for publishing this account of his victory in this dispute was to be appointed Professor of Anatomy in the city of Amsterdam, a position he then held for sixty-five years.

A plagiarism case: Bidloo vs. Cowper 1698

In 1698 Govard Bidloo, Professor of Anatomy at Leiden University, heard via a friend, the chief physician to King William, John Hutton, that a London surgeon, William Cowper, was going to publish an English translation of his 1685 book, *The anatomy of the human body* (*Anatomia corporis humani*). Bidloo immediately wrote to Cowper, offering all necessary help and even offering some new anatomical plates he had made since his book had been published. His letters went unanswered. Then he started hearing rumours that Cowper's book was in print, but that his own name 'can scarcely be found' in it. Bidloo mustered what support he could in the Royal Society of London to get into contact with Cowper, but without result. Finally in 1700 he published a 58-page booklet in Latin accusing Cowper of plagiarism before the 'tribunal' of the Royal Society, of which Cowper was a

[131] See on this dispute, and especially for the role of Johannes van Horne, Luyendijk-Elshout's excellent introduction to Ruysch, *Dilucidatio valvularum*, 1964 reprint.

member.[132] Cowper, said Bidloo, was nothing but a highwayman. Bidloo criticised Cowper's anatomical knowledge in great detail.

It was indeed a bit daring on Cowper's part, and actually the plagiarism that Bidloo charged him with. All Cowper says in his book about the former history of these plates is this: 'These figures were drawn after the life by the masterly painter G. de Lairess, and engraved by no less a hand, and represent the parts of *humane bodies* far beyond any extant; and were some time since published by Dr Bidloo, now Professor of Anatomy in the University of Leyden' (To the Reader). Cowper even used the original title-page, but had a little scrap of paper printed asserting his authorship, and stuck over Bidloo's name!

The Royal Society was the best tribunal Bidloo could find, since it was involved in more than one way, though it was not the kind of body that could rule on intellectual property. What had happened was that Samuel Smith and Benjamin Walford, publishers to the Royal Society, had acquired hundreds of impressions of each of Bidloo's plates, and had approached Cowper to write an English text for them. This Cowper had done, and to such a high standard that for many years his book was regarded as the best anatomy text in English. But had he plagiarised Bidloo? Cowper said no: his right to use the plates was as good as Bidloo's, since the plates had in the first place been prepared by Gerard de Lairess for Swammerdam, not Bidloo anyway. Moreover, Cowper claimed that his text was not only completely different from Bidloo's (Bidloo had simply named the parts), but more correct.[133]

Facticity challenged: 'VENITE! ET VIDETE!' Bidloo vs. Ruysch, 1690s

Ruysch's favourite motto, which he constantly used in his printed works, sometimes in almost every paragraph, and usually printed in capitals with exclamation marks on each important word, was 'COME! AND SEE!', 'VENITE! ET VIDETE!' The evidence of the eyes of the body should be preferred to the hallucinations of the eyes of the mind. He used his motto in a long-running dispute over anatomical facticity with Govard Bidloo, one of his favourite opponents, the Professor of Anatomy at Leiden University: do some of the structures Bidloo claimed to have discovered, and portrayed in his *The anatomy of the human body* of 1685, actually exist? Ruysch said no, and constantly challenged Bidloo to the test of sight. However, although Ruysch never admitted it, the visual test is not in fact finally decisive, because people can see different things, even when handling and looking at the same anatomical preparation; this was the case with Ruysch and Bidloo, as we shall see.

The case as Ruysch first put it in his 1697 *Response of F. Ruysch to the little book of G. Bidloo, called 'Vindications'*, is totally one-sided of course, but is better

---

132 Bidloo, *Gulielmus Cowper, criminis literarii citatus*, 1700; quotation from p.6.

133 Cowper's reply (in Latin) is *ΕΥΧΑΡΙΣΤΙΑ*, 1701.

than any commentary I might offer on it.[134] The anatomical matters in dispute were as trivial or tiny as one could ask for. Ruysch writes:

> I have, as is well known, published some 'epistolary replies', where I answer questions posed to me by others. I have judged in those, that it is the duty of the office of teacher to unravel and explain some anatomical truths, and at the same time to warn the practitioners of anatomy, that they should arm themselves against the things which are falsely described by various authors. Among others, I have pointed at some errors made, in my opinion, by Professor Bidloo.

Bidloo had responded with a little book against these letters, with the title of *Vindications*, where he promised on pages 3 to 4:

> First, *that he would vindicate from insults the images of the human body that he had shown to the learned world thirteen years earlier*; whence one gathers that he believes that these images were wrongly *castigated* by me *as Chimeras*; on the contrary, he promises that he will prove that *his images are in agreement, conformity and correspond to the Animal Œconomy and the arrangements of the parts making up the whole living body*.
>
> Secondly, he objects to errors, made by me, so he thinks: for he states in the same place, *that he will present a sketch of the Ruyschian art, learning and work*; and afterwards he brings forward several passages from our writings …
>
> Furthermore, as is his habit, he adds despicable scorn and slander, and worse, namely he adds that *Ruysch is shameless and dishonourable*, and he vomits up more of this kind of slander, and with these things said, he judges that I am *a disgrace to the study of anatomy, the worst anatomist, an excellent butcher, a scandalous and disgraceful orator*, and furthermore he pours out more words that are too shameful to repeat.
>
> Let us assume that he has done what he promised. But if I can demonstrate that in every respect the contrary is true; if I can demonstrate that he does not reply to the dispute, but rather that he prefers to distort and mutilate my writings; if I can show that he has committed bigger blunders than I have so far listed … if I can reveal that he has twisted and falsely quoted my writings in bad faith; and that Bidloo publicly writes mad figments of his imagination, and fallacies (more properly I'd call them lies); if I make public his dishonest disposition; then is one not justified in asking the fair-minded reader, whether I can throw back his words, or rather slander, in his own face, even whether if Bidloo was his own

[134] All of this is Ruysch's *Responsio ad Godefridi Bidloi Libellum, quem Vindicias inscripsit*, first published in 1697, and reprinted in 1721, and then in 1738, in *Opera*, 1737, vol. 1. My warmest thanks to Dr Rina Knoeff for first drawing these disputes to my attention, and providing me with her own English translation from the 1744 Dutch version, itself translated from the original Latin, and for giving me permission to use it here. I give reference to the Latin text of 1697.

> judge one could say that he is what he declares me to be; whether one should not rather say of him that he is *a man, shameless, who is disgraceful, scandalous and without honour, shameless and mischievous, frivolous, dirty, lascivious, a moral cancer, an utter enemy of peace and learning*?
>
> This I shall now fully prove ....[135]

This sort of fruity insult was typical of Ruysch. Bidloo had been looking down the microscope at the glands of the human skin. The first matter of dispute between Bidloo and Ruysch was about the shape of these glands (*glandulae cutaneae*). Bidloo maintained that they have a needle-like shape (*pyramidales*), while Ruysch held that they are round (*orbiculares*). Fig. 4.25 is the figure of the skin glands from Bidloo's book that Ruysch found so offensive. Ruysch lays down the challenge to a public duel:

> Thus I ask, modest reader, that you pass a fair verdict on the first anatomical dispute, please judge whether this demand is unreasonable, namely, that Bidloo in the next public demonstration in the Anatomical Theatre of the Academy, prove to the onlookers that the cutaneous glands are shaped like needles, as he has illustrated them in his Tables, and as he says they are in his *Vindications*; our request will seem all the more reasonable, because I myself feel no objection to undergoing these kind of tests; I myself would not refuse to subject myself to this stipulation ....[136]

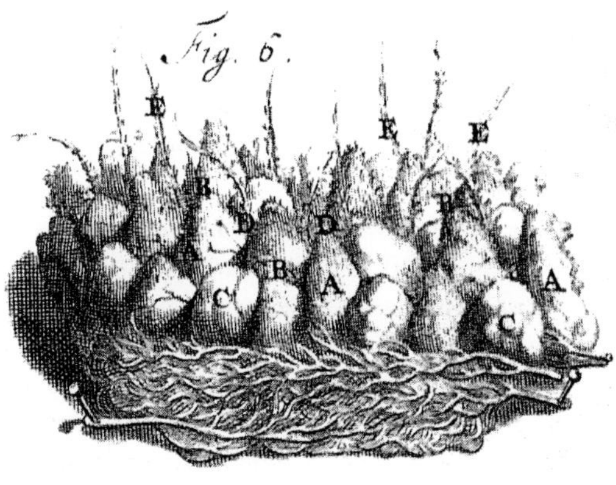

Fig. 4.25

The second issue is something else that Bidloo had illustrated and that Ruysch thought did not exist: the blood vessels in the *membrana arachnoidea*, 'the delicate serous membrane or membranous sac lining the *dura mater*, and enveloping the

135 Ruysch, *Responsio ad Godefridi Bidloi libellum*, 1721 (1697), pp. 3–4.

136 Ibid, pp. 6–7.

brain and spinal cord'.[137] Ruysch doubts that it is possible to separate the arachnoid- from the cerebral membrane, as Bidloo had illustrated it, and impossible to see the blood vessels. It is all (Ruysch claims) a fantasy of Bidloo, not based on sense perception.

> I have recently also shown in a public dissection, where Bidloo was present, how thin this arachnoid is, and how tightly it is fixed to the thin cerebral membrane, and how impossible therefore is the separation and demonstrations of the vessels; but in order to make this all more clear, let Bidloo in his next public dissection of a human show those things that he has depicted in his figures …[138]
>
> …
>
> Finally he [Bidloo] tells a story, which happened in our previous public dissection; but he tells it in such a way, as he is does in other cases, namely by partly hiding and partly distorting: but so that the reader understands the story fully, just like it happened, I shall tell it from the beginning, and judge from this the shamelessness of Bidloo!
>
> When I was allowed to dissect a dead body in public, among those present in the afternoon there was Mr. Bidloo the Apothecary, and as such head supervisor of the Collegium Medicum, brother of the Professor; among other things I demonstrated some cutaneous glands, adding these words: 'These glands, as the audience can see, are round, and are by no means in the shape of a needle, as Professor Bidloo has wrongly depicted in his book'. The next day the aforementioned Professor from Leiden himself came to the anatomical theatre; with few words I repeated what I had said and shown the day before, adding that yesterday I showed the audience the cutaneous glands, and demonstrated that these have a round figure and are not shaped like needles, as they are wrongly depicted in the illustrations of Professor Bidloo. A bit later during the demonstration I spoke about the cerebral membranes, adding these words: 'This membrane is called the arachnoid, which was found many years ago in a Society of learned gentlemen, among whose members were the celebrated gentlemen Blasius, Sladus, Swammerdam, and the famous Doctor Quina who is present here; these gentlemen gave this membrane its name'.
>
> That is why I cannot be surprised enough about the audacity of some people, who dare to claim to be the discoverers of this membrane: 'Please examine its thinness', I said (holding it in front of Bidloo's face and of the person sitting next to him). 'I judge that it is impossible, however hard one tries, to separate a part from it [because of its thinness], as some reckless person has shown in his figures, and shown it to the learned world, and woven through with visible blood vessels, which up till now no one has ever seen; with the help of my skill and instruments, vessels have been discovered in several parts of the body, which no one has seen before, but I say with good reason that it is an impossible task in

137 *Oxford English Dictionary*.

138 Ruysch, *Responsio ad Godefridi Bidloi libellum*, 1721 (1697), p. 8.

> this part, i.e. in the membrane'. Most of the public were not unaware that Bidloo had illustrated this membrane in such a way in his book, and that is why all eyes were on him. On all things Bidloo remained silent; but after the public had left, with the exception of the first row, Bidloo, on his way out, said in Dutch, 'Sir, I thank you for your courtesy, I am prepared to do you a similar favour, when you happen to be in Leiden'. Upon which I answered, 'It could easily happen, that I come there on some occasion'; Bidloo replied, 'When I have a demonstration, I shall let you know', after which I answered, 'As you wish'. Such was the end of the business; but our friend [viz. Bidloo] added 'that I answered with a trembling voice'; if this were true, what conclusion can be drawn from this? Namely that I am not as shameless as Bidloo is! That's fine with me, and for that reason, if I were to write like Bidloo does, would I not then call him the Professor of Impudence, for our imperturbable friend does not back down when he loses a case; as he says, I tremble when I demonstrate that Bidloo wrongly illustrated the cutaneous glands, and makes other blunders, which I can easily correct. Here you have, Reader, a faithful account of the incident, which I can prove with honourable witnesses at any time of the day.[139]

Much later – forty years later! – in 1727, Ruysch wrote about the incident to his friend and last pupil J. Christopher Bohlius, professor at Königsberg –

> As for those so-called olive-like glands [in the cerebral cortex], the truth is these perverse things are without any real existence. I still remember what happened, when Professor Bidloo, in defence of his views, wanted to show to me, in a public Anatomy, that the cerebral cortex is composed of olive-shaped glands. He was holding the cortex of a brain in his hands, and he was asking his audience if they could not see that the cortex of the brain is made up of such glands as he had described in his work? But when he came to Professor De Volder (who does not lightly believe things), and asked him the same thing, he got the answer: 'No, sir!' … If I now also asked you, learned sir, to show me these supposed olive-shaped bodies, would the same not happen to you? Soldiers, when they hear the approach of the enemy, shout, To arms, To arms! I say here likewise, Ad visum! Ad visum! *To looking, To looking!*[140]

A case of theft? Hunters vs. Monros: The testis and the lymphatics 1752–64

Who was the first to successfully inject mercury into the internal tubules of the testicle (*tubuli testis*), and thus show the structures wherein the semen is made? Was it William Hunter in November 1752, working with his brother John in

[139] Ibid., pp. 16–17.

[140] Ruysch, *Responsio ad dissertationem epistolicam … Bohlii* (1727), in *Opera*, 1737, vol. 2, pp. 11–12.

London? Or was it Alexander Monro *Secundus*, in January 1753, working in Edinburgh?[141] And who first proved that the lymphatics were an independent system of vessels and acted as the absorbent vessels for the body? Was it William Hunter in London, who from 1746 taught his students that this was the case? Or was it Alexander Monro *Secundus*, experimenting on live animals in Edinburgh in 1753 and publishing the results in Berlin in 1757? The two questions were so closely linked because both anatomists made their main experiments on the lymphatics of the testicle, either of a man or of a pig (in which all is more clearly visible) or other animal.

It was a curious situation, which pitted families against each other for many years. The Monros, father and sons, appear to have bonded over their exploration of the testis, so Alexander Monro *Primus* weighed in to support his son, and Dr Donald Monro joined in to support his brother. John Hunter in turn carried out numerous dissections to support his brother, William. But the families were not only at war with one another; they were also doubly and intimately linked – and opposed to each other – by anatomical teaching. For William Hunter had studied under Alexander Monro *Primus* at Edinburgh, and referred to him (sometimes sarcastically) as 'my old master'. By 1755, when the two disputants first met, William Hunter had been teaching in his own private anatomical school in London for some years. Alexander Monro *Secundus*, the son of William's 'old master', already appointed joint Professor of Anatomy with his father at Edinburgh (July 1755) and just in receipt of his MD (October 1755), now arrived in London and enrolled to attend William's lectures. William later claimed that he thought that the arrival of Alexander Monro on his course 'was surely a particular honour conferred upon me'.[142] However, these successive master–pupil relationships did not make either side like or trust the other.

The testis and the lymphatic system were of great interest to many anatomists in the mid-eighteenth century largely because of the success of injections in making preparations. For both the testis and the lymphatics seemed to *resist* injections: not even the master of injections, Ruysch, had been able to inject the testicle. Hence they became a particular challenge and a focus of investigation. No one could succeed in injecting far enough along the great length of the epididymis into the testis to reveal its fine structure and thus identify precisely where and within what structures the semen was made. And as for the lymphatics in general, their elusiveness, transparency, fragility, their numerous valves and, in many animals, their apparent total absence, continued to defeat all the skills of the anatomists to discover either their distribution or their function. At mid-century 'it still remained doubtful', wrote the contemporary historian of work on the lymphatics, 'whether the lymphatic system were universally extended over the human body … Neither could any part of [this] system be discovered in birds, amphibious

141 Alexander Monro *Secundus*, 'A description of the seminal vessels', 1754.

142 William Hunter, *Medical commentaries*, 1777, p. 19.

animals, and fishes'.[143] The prevailing opinion was that, whatever their function, the lymphatic veins were like the veins of the blood, and that there must either exist an as-yet-undiscovered complementary system of lymphatic arteries, or that the lymphatic veins are somehow attached to the blood arteries. So the questions were important: what does this system do; where is it present; is it independent of the blood vessels; and where does it originate in the body? Inflating the lymphatics with air, and failing to fill them via injections, was the means employed to answer these questions. It was only towards the end of the century that Paolo Mascagni, Professor of Anatomy in the Royal College of Siena, building on such earlier work, had by his mid-thirties produced an exhaustive illustrated account of these vessels throughout the human body.[144]

Alexander Monro *Secundus*'s inaugural medical dissertation in Latin as professor at Edinburgh (1754) was 'On the testicles and on the semen in various animals'.[145] He gave credit to Albrecht von Haller, whose own recently published experiments recorded that he had managed to inject mercury through the length of the epididymis – the much-coiled efferent duct from the testicle, which ultimately becomes the vas deferens – and even just into the testis itself. By contrast, William Hunter was not mentioned by young Monro. It is clear that Alexander was ambitious and keen to have some important new discovery to his name early in his career. In this dissertation the young Monro had also claimed, almost in passing, 'that the valvular lymphatic vessels, through the whole body, were a system of absorbent veins'.[146] For when injecting mercury into the spermatic artery, Monro had found that the mercury burst the walls of the artery and extravasated (that is, escaped out of the vessel into the tissues), but then it suddenly appeared in the lymphatics nearby, thus suggesting to him that the lymphatics take their origin in the cellular membranes (that is, outside of any vessels), and that they suck up or *absorb* free liquids from them, as here with the mercury. This is the opinion he put forward more fully in Latin in his 1757 Berlin publication 'On the valvular lymphatic veins and their true origin'.[147]

From William Hunter's point of view, it was clear that he himself had made all these discoveries first, and that the young Monro must have initially heard about them from some of William's Scottish pupils who had returned to Edinburgh; that Monro had there repeated William's experiments; and that he had then, at the urging of his father, used William's discoveries in his MD thesis of 1755, having earlier published them under his own name in a contribution to *Essays and Observations, physical and literary. Read before a Society in Edinburgh, and*

143 Sheldon, *The history of the absorbent system*, 1784, p. iii.

144 Mascagni, *Vasorum lymphaticorum corporis humani historia*, 1787.

145 Alexander Monro *Secundus*, *Dissertatio medica inauguralis, de testibus et de semine*, 1755.

146 Alexander Monro *Secundus*, *Observations, anatomical and physiological*, 1758, p. 36.

147 Alexander Monro *Secundus*, *De venis lymphaticis valvulosis*, 1757.

published by them in 1754.[148] In other words, he had stolen William's discoveries! William believed that Monro had then attended his lectures in London in order to learn more, and then the following year had published what he had learnt in his Berlin pamphlet. But in young Alexander Monro's opinion, it was Hunter who had stolen the idea from his own Edinburgh MD dissertation, a copy of which Monro had presented to Hunter when he arrived in London! When challenged, Monro accused Hunter of not having performed the experiments he claimed; of not having made any anatomical preparations to prove the case; and of having taken all his other 'facts' out of other people's books!

William Hunter was certainly a man to nurse a grievance, and he continued obsessed with the behaviour of Monro *Secundus* and *Primus* for many years, publishing his major assault on them as late as 1764.[149] He laid aside the dispute only grudgingly, because he was engaged in finishing his book on the human gravid uterus, but nevertheless signed off with a splendidly sarcastic insult to Monro *Primus*:

> At last, on the eleventh of February [1764], I was so fortunate as to meet with a gravid *Uterus*, to which, from that time, all the hours have been dedicated which have been at my own disposal. I have been busy in injecting, dissecting, preserving, and shewing it, and in planning and superintending drawings and plaister casts of it ... When the additions which have been made [to the book] shall be published to the world, I shall have an opportunity of learning whether, for the future, I ought to be directed by your consummate wisdom and prudence, or go on as well as I can, in my own simple and blundering manner.[150]

One of the problems about proving priority here was that according to one of the combatants – William – the act of 'publishing' (that is, making public) his ideas consisted in his lecturing on them privately to paying students, while according to the other – young Monro – 'publishing' meant publishing in print. So the precise publication date was not decisive in itself to both parties. 'I was the first who published this doctrine about the lymphatics', Hunter wrote, 'and Dr Alexander Monro jun. was the first who printed it'.[151] It has been argued that Hunter took this position because he was an entrepreneurial private anatomy lecturer. He had to have special knowledge that would attract students: 'reserving his teachings for the lecture-hall was for the hard-headed business reason of preserving the trade mystery'.[152] If he were to publish every discovery in print, then his lectures

148 William Hunter, *Medical commentaries*, 1777.

149 Ibid.; first published 1764.

150 Ibid., Supplement, pp. 32–3.

151 Ibid., p. 101.

152 Porter, 'William Hunter: A surgeon and a gentleman', 1985; see pp. 24–6.

would have no special appeal. This may be why he put that curious construction on 'publish' to mean 'made *public* in a *private* course'!

An internal job? Hunter versus Hunter: The structure of the placenta, 1755–80

As we just saw, William Hunter was a man who enjoyed nursing a grudge, and the greatest grudge of all was the one he held against his own brother – and it was a grudge he held until his death. The grudge was held so strongly by William both because it looked like betrayal from one he had loved and nurtured, but also because the discovery in question promised as much fame and respect as any in the long eighteenth century. It was the widely raised question of whether the mother and the fetus have each their own independent blood circulation, or whether up until birth the fetus is dependent on the blood circulation of the mother.

At birth the dark red placenta – so called because in humans it is 'cake-shaped' – peels away from the wall of the uterus and is spontaneously expelled a short time after the birth of the infant. The umbilical cord, containing three blood vessels (two of them conventionally but confusingly referred to as arteries, the other equally confusingly as a vein), comes to the fetus from the middle of the placenta, which itself looks like a mass of blood vessels. Since William Harvey's discovery of the circulation of the blood it was commonly accepted that the developing human fetus is nourished and respires through this connection, but it was not resolved whether the mother's circulating blood was extended to the fetus, or was separate and independent. Harvey himself had maintained that 'the extremities of the umbilicall vessels are in no way conjoined to the extremities of the uterine vessels by an anastomosis, nor do extract blood from them',[153] but the position still required experimental proof. This was what the Hunters provided. But which Hunter? And which Hunter did what? To whose eternal credit should the discovery be attributed? Should it be to the credit of the elder brother, Dr William, now a physician, or of the younger one, Mr John, still a surgeon?

The controversy arose long after the time at which the discovery had been made – a full quarter of a century after.[154] The brothers had not been on speaking terms for some time. Both brothers were Fellows of the Royal Society. In a talk that John gave there, probably in January 1780 (the talk was not printed in the *Philosophical Transactions*), on the subject of the human placenta, he claimed that he and a Dr Mackenzie had discovered the relation of the circulation of the mother and the fetus in utero some time around 1755.[155]

153 Harvey, *Anatomical exercitations, concerning the generation of living creatures*, p. 439.

154 See the excellent discussion of the issues in Brock, *William Hunter … A memoir*, 1983, pp. 47–8, on which I rely.

155 What follows comes from the letters William and John wrote to the Secretary and the President of the Royal Society, and that have been reprinted in Peachey, *A memoir of*

> Dr Mackenzie had injected the subject, and being unable, as I conceived, to explain an appearance which he had found in dissecting it, sent for me. I came to him, and having examined it further, explained the appearance in question, then, for the first time, to my own satisfaction and that of Dr Mackenzie; and in the evening of the same day, full of the discovery, I came to Dr Hunter, and brought him with me to Dr Mackenzie, to see and judge of the explanation I had given and Dr Mackenzie had agreed to. This is my state of the fact upon which I ground my belief of myself being the author of this anatomical discovery.

That is how John defended his claim after William wrote to the Secretary of the Royal Society an angry letter that he wanted to be read out to the Fellows. William claimed:

> First. That the doctrine has been many years ago published in printed books as my discovery, and had been communicated as such by myself ... Secondly. Besides treating of it as my own discovery in my lectures on the subject, I have always done so, for many years past, in the very first lecture of my course, which is the most public of all, because the door is then open to every person whose curiosity prompts him to be present.

William's third 'fact' was a patronising attempt to put Mr Hunter in his place, as a (mere) assistant to William in this and many other pursuits. He was proud of his little brother – 'I have given him all the little anatomical knowledge that I could communicate' – and helped his career, the Doctor wrote. 'May it be presumed then', William asked, 'that I stand possessed of the discovery in question, till proofs shall be brought to dispossess me?' While William did not go into any detail about the circumstances or precise date of his discovery, he said that his claims 'arise out of a long series of observations and experiments made with a view to the discovery in question; that it was not a random conjecture, a lucky thought, or accidental occasion, but a persevering pursuit for twelve or thirteen years at least'. So William presents his discovery – curiously never published, despite its importance – as that of a professional researching anatomist, while he portrays his brother merely as his assistant, who may have had a lucky break.

It looks as though the President of the Royal Society notified John of this letter and sought his response. Mr Hunter replied to this attack relatively calmly, given that (as he wrote), 'I am as tenacious as he [viz. William] is of anatomical discovery'. Having laid out his memory of how the discovery was made (as recited above), and that it was he who made it, John wrote, 'but as my brother thinks differently, after a period of twenty-five years, I am content to abolish all remembrance of the successions of time in the course of that day, and to suppose that Dr Mackenzie, Dr Hunter, and myself inspected the parts together,

---

William and John Hunter, 1924, pp. 267–71.

and made the discovery, by which means the honour of it will be divided into three'. It would certainly be strange that John would not know about such an important discovery if his brother had made it, for William was always extremely tenacious of anatomical discoveries. It is unclear whether the Fellows of the Royal Society were presented with these letters from the two brothers, and the matter of controversy – as so often – was unresolved.

Chapter 5

Animal Bodies and Comparative Anatomy

1. The animals of the anatomist: Vivisection and dissection

As will already be abundantly clear, the anatomist depended on being able to dissect and vivisect animals of all kinds, without constraint. William Harvey, for example, in the London of the first decades of the seventeenth century, had been able freely to dissect and vivisect a great range of domestic creatures as they came to hand. His own list included hens, geese, pigeons, ducks, fishes, shellfish, molluscs, frogs, snakes, bees, wasps, butterflies, silkworms, sheep, goats, dogs, cats, deer and cattle.[1] When Thomas Willis and Richard Lower investigated the brain in Oxford in the early 1660s, they dissected and vivisected 'whole Hecatombs almost of all animals, in the Anatomical court', including horses, sheep, calves, goats, pigs, dogs, cats, foxes, hares, geese, turkeys, fish, and a monkey.[2] When the Scottish anatomist John Hunter was called as an expert witness in a trial in 1781 at Warwick in England over whether someone had been deliberately poisoned, he was asked if he had ever poisoned animals. 'I have poisoned some thousands of animals', he replied.[3] All told, there were hundreds of thousands and possibly millions of animals vivisected or experimented on in one way or another by anatomists in the course of the long eighteenth century.

There was no opposition to this practice, although anatomists occasionally expressed regret at the pain they had necessarily caused dumb animals in the course of their experiments. The young Frederick Ruysch, investigating in Leiden in 1661 and looking for the route of the chyle, vivisected dogs yet wrote that 'it is not without abhorrence that I torture them with such prolonged pain. The Cartesians [that is, the followers of Descartes] take great pride in the truth of their philosophical system, but I wish they could convince me as thoroughly as they are themselves convinced of the fact that animals have no souls!'[4] Albrecht von Haller, exploring the anatomical phenomena of irritability and sensibility in Göttingen in the early 1750s, described some of his horrific animal experiments on a hundred and ninety animals as 'a species of cruelty for which I felt such a reluctance, as could only

---

[1] Harvey, *Anatomical exercitations, concerning the generation of living creatures*, 1653, Preface, b7v.

[2] Willis, *The anatomy of the brain and nerves*, 2 vols, 1664, vol. 1, Dedication.

[3] Hunter, *Lectures on the principles of surgery*, 1835, p. 195.

[4] Letter to Bartholin, quoted by Luyendijk-Elshout in her introduction to Ruysch, *Dilucidatio valvularum*, 1964 reprint, p. 36.

be overcome by the desire of contributing to the benefit of mankind'.[5] Sometimes – bizarrely, to a modern sensibility – the anatomist expressed regret at the offence to their *own* feelings that vivisection incurred. For instance, Thomas Pole wrote in 1790 that, 'Unless the man of feeling will submit to derive his knowledge from the testimony of the less compassionate, and, perhaps, more industrious Anatomist, he must condescend to hurt his sensibility by adopting those means, from which he can alone acquire substantial information: the dissection of living animals is here meant … necessity will occasionally require a reluctant acquiescence with such means; but the feelings of human nature lead us to hope that this employment will never be wantonly indulged in'.[6] The sensibility that vivisection is cruel did not appear widely in Europe until the nineteenth century, and even in some European countries has still not appeared today. Moreover, people in western Europe, even in the great cities, lived more closely with animals than most of us do today and were in general unsentimental about them. The horse was the primary means of transport and power; pigs and cattle were kept in almost every quarter of every town. While dogs and cats were sometimes kept as pets, as were exotic birds, they also roamed freely, scavenging a livelihood from the refuse produced by human life and industry. Attitudes to animals were far less sensitive than they are today.[7]

In addition to the use of domestic and easily accessible animals for vivisectional experiments, there was also widespread interest, both among anatomists and other people, in the anatomical structure of an enormous range of different animals in this period, and these animals were dissected after their death. These included large and exotic creatures such as elephants and whales, but also small and familiar ones such as caterpillars and silkworms. The number of different animals that were dissected, and whose anatomies were put into print, was so great, so apparently exhaustive, that looked at as a whole it almost seems as though there was a co-ordinated international campaign in the long eighteenth century to explore and publish the anatomy of every kind of animal, bird, fish and insect that could be discovered.

In fact these developments were piecemeal and largely without co-ordination, but formed part of the widespread practice of the discipline of 'natural history'. Such anatomies were for the most part conducted by people calling themselves 'naturalists' or 'natural historians', and occasionally 'zoologists' (but not yet 'biologists', since this modern coinage from the Greek had not yet been made). Different aspects of animals were studied by such people under different headings. Our interest here must be limited to the subjecting of animals to being anatomised, putting them to the knife to explore their structure and functioning. It will be possible only to give a mere sketch of the myriad encounters that the anatomists

5 Haller, *A dissertation on the sensible and irritable parts of animals*, 1755, p. 657.

6 Pole, *The anatomical instructor*, 1790, pp. xxv–xxvi.

7 The gradual changes in attitude are documented by Maehle, 'The ethical discourse on animal experimentation', 1993. For a brief but useful history of vivisection, see Guerrini, *Experimenting with animals and humans*, 2003.

had with their investigative and experimental animals during the period of the perfection of anatomy. My guiding questions in this chapter are these:

1. What kinds of animals were being dissected?
2. Where did these animals come from, especially the exotics?
3. Why were the anatomists dissecting them?

Although natural historians had been interested in animals and many things about them since the early 1500s, such as is recorded and illustrated in the five-volume work by Conrad Gesner, the *History of Animals* of 1551, it was not until the end of the sixteenth century that any of this interest was turned towards the actual anatomy of the animals being studied. In 1598 Carlo Ruini first published in Bologna his *Anatomy of the Horse, its Infirmities and Remedies*, many editions of which were issued in the next fifty years, in which he studied and illustrated the skeleton, internal organs and muscles of the horse (for which see later in this chapter).

Most of the earlier work on the anatomy of animals in our period was done under the aegis of the Académie royale des sciences in Paris. Unlike its English counterpart, the Académie royale was structured in a way that followed suggestions by Francis Francis Bacon, with a select group of pensioned professionals investigating nature and advancing knowledge as a group activity.[8] One such early project was on plants, and the result was co-ordinated by Claude Dodart, and published in 1676 as *Mémoires pour servir à l'histoire naturelle des plantes*. Dodart announced, 'Ce livre est l'ouvrage de toute l'Académie' ('This book is the work of the whole Academy'). The Académiciens are seen in action in the frontispiece of the resulting volume, in Fig. 5.1. Indeed, the Académie remained officially committed to joint labours up to 1699. According to Dodart, the Great Prince (that is, Louis XIV) had given the Académie the means to pursue this project in new ways. The Great Prince appears to have found the only proper way to advance the Sciences: by assembling several people to work in the same spirit and to the same plan, without slacking or undue haste, to assess the opinions of Philosophers by experiment, and to assess experiments by their own eyes – so runs the *Avertissement* of the volume. The Académiciens investigated the description of plants, their forms, their virtues – with much chemical work to discover their virtues by fire – and reported their research on the constitution of plants.

The Académie's project on animals was, as it were, a sister project to the one on plants. Both volumes have as a frontispiece a picture of Louis XIV visiting the Académie, which was then housed at Versailles (it did not move to the Louvre until 1699)[9] – see Fig. 5.2.

The findings of this second project were published in two volumes in 1671 and 1676 as *Mémoires pour servir à l'histoire naturelle des animaux* co-ordinated by Claude Perrault: *Memoirs to aid the natural history of animals*. The title-page is

[8] On the Academy and its projects see Stroup, *A company of scientists*, 1990.

[9] Schiller, 'Les laboratoires d'anatomie', 1964, pp. 104–7.

Fig. 5.1

Fig. 5.2

shown here in Fig. 5.3.[10] Perrault, who is more commonly remembered today as an architect, had also been fully involved in the plant project. The grand scale of this work on animals, and the fact that the book seemed only to be available if offered as a gift from the French king, led the members of the Royal Society of London to pirate it and republish it in a reduced format in 1688 in English, and later the work was extended and Latinised for a wider world by a Prussian professor, who published his edition in 1720.[11]

Fig. 5.3

More than fifty animals and birds are treated in the book, each of which had been systematically dissected. They included the lion, chameleon, dromedary, bear, gazelle, mountain cat, seafox, lynx, beaver, otter, civetcat, elk, sea calf, barbary cow, cormorant, chamois, porcupine, monkeys, pintado, eagle, Indian cock, bustard, ostrich, cassowary and tortoise. What was special about this great work, according to Perrault, was that it was a collaborative exercise, and that 'it contains no facts that have not been verified by the whole company, composed of people who have eyes to see such things, unlike most of the rest of the world, and who also have hands to seek them with the greatest dexterity and success, who see clearly what is the case, and who can only with difficulty be persuaded to see what is not the case'.[12] The accuracy and the facticity of the work, Perrault hoped, would make these *Mémoires* 'the material of a Natural History not unworthy

[10] Schiller dates this image to 1668 when the Académiciens dissected a fox, and places the event in the Bibliothèque du roi: ibid., p. 102n. Members of the court wear broad-brimmed hats; see Watson, 'The early days of the Académie des sciences', 1939, p. 567.

[11] Perrault, *Mémoires pour servir à l'histoire naturelle des animaux*, 2 vols, 1671, 1676; [Perrault], *The natural history of animals*, 1702; Valentinus, *Amphitheatrum zootomicum*, 1720. On the convoluted printing history of this work, see Cole, *A history of comparative anatomy*, 1949, pp. 393–400.

[12] Perrault, *Mémoires pour servir à l'histoire naturelle des animaux*, 2 vols, 1671, 1676, Preface.

of the greatest King who has ever lived', that is, of Louis XIV. The dissections themselves were made not by Perrault, who was a physician, but by a series of surgeons: Gayant, Dionis and Joseph-Guichard Duverney (Professor of Anatomy at the Jardin from 1679 to 1730).

Similarly, and at the same period in the late seventeenth century, Edward Tyson, a Fellow of the Royal Society of London, had busied himself in dissecting and recording all the unusual animals that came to London that he could get his hands on. In 1679, he dissected a porpoise; in 1681, an ostrich; and in later years a lion, tape-worm, dogfish, rattle-snake, chimpanzee (which he called the orang-outang), and an opossum.[13] The front view of the musculature of the orang-outang is shown in Fig. 5.4, as dissected by Tyson.

2. The source of exotic animals for dissection

The giving and receiving of exotic animals had long been a gesture of power and honour among rulers, both in Europe and elsewhere. The lion was the most symbolically important and the most frequently given – and the lion, naturally enough, appeared as the first animal in the *Mémoires pour servir à l'histoire naturelle des animaux* (see Fig. 5.5) – but other exotic animals were also considered suitable presents for princes. Emperors and kings thus often accumulated a collection of lions and other unusual animals, such as those kept in the Tower of London from the thirteenth century. The tradition was still alive and well in the seventeenth century when, for instance, the King of Morocco gave Charles II, King of England, a present of two lions and thirty ostriches in 1681,[14] and the tradition survives even today among the royal families of the world.

The dedicated creation of a collection of exotic animals for private or public viewing seems to have been first undertaken in early modern Europe by Louis XIV of France, whose purpose-built building for his collection at Versailles, erected in 1662, was the first to be referred to as a *ménagerie*, a word whose basic meaning is the management of livestock.[15] In the view we have of it in about 1662 the elephants are clearly visible (Fig. 5.6). The royal menagerie was the source of most of the animals in Perrault's *Mémoires*, the whole series of which had begun from the moment in 1667 when the king gave the Académie a recently deceased lion and a thresher shark to dissect. Louis XIV loved his menagerie so much that he was persuaded to give it to his young daughter-in-law, whom he

---

13 Ashley Montagu, *Edward Tyson*, 1943, pp. 126–7, 135.

14 Ibid., p. 132.

15 See Loisel, *Histoire des ménageries*, 2 vols, 1912, for all aspects of the history of menageries. For the structure of the menagerie, but not its animals, see *Almanach de Versailles*, 1782. Loisel's marvellous study is now complemented by the wonderful Baratay and Hardouin-Fugier, *Zoo: A history of zoological gardens in the West*, 2002. See also Robbins, *Elephant slaves and pampered parrots*, 2002.

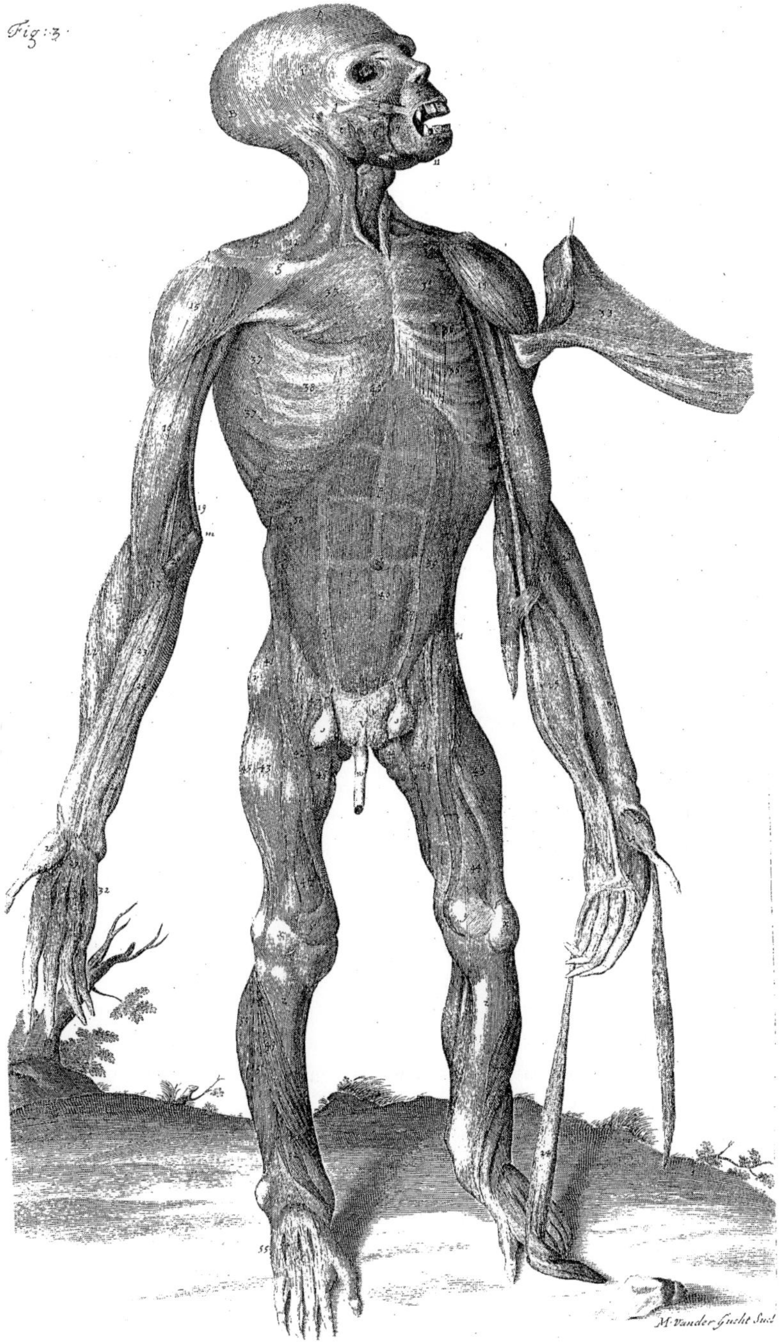

Fig. 5.4

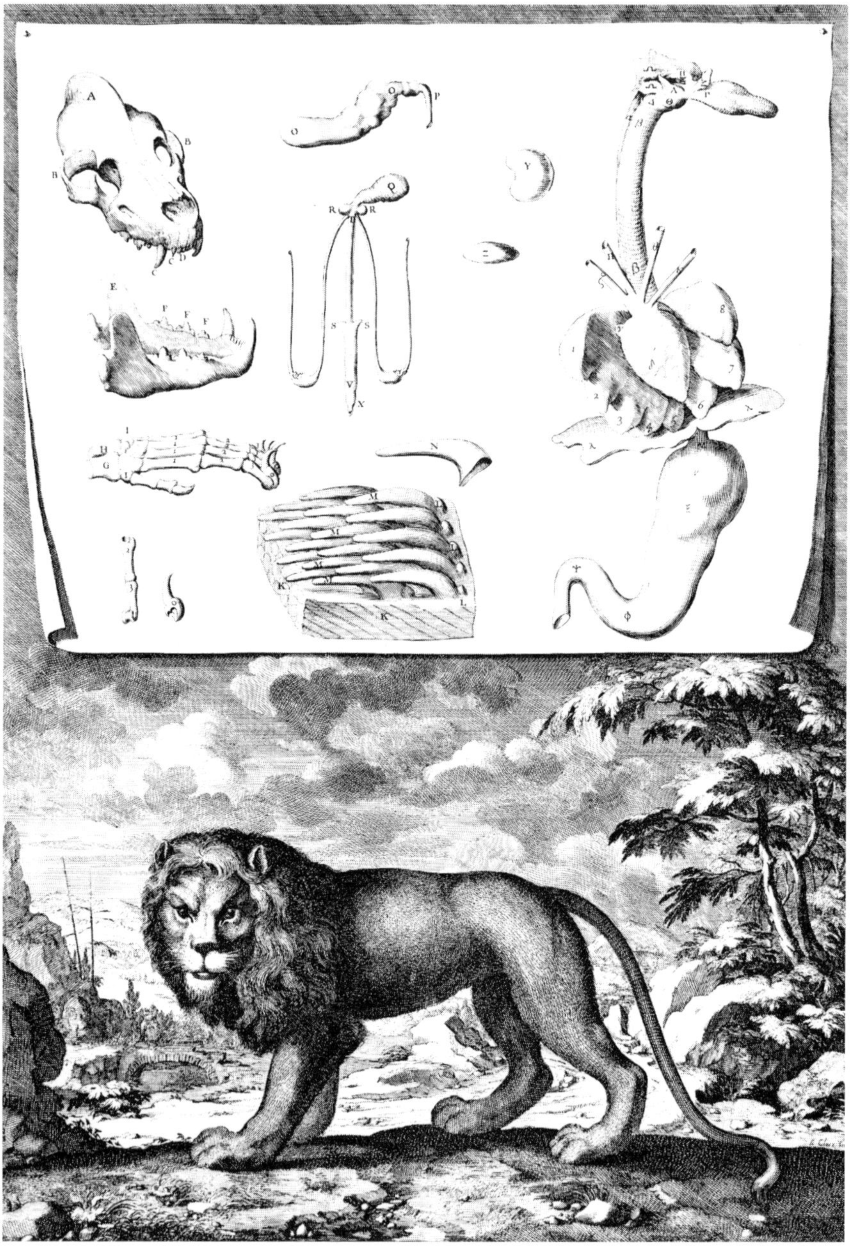

Fig. 5.5

adored even more, as a very special present. But Louis XV and Louis XVI showed no interest in the menagerie, which went into gentle decay in the course of the eighteenth century. But 'menagerie' as a term had become so familiar by the end of the eighteenth century as one with royal associations, that it was frequently used by those opposed to the French royal family in the French Revolutionary years to mean exotic, expensive and useless people: that is, the royal family and their hangers-on.

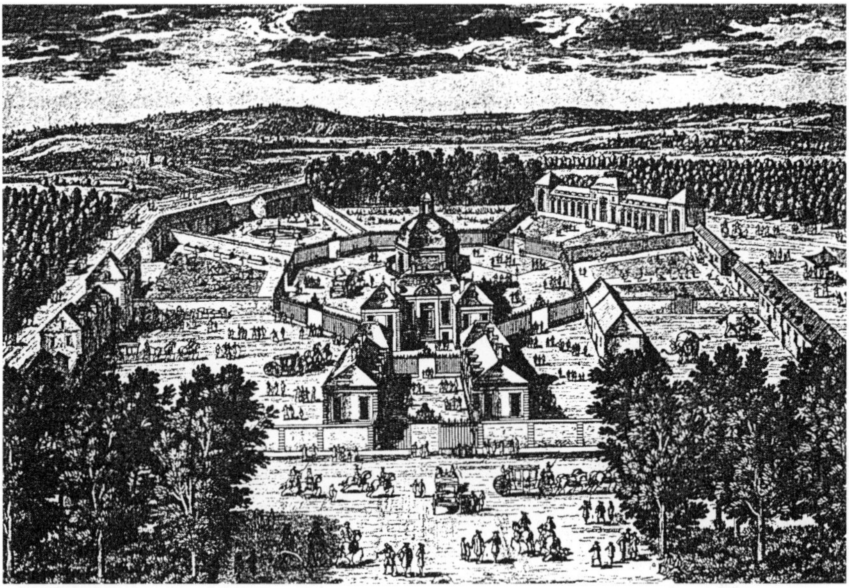

Fig. 5.6

In Vienna there was a menagerie in the Shönbrunn imperial palace, and several elsewhere in Europe, such as at Potsdam, and the royal one in St James's Park in London.[16] Curiosity and spectacle seem to have been the point of such menageries and, like the palaces they adorned, they were all based on the Versailles model.

The Versailles menagerie continued in existence until the French Revolution, when the less ferocious animals were freed by the Jacobins in 1792. The five surviving large animals – a rhinoceros, a lion with its constant friend a dog, a couple of antelopes, and a cougar – were offered to the Jardin des plantes in 1794, where they came under the care of Etienne Geoffroy Saint-Hilaire. This little group of animals was the origin of this most important of all menageries. To increase the size of this new menagerie, all the travelling menageries in Paris were closed down and the animals transferred to the Jardin. It had early problems: on one occasion during the course of the Revolution the least valuable animals had to be killed to

[16] See Loisel, *Histoire des ménageries*, 2 vols, 1912, passim.

feed the more valuable ones.[17] However, these problems were overcome and the menagerie of the Jardin des plantes still exists today in the form of a zoo.[18]

In addition to such royal menageries, there were in the eighteenth century a number of travelling commercial menageries, or collections of exotic animals, presenting the exotic to a paying public with shows that would include dwarves, giants and freaks, as well as lions, tigers and elephants.

Some erstwhile travelling menageries took on a more settled existence, such as that of Mr. Gilbert Pidcock in the 1790s at Exeter-'Change in the Strand in London, which had previously spent many years touring the country. The celebrated 'horse painter' George Stubbs (of whom we shall be hearing more below) got hold of a dead tiger from this menagerie. At 10 o'clock one evening, the story goes, probably in the 1780s, a messenger told Stubbs that a tiger had just died at the menagerie. Stubbs 'flew towards the well known place and presently entered the den where the dead animal lay extended: this was a precious moment; three guineas were given to the attendant, and the body was instantly conveyed to the painter's habitation, where in the place set apart for his muscular pursuits, Mr. S. spent the rest of the night in carbonading the once tremendous tyrant of the Indian Jungle'.[19] Why Stubbs should have grilled ('carbonaded') the tiger in making his anatomy is not clear; let us hope it is a misprint. This tiger appeared in Stubbs's 'comparative anatomy' that he prepared in the last years of his life, and that will be dealt with later in this chapter.

In the course of the nineteenth century menageries for the most part came to be turned into or supplanted by 'zoological gardens'. Zoology and zoological were effectively new terms at the beginning of the nineteenth century, first brought into popular use in the years around 1800, especially by Lamarck in his *Philosophie zoologique* of 1809. In these new 'zoos' the primary aim was that animals should be brought from every part of the globe 'to be applied either to some useful purpose, or as objects of scientific research, not of vulgar admiration', as the prospectus of the first zoo, that of the Zoological Society of London, put it in 1825.[20] Among the intended useful purposes were selective breeding and acclimatisation, so that exotic economically valuable animals could be farmed in Europe, and European animals be farmed in the overseas colonies of the European nations. The curiosity

---

[17] Deleuze, *History and description of the Royal Museum of Natural History*, 2 vols, 1823, p. 81.

[18] On its collection in 1801 see *Notice des animaux vivants, actuellement à la Ménagerie du Muséum National d'Histoire Naturelle*, 1801, and Lacépède and Cuvier, *La Ménagerie du Muséum d'Histoire Naturelle*, 1801; for its condition later see *Description de ce qu'il y a de remarquable à la Ménagerie*, 1818.

[19] N., 'The lion and horse', 1808.

[20] Chalmers Mitchell, *Centenary history of the Zoological Society of London*, 1929, p. 13. Even specialised historians of menageries and zoos continue to be quite sloppy in their use of terminology, and do not recognise the linguistic, historical and functional distinctions between the terms. See for example Hoage, *New worlds, new animals*, 1886.

and spectacle typical of the menagerie was intended to be replaced by education and scientific research.

At all events, it can be seen that there were many exotic animals available in eighteenth-century Europe, especially in the greatest towns, including lions, tigers, elephants and bears. The anatomists pounced, when they could, on dead animal bodies in the royal and commercial menageries, and dissected them. For instance, the members of the early Royal Society of London had had 'the design of getting the bodies of all exotic animals as should chance to die in St James's Park in order to have them anatomized and described'. They made an approach to the Keeper of the Tower of London in 1680, and they swooped on the troupe of ostriches that the King of Morocco had given to Charles II in 1681, dissecting them as they died in quick and sad succession.[21] In the same year Duverney and Perrault dissected the famous friendly elephant at Versailles, who had entertained people by patiently undoing knots with his trunk. Louis XIV turned up to watch the dissection and asked where the anatomist was, whereupon Duverney rose up from the bowels of the elephant's corpse to greet his king.[22] The skeleton of this elephant, seen here in Fig. 5.7, was one of the earliest that went to make up the royal anatomical collection and is still on show in Paris today. In the eighteenth century in Paris, Buffon was able to dissect all the animals that died in the royal collection, and at the end of the century Cuvier was able to dissect all those that died in the menagerie of the Muséum d'histoire naturelle (as the Jardin des plantes was by then called).

The elephant was fascinating to anatomists: 'a prodigious creature', 'the hugest of all that tread the ground, a mountain of animated matter', in William Stukeley's words.[23] And there were lots of them: I know of at least twelve elephants that happened to die near friendly anatomists in Europe in this period, and were therefore anatomised after their deaths – goodness knows how many died safe from the anatomist's knife; the results of some of these dissections were published. After Duverney in 1681 (above), Patrick Blair dissected one in Scotland in 1690 and sent a picture of it to the Royal Society in London (Fig. 5.8); Monsieur Suply opened one in Fort St George in 1715; William Stukeley one in Sir Hans Sloane's garden in London in 1720; J. G. Duvernoy one in Tübingen in 1726; Petrus Camper one in Holland in 1774; and the Hunter brothers dissected no less than three in London in the 1770s.[24] It was for a long time a great mystery to the anatomists

---

21 Ashley Montagu, *Edward Tyson*, 1943, p. 90.

22 Loisel, *Histoire des ménageries*, 2 vols, 1912, vol. 2, p. 298.

23 Stukeley, 'An essay towards the anatomy of the elephant', 1723, in his *Of the spleen*, 1723, p. 89.

24 This is a list assembled non-systematically: **1**. Perrault and Duverney, 1681 in Paris; **2**. Blair 1690s, Scotland; **3**. M. Suply at Fort St George, 1715 [see Stukeley, 'An essay towards the anatomy of the elephant', 1723, in his *Of the spleen*]; **4**. Dr Moulins in Dublin 'lately', says Stukeley (p. 106), without further reference; **5**. Stukeley, London 1722; **6**. J. Georgius Duvernoy (or Duvernoi) of Tübingen (not the famous Duvernoy of

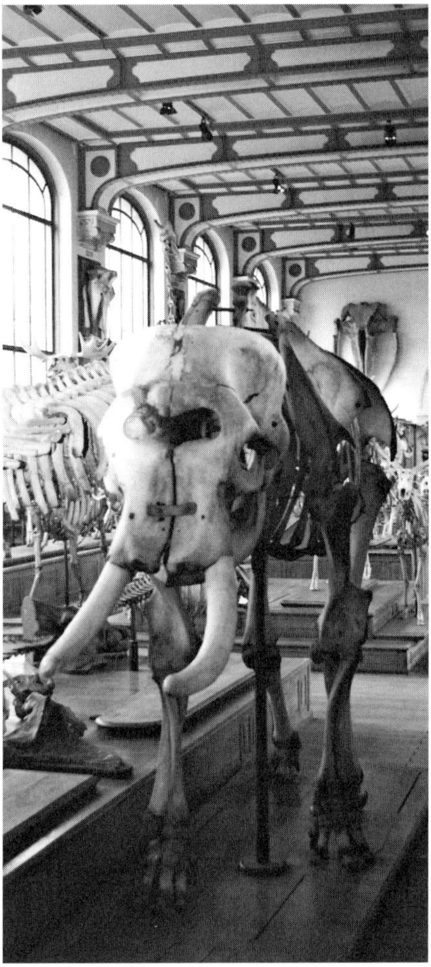

Fig. 5.7

Paris) published on the thoracic duct, heart and penis of an elephant in the *Memoirs* of the St Petersburg Academy – see Portal, *Histoire de l'anatomie*, 6 vols, 1770–73, vol. 4, pp. 641–2, and Haller, *Bibliotheca anatomica*, 2 vols, 1774, 1777, vol. 2, pp. 172–3; **7**. an Italian, Serrao, Naples, *c.* 1750, referred to in Camper, *Description anatomique d'un éléphant mâle*, 1892, p. ix, note; **8**. Camper *père*, 1774, Netherlands; **9**, **10** and **11**. William and John Hunter: one in 1773 in London, then two in London in 1775 – see Home, *Lectures on comparative anatomy*, 1823–28, p. 250; **12**. Soemmering, 1780 at Cassel (didn't publish) – see Camper, ibid., p. ix. Despite its title the very curious work Hartenfels, *Elephantographia curiosa*, 1715, did not celebrate the dissection of an elephant. Hartenfels had only once seen an elephant himself, and knew of no German who had ever dissected one. The illustrations indicate that he did not remember much of the one elephant that he had seen. The earliest dissection of an elephant in the early modern period of which I am aware took place in Verroia in Greece in the early 1610s, though the dissectors were not anatomists.

to determine the sex of the elephant they were dissecting, or to ascertain how elephants had intercourse.

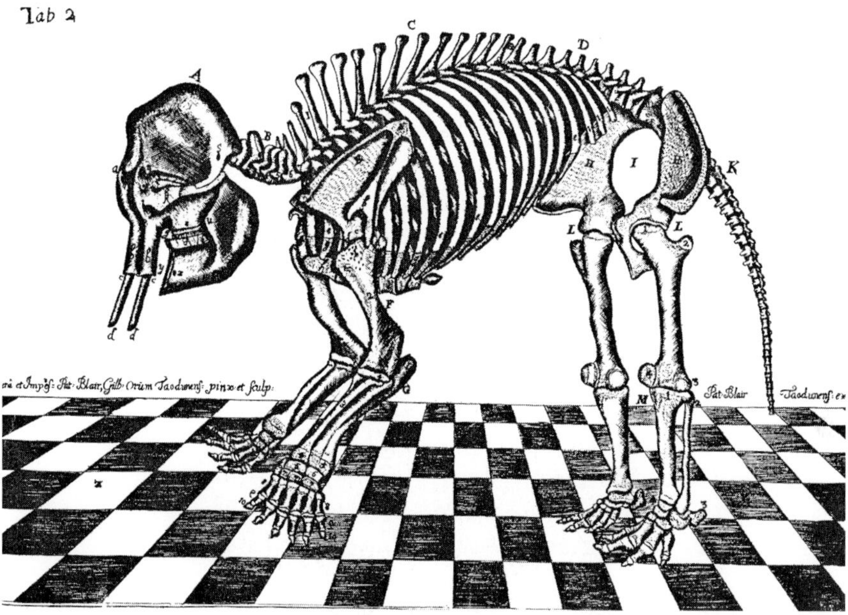

Fig. 5.8

But to eighteenth-century eyes the rarest, and hence most fascinating, large exotic animal was the rhinoceros. The famous so-called 'Dutch' rhinoceros, which came from India, was toured in a wagon by a Dutchman around all the major towns of Europe for almost twenty years from 1741 to 1758, beginning from the Netherlands. It was a female called Clara, and she was taken to Germany and Switzerland and then to France and Italy, being later exhibited in Vienna, London, Warsaw, Danzig, Copenhagen and many other towns. Visitors to the rhino's booth paid more the nearer they could approach to the rhino. People clearly believed that the rhino was a fierce and dangerous animal, though in fact the Indian rhino is generally docile and only fierce when it is deliberately antagonised. It was popularly believed that a rhinoceros would 'lick a man to death by raking away the flesh to the bone with his rough and sharp tongue'.[25] Clara was so striking that local artists portrayed her in many of the places that she visited. With respect to the history of anatomy, the most famous such portrayal of her are the pictures by the anatomist Albinus and his artist Jan Wandelaar, who saw and drew her in Leiden at the beginning of her tours. We saw these images in Chapter 4. The anatomist Peter Camper drew and modelled her in clay on a return visit of the beast to Leiden

25 Grew, *The comparative anatomy of stomachs and guts begun*, 1681, p. 30.

in 1748. A particularly remarkable set of images of her was produced at Venice in 1751 by Pietro Longhi, a Venetian artist whose 'whimsical and capricious skill' was said by contemporaries to have 'forged for himself a new and individual style of painting conversation pieces', and 'with this great talent he rose to high fame, and his works fetch high prices'. Longhi's sense of humour is always mentioned by commentators. For a period of his career he was obsessed with the portrayal of the Venetian nobility in their carnival masks; his son wrote that 'he painted pictures of masked figures which were so realistically depicted in their natural attitudes that they were recognisable even behind the mask'.[26] It was at this period that Clara arrived on the Venetian scene where, unfortunately, her horn fell off. There are several versions of Longhi's representation of Clara. We see both humour and mask in this particular version: every reveller present at an exhibition of the rhino is masked, but they do not watch the rhino, looking instead directly at the viewer (Fig. 5.9). The keeper, however, holds the horn of the rhinoceros in his hand. The rhino has been 'unmasked'!

The Versailles menagerie had a rhinoceros, which arrived in 1770 and died (or was killed) shortly before it was to be transferred to the new menagerie of the Muséum d'histoire naturelle in 1793. It was nevertheless taken there for dissection. It was dissected – the first rhinoceros ever to be dissected – by Mertrud and Vicq d'Azyr, in the presence of Daubenton. A series of thirty-eight coloured drawings of the organs of the beast were made by Nicolas Maréchal and Pierre-Joseph Redouté, and the re-articulated skeleton was placed in the Museum.[27]

There were no less than two rhinoceroses at the largest London animal dealer's premises at the end of the century, a male in 1790–93, which the 'horse painter' George Stubbs painted, and, when Pidcock took over the menagerie, another male in 1799. The first occasion when a rhinoceros was dissected in England occurred in 1800.[28]

3. What did 'comparative anatomy' mean in the long eighteenth century?

We have seen something of the vast array of animal species to which European anatomists had access in the course of the eighteenth century. But why were anatomists dissecting animals at all, given that their primary interest in anatomy was the structure of the human body? And why were they interested in anatomising such a very wide range of different animals, domestic and exotic? The term 'comparative anatomy' was widely used by anatomists from the mid-seventeenth century to the end of the eighteenth century in connection with their anatomising

[26] Pignatti, *Pietro Longhi*, 1969; quotations are from pp. 58, 60. See now also Ridley, *Clara's grand tour*, 2005.

[27] Saban, 'La dissection au Muséum du rhinocérus', 1983; Rookmaaker, 'Histoire du rhinocérus de Versailles', 1983.

[28] Thomas, 'An anatomical description of a male rhinoceros', 1801.

Fig. 5.9

of animals. What did it mean? Or rather, what did the anatomists mean when they used the expression? Were these anatomists practising comparative anatomy in the modern sense? What relation, if any, did their 'comparative anatomy' have to modern comparative anatomy? If it had no direct relation to modern comparative anatomy, then what did the anatomists of the long eighteenth century mean by their use of the expression? Was it a discipline, as anatomy was? If not, what kind of enterprise was it?

A definition of the modern discipline or sub-discipline of comparative anatomy compares it to ordinary modern anatomy:

> Comparative anatomy is a more dynamic approach to the study of anatomy than the mere learning of the location and structure of organs and systems. Homologous structures in different animals are compared with each other to determine how these structures have originated and become modified. Comparative anatomy seeks to show the phylogenetic origin of the various structures in different groups of animals by comparison of the form, location, and development of these structures. It seeks to explain why animals are constructed as they are and constitutes important evidence for the evolution of animal life.
>
> The comparative method is based on a critical comparison of structures that are truly homologous rather than analogous. Homology deals with those structures which are alike in origin and fundamental structure.[29]

We can see from this that comparative anatomy today is a discipline that is completely based on Darwinian evolutionary theory. That is to say, evolutionary theory is the *explanatory basis* of the accounts given by comparative anatomists of why animals are as they are and why they differ from each other: the answer lies in differing development over time from common ancestors. As another modern comparative anatomist has put it, 'Organs do not arise from nothing at all. They are modifications of structures which already exist'.[30] But as the definition above claims, comparative anatomy is also used as *evidence for* the soundness of the Darwinian theory, as presenting evidence of evolution having taken place. So comparative anatomy is crucially connected with evolutionary theory, both as explanation and as evidence. In addition we may note that modern comparative anatomy is almost exclusively concerned with the study of the chordata (animals with spinal cords), and especially the vertebrates (animals with the spinal cord contained within a flexible bony column).

Obviously any comparative work undertaken on animal anatomy in the long eighteenth century will not qualify as a meaningful predecessor of the modern discipline of comparative anatomy, because it all precedes Darwin's theory, which was only published in 1859. This modern, Darwinian, version of comparative anatomy itself replaced a new discipline of comparative anatomy that was established in the early years of the French Revolution by Georges Cuvier, on which more will be said in Chapter 6 below.

Yet the claim has certainly been made by historians that there was a discipline – at the very least an intermittent enterprise – of comparative anatomy from Antiquity onwards. So we need to ask here whether in the long eighteenth century

[29] Adams and Eddy, *Comparative anatomy*, 1949, p. 105.

[30] Atwood, *A concise comparative anatomy*, 1947, p. 20.

there were indeed other disciplines of comparative anatomy, but that were based on different, and now superseded, principles and aims. Before we can be in a position to answer this, we have to first dispose of the tradition of modern historical writing on comparative anatomy that has given a long but ambiguous history to the modern discipline.

The history of comparative anatomy, as we currently have it from the historians, is problematic, and needs to be closely inspected rather than just being adopted as the basis for our own account here. For comparative anatomy has been given many founders and contributors, from Aristotle onwards. F. J. Cole, who was Professor of Zoology at the University of Reading from 1907 to 1939, is the author of the most important modern English-language work in this respect: *A history of comparative anatomy from Aristotle to the eighteenth century* (1944). He begins his account with Aristotle and Galen, picks his story up again in the fifteenth century with Leonardo da Vinci, and then continues through virtually every anatomist of note until the period of what he calls 'The new comparative anatomy', beginning with Malpighi in the 1660s and continuing up to Ruysch in the early eighteenth century, but stopping before he reaches Cuvier. According to Cole's account, then, it would appear that virtually every anatomist who ever practised between the time of Aristotle and Ruysch was also a comparative anatomist! But what is the criterion that Cole is using to identify such supposed comparative anatomists? The nearest Cole comes to giving a definition of the practices he is talking about is in a question-begging sentence in the Introduction. Cole asks whether there can ever be comparative anatomy without the concept of evolution, and replies: 'The elucidation of an organ in one animal by comparison with the corresponding feature in another was in fact practised in the earliest days of comparative anatomy. The *explanation* of those homologies is another story, though a closely related one'.[31] Thus Cole asserts that there was indeed a practice of comparative anatomy in the many centuries before Darwin, where the establishing of comparative-anatomical facts was not yet linked to an adequate explanatory theory, and the thrust of his whole book is that this practice had some contributory role in the making of modern (that is, Darwinian) comparative anatomy, though what this role was he does not specify. Cole's definition of this practice is actually extremely vague and never clearly specified. It seems to amount to this: if an anatomist dissected or vivisected *animals* at all, or if he made a comparison *of any kind*, no matter how trivial, between the anatomy of man and that of any other animal, or between the anatomy of one animal and another, then he was a practitioner of this supposed old version of 'comparative anatomy'.

This very loose form of definition of 'what counts' as instances of the practice of investigative disciplines is typical of the historian of science who seeks in the past forebears for his modern discipline. In this way every 'great name' can be included as a practitioner of the discipline in question, and the discipline can be given a long and glorious past, however new it may actually be. But this approach

[31] Cole, *A history of comparative anatomy*, 1949, p. 6.

is simplistic in the extreme and actually produces a false picture. For it sidesteps the question, 'What counts as a particular discipline and as the practice of that particular discipline?' Even Cole himself concedes that, 'In the pre-evolutionary period [i.e. before 1859, when Darwin published *On the origin of species*] a comprehensive system of comparative anatomy was scarcely to be expected'.[32] His book, therefore, although presented as a history of comparative anatomy, is actually simply a history of interest in animals and their anatomy, and of the occasional use of the anatomy of animals to elucidate the structure of humans. Even in the brief 'chronology of comparative anatomy' that prefaces his book, Cole notes that the term itself is not used before Francis Bacon, and even then 'not in the modern sense'.[33] Cole himself shows that many of the seventeenth-century usages of the term merely mean the dissection of animals. So despite all the years of work that Cole's book represents, we cannot take his *History of comparative anatomy from Aristotle to the eighteenth century* as much of a guide here as to whether there was in the eighteenth century in fact one or more authentic – authentically eighteenth-century, that is – practices or disciplines of 'comparative anatomy'.

It might be said in Cole's defence that he was himself only following in a tradition established by earlier comparative anatomists in making this long and ambiguous history for the subject. Early developers of a discipline of comparative anatomy in the nineteenth century, such as Georges Cuvier, looked back in a similar way to a similar list of heroes.[34] But this is how it is for founders of a new discipline: they think they need to find some grand forefathers for it, however tenuous or spurious.

However, I think we can safely say that, in the period 1650–1800, the period covered by the present book, there were three main senses in which anatomists used the expression 'comparative anatomy' to describe particular aspects of their activity, and these meanings were deployed in parallel.[35] It will become clear from an analysis of these senses that, while a fair amount of comparison of one sort or

---

32 Ibid., p. 23.

33 Ibid., p. 10.

34 However, this is not the case with the transcendentalist comparative anatomist such as Carl Gustav Carus, whose historical account starts quite rightly with Goethe; see Carus, *Traité élémentaire de l'anatomie comparée*, 3 vols, 1835, vol. 3. This history is not present in the English translation: cf. Carus, *An introduction to the comparative anatomy of animals*, 2 vols, 1827, vol. 1, p. 5 – 'A general view of the literature of comparative anatomy'.

35 I omit from discussion very individual claims about 'comparative anatomy'. For instance, Bernhard Siegfried Albinus, who was to become one of the foremost anatomists of the eighteenth century, presented his inaugural oration at Leiden in 1719 as Professor of Anatomy as being 'Oratio Inauguralis de Anatome Comparata'. However, what he presented under this title was the curious claim that whatever exists in the realms of Animal, Vegetable or Mineral can be compared together, because they are all made by God.

another was going on, nevertheless there was no eighteenth-century discipline or sub-discipline of comparative anatomy. There very nearly was such a discipline as early as the 1660s (as we shall see), but not in fact. Such a discipline did not exist until the very last years of the eighteenth century. To guide us, we need to note that in the eighteenth century the basic meaning of 'compare' and related words was (as Dr Johnson put it in his *Dictionary* of 1755) 'to make one thing the measure of another'. This is also the meaning of 'comparo' in Latin. Neither term implies any necessary systematic work.

The first sense of 'comparative anatomy' was in this sense: as simple comparison. For instance, an anatomist might say that some part or organ is smaller than this in a fish, larger than this in an elephant, the same size or shape in a horse, different in a man, and draw no conclusion from this comparison nor invoke any law or principle underlying these apparent similarities or differences, nor use it to discover any law of nature underlying their similarity or difference. Ruini's work on the anatomy of the horse (1598) was like this because Ruini directly compared features of the horse, such as the hemispheres of the brain, to those of the human.[36]

The curious and elaborate *Systeme of anatomy* by Samuel Collins, published in two volumes in 1685 in London, is also like this. Collins had obviously performed masses of anatomical work on animals, fish, insects, and even vegetables. His Proposals, with which he sought financial support for publication, refer to the work as 'two large folios of comparative anatomy (written in the Latin tongue) consisting of many hundred sheets … in which all parts of the body of man, are compared with the Bodye of Beasts, Birds, Fish and insects'.[37] And yet when one looks in detail at the text (which is actually in English), the comparative dimension is very limited. For instance, Collins writes: 'Thus I have given you a short view of Humane Skin, that we may fitly make a comparison of it, with that of other Animals'.[38] He proceeds to do so, right down to insects and vegetables, yet he draws no conclusion from all this. Or, to give a further example, when talking about the abdominal muscles, Collins writes:

> And here I beg Pardon, for making a Digression (it being my design to Treat of Comparate Anatomy) in speaking of the Muscles of Insects, which hold some Analogy with those of Mans Body, and of the Woody parts of Trees, which may seem in some manner in their thin flaky Expansions, to resemble the Muscles of the Belly ….[39]

The key to Collins's approach is given in the Preface to the Tables, at the end of the second volume. 'I humbly conceive the great Use of Comparative Anatomy', writes

36 Ruini, *Anatomia del cavallo, infermita, et suoi rimedii*, 1618.

37 Russell, *British anatomy 1525–1800*, 1987 (1963), p. 49.

38 Collins, *A systeme of anatomy*, 2 vols, 1685, vol. 1, p. 49.

39 Ibid., vol. 1, p. 91.

Collins, 'is to illustrate the Structure, Actions, and Uses of Man's Body, which are sometimes more clear in that of other Animals, than in ours; as I have discovered in frequent Dissections to my great satisfaction, pleasure, and admiration'.[40] The set of images of the cerebellum of different creatures in Fig. 5.10 shows to what a short distance Collins takes such comparison.

This leads us on to the second tradition of invoking some sort of comparative anatomical element, insofar as we can fully distinguish it from the first tradition. This was one practised and mentioned by just about every anatomist of the long eighteenth century, and of many centuries before. The tradition was as old as Galen.[41] It is the view that the anatomist can see some internal parts more clearly in some animals than in others, or can use certain animals to anatomise in the absence of human cadavers. It was not at all controversial as a claim, though its repeated appearance in the literature suggests that some anatomists thought that other anatomists were not exploiting it as fully as they might.

This claim only mattered when the anatomy of man was under discussion. The use of the term 'comparative anatomy' in this sense meant that, by comparing the anatomy of certain animals to that of man, one would be able better to see and understand the anatomy of man. The argument was not used in the opposite direction – from man to animals – nor were animals regularly compared to each other in order to bring light to the anatomy of one from the anatomy of another. This claim bears no relation – *pace* F. J. Cole – to modern comparative anatomy. But because this claim and practice were so common, Cole was able to call on many instances of it in order to fill out his suppositious history of comparative anatomy before Darwin.

A few examples of this very common practice should suffice. In Amsterdam in the 1670s Gerard Blasius was pursuing anatomy along these lines as Fig. 5.11, the title-page of his little book, indicates. It is the lack of human bodies, Blasius writes, that means that anatomists often have to use animals as substitutes, especially the dog; this makes for an adequate basis for arriving at knowledge of the parts of the human body, but there is always the danger of transferring to humans things one has only observed in brutes. To avoid this danger, Blasius has noted what is distinctive about the usual substitute animals. Among the varied animals whose anatomy he explores in this way are the monkey, bull, sheep, tortoise, hedgehog, dormouse, snake and heron.[42]

Similarly, in London at the beginning of the eighteenth century, the anatomist William Cowper, author of the *Muscle dissection reformed* (*Myotomia reformata*)

---

40 Ibid., vol. 2, unnumbered page.

41 Although Aristotle certainly compared many species of animal, it was not to highlight the anatomy of man. See Cunningham, 'Aristotle's animal books', 2000.

42 Blasius, *Miscellanea anatomica*, 1673; Blasius, *Observata anatomico practica*, 1674; Blasius, *Zootomiae seu anatomes variorum animalium*, 1676. These are all developments of the same work.

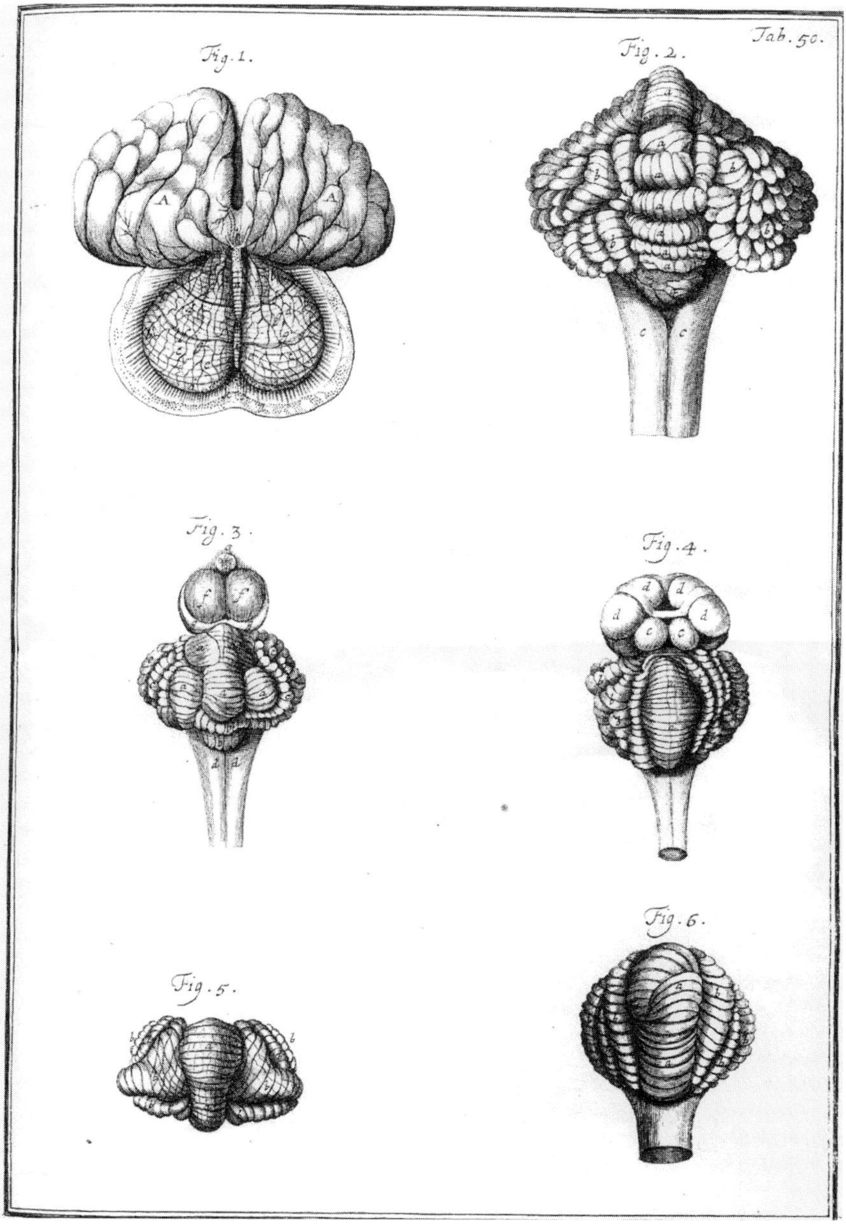

Fig. 5.10

of 1694, and with whom Tyson collaborated, wrote to Tyson in 1704 to the same effect:

> *Comparative anatomy* (as instructive as it is) does not escape the Censure of the Vulgar: though you know the greatest illustrations of the use of Parts are not only to be had from thence, but the very existence of divers Organs in Human Bodies have been made known to us by Discoveries first made in the Bodies of Quadrupeds. The Circulation of the Blood, and the Passages for the Chyle and *Lympha*, had been as little known to us as our Predecessors, were it not for Dissections made on the Bodies of several Animals.[43]

Fig. 5.11

The same assumption is the basis of James Douglas's book of the same period, *A Specimen of comparative myology* (*Myographiae comparatae specimen*, 1707). The 'comparison' being made is between the muscles of man and those of the dog.

43 Quoted in Ashley Montagu, *Edward Tyson*, 1943, p. 243, citing a letter from Cowper to Tyson printed in *Phil. Trans.*, vol. 24, 1704, pp. 1565–90 – see p. 1576.

'As for the Comparative Part of this Treatise, or the Interlacing the Descriptions of the human Muscles with those of the canine', writes Douglas, paraphrasing Cowper,

> … that, I presume, needs no Apology. The many useful Discoveries drawn from the Dissection of Quadrupeds, the Knowledge of the true Structure of divers Parts of the Body, of the Course of the Blood and Chyle, and of the Use and proper Action of the Parts, that are chiefly owing to this Sort of Dissection; these, I say, give a very warrantable Plea for insisting upon it, though it may be censured by the Vulgar.[44]

This work, intended as a ready reference that students could take with them to dissections, was comparative in only one way: 'I only mention when the origin or insertion of a muscle in a dog is different from man … it is there (and there only) that my comparative Remarks take Place'.[45]

Alexander Monro *Primus* wrote a celebrated essay on this same meaning of comparative anatomy, first published in 1744, and then reprinted by his son in 1783. 'The principal Advantages of Comparative Anatomy', wrote Monro,

> … are the following: First, It furnishes us with a sufficient Knowledge of the different Parts of Animals, to prevent our being imposed upon by such Authors who have delineated and described several Parts from Brutes as belonging to the human Body. Secondly, It helps us to understand several Passages in the ancient Writers in Medicine, who have taken many of their Descriptions from Brutes and reasoned from them … The third and great Use we reap from this Science, is the Light it casts on several Functions in the human Œconomy about which there have been so many Disputes among Anatomists: These will be in a great Measure cleared up by exhibiting the Structure of the same Parts in different Animals, and comparing the several Organs employed in performing the same Action, which in the human Body is brought about by one more complex.[46]

The whole focus is on ensuring that one has a good account of *human* anatomy. To this end Monro dissected many quadrupeds (dogs and cows), birds and fish. There is actually a properly comparative question being deployed here, though Monro does not make much of it. It concerns the *reasons* for the differences in the anatomies of different creatures. It all depends on the way of life of the animal: where their anatomies differ, it is because their ways of life differ. In comparing man and dog and the differences in their anatomies, Monro writes: 'that one Circumstance, the *Difference* of an erect and horizontal Posture, explains all the Variations. This Reasoning then gives Solution to many Difficulties in human

---

44 Douglas, *Myographiae comparatae specimen*, 1750, p. v.

45 Ibid., pp. iii–iv.

46 [Alexander Monro, *Primus*], *An essay on comparative anatomy*, 1744, pp. 1–2.

Anatomy; why the *Spleen* so firmly attached to the *Diaphragm*, why the *Omentum* reaches only so far, why the posterior Part of the *Bladder* only covered by the *Peritonaeum*, &c.' (prefatory letter, p. 13). But even in pursuing this question, as can be seen, everything still revolves for Monro around the anatomy of the human.

This too is how comparative anatomy is defined in the *Encyclopédie* of 1751, in the article on anatomy written by Albrecht von Haller:

> *Comparative Anatomy* is that branch of *Anatomy* concerning the investigation and examination of the different parts of animals, considered in relation to their particular structure and to the form that best suits their way of life and satisfies their needs. For example, in the *Comparative anatomy of stomachs* [by Grew, see below], one notices that the animals who need to eat frequently have very small stomachs, in comparison to certain animals who, troubled by the other animals that they eat, need to fast frequently, and to whom it appears Nature has for this reason given a stomach capable of retaining the nourishment for a long time.
>
> In *Comparative Anatomy* one examines animals and even vegetables in order, by comparison of what happens in them to what happens in us, to reach a more perfect knowledge of the human body. It is the method that Aristotle followed. One might say that the only reason he sacrificed so many animals was in order to compare the structure with that of man.[47]

We now turn to the third use of 'comparative anatomy'. The earliest work I know of that was actually entitled a 'comparative anatomy' of animals in any language by its own author – rather than by some later historian – is that by Nehemiah Grew, *The comparative anatomy of stomach and guts begun*, published in London in 1681. This work with its rather unpromising title was meant to be the first part of a whole *Comparative anatomy of animals begun*, which Grew projected in February 1676/77 but never completed. The 'begun' part of these titles is related both to Grew's awareness that nothing like this had been attempted before, so that this was a first step, and also to his view that, following his approach, 'a better History of [animals] might be written in five years, than hath hitherto been done in two Thousand', so that the work here 'begun' would soon be developed and completed.[48] The comparative anatomy of animals was to have been modelled on Grew's own work on the anatomy of plants, which had itself been based on certain anatomies of humans.

If we trace the personal route that Grew took to arriving at his 'comparative anatomy' of animals, we will be able to see its source and nature. Grew in 1682

---

47 Diderot, *Encyclopédie*, 1751–65, vol. 1, p. 416. The discussion of stomachs of course draws on the work of Grew (on which more below).

48 Grew, *The comparative anatomy of stomach and guts begun*, 1681, Preface.

wrote his own account of how he came to take up the anatomy of plants when he was about twenty-three years old:

> The first occasion of directing my Thoughts this way, was in the Year 1664, upon reading some, of the many and curious Inventions of Learned Men, in the *Bodies* of *Animals*. For considering, that both of them [viz. plants and animals] came at first out of the same *Hand* [viz. God's]; and were therefore the *Contrivances* of the same *Wisdom*: I thence fully assured my self, that it could not be a vain Design, to seek it in both. And being then newly furnished with a good stock of *Seeds*, in order to raise a *Nursery* of *Plants*; I resolved, besides what I first aimed at, to make the utmost use of them for that purpose: so that I might put somewhat upon that side the *Leaf* which the best *Botanicks* had left bare and empty … But notwithstanding the reasonableness of the Design; yet I did not forget, that, in respect of the Undertaker, there might be *Impar congressus*. And therefore, before I had ventured very far, in the Year 1668, I imparted it to my Brother-in-Law, the Learned Dr. *Henry Sampson*, now *Fellow* of the *College* of *Physicians* in *London*. Who not only very well liked the same, but also excited me to a vigorous and accurate prosecution of it. Which he did, partly, by mentioning a very pertinent passage of Dr. *Glisson*, in the *Preface* to his *Book de Hepate*, *ch. 1*, which I had not then read.[49]

The passage in question appears in the Prolegomena of Francis Glisson's book *The Anatomy of the liver*, published in Latin in 1654. Glisson wrote:

> Again in this sense [i.e. taking 'anatomy' in its largest sense] plants may be the object of anatomy, for they have texture and differences of parts; and without question, by the accurate retexture [= unweaving] of those parts, good observations might be made, and though the object be mean, yet I conceive men might better improve their industry even in these, than in mere transcribing one out of another, like bees robbing each others' hives, by which nothing is added to the common stock.[50]

So Grew, who was at this time a young Leiden-educated physician practising in Coventry, took one of his cues from a suggestion made in the context of the discussion of *human* anatomising, and then transferred his attention to *plants*, thereby becoming one of the first persons ever to act on the view that plants have internal structure amenable to being anatomised.[51] He sent his first efforts to the

---

49 Ibid., a1r–v.

50 Cunningham, ed., *English manuscripts of Francis Glisson (1)*, 1993, p. 21. This is Glisson's original English version; Grew quotes from the published version, which had been translated into Latin by George Ent.

51 Marcello Malpighi was working in Italy on a similar subject at the same time, and by chance the published book of Grew, *The Anatomy of Vegetables begun*, and the

Royal Society, where they received the patronage of John Wilkins, the Bishop of Chester, who persuaded the Royal Society in 1672 to appoint Grew their 'Curator for the Anatomy of Plants', on £50 a year, an appointment that lasted for six years. Like the Curator of Experiments, Robert Hooke, Grew was expected to bring in regular demonstrations and reports for the meetings of the Society, which he duly did.

The result of his work, both privately and in his Royal Society capacity, enabled Grew to build up *The anatomy of plants*, which he published in 1682, and which covered the history of 'perfect' plants. He chose, he said, 'the Method of Nature her self, in her continued Series of *Vegetations*; proceeding from the *Seed sown*, to the formation of the *Root*, *Trunk*, *Branch*, *Leaf*, *Flower*, *Fruit*, and last of all, of the *Seed* also to be *sown again*'.[52] For his own guidance in pursuing this anatomy of plants Grew had drawn up 'An Idea of a Philosophical History of Plants' (which he also called 'An Idea of a Phytological History Propounded'), published in 1673. 'In this *Idea*,' Grew wrote, 'one principal Thing I insist upon, for a *Philosophical History* of *Plants*, is *Anatomy*'.[53] And it is here that his novel *comparative* approach was first laid out. He proposed 'Five General *Heads* of Enquiry':

> *First*, of those Things, which are of more *External* Consideration about *Plants*, as their *Figures &c. Secondly*, Of their Compounding *Parts*, as *Vessels, &c. Thirdly*, Of their *Liquors,* and other *Contents*. *Fourthly*, Of their *Principles*, as *Salts &c. Fifthly*, Of their *Aliment*, as *Water*, and other Means of Growth.
>
> And therefore, first of all, To go through all the *Parts*, with equal care; examining the *Root*, *Trunk*, *Branch*, *Leaf*, *Flower*, *Fruit*, and *Seed* ... To examine, again, not only all the *Parts*, but *Kinds* of *Vegetables*, and comparatively, to observe divers of the same *size*, *shape*, *motion*, *age*, *sap*, *quality*, *power*, or any other way the *same*, which may also agree, in some one or more particulars, as to their *Interiour Structure*: and to make this comparison, throughout all their *Parts* and *Properties* ...
>
> ... From all which, we may come to know, what the *Communities* of *Vegetables* are, as belonging to all; what their *Distinctions*, to such a Kind; their *Properties*, to such a Species; and their *Peculiarities*, to such Particular ones. And as in *Metaphysical*, or other Contemplative Matters, when we have a distinct knowledge of the *Communities* and *Differences* of Things, we may then be able to give their true *Definitions*.[54]

manuscript of Marcello Malpighi on plant investigation were presented to the Royal Society of London on the same day, 7 December 1671. Malpighi's work (begun in the early 1660s) was published in London in 1675 as *Anatome plantarum*, 1675.

52 Grew, *The anatomy of plants*, 1682, p. 1.

53 Ibid., a3v. For the long-term relation of anatomy to the study of plants, see Delaporte, *Nature's second kingdom*, 1982.

54 Grew, *The anatomy of plants*, 1682, pp. 9–10.

By comparing the plants and their parts with one another we will be able to come to a true understanding of them, 'their true *Definitions*'. We will, in other words, have true 'scientific' knowledge of plants in the sense in which seventeenth-century philosophers used this term: we will know their 'essences'. In fact this 'Idea' bears an extraordinary similarity to what Francis Glisson describes as the role of the anatomist of the *human* body in *On the liver*, including seeking the true 'definition', which is equivalent to the essence or soul.[55]

Now, at some point Grew came to apply this comparative approach to the anatomising of animals, resulting in his *Comparative anatomy of stomach and guts begun* of 1681. He explained his intentions extensively in his original presentation to the Royal Society. As only a little of this eventually got printed, I shall quote it at length here from the manuscript original.[56] In introducing his *comparative anatomy* at the Royal Society meeting, he said that he was well aware of the work on the anatomy of animals that had been performed by the moderns over the previous two centuries, yet nevertheless he believed the comparative anatomy of animals had not yet been begun.

> That then which I would design, or at least propose to be performed is this, viz. First, That all the parts not only of several men, but of all sorts of animals be dissected and examined, both by the knife and by the glass or any other way as shall be necessary.
>
> 2. That in the doing hereof, there be a comparison made betwixt all their several parts. Of which to note where they are the same in genere and where they are different. If the same in genere, what special differences they may have, either for their position, colour, size, shape, texture, or constituent parts, or in any other property. But if they be of different kinds, then to note whether they be supernumerals or equinumerals, if I may have leave to use those words. By supernumerals I mean when one animal hath all the same parts with another, but one or more besides. By equinumeral I mean when the number of their viscera or other organical parts is the same, but one or more of them is of a different kind.
>
> 3. That this comparison be not made confusedly betwixt any animals at the first, as betwixt a dog and a fish. But in this method: first, betwixt the several animals of one species or subgenus, as a dog and a fox; next betwixt those of one kind, as all quadrupeds; and then last of all betwixt those of several kinds, as beasts, fowls and fishes. By which I mean not that they should be anatomised precisely in this order (though if they could always be procured, that would be wished), but that the consideration, and the history of them should as near as may be, be made and delivered in this method.

---

[55] Cunningham, ed., *English manuscripts of Francis Glisson (1)*, 1993, see esp. pp. 17–19.

[56] Royal Society Ms. RBC 5, pp. 37–40. My thanks to the Royal Society for permission to consult and quote this material.

> 4. That as not the animals, so neither their several parts be confounded, but all sorted together according to their several species, and the history of all of one species to be set down so that they may all be taken in, as it were at one view, as for instance all kinds of hearts, or all kinds of livers at one view, and so for all the other parts, without doing of which the foresaid comparison cannot be well made.
>
> 5. The comparison being thus made betwixt the several parts, we are lastly to compare these parts with the actions of those animals whereto they belong, and with all the other properties whereby their natures are defined.
>
> Having done all this, we may hence make a clear judgement of the uses of the parts, according to these three general conclusions. First, that any part which is common to divers animals hath some one common use in all those animals. 2. That the same part so far as it is diversified in divers animals, so far the use of it in those animals is also different. 3. So far as we see a relation betwixt the differences of any more parts than two, we may also collect what are the special uses of every part.
>
> Having thus proposed the work to be done, I will begin with it. And first of all, that I may rid them out of my way, I will begin with stomachs and intestines, and in the first place with those of quadrupeds. In describing of which, omitting most of what is already noted by anatomists, I shall principally speak of those things as have hitherto been unobserved.

The comparative anatomy of stomach and guts begun was published as a sort of appendix to Grew's catalogue of the museum of the Royal Society, and in the introduction to that he speaks of some of his goals that are also pertinent to his comparative anatomy of animals:

> … it were also very proper, That not only Things strange and rare, but the most known and common amongst us, were thus described. Not merely, for that what is common in one Country, is rare in another: but because, likewise, it would yield great abundance of matter for any Man's Reason to work upon … He that shall observe, That a Horse, which ought to have many and strong Teeth, and large and thick Hooves, hath no Horns; and that an Ox, with Horns, hath fewer Teeth, and weaker Hoofs: cannot but at the same time see the Providence of Nature, in disposing of the same Excrementitious parts of the Blood, either way, as is most suitable to the Animal … Together with such Notes as these, arising from the Description of the *outward Parts*; how largely and usefully might that of the *Inner*; his [i.e. the horse's] Generation, Breeding and the like, be also insisted upon. As so the like of other Animals.[57]

What Grew does in the *Comparative anatomy of stomach and guts begun* is to take specimen animals with a particular diet, and look at their stomach and

[57] Grew, *The comparative anatomy of stomach and guts begun*, 1681, Preface.

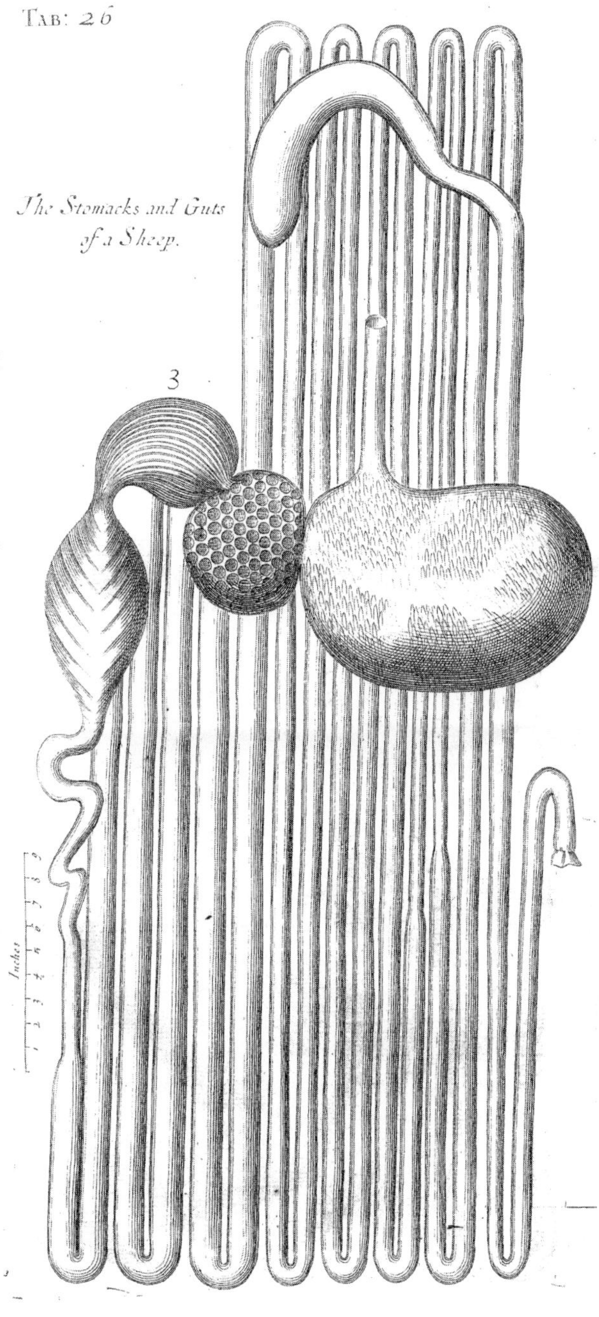

Fig. 5.12

guts, and ascertain how these fulfil the needs of their diet, whether they eat meat, fruit or grasses. He deals in turn with the structure of the stomachs and guts, and then of their uses. So there are the weasel, stoat, polecat, cat, dog and fox as instances of carnivorous quadrupeds; the mole, hedgehog, squirrel and rat as insectivorous quadrupeds; the rabbit, horse and pig as both frugivorous and graminivorous quadrupeds; a sheep and a calf as graminivorous quadrupeds; and then the stomach and guts of birds and fishes. Among the kind of conclusion he can reach by adopting this method is the following: 'all *Carnivorous Quadrupeds* have the smallest *Ventricles*; flesh going furthest. Those that feed on Fruits and Roots have them of a middle size … *Sheep* and *Oxen*, which feed on *Grass*, have the greatest'.[58] This is all illustrated in Fig. 5.12.

This comparative anatomical work by Grew is very theoretically sophisticated and well thought through. He had drawn up a plan or method, which he systematically followed. The work is also truly comparative: we can learn more about *all* stomachs and guts, or stomachs and guts *in general* (for instance) by comparing those of different animals in relation to their respective styles of life, than we can by just dissecting one kind of animal, however diligently. It is noteworthy that Grew's scheme was first inspired by some comments by the anatomist Francis Glisson, and that (unknown to Grew, or to Glisson) it matches almost exactly the programme for the investigation of animals that Aristotle had originated in ancient Greece, in which he had had no immediate followers![59] Nothing quite so systematically comparative as Grew's work was to be produced again for almost a century.

This next authentic work on comparative anatomy was by Alexander Monro *Secundus*, Professor of Anatomy at Edinburgh University, who was possibly the greatest anatomist of the dynasty, though nowadays forgotten in the shadow of his father. From at least as early as 1759 he gave, he later reported, 'Lectures on Comparative Anatomy when I accompanied my descriptions with dissection and demonstration'. These were part of his larger course on anatomy.[60] He made a systematic study of fishes, which was published in 1785 as *The structure and physiology of fishes explained and compared with those of man and other animals*. This is a most interesting work because it compares both physiological *functions* (how is function *x* carried out in this animal?) and anatomical *structures* (what are the distinctive anatomical structures in this animal?) together, to reach comparative conclusions about both function and structure. In his exposition his argument is both anatomical and physiological. Thus he says,

[58] Ibid., p. 23.

[59] Cunningham, 'Aristotle's animal books', 2000.

[60] Alexander Monro, *Secundus*, *The structure and physiology of fishes*, 1785, pp. 40, 38.

> I shall begin with tracing the blood from the heart, and its return to that organ.
>
> I shall next make some cursory observations on their [fishes'] organs of secretion.
>
> I shall afterward give an account of their absorbent system; and shall conclude with some Observations on their Brain, Nerves, and the Organs of their Senses. (pp. 13–14)

Monro *Secundus* shows that it was possible not only to reason from anatomical structure to physiological action, but also in the other direction, viz. from (comparative) physiological function to anatomical structure. Here, for instance, he draws conclusions about the structure and function of certain arteries in man from his investigation of the circulation of the blood in the skate (Fig. 5.13):

> At the beginning of the branchial artery [of the skate] there are three semilunar valves, the middle parts of which, analogous to the Corpuscula of Morgagni, are much thicker than in man, and illustrate the use of these organs in him, as they very evidently assist in preventing the return of the blood when the artery is in action. Between these valves and the principal cavity of the ventricle in the skate, there is a cylindrical canal interposed, the coats of which have the same muscular texture as the ventricle itself; and their contraction co-operating with that of the ventricle, we are thence led, more evidently than in man, to perceive the analogy between the structure of the arteries and those of the heart.
>
> Applying to man what we have observed of the vessels and circulation in fishes, we in the first place receive strong confirmation of an opinion I have always taught, That our arteries are of a muscular nature: and that their activity is essential in circulation, secretion, and other important offices. (pp. 19–20)

This is perhaps the most sophisticated kind of comparative reasoning that we see under the label of 'comparative anatomy', an attempt to make 'comparative anatomy' meaningfully comparative.

The fourth and most important tradition of talking about 'comparative anatomy' – sometimes also called zootomy – in the long eighteenth century lies in the *history of animals*, which was itself part of a larger enterprise, *Natural History*. Natural History of this kind is no longer really a scientific subject, having been divided up into different scientific specialities, and surviving today mainly as the title and remit of certain museums. 'History' here meant the same as the Latin term *historia*, a full, rounded account, in this case of the natures of things. Those 'things' were the productions of Nature; or, in other words, the created sub-lunar world. Hence the remit of the Natural Historian covered man, animals, birds, fishes, insects, creatures between animals and plants, all plants, minerals, and sometimes even different kinds of stones. All these things had been created by the Creator.

Natural History as a subject area may perhaps be dated back to the 1550s, with the works of Belon and Coiter on birds and other animals. It was certainly

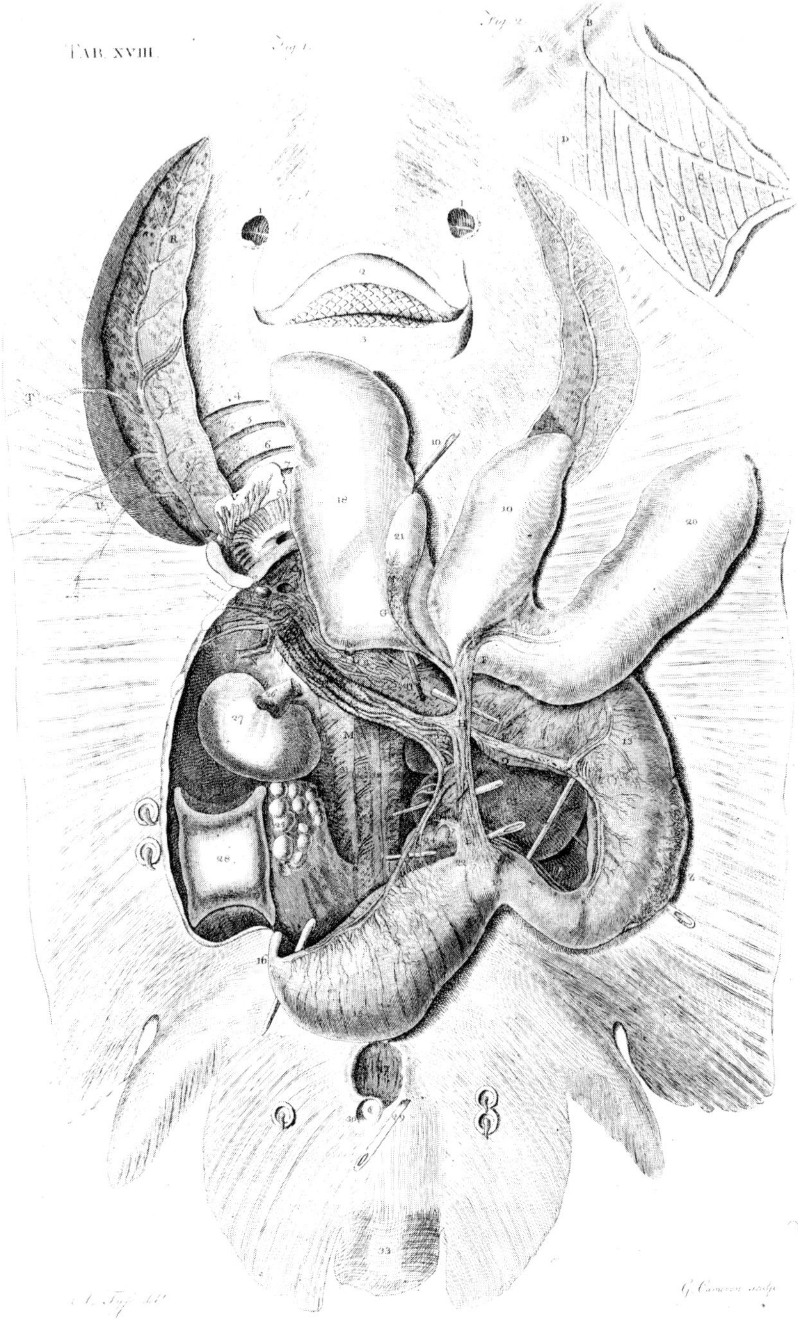

Fig. 5.13

being practised from the 1650s, with many investigators active across Europe. A modern historian has remarked, 'The taste for Natural History and for scientific research on it was spread across Europe in the 18th century, above all in France, and particularly in Paris'.[61] As with any discipline investigating nature actively, Natural History was a contested area, with different practitioners claiming diverse directions and bases for it.

Most of the anatomising of the great range of different animals in the eighteenth century (as opposed to the vivisection of domestic animals) was done in this tradition. For instance, the grand volumes edited by Claude Perrault as the *Mémoires pour servir à l'histoire naturelle des animaux* (1671, 1676), which we have already seen, are a contribution, as the title of the book announces, to the natural history of animals. In this case, Perrault says in the Preface, he and the other members of the Académie had limited themselves to the different peculiarities of the insides of these animals 'because Natural History lacks virtually nothing except this exact description of the internal parts'. So this is animal anatomy as part of Natural History, not as comparative anatomy since, notwithstanding such a wealth of material of different animals, nothing is being directly compared or contrasted; the structure or function of no part or organ of any of these animals is being illuminated by its being compared to those of any other animal.

Edward Tyson, working over the last three decades of the seventeenth century, talks repeatedly about making 'a *comparative* survey' in his extraordinary anatomies of the porpoise, orang-outang (chimpanzee), and other animals. He himself placed this anatomising within Natural History rather than some new or old discipline of comparative anatomy. A 'history of animals', he wrote, 'can't rationally be writ without their Anatomy ... [and] the Anatomy of one animal, will be a key to open several others; and until such time as we can have the whole completed, 'tis very desirable to have as many as we can of the most different, and anomalous'.[62] In his book of 1680 on the porpoise, Tyson said that he was putting forward –

> ... a rude Draught or Sciagraphy [i.e. basic plan] of a Natural History of Animals, such as hastily occurred to me ... [constructed] not, like some people do, from books, but from the Book of Nature ... I could therefore wish we had a good History of the Animals of our own Country ... This History I could desire likewise to commence from the lowest degree of Animation in Zoophyta's or Plant-animals; inquiring here into Nature's first Rudiments and obscurer Vestigia of forming the Organs of an Amphibious life in them; then gradually ascending by her clue to run through all the various Tribes of Animals; carefully observing all along the Harmony she keeps, or the Liberty she takes in the different formation of them, and from the whole to give a general Prospect of her workings.

[61] Lamy, *Les cabinets d'histoire naturelle en France*, 1931, p. 5.

[62] His Preface to the translation of Swammerdam's *Ephemeri Vita: or, the Natural History and Anatomy of the Ephemeron A Fly that lives but five hours* [i.e. mayfly], 1681. Quoted in Ashley Montagu, *Edward Tyson*, 1943, p. 125.

The physiological functions and the anatomical structures should be examined between these 'tribes', which 'by such a comparative survey may be rendered more intelligible; and from a clearer knowledge of them in Brutes, at length we may come the better to know ourselves'.[63]

What Tyson actually had in mind was an investigation of the 'scale of nature' or 'chain of being', the supposed continuous gradation along which God had created all creatures, looking first at the '*amphibious* or Hermaphroditical Nature of Animals' from the zoophytes to 'the Testaceous types and Worms', then in more perfect creatures, 'at last in Man and Intelligent Beings, that are a boundary between Divinity and the Creation', that is to say angels.[64] This was still his quest almost twenty years later in 1699, when he wrote about the orang-outang (chimpanzee):

> My chief Design in this Undertaking is the Improvement of the Natural History of Animals … and the Orang-Outang (whose Anatomy I here give) being a Creature so very remarkable, and rare … To render this Disquisition more useful, I have made a Comparative Survey of this Animal, with a Monkey, an Ape, and a Man. By viewing the same Parts of all these together, we may the better observe Nature's Gradation in the Formation of Animal Bodies, and the Transitions made from one to another; than which, nothing can more conduce to the Attainment of the true Knowledge, both of the Fabric, and Uses of the Parts. By following Nature's Clue in this wonderful Labyrinth of the Creation, we may be more easily admitted into her Secret Recesses, which Thread if we miss, we must needs err and be bewildered.[65]

Tyson concludes that the chimpanzee is neither an ape nor a man, and he draws up tables of similarities and dissimilarities among the three of them. And, with respect to the chain of being, Tyson concluded that the chimpanzee was 'more resembling a Man, than any other *Animal*; so that in this *Chain* of the *Creation*, as an intermediate Link between an *Ape* and a *Man*, I would place our *Pygmy*'.[66] This view of what a 'comparative survey' is undertaken for is very specific: it locates, by similarities and differences, each creature or group of creatures on the chain of being, through which 'we may find the knowledge of a Deity and our selves'.[67] While this is a very interesting (and extremely traditional) motive for the investigation of animals,[68] it does not make Tyson's enterprise comparative anatomy as we know it today, despite the strenuous protestations of his modern biographer that Tyson's 'leading interest lay in drawing parallels and in making

63 Tyson, *Phocæna, or the anatomy of a porpess*, 1680, pp. 4–5.

64 Ibid., p. 11.

65 Tyson, *Orang-outang, sive homo sylvestris*, 1699, Preface.

66 Ibid., p. 5.

67 Tyson, *Phocæna, or the anatomy of a porposs*, 1680, p. 3.

68 Lovejoy, *The great chain of being*, 1965.

comparisons, and not merely in accurate description for the sake of accurate description'.[69]

Fig. 5.14

In his massive *Histoire naturelle* Le Comte de Buffon, the most influential natural historian of the eighteenth century, gave a great role to the anatomy of every creature he dealt with.[70] Buffon was Intendant at the Jardin royal, with access to all the animals that died in the Versailles menagerie, and he also built up a menagerie of his own on his estate at Montbard in the Côte d'Or in Burgundy. *L'Histoire naturelle*, 36 volumes of which had been published at the time of his death in 1788, deals with all aspects of Natural History, beginning with many volumes on animals.[71] Working with the anatomist Louis Daubenton, using both dissection and the microscope (Fig. 5.14), Buffon offered a description of the anatomy of each animal, linked to a generous number of engraved pictures of the anatomised animal. For instance, in the case of the lion and lioness (vol. 9, 1761), there is one view of the lion from the outside, and then nine plates of the lion in

---

69 Ashley Montagu, *Edward Tyson*, 1943, esp. p. 243.

70 Buffon and Daubenton, *Histoire naturelle*, 54 vols, 1749–1804. The first edition needs to be consulted, since later editions and translations omit crucial texts, discourses and illustrations. The anatomical dissections and their illustrations were particularly liable to be omitted from later editions, partly on account of cost. For our present purposes the English translation by William Smellie (not the anatomist of that name) is completely useless, since it omits so much: Buffon, *Natural history, general and particular*, 9 vols, 1785. For a modern translation into English of the 'Initial Discourse', see Lyon and Sloan, eds, *From natural history to the history of nature*, 1981.

71 On Buffon see the modern biography, Roger, *Buffon: A life in natural history*, 1997, and the essays in Gayon, ed., *Buffon 88*, 1992.

various states of dissection, ending with the articulated skeleton, but not dealing with the musculature at all, nor with the systems of vessels such as the arteries, veins and nerves. Rather, what is illustrated are the internal organs together (plate 3), the intestines (plate 4), the colon and the external female sexual organs (plate 5, Fig. 5.15), the stomach (plate 5), the male sexual organs (plate 6), the membranes covering the fetus (the amnion, chorion and allantois) (plate 7), the gravid uterus (plate 8), the sexual organs of the female (plate 9), and finally the skeleton (plate 10). This is the general pattern for the internal organs illustrated of every animal in the *Histoire naturelle*: they show what is distinctive about the arrangement of the organs in general in the animal being investigated, and of the organs of digestion and of generation in particular, paying attention also to the formation of the fetus, and any features that are unique to the particular animal, such as its horns.

The rationale for this Buffon explained in an essay, 'On the Description of Animals' (in volume 4), that prefaced the accounts themselves. This is for the most part an attack on Linnaean binomial nomenclature, compared to the kind of account of each animal that Buffon thought necessary for the accomplishment of good Natural History. 'What characterises an animal in our eyes', writes Buffon, 'is the ensemble of the shape, the attitude, the bearing, the movement, and the proportion of the different parts to each other'. But 'Natural History is not limited to the exterior of the animal, but extends further; its principal object is to open up the interior, and to discover, by the inspection of the insides, the mechanism of the movements that appear on the outside, and the causes of the appetites and inclinations that are peculiar to each species of animal'.[72] It is his pursuit of the 'appetites and inclinations that are peculiar to each species of animal' that led Buffon to concentrate so much on the digestive and generative organs. But although anatomising of animals is part of Natural History, Buffon continues, the aims of the anatomist and the Natural Historian differ. The anatomist gets obsessed by the detail of the particular creature he is dissecting; by contrast, the Natural Historian just observes. The anatomist particularises his observations, the Natural Historian generalises his, since the limits of his field of study are as extensive as those of Nature itself.[73] The Natural Historian wants to find the *essential qualities* of each kind of animal, and can only reach this goal by comparing the resemblances and differences of animals. Obviously a 'plan of description' is necessary to facilitate this. According to Buffon, this is where Perrault and others had gone wrong: they had claimed to be doing comparative anatomy, but they had had no *plan of description*, 'without which all description is virtually useless for comparative anatomy'.[74] As far as Buffon's own 'plan of description' is concerned, he does not need the detail that the anatomists normally pursue:

[72] Buffon, *Histoire naturelle*, 54 vols, 1749–1804, vol. 4, pp. 118–19.

[73] Ibid., vol. 4, pp. 128–9.

[74] Ibid., vol. 4, p. 133.

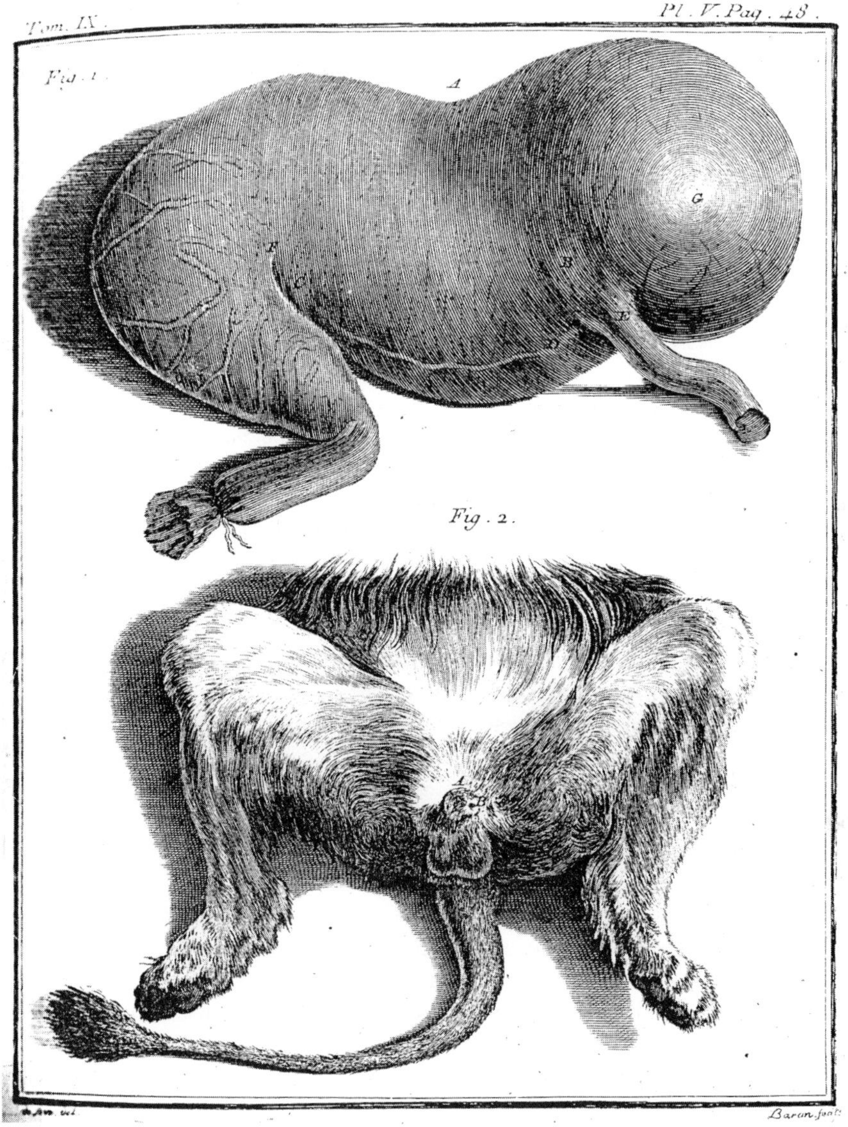

Fig. 5.15

> The Natural Historian needs to abandon these details, to ensure that he does not lose from view the general connections [*rapports*] that exist between the different species of animals, the resemblances between them, and the essential differences that can be found in the mechanism of their body; also I insist on the examination of the principal parts, and I look only at the situation, figure, dimensions and proportions of the bones, of the skull, of the heart, the lungs, the diaphragm, the stomach, the intestines, the liver, the spleen, the pancreas, the kidneys, the bladder, and the generative parts of the male and the female, and the embryo and its envelopes. No mention will be made of the muscles, arteries, veins or nerves, because I believe that the insight one can gain from the description of these parts has less influence on the knowledge of the animal economy, than the results that the comparison of the viscera and bones produce.[75]

Buffon's written account of the anatomy of animals was complemented in the Jardin by the skeletons and other animal parts in the Cabinet du Roi, of which Daubenton was the keeper, and which had been in existence from the very beginnings of the Jardin in 1635. In the volumes of the *Histoire naturelle* Daubenton inserted a section after Buffon's treatment of each animal, which detailed what pertinent dry or wet specimens of the animal and its parts were kept in the Cabinet.[76]

Buffon's view of what comparative anatomy was, and its role in Natural History, is very distinctive and precise. Its role was to extend the capacity of the Natural Historian to view 'the general connections' between animals, and that is the significance of any comparisons between animals that he made, whether to do with their outward appearance, their way of life, their use for humans, or their internal parts as revealed by dissection. This whole approach of Buffon's originated in the 1740s, but by the time the last of the dozens of projected volumes gradually appeared over the later decades of the century and into the next, this version of Natural History was beginning to look increasingly old-fashioned. Meanwhile the nature of Natural History, and the role of comparison between animals within it, was undergoing great change in France, as we shall see later (Chapter 6).

One particular practitioner of comparative anatomy in natural history (in yet another sense) in France was able to get a professorship for himself under this title. This was Félix Vicq d'Azyr (1748–94). A follower of both Petit and Daubenton, Vicq d'Azyr claimed that man's place in the order of nature cannot be understood if man is considered in isolation from the other animals:

> Man must not be considered alone; he must be compared with the other animals: when thus brought together, they will form a tableau that is imposing in its extent and stimulating by its variety. Man alone does not seem so great, one does not

[75] Ibid., vol. 4, pp. 137–8. For a different version of the same sentiments, see the First Discourse in volume 1, 'De la manière d'étudier et de traiter l'Histoire Naturelle', esp. pp. 3–4 and 29–30.

[76] Daubenton, 'La description du Cabinet du Roi', 1749.

> see as well what he really is; the animals, without man, cannot be separated from their type and one does not know to what centre to relate them.[77]

It may have been his interest in the anatomy of animals that made Vicq d'Azyr a candidate to investigate a rinderpest epidemic in 1774–76. His experience here, in turn, led Vicq to found a Commission for Epidemics and Epizootics under royal patronage and then, with the patronage of Turgot, the comptroller general of finance of the kingdom, a professorship in human and comparative anatomy was established for him in this Commission, though it had neither premises nor students. Finally Vicq was able to become Professor of Comparative Anatomy at the royal veterinary school at Alfort for a few years in the 1780s, until the professorships there were closed down in 1788.[78]

Another of the important anatomists who wrote about 'comparative anatomy' as part of natural history (but meant something different again by it) is the Scot, John Hunter. Like several other eighteenth-century anatomists, Hunter has often been credited with being an early comparative anatomist in the modern sense, a claim we can put aside. Hunter worked in London from 1748 to his death in 1793, with the exception of some years in the 1760s spent working abroad as an army surgeon. He acquired the right to dissect all the animals that died in the Tower of London, and he had his own menagerie at Earls Court, where he kept leopards and other exotic animals. He dissected many animals and performed thousands of experiments on them. His comparative anatomy, like that of Buffon, is part of *the history of animals*, which in turn was for both of them part of Natural History, though Hunter's differed from that of Buffon in aim and approach. At one time, in 1775, Hunter even planned to open a school of Natural History, of which comparative anatomy would have been a part. He wrote in a letter to his former pupil Edward Jenner:

> I have a great scheme to communicate to you, and I want you to take part in it; but remember, it is as yet a most profound secret. My scheme is to teach natural history, in which will be included anatomy, both human and comparative. The labour of it is too much for one man, therefore I must have someone to assist.[79]

John Hunter's own account of his comparative anatomy was not published in his lifetime or for many years after. It escaped the fate of so many of his unpublished papers – that is, the fate of being plagiarised and then burnt by his executor, Everard Home – for Hunter's disciple William Clift had discreetly made a copy of it, and kept it in his own possession. Virtually on his own deathbed in 1849, Clift handed these papers to his own son-in-law, the famous comparative anatomist

---

[77] Hannaway, 'Vicq d'Azyr, anatomy and a vision of medicine', 1994, pp. 282–3, citing Vicq d'Azyr, *Traité d'anatomie et de physiologie*, 2 vols, 1786, 1789, vol. 1, p. 7.

[78] Hannaway, 'Vicq d'Azyr, anatomy and a vision of medicine', 1994.

[79] Quoted in Ottley, 'The life of John Hunter, F. R. S.', 1835, p. 55.

Professor Richard Owen. Owen then published Hunter's comparative anatomy in 1861. Thus this comparative anatomy was unknown to anyone but these three until 1861. It could not have been seen as the model for a truly Hunterian form of comparative anatomy by anyone apart from Home or Clift until 1849, or Owen until 1861, by which time Home and Owen had anyway developed and published their own modern versions of comparative anatomy. The text by Hunter that was published by Owen in 1861 runs to almost five hundred pages of printed text, but, given Hunter's perfectionism about his publications, it is probably not in a form in which Hunter would have been ready to publish it.[80] And it is certainly not in the order that Hunter would have given it.

What does John Hunter's practice of comparative anatomy look like? Before we start, we have to remove Richard Owen from the equation. In preparing these manuscripts for the press Owen had felt himself obliged to modernise all of Hunter's terminology, though thankfully he put such revisions in square brackets. Moreover, from Owen's point of view, Hunter's notes needed to be put into a particular systematic order: that of the sub-kingdoms of vertebrates, molluscs, articulata and radiata, each with their appropriate classes and sub-classes, orders and sub-orders, families and species. Unfortunately, this intervention by Owen gives us a presentation of Hunter's comparative anatomy that makes it look finished and complete (which it wasn't – it was always 'in progress' from Hunter's point of view), and it is also a presentation that not only insists that Hunter was concerned primarily with *classifying* animals but also that his intended classification was Cuvierian. For the classificatory terms 'vertebrata', 'articulata' and 'radiata' were all coined by Cuvier at the end of the eighteenth century, and while 'mollusca' had been coined by Linnæus in 1758 it was redefined by Cuvier. As Hunter had died half a dozen years before Cuvier published his classification, it is absolutely impossible that this presentation represents Hunter's intentions!

Hunter did write some general comments about comparative anatomy, and these Owen placed at the front of the volume. Here are the crucial passages:

> Comparative Anatomy [Hunter writes] is fit only to be read by those that understand a good deal of the human body and the general animal economy; for if they do not understand the standard, they cannot understand the variations from it. It is too great a task for any individual to make it any way perfect, for he must dissect most animals to be able to class them, and arrange them in such order as to make them intelligible and easier for others to pursue this species of knowledge. For example, if we find that all the lion-tribe are the same, it will save others from dissecting all this class, or it will enable us to dissect or examine them with much more accuracy.
>
> When I compared the human with the quadruped (not considering the intended uses of each, which would make me admire them equally), it always put me in

[80] John Hunter, *Essays and observations on natural history, anatomy*, 2 vols, 1861. The so-called 'comparative anatomy' takes up the whole of volume two.

> mind of two machines of the same kind, one made by an artist, the other only an imitation of it made by a novice.
>
> The human body is not a standard in every part for Comparative Anatomy; for though the brain may be a standard, yet the teeth are not; for they are of the mixed kind in the human; therefore we must either begin with the true herbivorous or carnivorous animals, and the human comes in between. The human, however, seems to have most of original forms, or fewer forms in common.
>
> In Comparative or Universal Animal Anatomy it would be, perhaps, the most natural way to begin with the most simple animal, and proceed to the most complex. But by considering that the most compound are either ourselves, or those we are most conversant with, and therefore know most of (the others being in a state of obscurity, much beyond their simplicity of structure), it becomes necessary that we should perfect ourselves in that we are most acquainted with: like the necessity of perfecting ourselves in our native language, which we know a good deal of, although it is very complex, before we enter upon another, although very simple in comparison, of which we know nothing.[81]

In these general remarks John Hunter speaks of 'Comparative or Universal Animal Anatomy', and it would appear from this that for him comparative anatomy was a project that would ultimately cover the dissection of every species of animal and insect. However, complete anatomical accounts of each animal would serve little purpose and be simply repetitive, and instead Hunter proposed a system of taking one or more animals as a *standard*, and then *comparing* the animal being dissected, noting down any *variations* from the standard animal. When he is writing about monkeys and apes, the standard is the human, and Hunter only remarks on differences from the human. For instance, he records:

> A large black monkey, from Lord Shelburne, of the spider-monkey kind, but has two thumbs. His hair is black and strong: it is not longer on one part than another, excepting on the top of the head, where it is also thickest. Without a tail. The liver is exactly similar to the human. The gall-bladder is smaller than in the human, and is somewhat contorted. The stomach is similar to the human, but rather more globular. The duodenum exactly similar to that of the human.[82]

And so on, with Hunter paying particular attention to the digestive and the generative organs. In the case of other animals, the standard was a different animal. For instance, when dealing with the fox Hunter says, 'I have compared this animal with the dog, so that what I have not taken notice of, is to be found in a dog'.[83] By adopting this approach, as he said (above), 'if we find that all the lion-tribe are the same, it will save others from dissecting all this class'. This is one of the features

81 Ibid., vol. 2, pp. 1–2.

82 Ibid., vol. 2, p. 8.

83 Ibid., vol. 2, p. 64.

that so distinguishes John Hunter's comparative anatomy in contrast to that of Buffon: from John's accounts it is very rarely possible to gain a mental picture of what the animal being discussed looks like!

Although he dissected over five hundred different kinds of animal,[84] for the most part John Hunter was *not* very concerned with classification of animals or insects (despite Richard Owen's later attempt to make him a Cuvierian and hence a classificatory comparative anatomist). The most he does is speak in a loose way of the 'tribes' and 'affinities', the 'classes' and 'gradations' of different animals.[85] Nor is he concerned with standardising or otherwise improving *nomenclature*, as was the case so often with his contemporary naturalists, and he always uses common English names for the animals. And finally we need to note that comparative anatomy was, for John Hunter, of a piece with his other great interrelated projects on anatomy, physiology and pathology. If they were of interest to him, John might spread the dissected and preserved parts of a given animal between the physiological, the osteological or the pathological series in the collection, depending on what particular features he wished to highlight or record.

Thus far John Hunter on his own sub-discipline of comparative anatomy. However, poor John Hunter has been represented as many other kinds of comparative anatomist since his death, for his comparative anatomy became a very flexible concept indeed, being constantly refashioned by well- and ill-disposed followers. First his brother-in-law and executor Everard Home turned it into a French enterprise, *à la* Cuvier:

> In the year 1800, Mr Cuvier, Secretary to the National Institute of Paris, an anatomist deservedly eminent in the eyes of Europe, published the first two volumes of his *Leçons d'Anatomie Comparée*, a most valuable work, so nearly on the same plan as Mr Hunter's Collection, that it might be considered as a descriptive catalogue of it on a more extensive scale.[86]

It is astonishing that Home, who had been a pupil of Hunter's as well as his co-worker, could have so mistaken the identity of Hunter's collection.

But then the collection received another identity crisis when the new conservator, Richard Owen, having once made it Cuverian, then at a later date turned it into

---

84 Richard Owen's Preface to *The works of John Hunter*, 4 vols, 1837, vol. 4, p. vii.

85 The nearest such universal division of animals into classes by Hunter is the one Owen reproduced from a manuscript in the Preface to *The works of John Hunter*, ibid., vol. 4, pp. xxxiii–xxxv, though the status even of this is not clear. One needs to bear in mind that, in this Preface, Owen had set out to redeem Hunter's reputation as a comparative anatomist in the eyes of continental historians of Owen's own day, and he therefore grasps at anything in Hunter's writings or collections that might 'count as' comparative anatomy seen from the 1830s.

86 Home, *Lectures on comparative anatomy*, 6 vols, 1814–28, pp. 7–8.

a German one, following the teachings of Oken and St.-Hilaire of whom he was himself a follower for a while. This is not the place to go into the intricacies of all these new identities that John Hunter's comparative anatomy received through the assembling and publication of catalogues, the reorganisation of the material in the collections, and the writings of friendly and hostile historians and others commemorating his memory, since to do so would take us up to the early twentieth century. L. S. Jacyna has recently shown how the Hunterian Orators at the Royal College of Surgeons in the first three decades of the nineteenth century made 'comparative anatomy' the centre of Hunter's supposed achievements. Jacyna points out that a significant motivation in their doing so was to raise the status of surgeons from that of being lowly hand-workers to being practitioners of a science.[87] However, none of these practices of comparative anatomy in the 1820s and '30s were that of John Hunter himself.

Finally here we might notice the meaning that a very minor practising anatomist, Sir Busick Harwood (?1745–1814) gave to the term 'comparative anatomy' at the very end of the eighteenth century. Harwood was Professor of Anatomy at Cambridge University from 1785, where his lectures frequently included demonstrations – apparently successful – of transfusing the blood of a sheep into a dog. A somewhat hostile contemporary recalled that in 1786 Harwood 'was at that time lecturing on comparative anatomy; and it was no unusual thing to see the turbot on which Mr. Orange [Harwood's demonstrator in anatomy] had exercised his skill one day, carved by the Professor on the following' at dinner.[88]

Harwood published a first (there was never a second) volume of *A system of comparative anatomy and physiology* in 1796. It was translated into German, but the English version sold so poorly that it was pulped.[89] In fact it is little more than a treatise on noses. It appears to reflect some of his teaching in the 1790s.[90]

In this solitary first volume Harwood begins by discussing the brain and nerves. Although we know the detailed *structure* of the brain, he says, we are unable with any assurance to allocate intellective *functions* to the different parts:

> Abandoning the vain attempt, we flatter ourselves, that some resource is left to us in comparative anatomy; and indeed at first view it seems not at all improbable, that in different animals there should be some marked peculiarities in the structure of the brain, corresponding with their different habits and sagacities. Whatever

[87] Jacyna, 'Images of John Hunter', 1983.

[88] Gunning, *Reminiscences of the University … of Cambridge*, 2 vols, 1854, vol. 1, p. 55.

[89] Russell, *British anatomy*, 1987 (1963), p. 97.

[90] Harwood, *A synopsis of a course of lectures on anatomy and physiology*, 1792, p. 82, item 12 'Of the NOSE of Quadrupeds. Of its internal structure, and of the passage, and distribution of the olfactory nerves'.

> our hopes or wishes may be, from this source little if any thing has hitherto been derived; nor have I advanced a single step beyond my predecessors.[91]

However, Harwood argues, we make more progress when we consider sensation and trace the nerves:

> We can trace them from the Sensorium itself, to the seat of sense; we can compare their various magnitudes, and distributions in various animals; we can illustrate the obscurity of one subject, by the reflected light of another; and by extending the boundaries of knowledge, exalt the mind to the most congenial of it's [*sic*] operations, the adoration of the Almighty Creator.[92]

One of the things he does is compare the structures serving the same function (in this case serving the sense of smell) in different animals. It is illustrated here in Fig. 5.16, where the skulls of three herbivores – the hog, horse and deer – are compared, revealing a gradation with respect to their similarity to carnivorous animals. By pursuing this route, Harwood came up with the novel finding that the sense of smell is located in a small area of the membrane lining the anterior nares, and that animals with a greater need for a good sense of smell (carnivores) have more complex olfactory organs and nerves than those that do not (herbivores).[93]

> In taking a view of the comparative anatomy of the Nose, in the classes, *Mammalia*, *Aves*, *Pisces*, and *Amphibia*, we are struck with the general analogy, observable in the composition and adjustment of the olfactory organs of each. And even, where they fail with respect to a close conformity, we can in most instances discover the necessity of the deviation, by examining their usual places of residence, and scrutinizing the manners of the inhabitants. The external nostrils in all animals are in the neighbourhood of the mouth. In the generality of Quadrupeds they are situated near the extremity of the snout, just above the upper lip, as if to perform the office of sentinel at the principal entrance.
>
> In birds they are placed at some distance from the end of the beak. Had they been at the extremity, the inconvenience, that would have arisen from such a situation, is evident.[94]

Over the years Harwood's interests and teaching moved more towards comparative anatomy and Natural History. The *Synopsis* of his lectures for 1807 shows that his teaching was by then arranged around comparative anatomy:

---

91 Ibid., p. 3.

92 Ibid., p. 4.

93 Williamson, 'Sir Buswick Harwood: A reappraisal', 1983, p. 430.

94 Harwood, *A system of comparative anatomy and physiology*, 1796, p. 51.

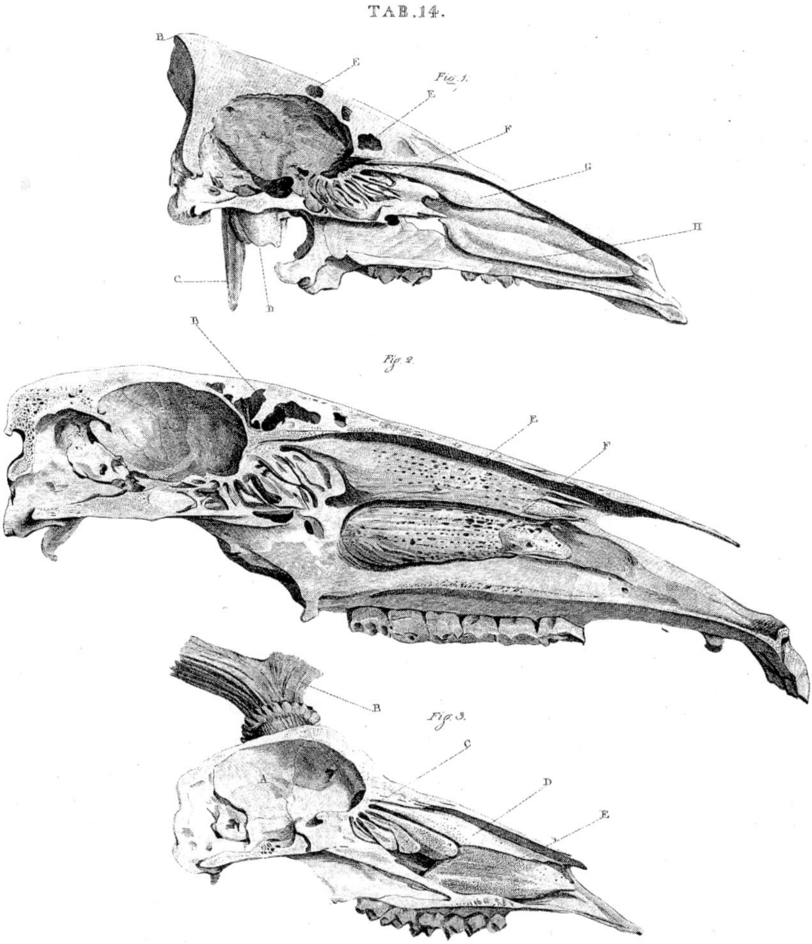

Fig. 5.16

> By the term C*omparative Anatomy* he [the Professor; that is, Harwood] means to include not only an enquiry into the Organization of Men and of other Animals, with their respective relations and distinctions, but also into such Analogies as are found to exist between the Animal and Vegetable kingdoms. It is moreover his intention that this plan of Lectures should comprehend a succinct account of the progressive scale of Animal and Vegetable existence, so as to impress the minds of his hearers with the sublime truths deducible from the *Philosophy of Natural History*.[95]

[95] Harwood, *A synopsis of a course of lectures on comparative anatomy and physiology*, 1807, p. vii.

By 1812, when he issued the last Synopsis that we know of, he had gone even further in this direction. It is called *A synopsis of a course of lectures on the philosophy of natural history and the comparative structure of plants and animals*.[96] The first part deals with the analogies one can make between plants and animals, such as structure, absorbent systems, respiration, circulation of fluids, generation. The second deals with the comparative anatomy of man and animals. In this the blood, bones, internal organs, senses, sense-organs and nerves are compared. This is the most thorough-going series of comparisons of the human with animals that we have yet seen, and it demonstrates a real interest in the behaviour and structure of animals. On the evidence of this Synopsis Harwood was no Cuvierian, for he still maintains the existence of the scale of nature, although in other respects he shows evidence of keeping up with the latest discoveries.

We can conclude from this variety of uses of the term 'comparative anatomy' in the eighteenth century that it was a term in widespread use among anatomists, and that in the cases of certain individuals – in particular Buffon and John Hunter – it applied to a particular sub-discipline of both anatomy and natural history, though in each case a sub-discipline unique to each investigator. In other cases it refers to a very limited procedure of looking at analogous structures in animals to elucidate structures in man, a procedure thus subservient to the discipline of human anatomising and not a discipline in its own right.

4. George Stubbs and the dissection of the horse

Following eighteenth-century usages of the term 'comparative anatomy' we have seen anatomists dissecting animals in order to get a better view of the anatomy of man, and we have seen anatomists acting as the handmaids of Natural Historians, who were seeking to get as complete an account of animal creation as possible, by offering them accounts of the internal anatomy of animals. But this is to overlook what was in some ways the greatest single-handed accomplishment in the anatomising of a particular species of animal, and which was carried out not by a professional anatomist but by a professional artist who had taught himself anatomy. This was the dissection of the horse by George Stubbs.

Of all the animals dissected for anatomy in the eighteenth century, the most important one economically and socially was the horse. Stubbs's anatomical atlas was produced in England. In eighteenth-century England the horse was being given new roles by the aristocracy and gentry, for it was the beginning of fox-hunting on horse-back, of horse-racing on a large scale, of the selective breeding of animals, and particularly the specialised breeding of hunters and racehorses. With respect to racing, the first racing calendar was produced in 1727; the Jockey

[96] Harwood, *A synopsis of a course of lectures on the philosophy of natural history*, 1812. Not listed in Russell, *British anatomy, 1525–1800*, 1987 (1963); copy in Cambridge University Library.

Club was founded in 1750; the first steeplechase was held in 1752; the first St Leger was run in 1778; the first Oaks in 1779; and the first Derby in 1780. In these circumstances it is not surprising that artists were offered a new form of employment now: portraying their patrons and their horses engaged in these country pursuits, in order to provide pictures to adorn the walls of the patrons' new country houses.

Surprisingly these consumer demands in England did not lead to any improvement in the horse's lot, nor to improvements in the veterinary art, nor to concern with the anatomy of the horse – knowledge that would necessarily underlie any significant development in pathology and therapy for the horse.[97] Indeed, England lagged very much behind the continental countries in its concern with veterinary knowledge and training. The French state established a veterinary school in Lyons in 1761, which was moved to Alfort (near Paris) in 1765, and another one in Montpellier. In the next thirty years veterinary schools were established in Vienna, Berlin, Copenhagen, Stuttgart, Dresden, Hanover, Budapest, Turin, Padua, Parma, Utrecht, and elsewhere. But it was not until 1788 that one was established in England – and even that one was on the initiative of a Frenchman unable to return to France for political reasons when the French Revolution was about to break out. This Frenchman, Charles Vial De Saint Bel, was a former student at the Lyons school and teacher both there and in Montpellier. In 1791 he became first professor of The Veterinary College of London (later The Royal Veterinary College) that he himself largely founded.[98] Until that time sick and injured horses in England received the rough and ignorant treatment and the folk remedies they had always been given at the hands of blacksmiths, farriers and self-proclaimed horse-doctors. Among the motivations of those behind the founding of the London College were that 'farriery as it is commonly practised is conducted without principle or science, and greatly to the injury of the noblest and most useful of our animals', and that 'the improvement of Farriery established as a study of the Anatomy, diseases and cure of cattle, particularly Horses, Cows and Sheep, will be an essential benefit to Agriculture and will greatly improve some of the most important branches of national commerce such as Wool and Leather'.[99] John Hunter was the first chairman of the committee that appointed professors to the Veterinary College. The anatomy of the horse was a central part of the teaching.

---

97 For a more positive assessment see Curth, 'The care of the brute beast', 2002.

98 Vial De Sainbel, *Elements of the veterinary art*, 1797; for details of the Lyons and Alford schools, see 'Preliminary Discourse', pp. 20–29. This account is the basis of Vines, *Article 1: On the origin of the veterinary science in Great Britain*, 1836; Vines is identified as the author in the British Library catalogue. There is no trace of a title-page to this document, and it may be a reprint from a publication of some kind.

99 Cotchin, *The Royal Veterinary College London*, 1990, pp. 15–16: motions of the Oldham Agricultural Society, 1785, which was behind the committee that founded the London College.

Anatomical monographs had been devoted to the horse before the eighteenth century, but the best of these had been published over two centuries earlier, Carlo Ruini's *The anatomy of the horse, its infirmities and their remedies* (Bologna, 1598), which went through several new editions and plagiarisms.[100] Ruini's book followed the anatomy of the horse through from the head (book one, on the 'animal' organs), to the chest (book two, on the 'spiritual' organs), to the organs of nutrition in the lower belly (book three), to the organs of generation (book four), to the muscles, skeleton, veins, arteries and nerves (book five), thus ending up with displaying the superficial muscles. The woodcuts are skilful and striking, as can be seen in Fig. 5.17.

The modern masterpiece on the anatomy of the horse was produced by an English painter, George Stubbs (1724–1806). Effectively, Stubbs's work is the opposite, or perhaps the complement, of Ruini's, though it is not known whether Stubbs knew Ruini's book. Where Ruini had worked systematically through all the inner organs of the horse, somewhat in the manner of someone anatomising the human, Stubbs simply took the inner organs out and they were 'cast away'. But where Ruini presented only one view of the entire skeleton, and very few of the bones and muscles, Stubbs dedicated all eighteen of his great illustrations to just these. Again, Ruini had devoted one of his two books to the illnesses of horses and their cures, a theme Stubbs merely alludes to almost in passing in his Address to the Reader.[101] So Stubbs's aims in producing his *Anatomy of the Horse* appear to be quite novel. This is in fact the case: his aims were closely linked to art and patronage, and to the need to advertise his talents as a painter.

Virtually all our knowledge of Stubbs's early life, and most of what we know about his later life beyond his publications themselves, comes from a set of manuscript memoirs drawn up by a friend and fellow artist, Ozias Humphry, from conversations with Stubbs in the 1790s, with annotations by Stubbs's common-law wife. Fortunately the original manuscript has at last been published in full, and will be the source quoted here (with page numbers in brackets). Over the last two centuries the biographers of Stubbs, with so little original material to work from,

---

100 Ruini, *Anatomia del cavallo, infermita, et suoi rimedii*, 1618; the first edition was published in Bologna in 1598, and there was another edition in Italian in 1599. A German edition appeared in Frankfurt in 1603, and two French editions in Paris in 1647 and 1655. Among the plagiarisms are: Andrew Snape, *An Anatomy of a Horse*, 1683; J. de Saunier, *La Parfaite Connoissance des Chevaux*, 1734. For a less ambitious essay just on the bones of the horse, see Héroard, *Hippostologie, c'est à dire, discourse des os du cheval*, 1599. This was intended as a groundwork for a complete treatment of the veterinary art.

101 'As for Farriers and Horse-Doctors, the Veterinarian School lately established in France shews of what importance their profession is held in that country; amongst us they have frequent opportunities of dissecting, and many of them have considerable skill in anatomy: but it were to be wished that this, as well as other parts of medical science, were as generally attended to by them, as by those gentlemen who treat the diseases and wounds of the human body. If what I have done may in any sort facilitate or promote so necessary a study amongst them, I shall think my labour well bestowed'.

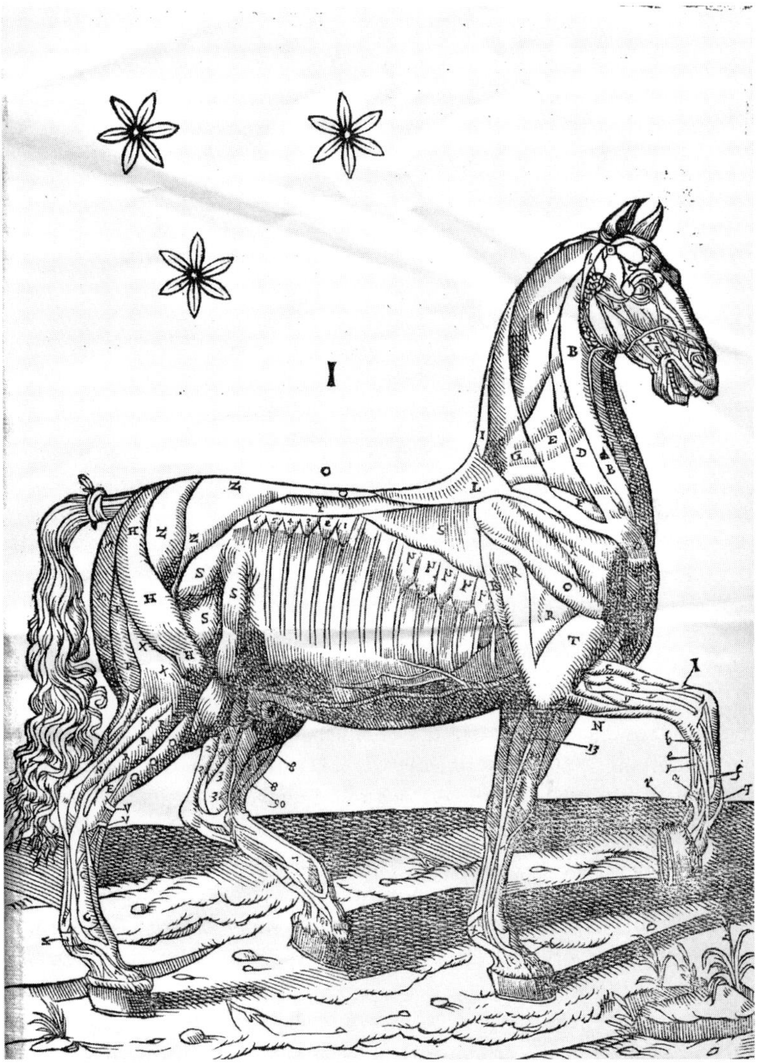

Fig. 5.17

have inevitably added embellishments to the story and in particular have ascribed to Stubbs motivations that are not at all supported by the original memoirs.[102] The

[102] The manuscript is in the Liverpool City Libraries, and was published in its entirety for the first time as an Appendix to Hall, ed., *Fearful symmetry: George Stubbs, painter of the English Enlightenment*, 2000, pp. 196–212; a paraphrase and abbreviation was made by Mayer, *Memoirs of Thomas Dodd, William Upcott and George Stubbs, R. A.*, 1879. All modern biographies of Stubbs are built on these, with or without acknowledgement:

memoirs show quite clearly how Stubbs approached his anatomy as a painter first and foremost, so I shall quote them extensively.

Stubbs became a horse anatomist because he was a painter. Indeed in his own eyes Stubbs was always a painter first, and an anatomist second: on the title-page of *The Anatomy of the Horse* (1766) he described himself as 'George Stubbs, Painter'. But, as he said to his friend Ozias Humphry, the book 'more than any other thing tended to throw him into Horse Painting, and to this he ascribes entirely his being a Horse Painter' (p. 202). Portraiture and historic painting would have been more conventional goals for a budding artist. As Humphry put it, 'without seeking assistance from any one, he [Stubbs] proceeded unassisted to make all his studies after nature, intending by every thing he did, to qualify himself for painting Rural, Pastoral and familiar subjects in History as well as portraiture' (p. 201). In fact, it appears that Stubbs came to anatomise the horse via his anatomies of humans, which he seems to have pursued in order to develop his artistic skills in human portraiture and other conventional areas.

The son of a leather-dresser in Liverpool, George Stubbs was engaged in both drawing and anatomising from a very early age: 'When he was scarcely eight years old he began his little studies in Anatomy, that he had Bones and prepared subjects dissected and lent to him by Doctor Holt, a benevolent neighbour, who wished to amuse him and to promote his improvement' (p. 211). Though it was an artistic commonplace to want to 'copy nature' rather than other painters, in Stubbs's case it seems to have been almost a fetish. He went to Italy aged about thirty, following many other aspirant painters to see the classical relics and the works of the Renaissance masters. But, 'let it not escape notice', Humphry recalls from Stubbs, 'that Stubbs's motive for going thither was to convince himself that nature was and is always superior to art whether Greek or Roman; – and having received this conviction he immediately resolved upon returning home' (p. 202).

There is, of course, no simple unmediated 'nature' for a painter to depict, nor is it possible for a painter to see 'nature' without some cultural spectacles (as it were) of one kind or another. So nor was it possible in Stubbs's case either, for his paintings clearly show that he was an eighteenth-century painter in style and mannerisms, and with eighteenth-century visual perceptions, and that he was also indebted for elements of his style to other painters.[103] However, the concern to produce his paintings 'from nature' certainly led him to undertake his horse anatomies in a particular way.

Sparrow, *British sporting artists*, 1922; Parker, *George Stubbs: Art, animals and anatomy*, 1971; Doherty, *The anatomical works of George Stubbs*, 1974. Some anecdotes, of unknown trustworthiness, were told in two short articles in *The Sporting Magazine* for 1808, pp. 55–7 and 155–7; some of the material here reproduces what is in Humphry's memoir.

[103] Hall, ed., *Fearful symmetry: George Stubbs, painter of the English Enlightenment*, 2000, argues the case that, notwithstanding all his originality, Stubbs was indebted to contemporary artistic philosophy and practice.

At the age of twenty Stubbs went to York 'to execute some commissions', that is of portraits, and it was here that 'he began a regular course of anatomical study, and of dissection for the first time of the human figure … Here he became acquainted with Mr Charles Atkinson a surgeon who procured him the first subject for dissection [i.e. a human body]. He was now prevailed upon to give anatomical lectures privately to the pupils of the Hospital' (p. 201). Human and animal dissection was still at that time an unusual but not unheard-of step to take in training to become a painter.

The hospital to whose medical pupils Stubbs gave these anatomical lectures was the recently founded York County Infirmary, set up in 1740 by (among others) Dr John Burton, who had studied in Cambridge and Leiden. Burton was engaged in writing *An essay towards a complete new system of midwifery*, and needed someone to produce the illustrations of the pregnant womb and the infant within it. Stubbs was recruited for this task (as we saw earlier in Chapter 3); 'But the better to enable him to perform this task it was necessary that he should make himself thoroughly a master of the human figure, and with that part of it in particular which concerned Midwifery. Fourteen or sixteen miles from York, a female subject who had died in childbed was found singularly favourable for the purpose of these studies and brought to York by Stubbs's pupils, where it was concealed in a Garret, and all the necessary dissections made' (pp. 201–2). As Burton could find no one willing to make the engravings either (probably because of the subject-matter), Stubbs taught himself to do this, and the book and its illustrations were published in 1751 (see Fig. 3.2 in Chapter 3 above).

So Stubbs had engaged enthusiastically in human dissection, and had even dissected the human gravid uterus and the fetus within it. He had been engaged as a professional portrait painter, and he had gone to Italy to confirm his prejudices. When he returned to York, the idea for a book on the anatomy of the horse represented itself to him.[104] 'In this neighbourhood it was that he resolved to carry on and complete what he had long since determined upon by the advice of his young chirurgical friends with whom he had formerly been connected at York, which was to publish the Anatomy of the Horse' (p. 202). That is precisely all we know directly about Stubbs's decision to undertake the anatomy of the horse.

However, the immediate connection of the horse anatomy with Stubbs's self-image as a painter is quite clear from the opening of his Address to the Reader in the finished book. Of the four motivations he reports, the first is concerned with painting:

> When I first resolved to apply myself to the present work, I was flattered with the idea, that it might prove particularly useful to those of my own profession

[104] According to Morrison, *The art of George Stubbs*, 1989, p. 21, 'The surgeon Charles Atkinson may have been important in the early stages of the project, for a reference exists which states that he taught Stubbs the anatomy of the horse'. Unfortunately Morrison does not identify this reference.

> [i.e. painting]; and those to whose care and skill the horse is usually entrusted, whenever medicine or surgery becomes necessary to him; I thought it might be a desirable addition to what is usually collected for the study of comparative anatomy, and by no means unacceptable to those gentlemen who delight in horses, and who either breed or keep any considerable number of them.
>
> The Painter, Sculptor, and Designer know what assistance is to be gained from the books hitherto published on this subject; and as they must be supposed best able to judge, how fitly the present work is accommodated to their purpose, any address to them is superfluous.

To carry out the work, Stubbs in 1756 'engaged a Farm House at Horkstow in Lincolnshire, that he might without inconvenience to others, have dead horses in his apartment and other subjects adapted to his purposes. In this retirement he continued with unremitting diligence to dissect for eighteen months (he was then in his 32nd year) and here also it was that he completed the Drawings and descriptions for his publication of the Anatomy of the Horse with no other companion or assistance than his female relation and friend Miss Spencer' (p. 202). 'Miss Spencer', Mary Spencer, was his common-law wife, with whom he had a child, but whose assistance he did not acknowledge at any time in print. Humphry describes her as Stubbs's 'constant companion, and principal assistant, both in his literary and his anatomical pursuits' (p. 199).

Having the idea to produce an anatomy of the horse was one thing, but how was he to go about it? What he did was to turn to the most recent and best book on the anatomy of the *human*: that of Albinus, which had been published first in Leiden in 1747, and then reprinted in English translation in London in 1749 (see Chapter 4 above). Every element of Albinus's approach to the human is evident in Stubbs's approach to the horse. The extensive Preface in which Albinus explained what he was doing was available to Stubbs in the English edition, so he could learn from this how to rig up and pose the succession of horses he used, and indeed even how to portray their anatomy. In fact, from the systematic way in which Stubbs follows Albinus (and shows no indebtedness to other anatomists of his day), it seems more than likely that it was while they were perusing Albinus's book that Stubbs and his surgical friends at York had the idea for the anatomy of the horse in the first place.

Just as he never acknowledged in print his indebtedness to Miss Spencer in the actual work of dissection and drawing, so neither did Stubbs ever acknowledge directly his use of Albinus as his model. All he says in this respect in his Address to the Reader was, 'what I would principally observe to the Reader concerning this my performance, is, that all the figures in it are drawn from nature, for which purpose I dissected a great number of horses; and that, at the same time, I have consulted most of the treatises of reputation on the general subject of anatomy'. But the role of Albinus's book as Stubbs's model would have been immediately evident to any contemporary anatomist, and indeed it was pointed out by the Dutch professional anatomist Peter Camper who, in a congratulatory letter in 1771, wrote to Stubbs: 'You have certainly had before you the scheme of the great Albinus'.[105] Stubbs's use of Albinus should probably not be thought of as plagiarism, and the fact that he does not acknowledge him should probably not be interpreted as a slight to Albinus either. We might prefer to say that Stubbs had adopted the very latest and best techniques and presentation in the world of human anatomy for his book on the horse, in order both to make it most accessible for learning and locating the bones and muscles, and also to make it most beautiful as a desirable object. Though he evidently had his own reasons for only acknowledging Nature as his source and for claiming all the work as his own, in practice both Nature and Albinus were his masters – and Mary Spencer was his assistant.

The ways in which Stubbs's book is modelled on Albinus's are manifold. In the first place, the horse that is drawn is a *generalised* or *idealised* horse, just as Albinus had portrayed the idealised human. It is not one particular creature, nor is it the mean or average of the many horses that Stubbs dissected. Rather, it was a standardised, or ideal, horse. So the proportions of the horse portrayed had to remain constant no matter what the variation in the horses that Stubbs had available to draw. To achieve this, Stubbs modified Albinus's device of a regular viewing frame through which to view and draw the skeleton. Stubbs's approach was more painterly: he effectively drew up a paper template by preparing a master drawing of the silhouette of the horse and of its skeleton from each of the viewing-points he wanted. He then made tiny pin-holes around these outlines. He placed one of these templates on a clean sheet of drawing-paper and tapped red chalk dust through the holes. On the clean sheet he then joined up the dots to re-create the outline identical in size and proportions.[106] In this way Stubbs was able to retain the desired continuity of size, scale and pose, while using lots of horses of different sizes as his models.

Second, Stubbs, like Albinus, had to start from the skeleton, onto which he then 'built up' the muscles. As he wrote to his Reader, 'the proportions which I have mentioned in several places of the book, are estimated from the length of the head, as is usually done by those who have treated the proportion of human figures; this length is taken from the top of the head to the ends of the cutting teeth, and is divided into four equal parts, each of which is again divided into twelve minutes'.

Third, in order to dissect and portray them, Stubbs suspended his dead horses in a manner very similar to that adopted by Albinus for humans. Where Albinus had constructed his tripod, and his rope running through a ring in the ceiling to place and keep the skeleton and then the corpse in the chosen fixed position, Stubbs too suspends his horses from a hook in the ceiling, attached to the upper spine. Albinus

[105] Mayer, *Memoirs of Thomas Dodd, William Upcott and George Stubbs, R. A.*, 1879, p. 13.

[106] Parker, *Mr Stubbs the horse painter*, 1971, p. 18; the pin-holes and chalk markings can apparently still be seen on the sketches at the Royal Academy.

places a board under the feet of his skeletons so that they are in a standing posture to be dissected and portrayed; Stubbs places one under the feet of his horses.

Fourth, Stubbs dissected and illustrated in the horse precisely what Albinus had dissected and illustrated in the human in the book of his that Stubbs had available to him: the skeleton and the muscles.

Finally, the format and appearance of the printed books of the two anatomists are remarkably similar. Albinus begins his portrayal of the anatomy of the human with three skeleton poses (front, rear and side); Stubbs does the same with his portrayal of the anatomy of the horse. Albinus then presents the tables of the muscles of the body, first with the intact body, and then a sequence of tables with the layers of muscles successively stripped off, and he does this for the body seen both from front and rear; Stubbs presents the musculature of the horse in the same sequential way, but from the perspective of front, side and rear. What is more, Stubbs follows Albinus exactly in presenting not one but two images on each table: one the image, the other the outline with the key to all the bones and muscles, which key is given a detailed description in the text (Figs 5.18–5.21). This system Albinus seems to have invented himself. Stubbs, by contrast, simply lifts the whole method from Albinus wholesale. However, Stubbs does not follow Albinus in his portrayal of all the detailed images of the individual muscles of the human body.

There are only a few significant differences in the approaches that Albinus and Stubbs took in producing their illustrations. One, the most notable, was that Stubbs was his own artist as well as being the dissector; he did not need to train someone else's eye and hand to portray what he wanted, as Albinus had had to train Jan Wandelaar. As Stubbs could claim for a later project, 'he has executed the whole himself, from Nature; that is, the Injection, Dissection, Drawing, Explanation, and Engraving. He has taken particular care to represent each thing in its proper place, that is, not loose or irregularly hanging; such as the Arteries, Veins, Nerves, Muscles, Viscera, &c. because their true situations and proportions are as necessary to be known as their forms'.[107]

Another difference between the performances of Albinus and Stubbs is that Stubbs gave no classical background to his figures as Albinus had done. We find here no rhinoceros or classical statuary in the background, no concession to the picturesque. Finally, for Stubbs the artist, the muscles and skeleton of the horse were all that he ever planned to dissect and portray, while for Albinus the muscles and skeleton were only the first part of his intended project of anatomical dissection illustration, which was to have gone on to treat all the internal organs as well.

It might sound as though with Stubbs we have an anatomist who was self-taught, who did not belong to the discipline of anatomy. I said in the Introduction that in order to practise anatomy (like any other discipline) one has to be initiated into it; one has to have a teacher. Was Stubbs the exception – the self-taught genius

---

[107] The subscription proposal by Stubbs for his *Comparative anatomical exposition of the structure of the human body* (1804–1806), as quoted in Lennox-Boyd, Dixon and Clayton, *George Stubbs: The complete engraved works*, 1989, pp. 313–15.

– that proved the rule? Did Stubbs have a teacher in his anatomies of the horse? The answer is in the affirmative: his teacher was Albinus, through his book.

We are fortunate that Stubbs gave his friend Ozias Humphry a detailed account of how he had gone to work in his Lincolnshire farmhouse, although Humphry omits the very first stage where Stubbs had to prepare and draw his basic skeletons on which to then portray the muscles. This reveals just how laborious was the work of following Nature, even with Albinus as one's silent guide:

> As these studies and operations were singular and very important, the manner in which they were conducted may not be uninteresting to relate. The first subject that he procured was a horse which was bled to death by the jugular vein; after which the arteries and veins were injected. – A Bar of Iron was then suspended from the ceiling of the room, by a Teagle to which Iron Hooks were fixed. – Under this bar a plank was swung about 18 inches wide for the horses feet to rest upon and the animal was suspended to the Iron Bar by the above mentioned Hooks which were fastened into the opposite side of the horse to that which was intended to be designed; by passing the Hooks thro' the Ribs and fastening them under the Back bone and by these means the Horse was fixed in the attitude which these prints represent and continued hanging in the posture six or seven lengths [?weeks], or as long as they were fit for use.
>
> His drawings of a skeleton were previously made, and then the operation upon this fixed subject were thus managed.
>
> He first began by dissecting and designing the muscles of the abdomen proceeding thro' five different layers of muscles till he came to the peritoneum and the pleura, through which appeared the lungs and the intestines – afterwards, the Bowels were taken out, and cast away. – then he proceeded to dissect the Head by first stripping off the skin, and after having cleaned and prepared the muscles &c. for the drawing, he made careful design of them, and wrote the explanations which usually employed him a whole day. – He then he took off another lay' of muscles, which he prepared, designed, and described, in the same manner as is represented in the work. – and so he proceeded till he came to the skeleton. It must be noted that by means of the Injection, the Muscles, the Blood vessels and the nerves, retained their form to the last without undergoing any change. – In this manner he advanced his work by stripping off the skin and cleaning and preparing as much of the subject as he concluded would employ a whole day to prepare, design, and describe, till the whole subject was completed. (p. 211)

This work took eighteen months of unremitting industry. But, 'so ardent was his thirst for acquiring experience by practical dissection, that he very frequently braved those dangers from putridity &c. which would have appalled most experienced practitioners; and such was his muscular strength, that he has more than once carried a dead horse on his back up two flights of a narrow staircase to the dissecting room on the attic floor' (pp. 198–9).

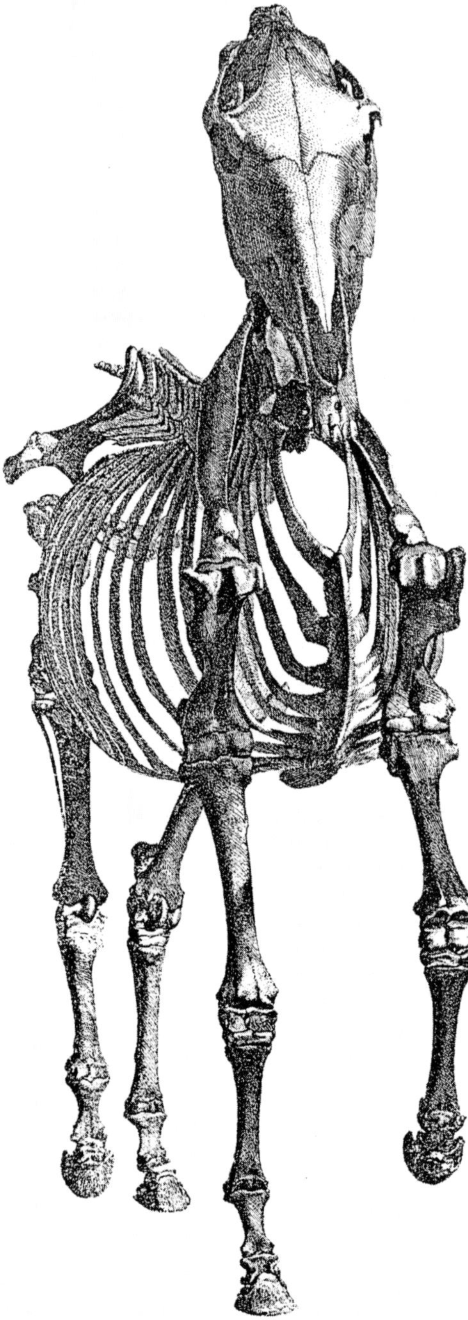

Fig. 5.18

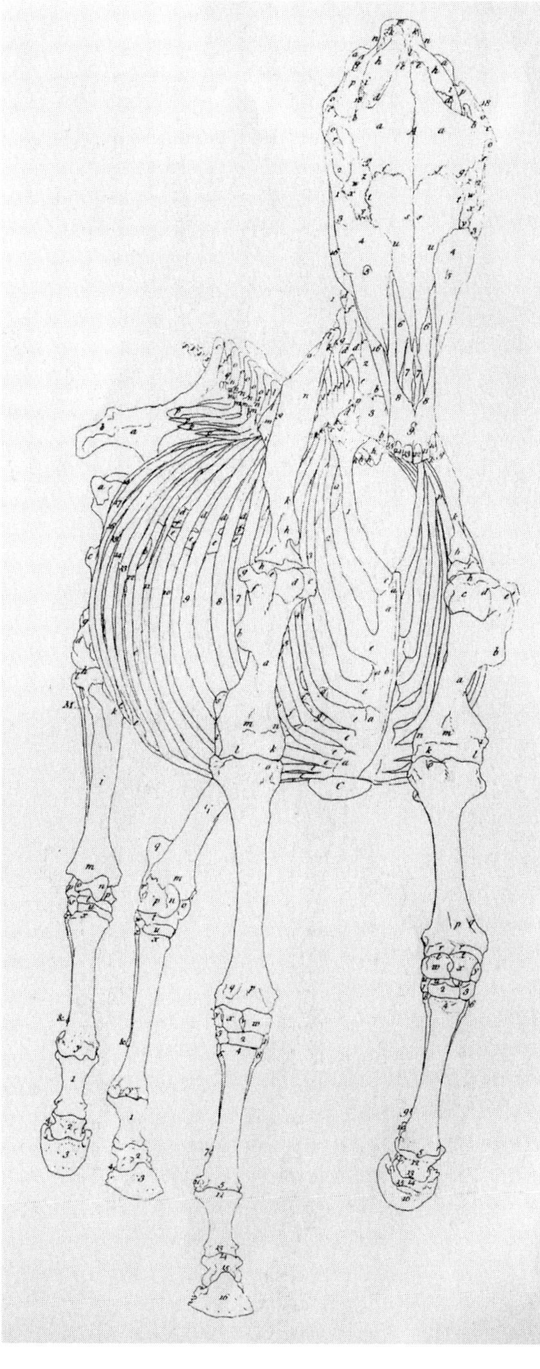

Fig. 5.19

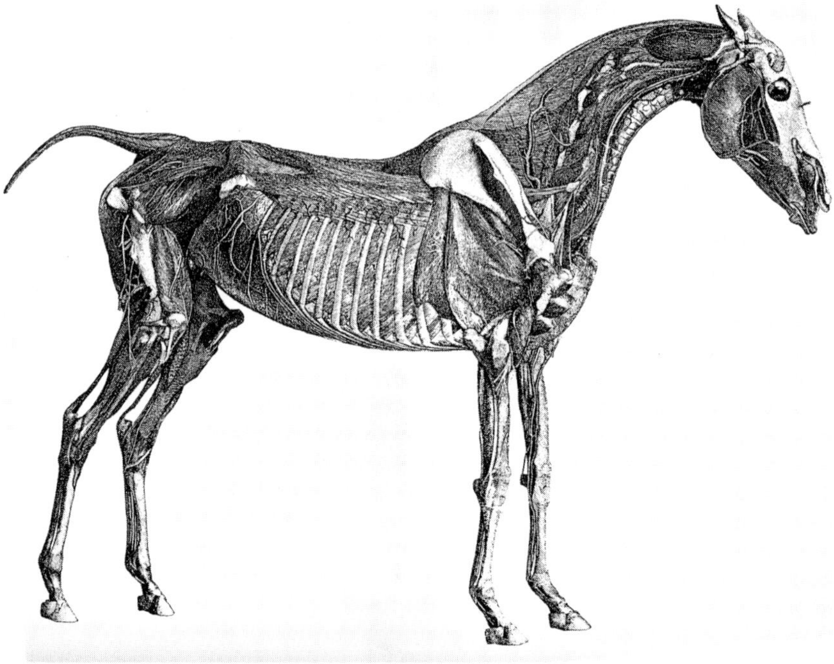

Fig. 5.20

Stubbs then in 1759 'took measures for bringing all his studies to London, in order to their being engraven and published' (p. 202). But no engraver, not even one who had been involved with engraving Albinus's plates for the London edition, was willing to take on the work. Stubbs had to do it himself, while earning his living as a painter. The thirty-six plates took him six years to engrave. He then sold the book by subscription.

The reception of the book seems to have been generally very good. The reaction of PetrusCamper, Professor of Anatomy at the University of Groningen, was unreservedly generous and charming. He wrote to Stubbs in 1771, incidentally describing a project of his own on the anatomy of the horse:

> Sir – If ever I was surprised to see a performance, it was surely when I saw yours on the 'Anatomy of the Horse'! The myology, neurology and angiology of men, have not been carried to such perfection in two ages, as these horses by you. How is it possible a single man can execute such a plan with so much accuracy and industry? You have certainly had before you the scheme of the great Albinus but even his plates have not that delicacy and fullness, nor the expression of yours. Give me leave to ask you, was you the engraver? for you do not mention the engravers name. I once had a plan to offer to the public, a subscription for the like; but I am sure I could not have obtained the elegancy and exactness of

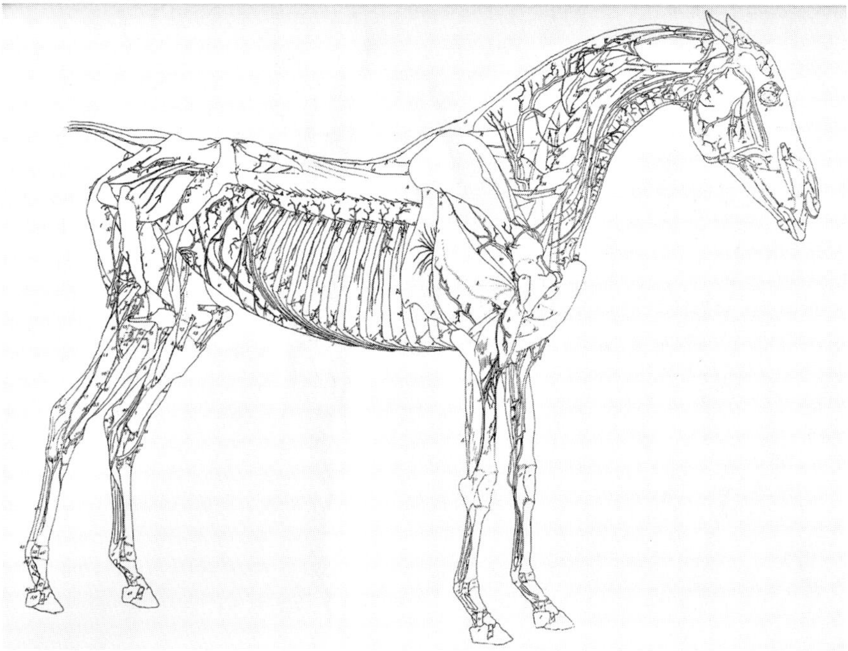

Fig. 5.21

yours. I dissected many horses; but I especially examined the head, and all the different sections of the inside, the bowels, and so on. I made figures as large as the life. I dare venture to say they are beautiful, mostly done by different means upon the life itself. My intention was to reduce them to one eighth, and to have them engraved, but after having seen and admired yours, I dropped all hopes of succeeding. This favour I hope you'll grant me, to inform me whether you still go on to finish this beautiful undertaking, and whether or not we may flatter ourselves to see the internal parts of this useful creature, and something about the disorders and internal diseases of the horse.

You will be curious to be acquainted with a Dutchman who admires with so much ecstasy your Tables. I am public professor of Medicine, Anat. and Surgery at Groningen; and I have published some figures of the human arm, pelvis, etc. I am actually publishing the Brain and the Organs of Hearing, Smelling, etc. in different animals. I dissect, but I do not love horses, though I keep them for proper use and for my family. I am sure my acquaintance can be of little use to you, but yours to me of great consequence. I desire to have two copies of your performance, one for me, and one for a gentleman who admires as well as I do your book …. (p. 210)

Stubbs's reply to this letter reconfirms his motives as being essentially painterly: 'What you have seen is all I meant to do, it being as much as I thought necessary for the study of Painting … I looked very little into the internal parts of a Horse, my search there being only matter of curiosity'.[108]

His dedication to the skeletal and muscular anatomy of the horse certainly enabled Stubbs to produce much more life-like portrayals of horses, especially with respect to their musculature, than had been achieved even by the great masters of painting. No one had portrayed horses so accurately before. In this he was more true to Nature, as he wished to be.

John Lawrence, in his rambling *History and delineation of the horse*, published in 1809, says that Stubbs's concept of following nature did not meet the taste of the times after his death:

> … the artists say that Stubbs saw all proportions through magnifying optics … a common objection to the works of our late celebrated horse painter Stubbs, whose superior genius and professional skill, have been acknowledged by all Europe … These opinions seem to be spreading beyond the professional line, and to have become, in a certain degree, fashionable. It has been lately discovered, that Stubbs was merely an anatomist, without any genius as a painter; that his Horses are all alike, and that after you have looked over his portraits of Marske, Protector, Shark, Gimcrack, and others, all that you shall have seen, is the anatomical figure of a Horse, by Stubbs, under different names ….[109]

This is a very topsy-turvy criticism to make: that all Stubbs's horses look like horses! But what was meant of course was that all Stubbs's horses look like idealised horses, and pay little concession to the small individualities of any particular horse that the owner of the horse was so aware of.

5. George Stubbs as comparative anatomist

Given the fame and celebrity of Stubbs the painter–anatomist, we need finally to look at his own version of 'comparative anatomy'. Late in life, when he had made his reputation with his horse paintings and other pictures, Stubbs somehow got the idea to produce a 'comparative anatomy'. He was already in his seventies, but threw himself into the work with as much energy as he had shown for the anatomy of the horse forty years earlier.

Different initial goals have been attributed to Stubbs in this project. An early commentator said that 'he most cheerfully began his Comparative Anatomy, after the plan of Professor Blumenbach, at the period of 80, promising a complete

[108] Quoted in *George Stubbs, 1724–1806. Exhibition catalogue*, 1984, p. 31, without further reference.

[109] Lawrence, *The history and delineation of the horse*, 1809, pp, 108, 274.

classification of the animal world'.[110] However, Stubbs's project does not seem to have much in common with Blumenbach's, nor is it concerned with classification, nor are the dates right.[111] Modern art historians have tended to assume that Stubbs was following out some programme for comparative anatomy put forward by John Hunter, since he is known to have made animal paintings for the Hunters.[112] However, this explanation must also be put aside since it does not in the least resemble John Hunter's comparative anatomy as we have seen it above. The best clue lies in the title of the book itself. The three completed numbers were published in 1817, with text, after Mary Spencer's death, as *The anatomy of the human body particularly adapted to the use of artists, the whole executed by George Stubbs Esq. R. A., author of the Anatomy of the horse, accompanied with fifteen large folio engravings*. That is to say, it was – like the *Anatomy of the Horse* – a comparative anatomy aimed primarily at artists rather than naturalists or anatomists.

For his exemplary animals Stubbs chose three: first, man, traditionally seen as the high-point of creation; second, the tiger, one of the great exotic animals; and third, a common chicken. Presumably the tiger was the one he had got from Pidcock's menagerie in the 1790s. Why he chose the chicken is not clear at all, except perhaps for its availability. Humphry's memoir of Stubbs records the event like this:

> In the year 1795 Mr S projected his plan of making a series of anatomical experiments [*sic*] in order to prove the analogy there is in the structure of the human frame with that of various animals, Quadrupeds (Bipeds) fowls and vegetables – and for this purpose having made some progress with human subjects, a Tiger and a Hen, it was his intention that the whole should be an explanation of the Bones, the Muscles, Fascias, Ligaments, Nerves, Arteries, Veins and Glands, in this undertaking he was laudably occupied and continued with unexampled exertion till the termination of his life. (p. 207)

The manuscript note by Mary Spencer, added to the Humphry memoir, reports:

> The last exertion of this valuable man was a Comparative anatomy, shewing the analogy between the Human frame, the Quadruped and the fowl, giving also an accurate description of the bones, cartilages, muscles, fascias, ligaments &c. &c. (which he intended to have carried on to the vegetable world) a work truly philosophical in which his Assiduity and perseverance are almost unexampled

110 *The Sporting Magazine*, July 1808, p. 22.

111 Blumenbach, *A short system of comparative anatomy*, 1807 (first published in German in 1805).

112 See Doherty, *The anatomical works of George Stubbs*, 1974, p. 17, where he speaks of a close relationship between Stubbs and Hunter; but compare p. 16, where Doherty says that the nature of the relationship between Stubbs and Hunter is unknown. See also Parker, *Mr Stubbs the horse painter*, 1971, p. 88.

> in the annals of mankind often employing himself from the dawn of day in summer till midnight so great was the Ardour with which he engaged in this grand undertaking. Indeed the enthusiasm which supported him with spirits and strength throughout this laborious task left him not till the latest period of his life … He had when he died published the three first numbers of his Comparative Anatomy. All the Drawings were finished and the explanation wrote to the remaining three Numbers, which only want the Engraver and Printer, to finish the work, which will be done. (pp. 207–8)

What kind of work was this? What kind of comparative anatomy did Stubbs intend? Although he published some of the engravings, Stubbs did not publish any explanatory text to go with them, nor is there a title-page or Preface to the published work.[113] As an artist of few words, it is unlikely that he would have offered much explanation even if he had completed the letter-press. However, the subscription proposal is more forthcoming, and it appears to have been written by Stubbs himself:

> An exact knowledge of the Bones is the foundation of Anatomy, because without it we can never have a just idea of the situation, disposition, connexion, and uses of the other parts of the Human or Animal Body; it is necessary an Explanation of the Skeleton should employ the first Number.
>
> The Author has, in the explanatory Plates, or Outlines of the Skeletons, given the proportions, marked by dotted lines from centre to centre of motion in each joint; because from those points the proportions will never vary, whether the joints be bent or strait: he has likewise marked, in dotted lines, the outlines of the complete or living Figure, to shew how the Bones lie within them. These proportions are meant for the use of Painters and Sculptors, to whom the knowledge of Anatomy is as necessary as to the Physician or Surgeon. Leonardo da Vinci and Michael Angelo de Bonaroti were excellent Anatomists.
>
> The Second Number is a View of the External parts of the Human Body, and an enumeration of the parts lying under them; with a description of the common integuments.
>
> The Third Number has the Common Integuments taken off, with the Membrana Adiposa and Fat.
>
> The Fourth Number has the First Lay of Muscles &c. taken off.
>
> The Fifth Number has the Second Lay of Muscles &c. taken off.
>
> The Sixth Number has the Third Lay of Muscles &c. taken off.

[113] The Royal Academy Library has copies of numbers one and two, but no text with them.

> The Author has not only endeavoured to give a full Explanation of those Muscles, &c. that come wholly in view, but likewise of those that appear but partially; and frequently of some which do not appear at all, that is, such as cannot be seen in complete Figures. And when the same thing is repeated in different Tables, he still gives the full explanation, as he thinks references are troublesome; and the oftener a thing is repeated the better it will be remembered.
>
> He has executed the whole himself, from Nature; that is, the Injection, Dissection, Drawing, Explanation, and Engraving. He has taken particular care to represent each thing in its proper place, that is, not loose or irregularly hanging; such as the Arteries, Veins, Nerves, Muscles, Viscera, &c. because their true situations and proportions are as necessary to be known as their forms.[114]

Stubbs was indeed very industrious about this project. He completed many pages of text for the book, some of which survive in manuscript, now to be found in the Paul Getty Museum of British Art at Yale University.[115] The four surviving manuscript volumes are in Stubbs's own handwriting, and someone with the initials 'F.M.B.' wrote an explanatory note:

> "A Comparative Anatomical Exposition of the Human Body, with that of a Tiger and a Common Fowl". This was the intended title of this unpublished Work, By Geo. Stubbs, an eminent Animal Painter and Anatomist. At his Death, in 1806, he left prepared these 124 Plates, with Three Parts in Manuscript, of the Six, of which he intended the Work to consist. Part of the Plates are coloured.

The manuscript consists of four volumes of fascicles bound together, some of which are fuller than others, so there are many blank pages. Two of the sets of fascicles are in English and two in French, presumably anticipating a French edition of the work, and cover the same anatomical material in each of the three chosen exemplars.[116]

---

114 This is the back of the subscription proposal, and the inside wrapper: see Lennox-Boyd, Dixon and Clayton, *George Stubbs: The complete engraved works*, 1989, pp. 313–15.

115 After Mary's death the manuscript was auctioned and bought by the anatomist Thomas Bell (1792–1880), who lectured in comparative anatomy at Guy's Hospital, and was Professor of Zoology at King's College London. Then the manuscripts got to the library of Dr John Green of Worcester, MA, who donated them in 1859 to the Free Public Library of Worcester, Mass., where they were rediscovered in 1957. Subsequently they were bought by the Getty Museum, where they complement the outstanding collection of Stubbs paintings held there in the Yale Center for British Art in New Haven, Connecticut.

116 Some pages of the manuscript are reproduced in Doherty, *The anatomical works of George Stubbs*, 1974, pp. 344–5, one English on the muscles of the human body, one French on the muscles of the tiger.

To show the nature of the commentary Stubbs gives, and the extent to which it is comparative across species, I offer the following two passages from the manuscript on the human and on the tiger.[117] Although Stubbs deals mainly with muscles and bones, for the sake of variety I have chosen what he has to say on the lungs. First the human:

> Pulmones; the lungs are two large spungy bodies of reddish colour in children, greyish in adult subjects, and bluish in old age; filling the whole cavity of the thorax, one being situated in the right side, the other in the left, parted by the mediastinum and heart; and of a figure answering to that of the cavity which contains them; that is, convex next the ribs, concave next the diaphragm, and irregularly flattened and depressed next the mediastinum and heart; they are distinguished into the right and left lung; and each of these into two or three portions called lobi, of which the right lung has commonly three or two and a half, and the left two: the right lung is generally larger than the left, answerable to that cavity of the breast, and to the obliquity of the mediastinum: the substance of the lungs is almost all spungy, being made up of an infinite number of membranous cells, in innumerable ramifications: this whole mass is covered by a membrane continued from each pleura, which is commonly double.[118]

Now the tiger:

> All the vessels of which the lungs are chiefly composed, that is the air-vessels or bronchia; and blood-vessels, that is, the pulmonary and bronchial arteries and veins, accompany each other through this whole viscus: they are disposed commonly in just such a manner even to the last ramifications, as that a subordinate trunk or branch of the bronchia lies between the like trunk or branches of the pulmonary artery and vein; the bronchial vessels being immediately joined to the bronchia; the bronchia are divided into a very great number of ramifications; and the last rami are the pedicles or footstalks of the small lobuli; all the lobuli are angular, oblong, broad, thin etc. the footstalks send out other small membranous pedicles, which are very short, and terminate in the bronchial vessels or cells, of which they are continuations; these subordinate trunks and rami detach a great number of these pedicles from their convex surface: when we blow into the lungs, their bronchial cells nearest their outer surface, like small portions of round vesicles; and from this appearance, all the bronchial cells have got the name of vesicles, though they are all angular, except those which have been now mentioned: when we examine a lung without blowing it up, we find that the cartilaginous segments of the bronchia lie so near as to be ingaged in each other; and in drawing out any portion of the bronchia by the two ends, these segments are parted, and the whole canal is increased in length; but it contracts again by

---

117 I thank the Getty Museum for permission to consult and quote these manuscripts.

118 Stubbs ms. vol. 3, p. 263.

> means of its elastic membrane as soon as that force is taken off: these three vessels lie in a sort of cellular vagina which accompanies all the ramifications, and is a continuation of the interlobular cells, or cellular substances in the interstices of the lobuli; the pelliculae which are, however, disposed there, in a more regular manner, and more longitudinally than in other places, thereby appear to form a true vagina.[119]

Nowhere does Stubbs make a direct comparison of any kind between humans and animals or animals and animals. Each account is complete in itself. It would presumably be for the reader to make such comparisons as he or she felt fit.

It is thus clear from these documents – the two accounts given by Humphry and Mary, the subscription proposal, and the Yale manuscripts – that the kind of comparative anatomy Stubbs had in mind was concerned with 'the foundation of Anatomy' in human and animal bodies, in their appropriate proportions, and 'These proportions are meant for the use of Painters and Sculptors, to whom the knowledge of Anatomy is as necessary as to the Physician or Surgeon'.

But more than this: Stubbs was also seeking *analogies* between the anatomical structure of the biped, the quadruped and the bird – functional resemblances between the structure of organs in the different species. The extensive text that he wrote to accompany the drawings was completely devoid (as we have seen) of any explicit comparison at this level. But nevertheless – at least according to those people closest to him – there lay behind all this an intention to produce 'a work truly philosophical', 'to prove the analogy there is in the structure of the human frame with that of various animals, Quadrupeds (Bipeds) fowls and vegetables (which he intended to have carried on to the vegetable world)'. It would have been truly philosophical because it would have been dealing ultimately in *causes*: the structure of part *a* in this animal (or vegetable) is like this *because* it is analogous to the structure of part *b* in this animal (or vegetable), carrying out a similar function.

By contrast, the definition of *modern*, post-Darwinian, comparative anatomy that we quoted at the beginning of this chapter stated: 'The comparative method is based on a critical comparison of structures that are truly homologous rather than analogous. Homology deals with those structures which are alike in origin and fundamental structure.'[120] Thus, in seeking to demonstrate analogies rather than homologies, Stubbs's work was certainly not akin to modern comparative anatomy.

---

119 Ibid., vol. 3, p. 465.

120 Adams and Eddy, *Comparative anatomy*, 1949, p. 105.

Chapter 6
The End of Old Anatomy

1. 'Dutch' anatomy in Japan

In the eighteenth century Japan was a closed country. Since 1639 all Japanese had been forbidden to leave Japan, and foreigners were excluded from the country in order to keep out the poison of Christianity. Only one group of foreigners were made exceptions to this isolation policy: the Dutch traders, who were seen not to be trying to spread Christianity. These traders were confined to Nagasaki, but made an annual visit to Edo (today's Tokyo) to pay homage to the ruler, the Shogun. A few Japanese were deputed to translate for the Dutch traders, but even they were forbidden to learn to write or read Dutch. There was thus a very limited amount of contact between a very few Japanese and a small number of Dutch traders, mostly just on the annual visit to Edo, when the Dutchmen, and particularly the surgeons on their ships, were questioned at their hotel on medical matters by some inquisitive Japanese.

Fortunately, the leading figure in first translating a work of anatomy from Dutch to Japanese wrote a short autobiography, which he finished in 1815, some forty years after the momentous events he there recalls. This was Genpaku Sugita, a practitioner of Chinese medicine, and his account was entitled *The beginnings of Dutch learning*.[1] It is a very modest account, but the difficulties faced by himself and his fellow Japanese scholars were immense. In 1771 a friend of Sugita visited the hotel in Edo where the Dutchmen were staying:

> He was shown the two illustrated books on the internal structure of the human body, the *Ontleedkundige Tafelen* [by Johann Adam Kulmus] and Caspar Bartholinus's *Anatomia nova* [first published in 1639]. The Dutchman who possessed these books was willing to sell them, if so requested. Jun-an [Sugita's friend] took them home and showed them to me. Of course, not a word in them could we read, but the structures of internal organs and the skeletal frames illustrated in them appeared very different from those we had seen in [Chinese] books or had heard of in the past. We concluded that these must have been drawn from the real things. And somehow a strong desire to possess them arose in me. (p. 24)

[1] The work was translated into English by Matsumoto in the volume edited by Sugita, *Dawn of Western science in Japan*, 1969. See also Ma, 'Japan's encounter with Western medical science: "The beginning of Dutch study"', 1959. My thanks to Dr Sachiko Kusukawa for calling this to my attention.

Sugita recalled that he and friends had repeatedly bewailed the lack of translations of such Dutch works: 'even just one volume of them, if rendered into Japanese, would be of immense value to the nation!' (p. 25)

Johann Kulm (Kulmus, 1689–1745), whose book Sugita now owned, was actually German, not Dutch, having been born in Danzig, where he first published his anatomical tables in German in 1722. The copy that Sugita acquired in Japan was of the 1734 edition in Dutch, published in Amsterdam. The work was also published in Latin in Amsterdam in 1732 (and in 1735 in Utrecht, 1744 in Amsterdam, 1765 in Rome), and additionally in a version translated from Latin into French published in 1732 and 1734, in Amsterdam. In the end it received twenty-three editions between 1722 and 1814, including the Japanese one whose origins we are discussing here. It was clearly being used by medical students as a straightforward and uncontroversial text for study.[2]

Shortly after acquiring his copy of Kulmus, Sugita was able to witness a dissection:

> It was, as I remember, on the night of March 3, 1771, I received a letter … inviting me to be present at a dissection … at Kotsugahara, Senju. (p. 26)
>
> …
>
> All of us together arrived at the designated place in Kotsugahara. The executed body to be dissected was of a female criminal about 50 years old who … had committed a heinous crime, we were told. An Eta [a kind of untouchable] was to perform the task … He said that he had performed a number of dissections ever since his youth [he was 90 years old]. In dissecting the human body, the custom till then was to leave everything to such outcast people. They would cut open the body and point out such organs as the lungs, the liver and the kidneys, and the observing doctors simply watched them and came away. All they could then say was: 'We actually viewed the innards of a human body'. With no sign tag attached to each organ, all they could do was listen to the dissector's words and nod.
>
> On this occasion too, the old man went on explaining various organs such as the heart, the liver, the gall-bladder and the stomach. Further he pointed to some other things and said: 'I don't know what they are, but they have always been there in all the bodies which I have so far dissected'. Checking them later with the Dutch charts, we were able to identify them to be the main arteries and veins and suprarenal glands. The old man also said: 'In my past experience of dissection, the doctors present never showed puzzle or asked questions specifically about one thing or another'.
>
> Comparing the things we saw with the pictures in the Dutch book … we were amazed at their perfect agreement. There was no such division as the six lobes

2 Sachs, 'Die "Anatomische Tabellen" (1722) des Johann Adam Kulmus', 2002. See also Myasita, 'Bibliography of Dutch medical books translated into Japanese', 1975, pp. 18–19.

> and two auricles of the lungs or the three left lobes and four right lobes of the liver mentioned in the old [Chinese] medical books. Also, the positions and the forms of the intestines and stomach were very different from the traditional descriptions. The Shogun's official doctors … had beheld dissections seven or eight times before, but always what they saw were different from what had been taught in the past thousand years … On this basis, I suppose, they had written that perhaps the Chinese and the Japanese were different in their internal structures. (pp. 29–31)

What is most striking here in this account is that human dissection was not unknown in Japan, that when it was performed it was carried out by low-caste people at the behest of the authorities, and that doctors – trained of course in the Chinese medical system – would sometimes attend, but either see no difference between the text and illustrations of the Chinese medical books with which they had been brought up and the dissected Japanese body in front of their eyes, or not know what to do with the difference between text and body that they did observe. Seeing a discrepancy does not, in itself, seem to have moved any of these Japanese doctors to any emotion beyond mild surprise or to any action beyond quiet nodding.

This is all in accord with Western experience in anatomy some two centuries earlier (as we saw in Chapter 1 above), when the onlookers, primarily learned doctors of medicine and medical students, also watched while a lower-status person – a surgeon – did the menial work of dissection. There had been a similar reliance on the authoritative nature of traditional texts (those of Galen and Mondino), which had meant that the onlookers did not feel called on to trust their own judgement about what they were seeing at anatomical demonstrations in the human body nor to question whether it differed from the canonical text. Texts always seem to prevail over eye-witness, unless there is some other, external, prompt to encourage people to make comparisons between text and nature, and then to take any discrepancy seriously. Vesalius in sixteenth-century Europe had needed such an external prompt quite unrelated to anatomy, and then – but only then – was he able to claim that the body was the true book that should be studied, and that the ancient text should be trusted only insofar as it accorded with the dissected body in front of one's eyes. Vesalius also showed the students of this discipline that they should ignore the head/hand distinction, and instead he applied his own hand to cutting the body.[3]

A strange inversion of this experience was now happening here in eighteenth-century Japan. For here bodies had been available for some years for dissection, but no one took seriously any discrepancy between what could be seen within them and the authoritative ancient texts. But Sugita and his fellow doctors had first become interested in the forbidden fruit of Dutch learning in general, believing that it would offer significant rewards to the Japanese nation. This was the external

[3] Cunningham, *The anatomical renaissance*, 1997, chapter 4.

prompt in their case. Only then had they obtained a Dutch book that they could not read, but to whose illustrations and text they now ascribed high authority. So now they could see the dissected body effectively for the first time, but on this occasion through what was illustrated in a *text*, that is to say through superior (as they thought) Dutch eyes. And only now that they could see the printed *text* were they able, at last, to see with their own *eyes* the true nature of the bodies that had been available in dissections for some years!

What was to be done?

> On our way home, three of us – Ryotaku, Jun-an and I – talked of what a startling revelation we had seen that day. We felt ashamed of ourselves for having come this far in our lives without being aware of our own ignorance. How presumptuous on our part to have served our lordships and pretended to carry out our duties as official doctors when we were totally without knowledge of the true make of our bodies that should be the foundation of the art of healing! …
>
> Thus we talked and sighed. Ryotaku, too, said all was very true, he was in complete agreement. I broke the spell saying. 'Even this one volume of *Ontleedkundige Tafelen* – suppose we translate it – many facts about the body will be clarified and the art of healing will be greatly benefited. I would like, by some way or another, to read this book without the aid of a Nagasaki interpreter'. (pp. 31–2)
>
> …
>
> Next day we gathered at Ryotaku's house. We talked over the experience of the day before. Then we faced the book. But it was as though we were on a boat with no oar or rudder adrift on the great ocean – a vast expanse and nothing to indicate our course. We just gazed at each other in blank dismay. (p. 33)

Sugita did not even know the letters of the Dutch alphabet, though his friend Ryotaku at least knew a little. So initially they began with the external parts of the body, as illustrated in Kulmus's book.

> But coming to such auxiliaries as 'de' (of), 'het' (it), 'als' (as) and 'welke' (which), we were completely at sea. We had a fragmentary knowledge of Dutch words, but their connections in a sentence were always puzzling to us. For example, we once came across the sentence: 'The wyn-braawen (eyebrows) are the hair growing above the eyes'. To us its meaning remained hazy in spite of a long spring day being spent on it. It was not rare that we would be thinking and staring together till sunset and yet unable to decipher a line or even a sentence one or two inches long. (p. 34)
>
> …
>
> After the passage of one year or so [meeting six or seven times a month], our vocabulary was enlarged, and naturally we became informed of Dutch things in general. By and by we found ourselves capable of translating as many as ten lines a day without much trouble when the language was not very complicated.

> Needless to say, we did take questions to the interpreters who came to Edo annually. Also, between times we attended the dissections of human bodies and oftener we opened animal bodies to confirm what we read. (p. 36)

Finally, after eleven re-writings, the text was ready for the printers, and it appeared in 1774 in Edo (Tokyo) as *A new book of anatomy* in five volumes (Figs 6.1 and 6.2), in the second of which the Vesalian origin of the image is still strongly visible.

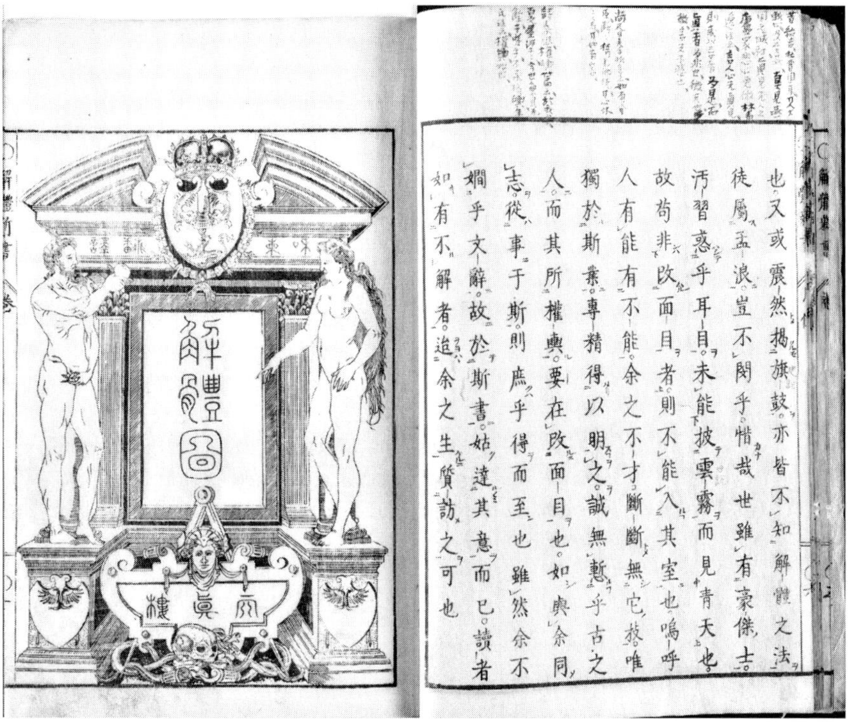

Fig. 6.1

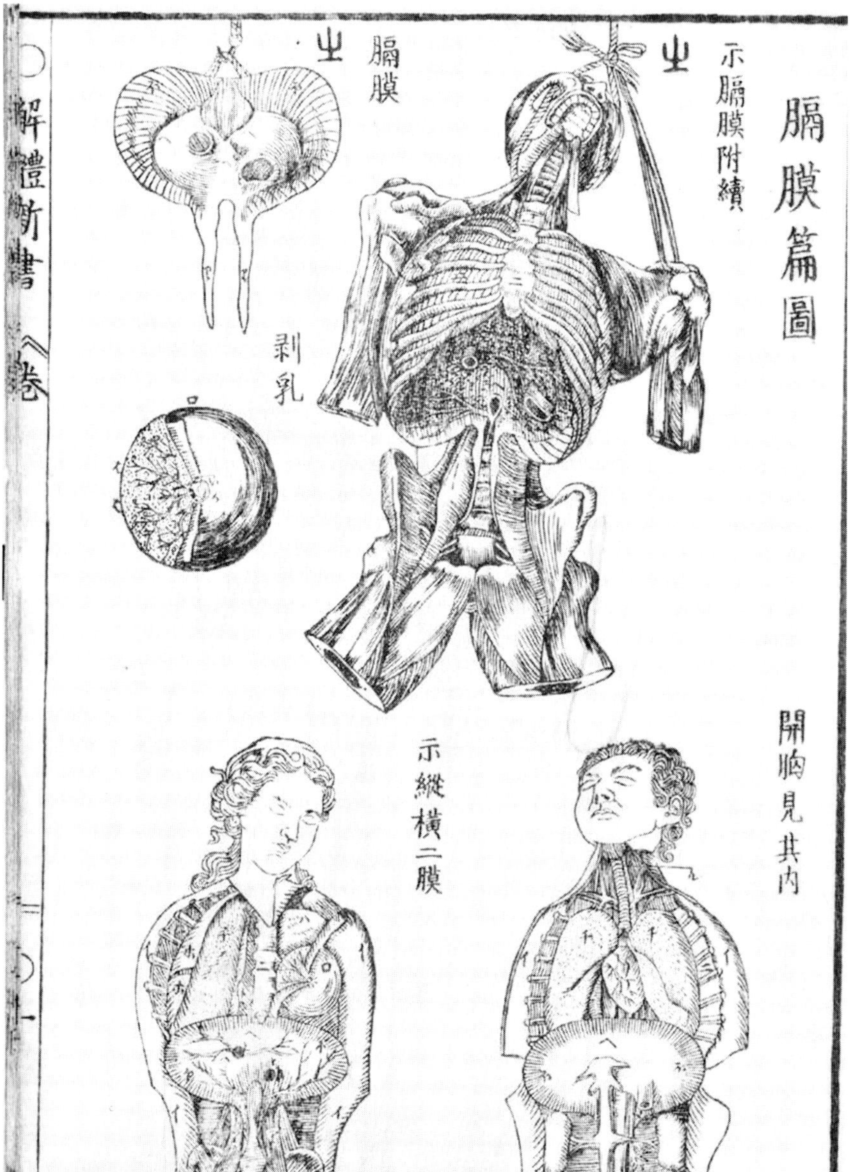

Fig. 6.2

2. 'The perfection of anatomy'

As these earnest Japanese scholars were trying, with agonising slowness, to catch up with the state of European and Dutch anatomical knowledge of fifty years before, the position of anatomy back there in the West, half a world away, was at a turning-point.

On the positive side it could be rightly said that the mid- to late eighteenth century was, as Albrecht von Haller, one of the greatest anatomists and teachers of that age called it in the 1770s, the period of 'the perfection of anatomy'.[4] It was a widespread sentiment among anatomists of the late eighteenth century that their discipline was at its peak, as we saw in Chapter 1. Indeed, so great was everyone's awareness of the achievements made in this period of the perfection of anatomy that only a few years later, in the early years of the nineteenth century, the extra-mural anatomy teacher at Edinburgh and celebrated wit, John Barclay, could use a wonderful image to describe to his students the current state of anatomical investigation. He was advising his students to beware of believing they had made anatomical discoveries and rushing with them into print:

> Anatomy may be likened to a harvest-field. First come the reapers who, entering upon untrodden ground, cut down great store of corn from all sides of them. These are the early anatomists of modern Europe, such as Vesalius, Fabricius, Fallopius, Malpighi, and Harvey. Then come the gleaners, who gather up ears enough from the bare ridges to make a few loaves of bread. Such were the anatomists of last century – Valsalva, Cotunnius, Haller, Winslow, Vicq d'Azyr, Camper, Hunter, and the two Monros. Last of all come the geese, who still contrive to pick up a few grains scattered here and there among the stubble, and waddle home in the evening, poor things, cackling with joy because of their success. Gentlemen, – we are the geese.[5]

In 1785 Jacques Tenon, Royal Professor at the College of Surgery and the Académie royale des sciences, put forward his 'Observations on the obstacles that hinder the progress of anatomy'.[6] But they were pretty trivial obstacles, concerning the pay and opportunities available to anatomists. Tenon argued that some people thought

4 Haller, *Bibliotheca anatomica,* 2 vols, 1774, 1777, vol. 2, book 9.

5 *The life of Sir Robert Christison*, 1885, pp. 71–2. The anecdote is not dated, but was probably told around 1815. My thanks to Wendy Moore for helping me relocate this reference. We have met all these anatomists in the present book except 'Cotunnius' (or Cottunni), who is probably Domenico Cotugno, who investigated the nervous system and the canals of the inner ear. The omission from this list of the third Monro, Barclay's contemporary in Edinburgh anatomical teaching, might be significant. Barclay's collections of 'comparative anatomy' were given to the Royal College of Surgeons of Edinburgh in 1828, where they may still be seen.

6 Tenon, 'Observations sur les obstacles', 1785.

that anatomy was so perfected that it was exhausted and had no more problems to solve; he claimed that it was in fact on the point of greater progress if only more people were to cultivate it and with more resources. But Tenon here makes no call for anatomising to take a different direction, nor for it to be freed from any intellectual shackles, nor for it to be made subservient to some other, more progressive, discipline. Basically Tenon is simply calling for more of the same anatomising tradition, but with improved career conditions. Publication of this discourse was approved by the members of the seven-man 'Class of anatomy' at the Academy, including Daubenton, Portal, Sabbatier, Vicq d'Azyr and Broussonet, who describe Tenon's discourse as 'suitable to excite the emulation of all those who cultivate anatomy'.[7]

Yes, anatomy was perfected, and all that hard-won anatomical *knowledge* would continue to be valid through to modern times. Indeed it is about the only thing concerning eighteenth-century medicine that would survive the changes in medicine brought about as a consequence of the French Revolution.

But on the negative side, just about everything about the *discipline* of anatomy and its status was about to be challenged and changed in the couple of decades around 1800. These changes were to constitute in effect the end of old anatomy. The discipline and the domain of anatomy that we have today is a mere shadow (I'd love to say a mere skeleton) of anatomising in the eighteenth century, since its content and relative position among the sciences of life have been changed to its detriment. It is no longer the star of the sciences of life. It would be nice to be able to claim that the enormous industry shown by the eighteenth-century anatomists of Europe, and particularly their singular obsession with how anatomising was the only way to establish *facts* about the body, laid the basis of the modern sciences of life. Unfortunately even this would not be justified. For while facts continued to be fetishised, anatomy was no longer to be the major science of life that generated them.

3. Fields of change

I shall deal successively here with some of the transformations that the discipline of anatomy underwent and which together comprised what I call 'the end of old anatomy'. Ideally each of these transformations deserves a book to itself, but here only a sketch of each can be given. I have chosen six fields to discuss, though these do not exhaust the subject, and I look at them here in differing degrees of detail.

---

[7] 'Sabbatier' is probably Raphaël Bienvenu Sabatier (1732–1811), who published on anatomy and surgery. I am not sure of the identity of 'Brousssonet'.

Physics and chemistry

In my first field I want to look at the effect on anatomical investigation of the new dominance of the two disciplines of physics and chemistry.

In year 4 (1796), the French Institut national pour les sciences et les arts (the old Académie royale des sciences reconstituted in the course of the Revolution), offered a prize for physics. The topic was the physics of the liver: 'The comparison of the nature, form and uses of the liver in the different classes of animals'.[8] Earlier in the eighteenth century the investigation of the functioning of the liver would have been a topic for the anatomists, but as will be clear from how the question is set up, it is now to be seen as a question primarily of physics and chemistry. For, according to the prize conditions, vegetables get 'the materials necessary for their organisation' from the air, water, and minerals. Then,

> ... animals eat vegetables (or other animals that themselves have eaten vegetables). Finally, fermentation, putrefaction and combustion return to the air and to the mineral world the principles that the vegetables and animals have borrowed from them. Through what process of nature does this circulation between the three kingdoms occur? How does nature manage to form fermentable, combustible and putrescible substances from materials that have none of these properties? It's clear that *la végétation* [the process of turning materials into vegetables] and *l'animalisation* [the same into animals] are the inverse processes to combustion and putrefaction. (p. 4)

The food taken into the mouth is subject to several conversions in the body, through saliva and gastric juice, through being taken into the blood stream, and being turned into evacuations. The prize essay will deal with just the role of the liver and its bile in the process of 'animalisation'. It is well known that the liver is important, that a part of the abdominal vascular system is dedicated to it, that the blood there is in a condition to secrete bile, that the bile is constant and regular and necessary for the integrity of all functions; that the liver exists in virtually every animal; that it does or does not have a gall-bladder; that there are essential relations between the spleen, the pancreas and the liver. 'These are the essential facts [*les premières données*] that anatomy has long offered to the speculation of physiologists; but until now they have been virtually sterile in application' (p. 8). The Institut knows this is a big subject: it requires extensive anatomical knowledge, comparative anatomy, chemical researches using the new methods of analysis, and so on: 'The Institut believes that the time has come to approach the complicated questions that the phenomena of the animal economy present, and that it is by the united efforts of physics, anatomy and chemistry that we can now look forward to the solution of these great questions' (p. 9).

---

8 Cuvier, *Programme des prix de l'Institut national des sciences et arts, proposés l'an IV de la République*, 1796 (year 4).

We can see here that the 'essential facts' that anatomy offers are merely ones of structure and connection. Anatomy is not presented here as an experimental discipline: the truly experimental sciences are now physics and chemistry, and it is through these that answers to 'the complicated questions that the phenomena of the animal economy present' are to be sought.

It is quite likely that such interpretations of anatomical function through physics and chemistry had been going on for a decade or more in France, especially following Lavoisier's work on respiration, beginning in the 1770s. But possibly the uniting of all the nature sciences into one Class of the Institut had some effect in hastening both this promotion of physics and chemistry as the pre-eminent experimental sciences, and the demoting of anatomy.

All this change of hierarchy among the sciences of nature was later confirmed by the 1808 report by Cuvier on scientific progress. Napoleon had declared himself Emperor in 1804 and a series of reports was made to him as Emperor on the progress of the arts and sciences since the Revolution. They were meant to be a declaration of the achievements and legacy of both the Revolution and Napoleon.

Cuvier wrote the report on 'les progrès des sciences naturelles', which included anatomy. The three parts of this report are on (1) chemistry, under which the whole of physics is subsumed, (2) natural history, which is presented as running from the natural history of the atmosphere, minerals and geology, to the natural history of living bodies, covering anatomy and physiology, and (3) applied sciences, meaning medicine, agriculture and technology. Cuvier was himself a very active agent in the reconfiguring of the old sciences and the making of new ones, so he is an interested party in what he says about anatomy and the other disciplines. But on the other hand, since Cuvier was a heavily involved party, so his views express both what had happened in the sciences and what he and others still wanted to happen to them.

Having dealt with the progress made in mineralogy and geology, Cuvier turned to living bodies:

> In a word, the *form*, which has no influence in the history of the atmosphere and waters … becomes, in the study of living bodies, the dominant consideration, and gives to anatomy a role just as important as that of chemistry; and these two sciences are becoming the necessary and joint [*simultanés*] instruments of all the researches that remain to be discussed.
>
> The first thing that strikes us about the study of life is this power of organised bodies for attracting foreign substances into their whirl, for retaining them for some time afterwards to assimilate them, and finally for distributing these substances, now made their own, into all the parts, according to the functions that they must carry out.
>
> This power presents three objects of study.
>
> It is necessary to inspect which materials these living bodies attract, and which they expel. The residue will form their own substance. This is the *chemical* part of the problem. Then it is necessary to describe the routes that these materials

> traverse from their entry into the living beings to their exit. This is the *anatomic* part of the problem.
>
> Finally it is necessary to examine by which forces these materials are attracted, retained, steered and expelled. One could call this research the *dynamic part* or more properly the *physiological* part.
>
> … The anatomic part of the general problem of life has long been worked out for animals … The routes of the chyle and blood are now clear; the eye can follow all the detours, and they encounter everywhere valves or other indicators that mark the direction of flow; the eye also notices how the routes, so complicated in man, simplify themselves by degrees in inferior animals, and end by reducing themselves to a uniform sponginess. The researches of Monsieur Cuvier have succeeded in assigning each animal its place on the great ladder of the complications of structure.[9]

So in this account of the proper study of living beings, put forward by Cuvier, anatomy still holds one role out of three, but only to provide information about the general structures and routes visible to the naked eye. The really interesting events now are what happens in these anatomical locations – and these events lie in the domains of chemistry and a physics of forces.

Here Cuvier is in the role of propagandist for the past and future progress of the natural sciences. We shall see him soon in other, more activist, roles in reshaping old anatomy.

Experimental physiology

Our next field is the creation of the discipline and practice of experimental physiology, which rapidly took over experimentation from anatomy, and whose early practitioners reclassified anatomy as a dead and sterile science, merely preliminary to physiology. Again this involved making physics and chemistry the models of true sciences of nature. It was François Magendie who claimed the credit for the creation of experimental physiology, and who wrote the first textbook of the subject in 1816, laying out the principles of the discipline and presenting his own experiments as exemplary of it.

How and why did Magendie invent this new discipline in the first place? He had been trained in medicine and as a young man was already giving classes in anatomy in Paris. Everyone thought he was going on to become a surgeon. But he was also very ambitious, and wanted to make his mark in the sciences. He mixed with some of France's most eminent investigators, and was welcomed into the Académie des sciences (the Institut reconverted), where one day he heard Pierre Laplace, the great astronomer, say 'the two sciences most worthy of the attention of superior intellects are physiology and astronomy – and physiology lacks its

---

[9] Cuvier, *Rapport historique sur les progrès des sciences naturelles depuis 1789, et sur leur état actuel*, 1810, pp. 152–3, 157–8.

Newton'.[10] Laplace thought he was the French Newton of astronomy, so this was a challenge indeed for Magendie: to be the Newton – or even the Laplace – of physiology! Obviously physiology would have to be rethought from the bottom up. It would have to be reformed, in Magendie's view, along the lines of physics and chemistry, which in the course of the eighteenth century had come to be centred round experiment. Magendie's object now was to do the same for physiology, 'and reduce it entirely to experiment: in a word, to impart to that science the happy renovation which has taken place in the natural sciences'.[11] No longer would physiology be 'a long and tiresome romance', but a proper science like physics and chemistry, and therefore, like them, it would have to have *experiment* at its core. And as the subject that physiology dealt with was the *live* human or *live* animal, then – for Magendie – *that* had to be the experimental object. The animal had to be *vivisected.* As Magendie used to warn his pupils when they first entered the laboratory, 'Les chiens ne s'amusent pas ici' – 'The dogs don't like it here'.

Actually Magendie's ambition for physiology was not just that it should be rebuilt on an experimental basis, but that in turn it should be the basis of a whole new *experimental medicine*. So for Magendie and his student Claude Bernard, what was required was an experimental physiology, to establish how the body works in health; an experimental pathology, to establish how disease occurs in the living body; and an experimental therapeutics, establishing what drugs and other means work in the living body. All this was to be built on the *sacrifices* – it's the technical word – of live experimental animals in the laboratory.

In 1865 Magendie's student and successor, Claude Bernard, published his famous book *An introduction to experimental medicine*, which laid out this position at length –

> Scientific medicine, like the other sciences, can be established only by experimental means, that is, by direct and rigorous application of reasoning to the facts furnished us by observation and experiment … Experimentation is undeniably harder in medicine than in any other science; but for that very reason, it was never so necessary … Nothing, I believe, is today so important for the progress of medicine.[12]

One of Magendie's earliest set of experiments was on the role of the stomach in vomiting, and must stand here for the rest. Is the stomach active or passive in vomiting? In his introductory textbook he briefly recounted some of his experiments:

---

[10] Quoted without source in Olmsted, *François Magendie*, 1944, pp. 19–20.

[11] Magendie, *Elementary compendium of physiology*, 1826, author's preface to second edition, p. xvii.

[12] Bernard, *An introduction to the study of experimental medicine*, 1927 (1865), author's preface, pp. 2–3.

> Vomiting was long believed to depend upon the rapid convulsive contraction of the stomach; but I have shewn, by a series of experiments, that, in this process, this organ is nearly passive; and that the true agents of vomiting are, on the one hand, the diaphragm, and, on the other, the large abdominal muscles; I have even succeeded in producing it, by substituting, for the stomach, in a dog, the bladder of a pig, which I afterwards filled with a coloured liquid.
>
> In the ordinary state, the diaphragm and the muscles of the abdomen co-operate in vomiting; but each of them can, nevertheless, produce it separately. Thus, an animal still vomits, though the diaphragm has been rendered immovable by cutting the diaphragmatic nerves; it vomits in the same manner, though the whole abdominal muscles have been taken away by the knife, with the precaution of leaving the linea alba and the peritoneum untouched.
>
> I never have seen the stomach contract in the instant of vomiting; we may conceive, however, that the motion of the pylorus may probably take place at this instant. This circumstance presented itself twice to Haller, and made that illustrious physiologist conclude, that the contraction of the stomach was the essential cause of vomiting.[13]

Magendie's experiments to establish the situation are typical of his approach. The first thing that is consulted is the experimental animal, not the anatomical facts, nor any previous authority. The methods Magendie uses are strikingly interventionist, especially if we remember that the animals were not anaesthetised. Each of the candidate parts of the body is successively removed or rendered immobile, and then vomiting is induced. The stomach itself is removed from one animal and replaced with a pig's bladder; the diaphragmatic nerves are cut in another animal; the abdominal muscles are cut out in yet another. The experiments are repeated several times, and in different combinations, in different animals. At the end the experimental physiologist can make what he believes is a Baconian induction about how vomiting is brought about: viz., by the diaphragm and the muscles of the abdomen, acting together or separately. The conclusion is precisely the opposite of the one that the 'illustrious physiologist' Haller had come to – who of course had been an old physiologist, not an experimental physiologist. The experiments Magendie carried out were actually far from random, but one of their most important characteristics is that they start from a position of presumed ignorance about the answer. The approach is true to Magendie's resolve, recorded by his first biographer, to 'admit no facts that do not find their confirmation in experiments competent for himself to repeat'.[14]

We can take the experiments of Pierre Flourens on the brain as another early instance of how the new experimental physiologist practised this new discipline. The young Flourens, a protégé of Cuvier, published in 1824 in Paris a report on his recent researches entitled 'Experimental researches on the properties and

---

13 Magendie, *Elementary compendium of physiology*, 1826, pp. 286–7.

14 Flourens, *Eloge historique de François Magendie*, 1858.

functions of the nervous system in vertebrate animals'. In the old system of things, experimental researches on the properties and functions of the structures of the body had been the domain of the old anatomist, not of the physiologist at all. In this report Flourens describes the experiments he had conducted on birds in order to discover the functions of the lobes of the cerebrum, and the functions of the cerebellum. He worked, like Magendie had done in investigating the action of the stomach, by removing the relevant organ in the living animal, on this occasion slice by slice. The assumption here is quite distinct from that of the old experimental anatomist: Flourens does not investigate the structure of the brain in order to deduce function, and then conduct experiments on the living animal to test his deductions. Rather, he begins by intervening and mutilating the organ in order to see what function is thereby *lost*, and from that makes a deduction as to the function of the removed parts. As he said, 'the end and goal of all physiology and pathology is to deduce the alteration of the parts from the alteration of the properties, and, reciprocally, the lesion of the properties from the lesion of the parts'.[15]

One experiment will illustrate these points. In his account of his experimental investigation of the function of the cerebellum (or 'hind-brain', as it was sometimes called in English), Flourens has been describing the effect of successive ablation experiments on three pigeons, and discovered that co-ordination of movement was progressively lost as the slices of cerebellum were removed. He now took his fourth pigeon:

> In a fourth pigeon I removed the upper layers of the cerebellum. When this mutilation was accomplished the animal saw and heard very well; it also kept itself upright, walked, and flew, but in an indecisive and very unsure manner.
>
> I continued my excisions; equilibrium was almost completely lost. The animal had very great difficulty in standing up and could do so only by supporting itself on its wings and tail. When it walked its staggering and unsure steps gave it the appearance of a drunken animal; its wings had to be used to help its legs, and despite this assistance it often fell over and rolled about.
>
> With the removal of the last layers [of the cerebellum], all semblance of equilibrium, that is all harmony between its efforts, disappeared. Ability to walk, fly, and be stationary were totally abolished ....[16]

This is how function is discovered by the experimental physiologist: by successive and repeated vivisection and mutilation of the experimental animal. Or, as Claude Bernard later expressed it, 'instead of proceeding from the organ to the function' as

---

[15] As translated by Young in *Mind, brain and adaptation*, 1970, p. 61, from Flourens, *Recherches expérimentales*, 1842 (1824), p. 57.

[16] Flourens, *Recherches expérimentales*, 1842 (1824), as translated in Clarke and O'Malley, *The human brain and spinal cord*, 1996 (1968), pp. 656–9.

old anatomists had done, experimental physiologists 'start from the physiological phenomenon and seek its explanation in the organism'.[17]

This new discipline crossed – or rather, ignored – the discipline boundaries of both the old disciplines. It had some characteristics of the one (the experiment of old anatomy) together with some characteristics of the other (the explanation of the 'grand functions' of old physiology), but it was actually a new and distinctive discipline, with new goals, standards, procedures, ideology and products.

Comparative anatomy and the remaking of Natural History

Our third field of transformation of old anatomy is in the realm of comparative anatomy. The increasing importance by the end of the eighteenth century of the domain of Natural History and the radical changes it was experiencing directly affected comparative anatomy, effectively making it a discipline and a profession, and with new goals and methods.

The major events were all centred in the Jardin royal in Paris as it was transformed in the course of the French Revolution into the Muséum d'histoire naturelle.[18] The Muséum was founded in 1793 and it took over the site and all the properties and collections of the Jardin. It was the only royal institution (apart from the Collège royal) that the revolutionaries did not demolish, dismantle or replace. Instead, they enlarged and extended the original institution while renaming it, giving it several more permanent professorships, and steering it towards new goals. In a word they transformed it from an *ancien régime* institution for the pursuit of *ancien régime* Natural History, into a Revolutionary institution for the pursuit of a revolutionised Natural History, including a new kind of comparative anatomy.

Of course the Revolutionary committee that made this decision did not dream up this new scheme for themselves. What happened was that they accepted a scheme that had been under development by the officers of the Jardin since 1790, under the leadership of Daubenton, who now put forward the Jardin as a place promoting the ideals of liberty typical of the revolutionaries. 'Shall the tree of liberty be the only one that cannot be naturalised at the Jardin des plantes?', asked the officers of the Jardin in their first address to the National Assembly.[19] In their scheme the officers promised a new form of education for a new generation, turning the garden into a school 'designed to form in the Nation and for the Nation men capable of improving all the arts'.[20] The new Museum, they claimed, 'will soon be able to become a kind of metropolis of all the sciences useful for agriculture, commerce, and the arts'.[21] It was a remarkable opportunity to seize

17 Bernard, *An introduction to the study of experimental medicine*, 1865, p. 106.

18 See the important discussion of these changes in Corsi, *The age of Lamarck*, 1988, chapter 1, 'Schemes for reforming "Histoire naturelle"'.

19 Hamy, 'Les derniers jours du Jardin du Roi', 1893, p. 100.

20 Ibid., p. 106.

21 Ibid.

the initiative and to reconstruct a grand discipline – Natural History – together with its newly appropriate sub-disciplines, all at once. Natural History was to be a discipline to serve the Revolutionary French nation, producing products and knowledge for use, and men educated to develop the nation. In an audacious move in their representations to the National Assembly, the officers of the Jardin claimed that theirs was now an institution that pursued not only the Revolutionary goal of liberty, but also that of equality. Therefore, they argued, they needed no Intendant in the new institution, but instead they should all be of one and the same status: professors! The professorships were all duly established on 10 June 1793, thus raising into formal chairs all the old officers of the Jardin.

The most important novelty in the concept and creation of the research and teaching posts consisted in the new positions for the study of animals. Human anatomy had of course been taught in the Jardin since the 1670s at least.[22] But now there were to be three (soon four) new additional anatomical professorships. The first of these, that of 'The Zoology of the Quadrupeds, Cetacea, Reptiles, Birds and Fish', was divided into two posts within less than two years. Of these, (1) the Chair of *The Natural History of Mammals and Birds* was held by Geoffroy Saint-Hilaire until 1841, while (2) that of *The Natural History of Reptiles and Fish* was held by Lacépède until 1825. The next new position (3) was *The Zoology of the Insects, Worms, Microscopic Animals etc.*, held by Lamarck until 1829. The Chair of *Human Anatomy* (4) was held by Antoine Portal (erstwhile historian of anatomy and anatomist to the Dauphin) until 1832. Finally there was the chair of greatest interest to us here: that of *The Anatomy of Animals*, held at first by the aged Mertrud, but from October 1802 by Cuvier, when it was renamed – presumably by Cuvier himself – (5) the Chair of *Comparative Anatomy*.[23] Cuvier held this professorship until his death in 1832.

From the specification of the duties of the holders of the chairs of human anatomy and of the anatomy of animals in the 1793 regulations, it is clear that an old (that is, eighteenth-century) view initially still prevailed of what constituted comparative anatomy. The professors 'will take care between them to ensure that the teaching of these sciences is as complete as possible; to acquaint the students with the most recent discoveries; to elucidate the structure of man by the structure of animals; to offer, in some detail, the facts of comparative anatomy that have not

22 Hamy, 'Recherches sur les origines de l'enseignement de l'anatomie humaine', 1895, and see above, Chapter 2.

23 For the professors, the chairs, and the changes to them over the first century, see *Centenaire de la fondation du Muséum d'histoire naturelle*, 1893, pp. iii–vii. The other new chairs, in addition to those of anatomy, were Mineralogy (Daubenton), General Chemistry (Fourcroy), The Chemical Arts (Brongniart), Botany in the Museum (Desfontaines), Botany in the Field (A.-L. de Jussieu), Plant Cultivation ('*Culture*') (Thoüin), Geology (Faujas de Saint-Fond), and Iconography [of plants and animals] (Van Spaendonk).

hitherto been gathered together in any course'.[24] More precisely the Professor of Comparative Anatomy was

> To give in the first lectures a general idea of the internal organisation of the different classes of animals; he will choose in each class those animals whose internal anatomy seems to him suitable for making an anatomical demonstration, and he will seize every occasion to put before the eyes of the students those animals whose organisation is less known. He will put particular emphasis on comparative anatomy, both between different animals, and between animals and man.[25]

The lessons were to take place in the amphitheatre of the Museum. It is clear that this kind of comparative anatomy was the eighteenth-century kind: the elucidation of the anatomy of man from that of animals, together with the anatomy of animals for their own sake and for the point of contrast. But that lasted just a couple of years, until Mertrud resigned, and the post was given to the young Georges Cuvier (1769–1832).

What was the role of the menagerie in this transformation? In the proposed regulations of 1793, such an institution was still only a project, and talk was still in terms of 'when the Museum will have the means to maintain in a menagerie living animals of different classes'. Come that day, it was anticipated that the care of the menagerie would be in the hands of the two professors of zoology, whose responsibilities would be to describe the animals, to study their habits, to try to acclimatise them, to breed them, and to distribute the useful species.[26] When the remnant animals of the menagerie of Versailles were hurriedly transferred to the Museum in April 1794, the new menagerie was actually put under the charge of Geoffroy St.-Hilaire, the professor of mammals and birds. One might say that it was the creation of the menagerie in the Museum that more than anything else validated the move by its professors to turn the Jardin into a Museum of Natural History, for now the final element, *live animals*, had been brought together with the living and dried plants.

Of course the dry and wet preparations of animals, together with the minerals, had already long been in the garden within the royal cabinet, as assembled by Buffon and Daubenton. But Cuvier now preserved the skeletons and soft parts of the animals that died in the menagerie. A building was acquired next to the amphitheatre, right next to the house of the Professor of Comparative Anatomy, with two great rooms to bring all the stuff together and display it. The preparations were now arranged according to 'les vues physiologiques', that is to say, not by animal, but by the organs whose structure they elucidate ('selon l'ordre … des

---

24 Hamy, 'Les derniers jours du Jardin du Roi', 1893, pp. 109–10.

25 Ibid., p. 150.

26 Ibid., p. 158.

organes dont elles éclaircissent la structure').[27] By 1803 there was a total of nearly three thousand preparations.

Cuvier it was who established a new science of comparative anatomy – not the Darwinian science of comparative anatomy, of course, but nevertheless one radically different from what had been pursued under this title in the long eighteenth century, which we discussed at length in Chapter 5 above. Given the chronological limits of the present study, this is not the place to give a full characterisation of this new science that in effect now became the basis (rather than the handmaid) of the Natural History of animals for more than half a century. However, it is necessary to outline what was so different about it. For it was indeed very distinct, so much so that it would be unwise of us even to see it as a direct successor of any of the eighteenth-century practices of comparative anatomy, or even of the animal studies of Buffon.

In 1817, in the Preface to *The animal kingdom*, the *Règne animal*, Cuvier pithily defined what comparative anatomy had meant for him for the previous thirty years:

> Having been devoted, from my earliest youth [i.e. the late 1770s and early 1780s], to the study of comparative anatomy, that is to say to the laws of the organization of animals, and of the modifications which this organization undergoes in the various species, and having, for nearly thirty years past, consecrated to that science every moment of which my duties allowed me to dispose, the constant aim of my labours has been to reduce it to general laws, and to propositions that should contain their most simple expression.[28]

I draw attention to this definition of comparative anatomy as the study of 'the laws of the organization of animals, and of the modifications which this organization undergoes in the various species'. None of our practitioners of eighteenth-century comparative anatomy had spoken in these terms, nor of having the goals of seeking *laws* underlying *animal organisation*, nor of systematically seeking the *modifications* that this organisation undergoes in the different species. Cuvier is explicit that this new comparative anatomy is a proper science and that it seeks general laws. While these criteria for what constitutes a proper science were new for comparative anatomy, they were becoming common to most of the sciences of nature at this time.[29] The uncovering of such natural laws for comparative anatomy (and indeed for Natural History as a whole) Cuvier believed would make the world of the organic as close to an experimental science as possible – and the model of experimental science for Cuvier was of course now physics or chemistry.[30]

---

27 Cuvier, 'Notice sur l'établissement de la collection d'anatomie comparée du Muséum', 1803, p. 412.

28 Cuvier, *Essay on the theory of the earth*, 1817, p. 1.

29 Cunningham and Williams, 'De-centring the "Big picture"', 1993.

30 Outram, 'Uncertain legislator: Georges Cuvier's laws of nature', 1986.

The idea that anatomy was itself an experimental discipline had been completely discarded.

Having created this new view of what the science of comparative anatomy ought to be, Cuvier was able to use it (1) to establish a new classification of living creatures, based on the comparison of their internal and external characters; (2) to establish the laws of 'the correlation of parts' and 'the subordination of characters', which showed which parts could and did go with which; (3) to reconstruct extinct species (for the first time ever) from as little as one fossil bone; and (4) to recognise that the earth itself had gone through a series of revolutions over time, that is that the earth itself had a history. This new comparative anatomy was clearly a most powerful discipline in Cuvier's hands, and he tried to make it the core discipline of Natural History.[31]

The spirit of emulation that was built into the teaching and research structure of the Museum had lasting effects. In fact the occupants of the anatomy posts were for over thirty years in constant rivalry with each other, each of them extremely active in constructing new theories about Natural History in general, about life, its origin, and the shapes and embodiments that it had taken since it originated, all these theories being built on the study of anatomy in its widest extent, but from different viewpoints. More particularly, each of their grand views of nature was built directly and literally on the interests of the particular anatomical chair that each professor held.

Thus Geoffroy St.-Hilaire saw *varied standards* (*archetypes*) everywhere in nature, all displaying many modifications; this view originated (in part) from his work on vertebrates and particularly the vertebrate skeleton. Lamarck's vision of organisms by contrast was *dynamic*. He saw nature as full of simple organisms striving to meet the changes in their environment, and hence developing in different ways, becoming increasingly complex; it was from his position as professor of worms and other 'invertebrates' (as he was the first to call them) that Lamarck looked up and up towards more 'complex' forms, and asked how this complexity developed from basic simplicity. Cuvier, in another contrasted version of Natural History, held to a view of the *stability* of the species, classes, families and genera of nature; this, and also his view of the revolutions that the earth must have undergone over time, he built on his own new speciality of comparative anatomy, which was the subject of his professorial chair, a comparative anatomy in which every creature was so exquisitely suited to its environment and way of life that it could not change an iota without becoming extinct. In one way or another these rival theories, created in the Museum anatomy chairs, shaped the work of many other naturalists elsewhere, including Charles Darwin, out of whose work the modern discipline of comparative anatomy was in time to be developed.[32]

---

[31] Coleman, *Georges Cuvier, Zoologist*, 1964; Outram, *Georges Cuvier*, 1984.

[32] For instance, Cuvier's approach influenced Carl Gustav Carus in his 'romantic' approach to comparative anatomy, in which he attempted to fulfil 'the principal object of Comparative Anatomy; viz. to give a history of organization, in its successive gradations of

Clinical medicine and a new pathology

Despite all the enormous effort and dedication put into the development of a properly anatomically based pathology by Morgagni and his circle and by others in the long eighteenth century, including Portal and Baillie, this sub-discipline too was turned into something quite different in the early decades of the nineteenth century, and this is my fourth field of transformation of old anatomy. Pathology based on the seats of diseases was transformed in the course of creating 'Paris hospital medicine', whereby pathology became a bedside – or, to put it in Greek, a *clinical* – subject, rather than a theatrical one. Indeed, this new discipline of pathological anatomy was to be so different from the sub-discipline we were looking at in Chapter 3 above, that its practitioners took almost no notice of the century or more of observations, dissections and reasonings that had been so laboriously gathered together and printed by the eighteenth-century practitioners.

For when we look at the era of 'Paris hospital medicine', recalled to fame in recent years by Erwin Ackerknecht and Michel Foucault,[33] Morgagni was certainly not treated as the 'father-figure' for this Revolutionary attempt to correlate the clinical signs and symptoms with the pathological changes during life that are otherwise invisible to the eye.

René Théophile Laennec, who invented the stethoscope in order to investigate the state of the organs of the thorax during life, defined the new pathological anatomy like this in 1812:

> Pathological anatomy is a science that has for its goal the knowledge of the visible alterations that the condition of illness produces in the organs of the human body. Opening of corpses is the means of acquiring this knowledge; but, to ensure it becomes of direct value and of immediate application to practical medicine, it is necessary to join to it the observation of the symptoms or the alterations of functions that occur together in each type of alteration of organs.[34]

One of the most important participants in the creation of 'Paris hospital medicine' was Jean-Nicolas Corvisart, who was the promoter of the technique of percussion (tapping the chest to hear the state of the organs within). Corvisart worked at the hospital of La Charité and had adopted the heart as his special organ of clinical research, seeking to co-relate the symptoms and course of diseases with the lesions found in the heart after death. He reckoned that 'the most frequent organic diseases ... are those of the heart'. In his classic book of 1806, *On the organic diseases and*

---

perfection, by a description of the varieties in the structure of individual parts of the animal creation' – Carus, *An introduction to the comparative anatomy of animals*, 2 vols, 1827. On Romanticism in comparative anatomy, see Rehbock, 'Transcendental anatomy', 1990.

33 Ackerknecht, *Medicine at the Paris hospital*, 1967; Foucault, *The birth of the clinic*, 1973.

34 *Dictionnaire des sciences médicales*, vol. 2, 1812, pp. 46–61, see pp. 46–7.

lesions of the heart and great vessels, Corvisart put forward many case-histories to connect together the *symptoms*, the *course* of the disease, and the *lesions* of the organ of the heart discovered after death, in order to be able to establish recognisable *signs*. Here he's giving an account of what he calls 'aneurysm of the heart' – by which he means a pathological dilation of the whole heart or one of its chambers beyond the normal size:

> A wheelwright, aged fifty-eight years, of a strong constitution, felt, after a violent effort, a sharp pain in the right side of the thorax. Immediately there ensued oppression, cough, and spitting of blood. The extremities then became oedematous [that is, swollen with the accumulation of fluid] … the strokes of the heart violent, dry and hurried, but regular. The region of this organ sounded feebly [when percussed] … The patient died three months after the violent effort which had originated the phenomena of the disease … On dissection … the heart was twice its natural size. Its length, from base to apex, was more than eight inches. The walls of the heart very thick ….[35]

As a result of such investigation Corvisart could describe the *signs* by which the physician could in future diagnose this 'aneurysm of the heart' –

> It is commonly in the second period of the disease that the patients enter hospital for relief … *External examination*. The face is now bloated, the cheeks and lips florid, or inclining to purple … there is a swelling of the feet and ankles while standing, which commonly disappears during the night. The chest, when percussed, sounds equally clear in every part, except that of the heart, where it is usually silent, for a considerable extent … The patient cannot breathe in a horizontal posture ….[36]

So if a clinically trained medic were to see a patient with these characteristics, he would know what is wrong with the patient, on the inside, because he can see the signs and read them.

While Corvisart, the new pathologist, offered Morgagni all due praise, he nevertheless thought Morgagni's day and approach were over. For the anatomists since the recovery of learning in the sixteenth century, Corvisart says,

> … entirely occupied in developing the detailed structure of the parts, almost all neglected that which concerned the organic lesions of the viscera. But several noted the lesions that by chance presented themselves to them; and these sparse observations have given birth to commendable works, of which some have even immortalised their authors; to prove what I have been saying it suffices to name

35 Corvisart, *An essay on the organic diseases and lesions of the heart and great vessels*, 1812, pp. 80–81.

36 Ibid., p. 120.

> Morgagni. But his splendid work, a monument of a wise and vast erudition, has rather furnished ornaments to the erudition of other physicians than encouraged or accelerated the art of recognising organic diseases.

We shall recall that Morgagni's great work was called *On the seats and causes of diseases investigated by anatomy*. Corvisart adds a significant footnote to this passage, indicating how the direction of research has now been reversed:

> In one of the memoirs that I read a long time ago to the Institut [de France], I put forward the idea of an analogous work, but rather in the opposite sense to that of Morgagni; it would have had as its title, *On the seats and causes of diseases investigated by diagnostic signs, and confirmed by anatomy*. But for such a work one would need at least another Morgagni![37]

There the difference is encapsulated. Morgagni's approach, and that of eighteenth-century pathological anatomists in general, had been absolutely *anatomical*, primarily concerned with inspecting the body when dead, and usually in an *anatomical theatre*, and seeking out the seats of diseases, working back from the proximate causes evident in dissection to the exciting and predisposing causes in the history of the patient, thus working out the particular ways of life and life-events that caused the disease and the signs that indicated it. Corvisart's approach, by contrast, an approach now held in common by the French school, was *clinical*, beginning from diagnostic signs observed at the *bedside* and working forwards, as it were, from them to confirm, in *the post-mortem room*, the hidden organic lesions that had provoked those symptoms. A sign always points to something. For Corvisart a clinical sign pointed forwards, to an internal lesion; for Morgagni, an internal lesion, discovered in post-mortem, points backwards to the symptom – back, that is, to what the patient complained of, not what the physician observed.

It will be evident that these transformations of anatomy in these four fields all took place in one country – France – and indeed in one town, Paris. Moreover, they all took place within a very short time, approximately thirty years. The thread that connects them all is of course the French Revolution and its aftermath, the change from the *ancien régime* to the modern world. The same was happening in all other learned fields too: it was all part of what I have elsewhere called 'The invention of science'.[38]

Soon these changes were taken up elsewhere in Europe, particularly in the German states and, after the defeat of Napoleon, in Britain too. The fracturing of large fields of knowledge into many new specialisations thus became Europe-wide, and France soon lost its leadership in these matters.

---

[37] Corvisart, *Essai sur les maladies et les lésions organiques du coeur et des gros vaisseaux*, 1818, p. ix.

[38] Cunningham, 'Getting the game right', 1988.

The rise of organisation and the death of the soul

One can say that the existence and location of the soul was still being contested at the end of the Enlightenment period. In 1752 the great Albrecht von Haller announced his findings about irritability and sensibility, and defined both in terms of the soul: 'The soul is a being which is conscious of itself, represents to itself the body to which it belongs, and by means of that body the whole universe. I am myself, and not another, because that which is called I, is changed by every thing that happens to my body and the parts belonging to it … Irritability [therefore] is independent of the soul and will'. Similarly with sensibility: 'I call that a sensible part of the human body, which upon being touched transmits the impression of it to the soul; and in brutes, in whom the existence of the soul is not so clear, I call those parts sensible, the irritation of which occasions evident signs of pain and disquiet in the animal'.[39]

The occasional daring *philosophe* might say in the 1760s that the soul, if it existed, was a mere irrelevance: 'As for the mutual commerce between the soul and the body, it is not only the most inconceivable thing in the world but also the most useless to the physician'.[40] But as late as 1796 the highly skilled anatomist of Mainz (previously of Kassel), Samuel Thomas Soemmering, under the influence of the new doctrines of *Naturphilosophie*, was still trying to find a location for the soul, and using his anatomical skills to do so. He saw 'the intraventricular cerebrospinal fluid [as] the seat of a *sensorium commune*', or common-sense (where all the nerves from the senses meet), and decided to locate the soul there.[41] But within twenty years or so the soul was to be completely discarded from anatomical or physiological consideration.

The anatomist, as we saw in Chapter 1, traditionally investigated the parts and the *organs* of the body, taking the organs ('instruments') to be the instruments of the soul. That is, it was the existence of the soul that created the role of 'organ' for the various working parts of the body. The relation of the soul to its organs was a bit like that of master to servant: each is defined in relation to the other. This sense of 'organ' and of its associated terms such as 'organic' and 'organised', continued to be the primary ones right through to almost the end of the eighteenth century, but then they were discarded, and the terms were all given new meanings.

---

39 Haller, *A dissertation on the sensible and irritable parts of animals*, 1755, pp. 678, 659 [*sic*].

40 Dalembert, *L'Encylopédie*, vol. 12, 1765, p. 538, on physiologie: 'Quant au commerce mutual de l'ame et du corps, c'est non seulement la chose du monde la plus inconceivable mais meme la plus inutile au médecin'.

41 Account in *Dictionary of scientific biography*; Sömmering, 'Uber das organ der seele', 1796. See also Riese, 'The 150th anniversary', 1946, who deals with Goethe's and Kant's responses to the treatise; Marino, 'Soemmering', 1994. Hagner, 'The soul and the brain', 1992, stresses how localising the soul was for Soemmering a quintessentially anatomical issue, concerning the size and structure of the brain.

Over the course of the eighteenth century, and culminating in the last two decades, the adjective 'organic' came to replace 'animate', but with 'organic' now simply meaning 'alive', and it was contrasted with the status of inert natural productions ('inorganic'). It lost its old meaning in the process. The *Oxford English Dictionary* gives 1778 as the first instance of 'organic' in this new sense in English and, strikingly, as late as 1842 for the earliest such use of 'organism' in this new sense.[42]

Magendie, as an experimental physiologist, could mock the earlier understanding of the role of the soul:

> The most sublime features of the human character are intelligence, thought, the passions, and that admirable faculty by which we are enabled to direct our movements, and communicate by speech. These phenomena are dependant upon the brain, and are designated by many physiologists as the *cerebral functions*. Other physiologists, sustained and inspired by religious creeds, regard them as belonging to the soul, a being derived from the Divine essence, of which immortality is one of the attributes. It would not be becoming in us to undertake to decide here between these two modes of contemplating this important subject; our object is science, not theology.[43]

Recently it has been argued that one of the defining features of the Enlightenment was 'the demise of the soul': 'against the backcloth of the Enlightenment's watchword, *sapere aude* ('dare to know'), individuals reformulated the problems of existence and made sense of the self, with a changing, and waning, reference to the soul'.[44] The soul, in other words, became a subject of religion, of cults and of art, and not of the investigation of the natural world. For most investigators of nature in the early nineteenth century on the European mainland (and for all of them today), the whole investigation had become secularised, and it simply became 'unscientific' to talk in terms of souls or creatures or Creators. Anatomising no longer revealed the wisdom of the Creator to the onlookers, only the complexity of the organisation of the organism.

Britain was a long way behind the continent in abandoning the soul in this way, a conservatism that was linked to the opposition of the British government to Revolutionary France, and the fact that Napoleon never conquered Britain. As a consequence a 'natural theological' interest in the soul and teleology continued until at least the publication of Darwin's *Origin of species* in 1859. So most

---

42 See also Williams, *Keywords*, 1976, the entry 'organic'.

43 Quoted by Young, *Mind, brain and adaptation*, 1970, p. 82, from Magendie, *An elementary treatise on human physiology*, 1843 English trans. by John Revere of fifth French edition of 1838, pp. 135–6 (first published in French in 1816–17).

44 Porter, *Flesh in the Age of Reason*, 2003, p. 27. As usual with Porter, the story is limited to the 'Anglophone Enlightenment' (p. xv). A Europe-wide account of this change would be very welcome.

anatomists there, such as Charles Bell, still saw the human body as the product of divine design and as the high-point of creation.[45]

With the rejection of the soul in any form from the discussions of the practitioners of the new discipline of experimental physiology, the organs were now simply treated as the necessary constituent complex functioning parts of live beings. An 'organised body' in the new sense would be alive because of the *inter-relationship* of its constituent parts, and not because of its possession of discrete instruments serving the soul. Thenceforth, life was to be regarded as organisation (or as the product of organisation), but with nothing having the role of doing the organising – or at least nothing that could be scientifically investigated.[46] As far as I can see, this usage first appeared in the late 1770s. Complexity was the marker: more complex forms of organisation, especially of the nervous system, were, supposedly, indicators of higher animal forms.

The editors of a dictionary of the medical sciences, published in 1819, could define 'organisation' as an

> … expression intended to designate the mode of structure proper to all living beings in general, and to each part of these living bodies in particular; a mode of structure so named, because it consists of an assemblage of parts which, differing completely in the form, the composition, and the office that they fulfil in the economy of the being, nonetheless unite totally to form it and make it live, and which are called *organs*.[47]

Karl Figlio has argued that there was a 'groping for a vocabulary in which to designate life', but 'organisation' became the preferred, if awkward, term generally adopted.[48] He quotes John Barclay, the Edinburgh anatomist, as pointing out in 1822 the difficulties and circularities in the term, while concluding that 'organisation' nevertheless had to be the crucial term for both anatomists and physiologists. Barclay is arguing against William Lawrence:

> Of this term [Lawrence] gives two definitions: the one is, that it denotes the possession of organs, or instruments for accomplishing certain purposes; the other is, that it means that peculiar composition which distinguishes living bodies. If we understand either of the two, Mr Lawrence means by organisation an organised structure. But, in accuracy of language, organisation is not a structure, nor is a structure organisation. Organisation is a general expression

---

[45] See for instance his 1838 contribution to the Bridgewater Treatises: *The hand: Its mechanism and vital endowments as evincing design*.

[46] On these changes see Duchesneau, *La physiologie des lumières*, 1982.

[47] As cited and translated in Figlio, 'The metaphor of organization', 1976, p. 39, from 'Organisation' in the *Dictionnaire des sciences médicales*, vol. 38, pp. 205–49, quotation from p. 205 (translation modified); the article is by Messieurs Chaussier and Adelon.

[48] Figlio, 'The metaphor of organization', 1976, p. 40.

Fig. 6.3

> for the process or processes by which an organised structure is formed, and therefore can never properly denote the possession of organs or instruments for accomplishing certain purposes, but only a series of operations by which organs or instruments are formed, with a view to accomplishing certain purposes. The truth is, organisation … conveys an idea which is utterly inseparable from that of life being actually present ….[49]

Ending the sacred ritual

Although new anatomical theatres were still being constructed in many places across Europe in the mid-eighteenth century – the one at Kassel, in the Collegium Carolinum, was built as late as 1776 (Fig. 6.3) – yet by the 1790s, for reasons we do not know, some people – including the anatomists themselves – had lost their individual and collective memory of what the anatomical *demonstration* was for, what its sacred ritual purpose was. It came to be judged *solely* as a teaching event, and as one that was not effective. At the London Company of Surgeons, the master John Gunning in 1790 gave a very sour retiring account of what he found wrong with the Company. Among his criticisms he held forth about the teaching of anatomy:

---

[49] Barclay, *An enquiry*, 1822, pp. 340–41, as quoted by Figlio, 'The metaphor of organization', p. 40.

> I am sorry to observe that you have instituted lectures neither in surgery nor in anatomy of any degree of importance … Even the feeble attempts you have already made, though found in the end to be totally inadequate and ineffective, yet have been shamefully expensive … I know it will be said that our Hall can never be a school of anatomy. The situation of the different hospitals, the extent of the town, together with its interference with private courses, prohibit it … From the construction of your Hall, as well as from the name and nature of the officers founded in your body, it plainly appears that one great object of your institution was to raise a school for anatomy, and why it failed in the beginning I cannot readily conceive. Why it fails now I can readily explain.[50]

His explanation was that the teachers were not properly qualified nor rewarded, and that the Company could not compete with the private anatomical teachers of London.

Similarly in Bologna, where anatomical demonstrations on human corpses had been started back in 1316 by Professor Mondino, the new Professor of Anatomy in 1818, whose name – by astonishing coincidence – was Carlo Mondini, wrote a *Memoir* complaining that there were no proper facilities for teaching anatomy at the university:

> No doubt I shall be told that there is the anatomy theatre, built so many years ago for this purpose, and likewise put to use. But its huge size, and the remoteness of the cathedra and of the students' benches from the anatomy table [see Fig. 6.4], however convenient it may be when there are many people attending to hear a learned and elegant anatomical dissertation and an ingenious dispute, is equally inconvenient for the observation and contemplation of the anatomical preparations arranged on the table. As a result, when these are to be displayed, the scholars crowd around the table in such a way that only a very few of them can actually view and profit from the demonstration.[51]

In defence of the positions taken up here by John Gunning and Professor Mondini, one can point of course to a number of new initiatives that had been brought in on the teaching side in the course of the eighteenth century, which supplied effective teaching resources and made the traditional anatomical theatre look outdated. As we have seen in London, for instance, there were many private teachers of anatomy, who followed the so-called 'Paris manner of teaching', in which the students had access to human cadavers to dissect. They learnt in a 'hands-on' manner, and their need for fresh dead bodies to dissect was supplied by a whole

---

[50] Cope, *History of the Royal College of Surgeons of England*, 1959, p. 14.

[51] Carlo Mondini, *Memoir*, 1818, on taking up post as Professor of Human Anatomy, Bologna University Archives. As cited and translated in Giovanna Ferrari, 'Public Anatomy Lessons and the Carnival: The Anatomy Theatre of Bologna', *Past and Present* 117 (1987): 50–106, at p. 93.

Fig. 6.4

secret trade of grave-robbers and, in Edinburgh, by a couple of murderers. As we have also seen, in other places, especially Florence, Bologna and Amsterdam, the art of the anatomical wax-modeller had suddenly appeared, and extremely realistic representations of dissections of any part or system could be made in wax for teaching purposes.

But the anatomical demonstration in the anatomical theatre had never just been about teaching, as we saw in Chapter 1. What happened now to the sacred dimension of that ritual? While evidence of its continued presence, and of anatomists' dedication to it, can be found through most of the eighteenth century, nevertheless by the 1790s it is no longer there. The ritual is not only over, but forgotten.

Its disappearance, as far as I can see, was of a piece with many other cataclysmic changes in the world of learning that occurred at this period. I can only just mention briefly a number of coincidental events – on the strict understanding, of course, that a coincidence is never just a coincidence – in order to contextualise the end of the sacred ritual a little. In the first place there is the coincidence of the great *secularisation* of the world-view in general, especially in the sciences, in the course of the Enlightenment and the French Revolution, and particularly on the

European mainland. Secondly, there is the coincidence of the demise of *natural philosophy* – that God-centred approach to understanding the created world – of which anatomy had been a part, and its replacement by secular sciences in this period. In the case of anatomy, it meant the secularising of an existing science. Thirdly, there is the coincidence of the rise of *expertise* in the sciences – that is to say, a new profession of men of science, or scientists, with the university as the prime domain of making new knowledge, especially the research laboratory, where the general public were not allowed. Fourthly, there is the coincidence of the end of certain *public* forms of event, and their replacement with *concealed* or hidden versions. One of many events that goes from being open to closed here is the public execution. Even Tyburn, the popular public London execution place, was abandoned in 1783. From the mid-nineteenth century all official executions in England have taken place behind the closed walls of a prison.

In the context of such great changes in the world of knowledge, in its getting and its giving and its showing, it is no surprise that the sacred ritual of anatomy came to look out of place, a relic of an earlier age with no relevance in the age of science.

I like to think that my final picture sums up these changes. It is a postcard of the vestibule of a Russian anatomical theatre ('Russia' is the only location given) in the mid-twentieth century (Fig. 6.5).[52] As a postcard it is a disaster since it seems to show nothing at all. But if one takes seriously the closed doors, the closed cabinets, one can recognise that it shows that anatomy, like so many other forms of knowledge, has been taken by the experts into the closet! Its public, ritual, sacred, elements are lost forever. For the last two hundred years it has been the domain of scientific and medical experts only. It has literally been locked away.

Fig. 6.5

[52] My thanks to Simon Schaffer for this postcard.

Bibliography of Reference and Illustrative Works

Reference

The best of the following are basic works of reference for historians of anatomy for the long eighteenth century, and I have used them constantly and often without giving a particular reference. The first one comes with a health warning.

Eloy, N. *Dictionnaire Historique de la Médecine Ancienne et Moderne*, 2 vols (Liège, 1755); 2nd edn, 4 vols (Paris, 1778). Convenient because the second edition has received a modern reprint (Brussels, 1973). However, Portal's opinion (*Histoire de l'Anatomie et de la Chirurgerie*, vol. 5, p. 547) of this work was scathing: 'The extract that the editors of the Leipzig journal have given of this work sheds great honour on them, and should cover the author of this new history of medicine with shame. They show that M. Eloy has ruined the names of the best-known authors, he has truncated their accounts, and he has omitted most of the works which are worth knowing'. But Eloy's real crime seems to be that he had not acknowledged the importance of Haller! Haller himself (*Bibliotheca Anatomica*, vol. 2, p. 510) also dismissed the first edition as 'thin on recent writers. Names corrupted, imperfect catalogue of the books of each writer, writers repeated, the whole full of errors'.

Hahn, A., P. Dumaitre and J. Samion-Contet. *Histoire de la Médecine et du Livre Médical à la Lumière des Collections de la Bibliothèque de la Faculté de la Médecine de Paris* (Paris, 1962). More wide-ranging and valuable than the title indicates.

Haller, A. von. *Bibliotheca Anatomica. Qua Scripta ad Anatomen et Physiologiam Facientia a Rerum Initiis Recensentur*, 2 vols (Zurich, 1774, 1777). Haller was a skilled anatomist and had read virtually every work he discusses, which includes everything written on anatomy from Antiquity to his own day, and his judgements on anatomists are very fair and valuable. The work has a straightforward index (unlike Portal). Obsessionally accurate. Indispensable.

Jenty, C. *A Course of Anatomico-Physiological Lectures on the Human Structure and Animal Œconomy*, 3 vols (London, 1757). I have found 'The historical compendium' in volume 1 most useful, written as it was by a practising anatomist. Jenty seems to have assembled this himself, as he complains about the big volumes one has to read in order to find the small discoveries of most anatomists.

Portal, A. *Histoire de l'Anatomie et de la Chirurgerie*, 6 vols (Paris, 1770–73). Valuable not only for the generally reliable bibliographic and biographical information on anatomists, but also for Portal's assessments of their anatomical work. The sixth volume attempts a chronology of discoveries about each and every part of the body. Inconvenient to use since the anatomists are discussed in order of the date of their first publication, so they are rarely in strict date order (e.g. in volume 3 Duverney, first publication 1675, comes before Dionis, first publication 1683, even though Duverney succeeded Dionis as lecturer in the Jardin royal).

Richter, G. *Das Anatomische Theater* (Berlin, 1936). Still the basic work on the topic.

Roberts, K. B., and J. D. W. Tomlinson. *The Fabric of the Body: European Traditions of Anatomical Illustration* (Oxford, 1992). This beautiful and invaluable book is more analytical than most, and also has exemplary comments by an anatomist on the anatomical depictions. Its focus is on illustrations, and it is arranged by geographical region and period, treating anatomists one at a time. The occasional very important anatomist who barely used illustrations, such as Morgagni, unfortunately falls through the net.

Russell, K. F. *British Anatomy 1525–1800. A Bibliography of Works Published in Britain, America and on the Continent*, 2nd edn (Winchester, 1987), first publ. 1963. This covers (1) works by British anatomists, wherever published and in whatever language; and (2) anatomical works published in Britain, in English or Latin, whatever the origin of the author. Despite this curious choice of coverage, the volume is indispensable for the wealth and accuracy of its bibliographic information.

Wolf-Heidegger, G., and A. M. Cetto. *Die Anatomische Sektion in Bildlischer Darstellung* (Basle, 1967). A bibliographic work, it is a thorough (but not exhaustive) collection of printed frontispieces and other illustrations of dissection taking place, though it does not deal directly either with anatomy or anatomising.

Illustrations

This has been the primary way in which recent historians have explored the history of anatomy in the early modern period. Over the last decade or so there has been a vogue for producing books on historic anatomical illustrations, and for exhibitions on similar themes, with catalogues, with many images reproduced in full colour, each volume seemingly more gorgeous than the last (if you like that kind of thing). Among those in English, French or Italian are:

Cazort, M., M. Kornell and K. B. Roberts. *The Ingenious Machine of Nature: Four Centuries of Art and Anatomy* (Ottowa, 1996).

Comar, P., ed. *Figures du Corps: Une Leçon d'Anatomie à l'École des Beaux-Arts* (Paris, 2008).

Encyclopaedia Anatomica. A Complete Collection of Anatomical Waxes [at La Specola, Florence] (Cologne, 1999).

Kemp, M., and M. Wallace. *Spectacular Bodies: The Art and Science of the Human Body from Leonardo to Now* (London, 2000).

Petherbridge, D., C. Ritschard and A. Carlino. *Corps à Vif: Art et Anatomie* (Geneva, 1997).

Petherbridge, D., and L. Jordanova. *The Quick and the Dead: Artists and Anatomy* (London, 1997).

Premuda, Loris. *Storia dell'Iconografia Anatomica* (1957, repr. Saronno, 1993). Still of value.

Rifkin, B. A., M. J. Ackerman and J. Folkenberg. *Human Anatomy: Depicting the Body from the Renaissance to Today* (London, 2006).

Bibliography of Works Cited

Abernethy, John. *Physiological Lectures, Exhibiting a General View of Mr. [John] Hunter's Physiology, and of his Researches in Comparative Anatomy. Delivered before the Royal College of Surgeons, in the Year 1817*. London: Longman, Hurst, Rees, Orme, and Brown, 1817.

Abrégé d'Anatomie, Accommodé aux Arts de Peinture et de Sculpture, et Mis dans un Ordre Nouveau … Mis en Lumière par François Tortebat. Paris: Tortebat, 1667.

Ackerknecht, Erwin H. *Medicine at the Paris Hospital, 1794–1848*. Baltimore, MD: The Johns Hopkins University Press, 1967.

Adams, Leverett A., and Samuel Eddy. *Comparative Anatomy. An Introduction to the Vertebrates*, 2nd edn. New York: John Wiley and Sons Inc., 1949.

Adelmann, Howard B. *The Correspondence of Marcello Malpighi*, 5 vols, *Cornell Publications in the History of Science*. Ithaca, NY: Cornell University Press, 1975.

Adelmann, Howard B., ed. *Marcello Malpighi and the Evolution of Embryology*, 5 vols. Ithaca, NY: Cornell University Press, 1966.

Alaux, Jean-Paul. *L'Académie de France à Rome*, 2 vols. Paris: Editions Duchartre, 1933.

Albinus, Bernhard. *Demonstrationibus Anatomicis in Corpore Foeminino Bernh. Albinus Perofficiose Invitat*. Frankfurt am Oder: C. Zeitleri, 1684.

Albinus, Bernhard. *Demonstrationibus Anatomicis in Corpore Masculino Invitat Bernh. Albinus*. Frankfurt am Oder: C. Zeitleri, 1684.

Albinus, Bernhard Siegfried. *Tables of the Skeleton and Muscles of the Human Body. Translated from the Latin*. Edinburgh: Andrew Bell, Engraver, 1777 (first publ. 1747).

Albinus, Bernhard Siegfried. *The Explanation of Albinus's Anatomical Figures of the Human Skeleton and Muscles: With an Historical Account of the Work. Translated from the Latin. To which is Added the Explanation of the Supplement to Albinus, Containing a Compleat Anatomical Description of the Blood-Vessels and Nerves, and the External Parts of the Human Body. The Figures of Albinus, and the Supplement, are Curiously Engrav'd on Fifty-One Large Copper-Plates, Fifteen Inches by Twenty-Two*. London: John and Paul Knapton, 1754.

Albinus, Bernhard Siegfried. *Tabulae VII Uteri Mulieris Gravidae*. Leiden: Verbeek, 1748–51.

Albinus, Bernard Siegfried. *Explicatio Tabularum Anatomicarum Bartholomaei Eustachii, Anatomici Summi. Accedit Tabularum Editio Nova*. Leiden: Johann Arnold Langerak and Johann Herman Verbeek, 1744.

Albinus, Bernhard Siegfried. *Index Supellectilis Anatomicae quam Academiae Batavae quae Leidae Est Legavit Vir Clarissimus Johannes Jacobus Rau. Rogatu Illustrissimorum et Amplissimorum Academiae istius Curatorum et Urbis Consulum. Confectus a Bernhardo Siegfried Albino, qui et Vitam ejus et Curationem quam Calculosis Adhibuit Instrumentorumque Figuras Addidit.* Leiden: Henricus Mulhovium and Franciscus Schuly, 1725.

Albinus, Bernhard Siegfried. *Oratio Inauguralis de Anatome Comparata. Publice Habita, D. 2 Octobris MDCCXIX. Cum Publicum Anatomen & Chirurgiam Praelegendi Munus in Academia Leidensi Auspicaretur.* Leiden: Henricus Mulhovius, 1719.

Almanach de Versailles, Année 1782. Contenant la Description de la Ville, du Château, du Parc; la Maison du Roi. Celles de la Reine et de la Famille Royale; Les Bureaux des Ministres &c. Versailles: Chez Blaizot, 1782.

Anon. *Eloge historique de M. Winslow*. [n.p.; n.d.].

Anon. 'Mémoire sur une Controverse au Sujet de la Respiration', *Nouvelle Bibliotheque Germanique, ou Histoire Litteraire de l'Allemagne, de la Suisse, et des Pays du Nord* (1748): 412–28.

Armaroli, Maurizio. *Le Cere Anatomiche Bolognesi del Settecento*. Bologna: Università degli Studi di Bologna, 1981.

Aselli, Gaspar. *De Lactibus, sive Lacteis Venis, Quarto Vasorum Mesaraicorum Genere, Dissertatio: Quae Sententiae Anatomicae Multae vel Perperam Receptae Convelluntur, vel Parum Perceptae Illustrantur*, 1627; as reprinted in Adrianus Spigelius, *Opera Omnia*, 2 vols, Amsterdam, Iohannus Blaeu, 1645, 'Tractatus eorum, qui post Spigelium, Anatomiam Invento Aliquo Majoris Usus Adauxerunt', in vol. 1.

Ashley Montagu, M. F. *Edward Tyson, M. D., F. R. S., 1650–1708, and the Rise of Human and Comparative Anatomy in England; A Study in the History of Science*, Memoirs of the American Philosophical Society Held at Philadelphia for Promoting Useful Knowledge, vol. 20. Philadelphia: The American Philosophical Society, 1943.

Atwood, William Henry. *A Concise Comparative Anatomy*. London: Henry Kimpton, 1947.

Aucoc, Léon. *L'Institut de France. Lois, Statuts et Règlements Concernant les Anciennes Académies et l'Institut, de 1635 à 1889*. Paris: Plon, 1889.

Baillie, Matthew. *A Series of Engravings, Accompanied with Explanations, which are Intended to Illustrate the Morbid Anatomy of some of the Most Important Parts of the Human Body; Divided into Ten Fasciculi*. London: G. and W. Nicol, 1798–1802.

Baillie, Matthew. *The Morbid Anatomy of Some of the Most Important Parts of the Human Body. The Second Edition, Corrected and Considerably Enlarged.* London: J. Johnson; G. Nicol, 1797 (first publ. 1793).

Bamforth, Iain. 'Splendid Cadavers: A Memoir', in Andrew Patrizio and Dawn Kemp, eds, *Anatomy Acts: How We Come to Know Ourselves*, 99–107. Edinburgh: Birlinn Limited, 2006.

Baratay, Eric, and Elisabeth Hardouin-Fugier. *Zoo: A History of Zoological Gardens in the West*, trans. Oliver Welsh. London: Reaktion Books, 2002.

Barclay, John. *An Inquiry into the Opinions, Ancient and Modern, Concerning Life and Organization*. Edinburgh: Bell & Bradfute etc., 1822.

Bartholin, Thomas. *De Anatome Practica, ex Cadaveribus Morbosis Adornanda, Consilium*. Copenhagen: Petrus Haubold, 1674.

Bartholin, Thomas. *Bartholinus Anatomy; Made from the Precepts of his Father, and from the Observations of All Modern Anatomists; Together with his Own. With One Hundred Fifty and Three Figures; Cut in Brass, Much Larger and Better than any have been heretofore Printed in English. In Four Books and Four Manuals, Answering to the Said Books. Book I of the Lower Belly. Book II of the Middle Venter or Cavity. Book III of the Uppermost Cavities, Viz. The Head. Book IV of the Limbs. The Four Manuals Answering to the Foregoing Books. Manual I of the Veins, Answering to the First Book of the Lower Belly. Manual II of the Arteries, Answering to the Second Book of the Middle Cavity or Chest. Manual III of the Nerves, Answering to the Third Book of the Head. Manual IV of the Bones, Answering to the Fourth Book of the Limbs. Also Two Epistles of the Circulation of the Blood* [*by Johannes Walaeus*]. *Being Part of the First Volume of the Physitians Library, Published by Nich. Culpeper Gent. And Abdiah Cole Doctor of Physick*. London: Peter Cole, 1663.

Bartholin, Thomas [Younger]. *Cista Medica Hafniensis, Variis Consiliis, Curationibus, Casibus Rarioribus, Vitis Medicorum Hafniensium, aliisque ad Rem Medicam, Anatomicam, Botanicam et Chymicam Spectantibus Referta. Accedit ejusdem Domus Anatomica Brevissime Descripta*. Copenhagen: Petrus Haubold, 1662.

Belgrado, Anna Minerbi, ed. *Guillaume Lamy. Discours Anatomiques* [*1675*]. *Explication Méchanique et Physique des Fonctions de l'Ame Sensitive* [*1677*], ed. Antony McKenna, *Libre Pensée et Littérature Clandestine*. Paris and Oxford: Universitas and Voltaire Foundation, 1996.

Bell, Charles. *The Hand. Its Mechanism and Vital Endowments as Evincing Design*, *Bridgewater Treatises*. London: William Pickering, 1833.

Bennett, Jim. 'Malpighi and the Microscope', in Domenico Berteloni Meli, ed., *Marcello Malpighi Anatomist and Physician*, 63–72. Florence: Leo S. Olschki, 1998.

Bernard, Claude. *An Introduction to the Study of Experimental Medicine*, trans. Henry Copley Greene. First publ. in French in 1865; first publ. in English in 1927; this edition, New York: Henry Schuman, 1949.

Berteloni Meli, Domenico, ed. *Marcello Malpighi Anatomist and Physician*. Florence: Leo S. Olschki, 1998.

Bidloo, G. *Anatomia Humani Corporis, Centum et Quinque Tabulis Per Artificiosiss. G. de Lairesse ad Vivum Delineatis, Demonstrata, Veterum Recentiorumque Inventis Explicata Plurimisque, Hactenus non Detectis, Illustrata*. Amsterdam: Someren, 1685.

Bidloo, Govard. *Dissertationes Anatomico-Physiologicae Quatuor, Quoniam in Academia, quae est Lugd. in Bat. Discussae*. [Leiden]: n.p., 1708.

Bidloo, Govard. *Gulielmus Cowper, Criminis Literarii Citatus, Coram Tribunali Nobiliss. Ampliss. Societatis Britanno-Regiae, per Godefridum Bidloo*. Leiden: Jordanus Luchtmans, 1700.

Biheron, la Demoiselle. *Anatomie Artificielle. (Annonce de l'Exposition Publique de Pièces d'Anatomie Artificielle, Exécutées par la Demoiselle Biheron)*. Paris: Impr. de P.A. Le Prieur, 1761.

Billing, Christian. 'Modelling the Anatomy Theatre and the Indoor Hall Theatre: Dissection on the Stages of Early Modern London', *Early Modern Literary Studies* 13 (2004): 1–17.

Bils, Louis de. *The Coppy of a Certain Large Act* [*Obligatory*] *of Yonker Louis De Bils, Lord of Koppensdamme, Boenen, &c., Touching the Skill of a Better Way of Anatomy of Man's Body*. London, 1659.

Blasius, Gerard. *Zootomiae seu Anatomes Variorum Animalium. Pars Prima* [half title: *Zootomia seu Anatome Humanis Brutorumque Variorum*]. Amsterdam: Abraham Wolfgang, 1676.

Blasius, Gerard. [Full title:] *Observata Anatomica in Homine, Simia, Equo, Vitulo, Ove, Testudine, Echino, Glire, Serpente, Ardea, Variisque Animalibus Aliis. Accedunt Extraordinaria in Homine Reperta Praxin Medicam aeque ac Anatomen Illustrantia* [half title: *Observata Anatomico Pratica in Homine Brutisque Variis*]. Amsterdam: Gaasbaack, 1674.

Blasius, Gerard. *Miscellanea Anatomica, Hominis, Brutorumque Variorum, Fabricam Diversam Magna Parte Exhibentia*. Amsterdam: Caspar Commelin, 1673.

Blumenbach, J.F. *A Short System of Comparative Anatomy, Translated from the German of J.F. Blumenbach, Professor of Medicine in the University of Goettingen. By William Lawrence, Fellow of the Royal College of Surgeons of London, and Demonstrator of Anatomy at St. Bartholomew's Hospital: With Numerous Additional Notes, and an Introductory View of the Classification of Animals, by the Translator*. London: Longman, Hurst, Ress and Orme, 1807.

Bonet, Théophile. *Sepulchretum sive Anatomia Practica, ex Cadaveribus Morbo Denatis, Propens Historias et Observationes Omnium Pene Humani Corporis Affectuum, Ipsorumque Causas Reconditas Revelans Quo Nomine tam Pathologiae Genuinae, quam Nosocomiae Orthodoxae Fundatrix, Imo Medicinae Veteris ac Novae Promptuarium Dici Meretur. Cum Indicibus Necessariis. Editio Altera, Quam Novis Commentariis Et Observationibus Innumeris Illustravit, ac Tertia ad Minimum Parte Auctiorem Fecit Johannes Jacobus Mangetus, Medicinae Doctor, Et Serenissimi Electoris Brandenburgici Archiatri*. Leiden: Cramer and Perachon, 1700.

Bonet, Théophile. *Mercurius Compitalitius: Or, a Guide to the Practical Physician*. London: Thomas Flesher, 1684.

Bonet, Théophile. *Sepulchretum sive Anatomia Practica, ex Cadaveribus Morbo Denatis, Propens Historias et Observationes Omnium Pene Humani*

Corporis Affectuum, Ipsorumque Causas Reconditas Revelans Quo Nomine tam Pathologiae Genuinae, quam Nosocomiae Orthodoxae Fundatrix, Imo Medicinae Veteris ac Novae Promptuarium Dici Meretur. Cum Indicibus Necessariis. Opus Omnium Medicinae et Anatomie Cultorum Votis Hactenus Expetitum Summoque Labore Decerptum ac Congestum. Geneva: Leonard Chovët, 1679.

Bonet, Théophile. *Prodromus Anatomiae Practicae, sive de Abditis Morborum Causis, ex Cadaverum Dissectione Revelatis, Libri Primi Pars Prima, De Doloribus Capitis ex illius Apertione Manifestis*. Geneva: Miege, 1675.

Bonnet, Charles. *Considerations sur les Corps Organisés: Où l'on Traite de leur Origine, de leur Développement, de leur Réproduction, &c. & où l'on a Rassemblé en Abrégé Tout ce que l'Histoire Naturelle Offre de Plus Certain & de Plus Intéressant sur ce Sujet*, 2 vols. Amsterdam: Chez Marc-Michel Rey, 1762.

Bonnet, Charles. *Essai Analytique sur Les Facultés de l'Âme*. Copenhagen: Frères Philibert, 1760.

Boyle, Robert. *Some Considerations Touching the Usefulnesse of Experimental Natural Philosophy. Propos'd in a Familiar Discourse to a Friend, by Way of Invitation to the Study of it*, 2nd edn (first publ. 1663). Oxford: Ri. Davis, 1664.

Bradford, C.A. *Heart Burial*. London, 1933.

Brazier, Mary A.B. *A History of Neurophysiology in the 17th and 18th Centuries: From Concept to Experiment*. New York: Raven Press, 1984.

Brisbane, John, MD. *The Anatomy of Painting: Or a Short and Easy Introduction to Anatomy: Being a New Edition, on a Smaller Scale, of Six Tables of Albinus, with their Linear Figures: Also, a New Translation of Albinus's History of that Work, and of His Index to the Six Tables: To which are Added the Anatomy of Celsus, with Notes, and the Physiology of Cicero: With an Introduction, Giving a Short View of Picturesque Anatomy*. London: T. Cadell, 1769; repr. London: 1811.

Brock, C.H., ed. *William Hunter 1718–1783. A Memoir by Samuel Foart Simmons and John Hunter*. Glasgow: University of Glasgow Press, 1983.

Brockbank, William. 'Old Anatomical Theatres and what took place therein', *Medical History* 12 (1968): 371–84.

Brown(e), John. *Myographia Nova: Or, a Graphical Description of All the Muscles in Humane Body, as they arise in Dissection: Distributed into Six Lectures; at the Entrance to every of which, are Demonstrated the Muscles Properly Belonging to each Lecture. Now in General Use at the Theatre in Chyrurgeons-Hall, London* (5th edn of the original *Compleat Treatise of the Muscles*, 1681). London: Printed by Tho. Milbourn, for the Author, 1697.

Brown, Theodore M. *The Mechanical Philosophy and the 'Animal Œconomy'. A Study in the Development of English Physiology in the Seventeenth and Early Eighteenth Century*, repr. of 1968 Princeton University PhD thesis. New York: Arno Press, 1981.

Bueltzingsloewen, Isabelle von. 'Die Entstehung des Klinischen Unterrichts an den Deutschen Universitäten des 18. Jahrhunderts und das Göttinger Accouchierhaus', in Jürgen Schlumbohm and Claudia Wiesemann, eds, *Die Entstehung der Geburtsklinik in Deutschland 1751–1850*, 15–30. Göttingen: Wallstein Verlag, 2004.

Buffon, Georges Louis Leclerc, Comte de. *Natural History, General and Particular*, trans. William Smellie, 9 vols, 2nd reprint. London: W. Strahan and T. Cadell, 1785.

Buffon, Georges Louis Leclerc, Comte de, and Louis Jean Marie Daubenton. *Histoire Naturelle, Générale et Particulière, avec la Description du Cabinet du Roi*, 54 vols. Paris: Aux Deux-Ponts, 1749–1804.

Burmeister, Maritha Rene. 'Popular Anatomical Museums in Nineteenth-Century England', PhD, Graduate School–New Brunswick Rutgers, NJ: The State University of New Jersey, 2000.

Burton, John. *An Essay towards a Complete New System of Midwifry* [*sic*], *Theoretical and Practical*. London: James Hodges, 1751.

Burton, Robert. *The Anatomy of Melancholy, What it is: With all the Kindes, Causes, Symptomes, Prognosticies, and several Cures of it … by Democritus Iunior*. Oxford: H. Cripps, 1621.

Butterton, Joan R. 'The Education, Naval Service and Early Career of William Smellie', *Bulletin of the History of Medicine* 60 (1986): 1–18.

Campbell, R. *The London Tradesman. Being a Compendious View of All the Trades, Professions, Arts, both Liberal and Mechanic, Now Practised in the Cities of London and Westminster. Calculated for the Information of Parents, and Instruction of Youth in Their Choice of Business*. London: Printed by T. Gardner, 1747.

Camper, Peter. *Optical Dissertation on Vision 1746. Facsimile of the Original Latin Text with a Complete Translation and an Introduction by G. Ten Doesschate, Dutch Classics on History of Science*. Nieuwkoop: B. de Graaf, 1962.

Camper, Peter. *Petri Camperi Itinera in Angliam, 1748–1785*, trans. B. W. T. Nuyens, *Opuscula Selecta Neerlandicorum De Arte Medica*. Amsterdam, 1939.

Camper, Peter. *The Works of the Late Professor Camper on the Connexion between the Science of Anatomy and the Arts of Drawing, Painting and Statuary, &c in Two Books. Containing a Treatise on the Natural Difference of Features in Persons of Different Countries and Periods of Life; and on Beauty, as Exhibited in Ancient Sculpture; with a New Method of Sketching Heads, National Features, and Portraits of Individuals, with Accuracy &c. Illustrated with Seventeen Plates Illustrative of the Professor's Leading Principles. Translated from the Dutch by T. Cogan, M.D*, Dutch original publ. 1791; 1st English edn, London: C. Dilly, 1794.

Camper, Peter. *Dissertatio Optica de Visu. Quam, Favente Deo Ter Optimo Maximo, ex Auctoritate Magnifici Rectoris D. Hieronymi Davidis Gaubii,*

Publico ac Solemni Examini Submittit Petrus Camper, Lugd. Bat. ad Diem 14 Octobris 1746 ab Hora 8 ad 10 Loco Solito. Leiden: Elia Luzac, Jun., 1746.

Camper, Peter. *Dissertatio Physiologica de Quibusdam Oculi Partibus. Quam Favente Deo Optimo Maximo, ex Auctoritate Magnifici Rectoris D. Hieronymi Davidis Gaubii; Publico ac Solemni Examini Submittit Petrus Camper, Lugd. Bat. ad Diem 14 Octobris 1746 ab Hora 10 ad 12 Loco Solito*. Leiden: Elia Luzac, Jun., 1746.

Camper, Pierre. *Description Anatomique d'un Éléphant Mâle. Publiée par son Fils, A.G. Camper. Avec Vingt Planches*. Paris: H. J. Jansen, 1802, An XI.

Canezza, Alessandro. *Gli Archispedali di Roma nella Vita Cittadina Nella Storia e nell'Arte*. Rome: n.p., 1933.

Carus, Carl Gustav. *Traité Élémentaire d'Anatomie Comparée, Suivi de Recherches d'Anatomie Philosophique ou Transcendante sur les Parties Primaires du Système Nerveux et du Squelette Intérieur et Extérieur, et Accompagné d'un Atlas de 31 Planches in-4o, Gravées. Traduit de l'Allemand, sur la Seconde Édition, Par A.-J.-L. Jourdan*, 3 vols. Paris: J.-B. Baillière, 1835.

Carus, Carl Gustav. *An Introduction to the Comparative Anatomy of Animals; Compiled with Constant Reference to Physiology, and Elucidated by Twenty Copper-Plates. Translated from the German by R.T. Gore, Member of the Royal College of Surgeons in London*, 2 vols. London: Longman, Rees, Orme, Brown, and Green, 1827.

Cassebohm, Johann Friedrich. *Methodus Secandi, oder, Deutliche Anweisung zur Anatomischen Betrachtung und Zergliederung des Menschlichen Cörpers*. Berlin: Johann Jacob Schütß, 1746.

Cavazza, Marta. *Settecento Inquieto. Alle Origini dell'Istituto delle Scienze di Bologna*. Bologna: Il Mulino, 1990.

Centenaire de la Fondation du Muséum d'Histoire Naturelle 10 Juin 1793–10 Juin 1893. Volume Commémoratif Publié par les Professeurs du Muséum. Paris: Imprimerie Nationale, 1893.

Chalmers Mitchell, P. *Centenary History of the Zoological Society of London*. London: Printed for the Society, 1929.

Charleton, Walter. *Enquiries into Human Nature: in VI Anatomic Praelections in the New Theatre of the Royal Colledge of Physicians in London: With a Large Preface of the Antiquity, Uses, Differences, &c. of Anatomy. Published by Order of the most Learned President*. London: Printed for J. Conyers, 1680.

Cheselden, William. *Osteographia, or the Anatomy of the Bones*. London, 1733.

Choulant, Ludwig. *History and Bibliography of Anatomic Illustration*, trans, and annotated by Mortimer Frank, 1920. New York: Hafner Publishing, 1962.

Choulant, Ludwig. *Bibliotheca Medico-Historica, sive Catalogus Librorum Historicorum de Re Medica et Scientia Naturali Systematicus*. Leipzig: W. Engelmann, 1842.

Clark, Sir George. *A History of the Royal College of Physicians*, 2 vols. Oxford: Clarendon Press for Royal College of Physicians, 1964, 1966.

Clarke, Edwin, and C. D. O'Malley. *The Human Brain and Spinal Cord. A Historical Study Illustrated by Writings from Antiquity to the Twentieth Century*, first publ. 1968 by University of California Press; rev. edn, San Francisco: Norman Publishing, 1996.

Cole, Francis Joseph. *A History of Comparative Anatomy from Aristotle to the Eighteenth Century*. London: Macmillan, 1949.

Cole, Francis Joseph. *Early Theories of Sexual Generation*. Oxford: At the Clarendon Press, 1930.

Cole, Francis Joseph. 'The History of Anatomical Injections', in Charles Singer, ed., *Studies in the History and Method of Science*, 285–343. Oxford: Oxford University Press, 1921.

Coleman, William. *Georges Cuvier, Zoologist. A Study in the History of Evolution Theory*. Cambridge, MA: Harvard University Press, 1964.

Collignon, Charles. *Tyrocinium Anatomicum: or, An Introduction to Anatomy*. Cambridge: Printed by J. Bentham, 1763.

Collignon, Charles. *Compendium Anatomico-Medicum Eorum, quae in Scholis apud Cantabrigienses, Mense Martio, Quotannis Explicat, Demonstratque Carolus Collignon, M. D. et in Academia, Professor Anatomicus*. Cambridge: Printed by J. Bentham, 1756.

Collins, Samuel. *A Systeme of Anatomy, Treating of the Body of Man, Beasts, Birds, Fish, Insects, and Plants. Illustrated with Many Schemes, Consisting of Variety of Elegant Figures, Drawn from the Life, and Engraven in Seventy Four Folio Copper-Plates. And after Every Part of Man's Body hath been Anatomically Described, its Diseases, Cases, and Cures are Concisely Exhibited. The First Volume Containing the Parts of the Lowest Apartiment of the Body of Man and Other Animals &c.* [*The Second Volume, Containing the Middle and Highest Apartiment of Man's Body (and Other Animals) with its Diseases, Cases, and Cures. To which are Added in the End Seventy Three Schemes, Beautified with Many Elegant Figures, Fully Explained by Letters and Notes, Placed in Pages Opposite to the Tables, for the More Ready Inspection and Clear Understanding of the Figures*], 2 vols. London: Printed by Thomas Newcomb, 1685.

Comar, P., ed. *Figures du Corps. Une Leçon d'Anatomie à l'École des Beaux-Arts*. Paris: Les Éditions Beaux-Arts de Paris, 2008.

Commentaires de la Faculté de Médecine de Paris, 1777 à 1786, 2 vols, *Documents pour l'Histoire de l'Université de Paris*. Paris: G. Steinheil, 1903.

Condivi, Ascanio. *The Life of Michelangelo*, trans. Alice Sedgwick Wohl. Oxford: Phaidon, 1976.

Cope, Sir Zachary. *The History of the Royal College of Surgeons of England*. London: Anthony Blond, 1959.

Corsi, Pietro. *The Age of Lamarck: Evolutionary Theories in France, 1790–1830. Revised and Updated. Translated by Jonathan Mandelbaum*. Berkeley, CA: University of California Press, 1988.

Cortona, Pietro da. *The Anatomical Plates of Pietro Da Cortona* [*1596–1669*]*: 27 Baroque Masterpieces; with a New Introduction by Jeremy M. Norman*. New

York: Dover Publications, 1986. (Originally published as: *Tabulae anatomicae a Celeberrimo Pictore Petro Berrettini Cortonensi Delineatae*. Rome: Fausto Amideo, 1741.)

Corvisart, J. N. *An Essay on the Organic Diseases and Lesions of the Heart and Great Vessels, from the Clinical Lectures of J. N. Corvisart, Published, under his Inspection, by C. E. Horeau*. Boston, 1812; repr. New York, 1962.

Corvisart, J.N. *Essai sur les Maladies et les Lésions Organiques du Coeur et des Gros Vaisseaux* (first edn Paris, 1806), 3rd edn. Paris, 1818.

Coschwitz, Georg Daniel. *Theatri Anatomici sub Auspiciis Felicissimis Augustissimi Regis Borussiae Friderici Wilhelmi Semper Magni Felicis, Gloiosi in Florentissima Fridericiana Halensi Exstructi Natalia. Declamatione Inaugurali de Studii Anatomici Praestantia, Utilitate, ac Necessitate, Die IV Dec MDCCXXVII. In Splendidissimo Procerum, Patronorum, Pariter ac Illustris, Generosae, Praenobilissimae Civium Academicorum Coronae Consessu Concelebrata a Georgio Daniel Coschwitz, Med. Doct. Anat. Chir. Botan. In Alma Fridericiana Prof. Ordin. ....* Halle an der Saale: Johannes Christian Hilliger, Acad. Typ., 1727.

Cotchin, Ernest. *The Royal Veterinary College London: A Bicentary History*. Buckingham: Barracuda Books Limited, 1990.

Coulter, J. L. S., with J. J. Keevil and C. Lloyd. *Medicine and the Navy, 1200–1900*, 4 vols. Edinburgh, 1957–63.

Cowper, William. *ΈΥΧΑΡΙΣΤΙΑ, in qua Dotes Plurimae et Singulares Godefridi Bidloo, M.D. Et in Illustrissima Leydarum Academia Anatomiae Professoris Celeberrimi Peritia Anatomica, Probitas, Ingenium, Elegantiae Latinatis, Lepores, Candor, Humanitas, Ingenuitas, Solertia, Verecundia, Humilitas, Urbanitas, &c. Celebrantur, et Ejusdem Citationi Humillime Respondetur*. London: n.p., 1701.

Crainz, Franco, ed. *The Life and Works of Matthew Baillie M. D., F. R. S. L. & E. F. R. C. P., Etc (1761–1823)*. Rome: PelitiAssociati [*sic*], 1995.

Crescimbeni, Giovanni Maria. *Vita di Monsignor Gio. Maria Lancisi Camerier Segreto, e Medico di Nostro Signore Papa Clemente XI*. Rome: Antonio de Rossi, 1721.

Creswell, Clarendon Hyde. *The Royal College of Surgeons of Edinburgh: Historical Notes from 1505 to 1905*. Edinburgh: Privately printed for the College by Oliver and Boyd, 1926.

Crooke, Helkiah. *Mikrokosmographia. A Description of the Body of Man. Together with the Controversies Thereto Belonging. Collected and Translated out of All the Best Authors of Anatomy, Especially out of Gasper Bauhinus and Andreas Laurentius. The Second Edition Corrected and Enlarged*. London: Michael Sparke, 1631 (1st edn, 1615).

Cruickshank, William. *The Anatomy of the Absorbing Vessels of the Human Body*. London: G. Nicol, 1786.

Cullen, William. *Synopsis Nosologiae Methodicae, Exhibens Clariss. Virorum Sauvagesii, Linnæi, Vogelii, Sagari, et Macbridii, Systema Nosologica. Edidit*

Suumque Proprium Systema Nosologicum Adjecit Gulielmus Cullen, 2 vols. Edinburgh: William Creech, 1785.

Cunningham, Andrew. 'The Pen and the Sword: Recovering the Disciplinary Identity of Physiology and Anatomy before 1800. Part 2: Old Anatomy: The Sword', *Studies in History and Philosophy of Science* 34 (2003): 51–76.

Cunningham, Andrew. 'The Pen and the Sword: Recovering the Disciplinary Identity of Physiology and Anatomy before 1800. Part 1: Old Physiology: The Pen', *Studies in History and Philosophy of Science* 33 (2002): 631–65.

Cunningham, Andrew. 'The End of the Sacred Ritual of Anatomy', *Canadian Bulletin of Medical History* 18 (2001): 187–204.

Cunningham, Andrew. 'Aristotle's Animal Books: Ethology, Biology, Anatomy or Philosophy?', *Philosophical Topics* 27 (2000): 17–41.

Cunningham, Andrew. *The Anatomical Renaissance: The Resurrection of the Anatomical Projects of the Ancients*. Aldershot: Scolar, 1997.

Cunningham, Andrew. 'Pathology and the Case-History in Giambattista Morgagni's "On the Seats and Causes of Diseases Investigated through Anatomy (1761)"', *Medizin, Geschichte und Gesellschaft (Jahrbuch of the Institut für Geschichte der Medizin, Stuttgart)* 14 (1996): 37–61.

Cunningham, Andrew. 'The Historical Context of Wharton's Work on the Glands', in Stephen Freer (trans.), *Thomas Wharton's 'Adenographia'*, xxvii–lii. Oxford: Oxford University Press, 1996.

Cunningham, Andrew. 'Medicine to Calm the Mind: Boerhaave's Medical System and Why it was Adopted in Edinburgh', in Cunningham and Roger French, eds, *The Medical Enlightenment of the Eighteenth Century*, 40–66. Cambridge: Cambridge University Press, 1990.

Cunningham, Andrew. 'Thomas Sydenham: Epidemics, Experiment and the "Good Old Cause"', in Roger French and Andrew Wear, eds, *The Medical Revolution of the Seventeenth Century*, 164–90. Cambridge: Cambridge University Press, 1989.

Cunningham, Andrew. 'Getting the Game Right: Some Plain Words on the Identity and Invention of Science', *Studies in History and Philosophy of Science* 19 (1988): 365–89.

Cunningham, Andrew. 'Fabricius and the "Aristotle Project" in Anatomical Teaching and Research at Padua', in A. Wear, R.K. French and I.M. Lonie, eds, *The Medical Renaissance of the Sixteenth Century*, 195–222. Cambridge: Cambridge University Press, 1985.

Cunningham, Andrew. 'The Medical Professions and the Pattern of Medical Care: The Case of Edinburgh, c1670–c1700', in W. Eckart and J. Geyer-Kordesch, eds, *Heilberufe und Kranke im 17 und 18 Jahrhundert*, 1–28. Münster, 1982.

Cunningham, Andrew. 'Sydenham versus Newton: The Edinburgh Fever Dispute of the 1690s between Andrew Brown and Archibald Pitcairne', in W. F. Bynum and V. Nutton, eds, *Theories of Fever from Antiquity to the Enlightenment*, 71–98. London: Wellcome Institute for the History of Medicine, 1981.

Cunningham, Andrew. 'Sir Robert Sibbald and Medical Education, Edinburgh, 1706', *Clio Medica* 13 (1978): 135–61.

Cunningham, Andrew, ed. *English Manuscripts of Francis Glisson (2). Lectures and Other Papers*. Cambridge: Cambridge Wellcome Unit, 1998.

Cunningham, Andrew, ed. *English Manuscripts of Francis Glisson (1): From 'Anatomia Hepatis', 1654*. Cambridge: Cambridge Wellcome Unit, 1993.

Cunningham, Andrew, and Perry Williams. 'De-Centring the "Big Picture": *The Origins of Modern Science* and the Modern Origins of Science', *British Journal for the History of Science* 26 (1993): 407–32.

Curth, Louise Hill. 'The Care of the Brute Beast: Animals and the Seventeenth-Century Marketplace', *Social History of Medicine* 15 (2002): 375–92.

Cuvier, Georges. *Histoire des Sciences Naturelles depuis leur Origine jusqu'à nos Jours, chez les Peuples Connus. Complétée, Rédigée, Annotée et Publiée Par M. Magdeleine de Saint-Agy. Troisième Partie, Comprenant la Première Moitié du 18e Siècle*, 5 vols. Paris: Fortin, Masson, 1841.

Cuvier, Georges. *Essay on the Theory of the Earth. With Mineralogical Notes, and an Account of Cuvier's Geological Discoveries, by Professor Jameson*, 3rd edn. Edinburgh: Blackwood, 1817.

Cuvier, Georges. *Rapport Historique sur les Progrès des Sciences Naturelles Depuis 1789, et sur leur État Actuel, Présenté à Sa Majesté l'Empereur et Roi* [i.e. Napoleon]*, et Son Conseil d'Etat, le 6 Février 1808, par la Classe des Sciences Physiques et Mathématiques de l'Institut, Conformément à l'Arrête du Gouvernement du 3 Ventôse an X*. Paris: L'Imprimerie Impériale, 1810.

Cuvier, Georges. 'Notice sur l'Établissement de la Collection d'Anatomie Comparée du Muséum', *Annales du Muséum National d'Histoire Naturelle* 2 (1803): 409–14.

Cuvier, Georges. *Programme des Prix de l'Institut National des Sciences et Arts, Proposés l'An IV de la République*. Paris: L'Imprimerie de la République, IV (1796).

D'Agoty, *see* Gautier d'Agoty.

D'Alembert, Jean Le Rond. *Preliminary Discourse to the Encyclopedia of Diderot* [1751], trans. Richard N. Schwab, publ. 1963; rev. edn, Chicago: University of Chicago Press, 1995.

Darlington, Anne Carol. 'The Royal Academy of Arts and its Anatomical Teachings: With an Examination of Art–Anatomy Practices during the Eighteenth and Early Nineteenth Centuries in Britain', 2 vols. PhD thesis, London University, 1991.

Daubenton, Louis Jean Marie. 'La Description du Cabinet du Roi et de la Partie du Cabinet qui a Rapport à l'Histoire Naturelle de l'Homme', in Georges Louis Leclerc, Comte de Buffon and Daubenton, eds, *Histoire Naturelle, Générale et Particulière, Volume 3*, 1–304. Paris: L'Imprimerie Royale, 1749.

Decreta Facultatis Medicinae Parisiensis circa Exercitationes Anatomicas et Chirurgicas a Baccalaureis Deinceps Propria Manu in Cadavere Celebrandas, lata M. Hyacintho Theodoro Baron Decano. Paris: Typis Quillau, Universitatis et Facultatis Medicinae Typographi, 1733.

DeLacy, Margaret. 'Nosology, Mortality, and Disease Theory in the Eighteenth Century', *Journal of the History of Medicine and Allied Sciences* 54 (1999): 261–84.

Delaporte, François. *Nature's Second Kingdom: Explorations of Vegetality in the Eighteenth Century*, trans. Arthur Goldhammer. Cambridge, MA: MIT Press, 1982.

Delaunay, Paul. *Le Monde Médical Parisien au Dix-Huitième Siècle*. Paris, 1906.

Deleuze, J.F.P. *History and Description of the Royal Museum of Natural History. Published by Order of the Administrators of that Establishment. Translated from the French of M. Deleuze* [*by M. Royer*]. *With Three Plans and Fourteen Views of the Galleries, Gardens, and Menagerie*, 2 vols. Paris: Printed for A. Royer, at the Jardin du Roi, 1823.

Description de ce qu'il y a de Remarquable à la Ménagerie et au Cabinet d'Histoire Naturelle. Paris: Chez J. Moronval, 1818.

Dictionnaire des Sciences Médicales. Par une Société de Médecins et de Chirurgiens: MM. Alard, Alibertm, Barbier, Bayle ..., 60 vols. Paris: Crepart, 1812–25.

Diderot, Denis. *Éléments de Physiologie. Édition Critique, avec une Introduction et des Notes par Jean Mayer*. Paris: Librairie Marcel Didier, 1964.

Diderot, Denis, and Jean Le Rond D'Alembert. *Encyclopédie, ou Dictionnaire Raisonné des Sciences, des Arts et des Métiers, par Une Société de Gens de Lettres*, 17 vols. Paris: Briasson, David, Le Breton, Durand, 1751–65.

Dionis, Pierre. *Cours d'Opérations de Chirurgie, Démontrées au Jardin Royal*, 4th edn (first publ. 1708). Paris: Houry, 1740.

Dionis, Pierre. *A Course of Surgical Operations, Demonstrated in the Royal Garden at Paris*. London: F. Tonson, 1710.

Dionis, Pierre. *The Anatomy of Humane Bodies Improv'd, According to the Circulation of the Blood, and All the Modern Discoveries. Publickly Demonstrated at the Theater in the Royal Garden at Paris, by Monsieur Dionis, Chief Surgeon to the Late Dauphiness, and to the Present Duchess of Burgundy. Translated from the Third Edition, Corrected and Enlarged by the Author, with an Ample Dissertation Upon the Nature of Generation; and Several New Systems*. London: H. Bonwicke, W. Freeman, T. Goodwin, M. Wotton, B. Tooke, and S. Manship, 1703.

Dodart, Claude. *Mémoires pour Servir à l'Histoire Naturelle des Plantes*. Paris: L'Imprimerie Royale, 1676.

Doherty, Terence. *The Anatomical Works of George Stubbs*. London: Secker and Warburg, 1974.

Douglas, James. *Myographiae Comparatae Specimen: Or, a Comparative Description of All the Muscles in a Man, and in a Quadruped: Shewing Their Discoverer, Origin, Progress, Insertion, Use and Difference. To which is added an Account of the Muscles Peculiar to a Woman. With an Etymological Table, and Several Useful Indexes*. Edinburgh: A. Kincaid and G. Crawfurd, 1750.

Draper, John William. *History of the Conflict between Religion and Science* [*abridged*], first publ. 1874. New York: Vanguard Press, 1926.

Duchesneau, François. *La Physiologie des Lumières: Empiricisme, Modèles et Théories*. The Hague: Nijhoff, 1982.

Dupont, Michel. *Dictionnaire Historique des Médecins dans et hors de la Médecine*. Paris: Larousse, 1999.

Düring, Monika von, Georges Didi-Huberman and Marta Poggesi. *Encyclopaedia Anatomica: A Complete Collection of Anatomical Waxes* [*of the Museo di Storia Naturale dell'Università di Firenze, Sezione La Specola*]. Cologne: Taschen, 1999.

Duverney, Guichard-Joseph. *Traité des Maladies des Os*, 2 vols. Paris: Jean de Bure, l'Ainé, 1751.

Eloy, Nicolas François Joseph. *Dictionnaire Historique de la Médecine Ancienne et Moderne ou Mémoires Disposés en Ordre Alphabetique pour servir à l'Histoire de cette Science et à celle des Médecins, Anatomistes, Botanistes, Chirurgiens et Chimistes de Toutes Nations*, 1st edn Liège, 2 vols, 1755; 2nd edn Paris, 4 vols, 1778, repr. Brussels: Culture et Civilisation, 1973.

Emerson, Roger L. 'The Founding of the Edinburgh Medical School', *Journal of the History of Medicine and Allied Sciences* 59 (2004): 183–218.

Eriksson, Ruben, ed. *Andreas Vesalius' First Public Anatomy at Bologna 1540, an Eyewitness Report by Baldasar Heseler Medicinae Scholaris, Together with his Notes on Matthaeus Curtius' Lectures on Anatomia Mundini*. Uppsala: Almqvist and Wiksells Boktryckeri AB, 1959.

Eustachi, Bartholomaeus. *Bartolomaei Eustachii Sanctoseverinatis Medici et Philosophi Opuscula Anatomica* [Contains *1. De Renum Structura, Officio, & Administratione*; *2. De Auditus Organis*; *3. Ossium Examen*; *4. De Motu Capitis*; *5. De Vena, Quae Azygos Græcis Dicitur, et de Alia, Quae in Flexu Brachii Communem Profundam Producit*; *6. De Dentibus*]. Venice: for the Author, 1563.

Eustachius, Bartholomaeus. *Tabulae Anatomicae Clarissimi Viri Bartholomaei Eustachii Quas e Tenebris Tandem Vindicatas et Clementis XI Pont. Max. Munificentia Dono Acceptas, Praefatione ac Notis Illustravit, ac Ipso Suae Bibliothecae Dedicationis Die Publici Juris Fecit Jo. Maria Lancisius Intimus Cubicularius, & Archiater Pontificius* (first edn Rome, 1712). Amsterdam: R. & G. Westenios, 1722.

Farrington, Benjamin. 'The Last Chapter of the *De Fabrica* of Vesalius entitled *Some Observations on the Dissection of Living Animals*. Translated from the Latin by Benjamin Farrington', *Transactions of the Royal Society of South Africa* 20 (1931): 1–14.

Fields, William S., and Noreen A. Lemak. *A History of Stroke: Its Recognition and Treatment*. Oxford: Oxford University Press, 1989.

Figlio, Karl. 'The Metaphor of Organization: An Historiographical Perspective on the Bio-Medical Sciences of the Early Nineteenth Century', *History of Science* 14 (1976): 17–53.

Findlen, Paula. 'The Scientist's Body: The Nature of a Woman Philosopher in Enlightenment Italy', in Lorraine Daston and Gianna Pomata, eds, *The Faces of Nature in Enlightenment Europe*, 211–36. Berlin: Berliner Wissenschafts Verlag, 2003.

Flint South, John. *Memorials of the Craft of Surgery in England, Edited by D'Arcy Power*. London: Cassell and Company, Limited, 1886.

Flourens, Pierre. *Recherches Expérimentales sur les Propriétés et les Fonctions du Système Nerveux dans les Animaux Vertébrés*. Paris: Crevot, 1824. As translated in E. Clarke and C. D. O'Malley,.*The Human Brain and Spinal Cord: A Historical Study Illustrated by Writings from Antiquity to the Twentieth Century* (2nd edn), pp. 657–9, first publ. Oxford, 1972; repr. San Francisco: Norman Publishing, 1996.

Flourens, Pierre. *Éloge Historique de François Magendie. Suivi d'une Discussion sur les Titres Respectifs de MM. Bell et Magendie à la Découverte des Fonctions Distinctes des Racines des Nerfs*. Paris: Ganier Frères, 1858.

Fontenelle. *Suite des Éloges des Académiciens de l'Académie Royale des Sciences, Morts depuis l'An MDCCXXII*. Paris: La Veuve Brunet fils, and Marc Bordelet, 1733.

Forbes, Thomas. 'To Be Dissected and Anatomised', *Journal of the History of Medicine and Allied Sciences* 36 (1981): 490–92.

Foster, Sir Michael. *Lectures on the History of Physiology*. Cambridge: Cambridge University Press, 1901.

Foster, W. D. *A Short History of Clinical Pathology*. Edinburgh and London: E. & S. Livingstone Ltd, 1961.

Foucault, Michel. *The Birth of the Clinic: An Archaeology of Medical Perception*, trans. A.M. Sheridan (first publ. in French in 1963). London: Tavistock Publications, 1973.

Fountain, R.B. 'George Stubbs (1724–1806) as an Anatomist', *Proceedings of the Royal Society of Medicine* 61 (1968): 639–46.

Fozzard, J.A. Fairfax. *Professors of Anatomy in the University of Cambridge: The First Two Hundred and Sixty-One Years of the Cambridge University Department of Anatomy, 1707–1968*. Cambridge: FOSSLIA, 1983.

Fraser, Andrew G. *The Building of Old College: Adam, Playfair and the University of Edinburgh*. Edinburgh: Edinburgh University Press, 1989.

Freer, Stephen, and Andrew Cunningham. *Thomas Wharton's* Adenographia. Oxford: Clarendon Press, 1996.

French, Roger. *Dissection and Vivisection in the European Renaissance*. Aldershot: Ashgate, 1999.

French, Roger. 'Sickness and the Soul: Stahl, Hoffmann and Sauvages on Pathology', in Andrew Cunningham and French, eds, *The Medical Enlightenment of the Eighteenth Century*, 88–110. Cambridge: Cambridge University Press, 1990.

Gascoigne, Bamber. *Milestones in Colour Printing 1457–1859: With a Bibliography of Nelson Prints*. Cambridge: Cambridge University Press, 1997.

Gasking, Elizabeth. *Investigations into Generation 1651–1828*. London: Hutchinson, 1967.

Gautier d'Agoty [Dagoty], Arnaud-Eloy. *Cours Complet d'Anatomie, Peint et Gravé en Couleurs Naturelles par M.A.É. Gautier D'Agoty, Second Fils; Et Expliqué par M. Jadelot, Professeur d'Anatomie de la Faculté de Médecine de Nancy, et de l'Académie des Sciences et Belles-Lettres de la même Ville. Le Prix de cette Premiere Partie est de 90 Livres*. Nancy: Chez Jean-Baptiste-Hyacinthe Leclerc, Imprimeur-Libraire, 1773.

Gautier d'Agoty [Dagoty], Jacques. *Anatomie Générale des Viscères en Situation, de Grandeur et Couleur Naturelle, avec l'Angéologie, et la Nevrologie du Corps Humain*. Paris: n.p., 1752.

Gautier d'Agoty [Dagoty], Jacques. *Anatomie de la Tête, en Tableaux Imprimés, Qui Representent au Naturel le Cerveau sous Différentes Coupes, la Distribution des Vaisseaux dans Toutes les Parties de la Tête, les Organes des Sens, & une Partie de la Névrologie; d'après les Pièces Disséquées et Préparées, par M. Duverney, Maître en Chirurgie à Paris, Membre de l'Académie de Chirurgie, et Démonstrateur en Anatomie au Jardin Royal; En Huit Grandes Planches, Déssinées, Peintes, Gravées, & Imprimées en Couleur & Grandeur Naturelle, par Le Sieur Gautier, Seul Privilégié du Roy pour cet Ouvrage; Avec Des Tables Relatives Aux Figures. Dediée au Roy*. Paris: Le Sieur Gautier, M. Duverney and Quillau, 1748.

Gayon, Jean, ed. *Buffon 88: Actes du Colloque International* (Paris–Montbard–Dijon, 14–22 juin, 1988). Paris: Libraire Vrin, 1992.

Gelfand, Toby. 'A "Monarchical Profession" in the Old Regime: Surgeons, Ordinary Practitioners, and Medical Professionalization in Eighteenth-Century France', in Gerald L. Geison, ed., *Professions and the French State, 1700–1900*, 149–80. Philadelphia: University of Pennsylvania Press, 1984.

Gelfand, Toby. *Professionalizing Modern Medicine: Paris Surgeons and Medical Science and Institutions in the 18th Century*. Westport, CT: Greenwood Press, 1980.

Gelfand, Toby. 'The "Paris Manner" of Dissection: Student Anatomical Dissection in Early Eighteenth-Century Paris', *Bulletin of the History of Medicine* 46 (1972): 99–130.

Genga, Bernadino. *Anatomy Improv'd and Illustrated with Regard to the Uses Thereof in Designing: not only Laid Down from an Examen of the Bones and Muscles of the Human Body, but also Demonstrated and Exemplified from the Most Celebrated Antique Statues in Rome. Exhibited in a Great Number of Copper Plates, with All the Figures in Various Views. Intended Originally for the Use of the Royal French Academy of Painting and Sculpture. And Carried on under the Care and Inspection of Charles Errard Director of the Same in Rome. The Dissections Made by Dr. Ber. Genga, Anatomist Royal. The Explanations and Indexes Added by Signr. Can. John Maria Lancisi, P.

Physitian to His Holyness Pope Innocent XI. A Work of Great Use to Painters Sculptors Statuaries and All Others Studious in the Noble Arts of Designing. First Published at Rome by Dom Di Rossi, and Now Re-Engraven by the Ablest Hands in England. London: John Senex, 1723.

Genga, Bernardino. *Anatomia per Uso et Intelligenza del Disegno. Ricercata non Solo su gl'Ossi, e Muscoli del Corpo Humano; ma Dimostrata Ancora su le Statue Antiche più Insigni di Roma. Delineata in più Tavole con Tutte le Figure in Varie Faccie, e Vedute. Per Istudio della Regia Academia di Francia Pittura e Scultura Sotto la Direzzione di Carlo Errard gia Direttore di Essa in Roma. Preparata su'i Cadaveri dal Dottor Bernardino Genga Regio Anatomico. Con le Spiegazioni et Indice del Sigr. Canonico Gio. Maria Lancisi gia Medico Segreto della Sanctissima Memoria di Papa Innocentio XI. Opera Utilissima à Pittori, e Scultori, et ad'Ogni Altro Studioso Delle Nobili Arti del Disegno*. Rome: de Rossi, 1691.

George Stubbs, 1724–1806. Exhibition Catalogue. London: Tate Gallery, 1984.

Geyer-Kordesch, Johanna. 'Georg Ernst Stahl's Radical Pietist Medicine and its Influence on the German Enlightenment', in Andrew Cunningham and Roger French, eds, *The Medical Enlightenment of the Eighteenth Century*, 67–87. Cambridge: Cambridge University Press, 1990.

Gilpin, William. *An Essay on Prints*, 3rd edn. London: R. Blamire, 1781.

Glisson, Francis. *Tractatus de Ventriculo et Intestinis. Cui Praemittitur alius, de Partibus Continentibus in Genere & in Specie, de iis Abdominiis*. London: Brome, 1677.

Goelicke, Andrea Ottomar. *Historia Anatomicae Nova Aeque ac Antiqua, seu Conspectus Plerorumque, si non Omnium, tam Veterum Quam Recentiorum Scriptorum, Qui a Primis Artis Medicae Originibus usque ad Praesentia Nostra Tempora Anatomia Opus Suis Illustrarunt. In Usum Philiatrorum Edita*. Halle: Renger, 1713.

Goldin, Grace. *Work of Mercy: A Picture History of Hospitals*. Ontario: Boston Mills Press, 1994.

Gombrich, E.H. *Art and Illusion: A Study in the Psychology of Pictorial Representation* (first publ. 1960). Princeton, NJ: Princeton University Press, 1972.

Green, Robert Montraville. *A Translation of* [*the Modena, 1792 edition of*] *Luigi Galvani's De Viribus Electricitatis in Motu Musculari Commentarius. Commentary on the Effect of Electricity on Muscular Motion*, trans. Elizabeth Licht. Cambridge, MA, 1953.

Greenhill, W.A. *Thomae Sydenham, M.D. Opera Omnia*. London: The Sydenham Society, 1854.

Grell, Ole Peter. 'Between Anatomy and Religion: The Conversions to Catholicism of the Two Danish Anatomists Nicolaus Steno and Jacob Winslow', in Grell and Andrew Cunningham, eds, *Medicine and Religion in Enlightenment Europe*, 205–21. Aldershot: Ashgate, 2007.

Grell, Ole Peter, Andrew Cunningham and Bernd Roeck, eds. *Health Care and Poor Relief in 18th and 19th Century Southern Europe*. Aldershot: Ashgate, 2005.

Grew, Nehemiah. *The Anatomy of Plants. With an Idea of a Philosophical History of Plants, and Several Other Lectures Read before the Royal Society* [separate and additional title-page: *The Anatomy of Plants, Begun. With a General Account of Vegetation, Founded Thereupon*], 1st edn 1672; 2nd edn; London: Printed by W. Rawlins, for the Author, 1682.

Grew, Nehemiah. *Musaeum Regalis Societatis. Or a Catalogue and Description of the Natural and Artificial Rarities Belonging to the Royal Society and preserved at Gresham College. Whereunto is Subjoyned the Comparative Anatomy of Stomachs and Guts*. London: W. Rawlins for the Author, 1681.

Grew, Nehemiah. *The Comparative Anatomy of Stomachs and Guts Begun. Being Several Lectures Read before the Royal Society. In the Year, 1676* [publ. as part of his *Musaeum Regalis Societatis … Whereunto is Subjoyned the Comparative Anatomy of Stomachs and Guts*]. London: W. Rawlins, for the Author, 1681.

Guerrini, Anita. 'Duverney's Skeletons', *Isis* 94 (2004): 577–603.

Guerrini, Anita. *Experimenting with Humans and Animals: From Galen to Animal Rights*. Baltimore, MD: Johns Hopkins University Press, 2003.

Gunning, Henry. *Reminiscences of the University, Town, and County of Cambridge, from the Year 1780*, 2 vols. London: George Bell, 1854.

Hagner, Michael. 'The Soul and the Brain between Anatomy and *Naturphilosophie* in the Early Nineteenth Century', *Medical History* 36 (1992): 1–33.

Hall, Nicholas H.J., ed. *Fearful Symmetry: George Stubbs, Painter of the English Enlightenment*. New York: Hall and Knight, 2000.

Haller, Albrecht von. *Albrecht Haller. Tagebuch der Studienreise nach London, Paris, Strassburg und Basel 1727 bis 1728. Mit Anmerkungen Herausgegeben von E. Hintzsche*. Bern: Verlag Paul Haupt, 1942.

Haller, Albrecht von. *A Dissertation on the Sensible and Irritable Parts of Animals. Translated from the Latin*. London: J. Nourse, 1755. (Reprint, with introduction by Owsei Temkin, in Supplement to the *Bulletin of the History of Medicine* 4 [1936], 651–99.)

Haller, Albrecht von. *Albrecht Hallers Tagebucher seiner Reisen nach Deutschland, Holland und England 1723–1727. Mit Anmerkungen Herausgegeben von Ludwig Hirzel*. Leipzig: S. Hirzel, 1883.

Haller, Albrecht von. *Bibliotheca Anatomica. Qua Scripta ad Anatomen et Physiologiam Facientia a Rerum Initiis Recensentur. Tomus I ad Annum MDCC. Tomus II ab Anno MDCCI ad MDCCLXXVI*, 2 vols. Zurich: Orell, Gessner, Fuessli, et Socc., 1774, 1777.

Haller, Albrecht von. *Albrecht von Hallers Erster Umriss der Geschäfte des körpoerlichen Lebens*. Berlin: Haude and Spede, 1770.

Haller, Albrecht von. *Opera Minora Emendata, Aucta, et Renovata. Tomi Primi, Pars Prima. Anatomica; Ad Partes Corporis Humani Vitales, Animales,*

Naturales [vol. 2 is Botanical, vol. 3 Pathological]. Lausanne: Franciscus Grasset, 1763.

Haller, Albrecht von. *Sur la Formation du Coeur dans le Poulet; sur l'Oeuil, sur la Structure du Jaune, &c. Second Mémoire*, 2 vols. Lausanne: Marc-Mich. Bousquet & Comp., 1758.

Haller, Albrecht von. *Disputationes ad Morborum Historiam et Curationem Facientes. Quas Collegit, Edidit et Recensuit Albertus Hallerus*, 7 vols. Lausanne: Marci-Michael Bousquet, 1757–60.

Haller, Albrecht von. *Elementa Physiologiae Corporis Humanae*, 8 vols. Lausanne: Marci-Michael Bousquet, 1757–66.

Haller, Albrecht von. *Pathological Observations. Chiefly from Dissections of Morbid Bodies*. London: D. Wilson and T. Durham, 1756. (A translation of *Opuscula Pathologica*, Lausanne, 1754. These cases are again published in Latin in *Opera Minora*, 1763, vol. 3, with others.)

Haller, Albrecht von. 'De Amoenitate Anatomes Oratio Dicta 17. Sept. 1742, Cum Tres Doctores Medicinae Crearet Auctor', in *Opuscula sua Anatomica. De Respiratione, De Monstris, Aliaque Minora Recensuit, Emendavit, Auxit*, 327–36. Göttingen: Jo. Wilh. Schmidt, 1751.

Haller, Albrecht von. *Semestris Hyberni Anatomen Primam in Cadavere Puellae Auditoribus suis Indicit*. Göttingen: Vandenhoek, 1738.

Haller, Albrecht von. *Omnes Rei Anatomicae Studiosus ad Dissectionem Cadaveris Humani Invitat*. Göttingen: Vandenhoek, 1737.

Hamilton, James, Junior. *Observations on the Seats and Causes of Diseases: Illustrated by the Dissections of the Late Professor Morgagni of Padua*. Edinburgh: For P. Hill, 1795.

Hamy, M. E. T. 'Recherches sur les Origines de l'Enseignement de l'Anatomie Humaine et de l'Anthropologie au Jardin des Plantes', *Nouvelles Archives du Muséum d'Histoire Naturelle*, 3rd ser. 7 (1895): 1–22.

Hamy, M. E. T. 'Les Derniers Jours du Jardin du Roi et la Fondation du Muséum d'Histoire Naturelle', in *Centenaire de la Fondation du Muséum d'Histoire Naturelle 10 Juin 1793–10 Juin 1893. Volume Commémoratif Publié par les Professeurs du Muséum*, 3–162. Paris: Imprimerie Nationale, 1893.

Hannaway, Caroline. 'Vicq d'Azyr, Anatomy and a Vision of Medicine', in Ann La Berge and Mordechai Feingold, eds, *French Medical Culture in the Nineteenth Century*, 280–95. Amsterdam: Rodopi, 1994.

Hansen, Julie V. 'Resurrecting Death: Anatomical Art in the Cabinet of Dr Frederik Ruysch', *The Art Bulletin* 78 (1996): 663–79.

Hartenfels, George Christopher Peter. *Elephantographia Curiosa, seu Elephanti Descriptio, Juxta Methodum et Leges Imperialis Academiae Leopoldino Carolinae Naturae Curiosorum Adornata, Multisque Selectis Observationibus Physicis, Medicis et Jucundis Historiis Referta, cum Figuris Aenis*. Erfurt: The Author, 1715.

Harvey, William. *Praelectiones Anatomiae Universalis. Edited with an Autotype Reproduction of the Original by a Committee of the Royal College of Physicians of London*. London: Churchill, 1886.

Harvey, William. *Opera Omnia: A Collegio Medicorum Londinensis Edita*. London: Bowyer, 1766.

Harvey, William. *Opera sive Exercitatio Anatomica de Motu Cordis et Sanguinis in Animalibus*. Leiden: Johannes van Kerckhem, 1737.

Harvey, William. *The Anatomical Exercises of Dr William Harvey, Professor of Physick, and Physician to the King's Majesty, Concerning the Motion of the Heart and Blood. With the Preface of Zachariah Wood Physician of Rotterdam. To which is added Dr James De Back his Discourse of the Heart, Physician in Ordinary to the Town of Rotterdam*. London: Lowndes, 1653.

Harvey, William. *The Anatomical Exercitations Concerning the Generation of Living Creatures: To which are Added Particular Discourses, of Births, and of Conceptions, &c.* London: Octavian Pulleyn, 1653.

Harwood, Busick. *A Synopsis of a Course of Lectures on the Philosophy of Natural History and the Comparative Structure of Plants and Animals*. Cambridge: Printed by Francis Hodson, 1812.

Harwood, Busick. *A Synopsis of a Course of Lectures on Comparative Anatomy and Physiology*. Cambridge: Printed by Francis Hodson, 1807.

Harwood, Busick. *A System of Comparative Anatomy and Physiology by B. Harwood, M. D., F. R. S. and F. S. A., Professor of Anatomy in the University of Cambridge*. Cambridge: W. H. Lunn etc, 1796.

Harwood, Busick. *A Synopsis of a Course of Lectures on Anatomy and Physiology*, 3rd edn (1st edn 1787). Cambridge: J and J. Merill, etc., 1792.

Heister, Lorenz. *Oratio de Incrementis Anatomiae in hoc Seculo XVIII, Publice Dicta. MDCCXX*. Wolfenbüttel: Widow of G. Freytag, 1720.

Helm, Jürgen, and Karin Stukenbrock, eds. *Anatomie: Sektionen einer Medizinischen Wissenschaft im 18. Jahrhundert*. Wiesbaden: Franz Steiner Verlag, 2003.

Héroard, Jehan. *Hippostologie, c'est à dire, Discours des Os du Cheval*. Paris: Mamert Patisson, Imprimeur Ordinaire du Roy, 1599.

Hewson, William, F. R. S. 'Experimental Inquiries. Part the First. An Inquiry into the Properties of the Blood, with Remarks on Some of its Morbid Appearances; and an Appendix, Relating to the Discovery of the Lymphatic System in Birds, Fish, and the Animals Called Amphibious' (1771), in George Gulliver, ed., *The Works of William Hewson, F. R. S.*, 1–112. London: The Sydenham Society, 1846.

Hewson, William. 'Experimental Inquiries. Part the Second. A Description of the Lymphatic System in the Human Subject, and in Other Animals. Illustrated with Plates. Together with Observations on the Lymph, and the Changes which it Undergoes in Some Diseases' (1774), in George Gulliver, ed., *The Works of William Hewson, F. R. S.*, 113–204. London: The Sydenham Society, 1846.

Hewson, William. 'Experimental Inquiries. Part the Third. A Description of the Red Particles of the Blood in the Human Subject and Other Animals; with an Account of the Structure and Offices of the Lymphatic Glands, of the Thymus Gland, and of the Spleen: Being the Remaining Part of the Observations and Experiments of the Late Mr William Hewson, F. R. S., and Teacher of Anatomy' (1777), [ed.] Magnus Falconar, Surgeon, and Teacher of Anatomy [and Brother-in-Law of Hewson], in George Gulliver, ed., *The Works of William Hewson, F. R. S.*, 205–316. London: The Sydenham Society, 1846.

Hilloowala, Rumy, Bendetto Lanza, Maria Luisa Puccetti and Antonio Martelli. *The Anatomical Waxes of La Specola*, trans. Julianne Hilloowala, Rumy Hilloowala and Joseph Renahan, supplement to the original *Le Cere Anatomiche della Specola*. Florence: Arnaud, 1995.

Hippocrates: Pseudepigraphic Writings, ed. and trans. Wesley D. Smith. Leiden: Brill, 1990.

Hoage, R. J., and William A. Deiss, eds. *New Worlds, New Animals. From Menagerie to Zoological Park in the Nineteenth Century*. Baltimore, MD: The Johns Hopkins University Press, 1886.

Home, Sir Everard. *Lectures on Comparative Anatomy in which are Explained the Preparations in the Hunterian Collection. Illustrated by Engravings. To which is Subjoined Synopsis Systematis Regni Animalis, Nunc Primum ex Ovi Modificationibus Propositi. In Four Volumes* (i.e. vols 3–6). London: G. and W. Nicol and Longman, 1823–28.

Home, Sir Everard. *Lectures on Comparative Anatomy in which are Explained the Preparations in the Hunterian Collection. Illustrated by Engravings. In Two Volumes*, 2 vols (6 eventually). London: G and W. Nicol and Longman, 1814–28.

Horn, D.B. 'The Anatomy Classrooms in the Present Old College', *University of Edinburgh Journal* 22 (1965–66): 65–71.

Houstoun, James. *Dr Houstoun's Memoirs of his Own Lifetime. Collected from the Author's Original Manuscripts, and Publish'd by Jacob Bickerstaff, M.A.* London: Lawton Gilliver, 1747.

Howard, Rio. 'Guy De La Brosse: The Founder of the Jardin des Plantes in Paris', PhD thesis, Cornell University, 1974.

Huard, Pierre. 'L'Enseignement Médico-Chirurgical', in René Taton, ed., *Enseignement et Diffusion des Sciences en France au XVIIIe Siècle*, 171–236. Paris: Hermann, 1964.

Hunter, John. *Essays and Observations on Natural History, Anatomy, Physiology, Psychology, and Geology. Being His Posthumous Papers on Those Subjects, Arranged and Revised by Richard Owen*, 2 vols. London: John van Voorst, 1861.

Hunter, John. *The Works of John Hunter, F. R. S. with Notes. Edited by James F. Palmer*, 4 vols; vol. 4 [*Observations on Certain Parts of the Animal Oeconomy (1786), Ed. Richard Owen; Other Papers Delivered to the R. S.; Croonian*

Lectures on the Muscles]. London: Longman, Rees, Orme, Brown, Green, and Longman, 1837.

Hunter, John. *Lectures on the Principles of Surgery. With Notes by James F. Palmer*. 1835.

Hunter, John. *A Treatise on the Blood, Inflammation and Gun-Shot Wounds, by the late John Hunter. To which is Prefixed, a Short Account of the Author's Life, by his Brother-in-Law, Everard Home*. London: George Nicol, 1794.

Hunter, John. 'Lectures on the Principles of Surgery', in James F. Palmer, ed., *The Works of John Hunter, F.R.S. With Notes. Volume 1*, 199–643. London: Longman, Rees, Orme, Brown, Green and Longman, 1786–87.

Hunter, William. *Two Introductory Lectures, Delivered by Dr. William Hunter, to his Last Course of Anatomical Lectures, at his Theatre in Windmill Street: as they were left Corrected for the Press by himself. To which are Added, Some Papers Relating to Dr. Hunter's Intended Plan, for Establishing a Museum in London, for the Improvement of Anatomy, Surgery, and Physic*. London: Printed by order of the Trustees, for J. Johnson, 1784.

Hunter, William. *Medical Commentaries. Part I. Containing a Plain and Direct Answer to Professor Monro Jun. Interspersed with Remarks on the Structure, Functions, and Diseases of Several Parts of the Human Body. The Second Edition* (1st edn, 1762). London: S. Baker etc, 1777.

Hunter, William. *Anatomia Uteri Humani Gravidi Tabulis Illustrata. The Anatomy of the Human Gravid Uterus Exhibited in Figures*. Birmingham: John Baskerville, 1774.

Hussey, Christopher. *The Picturesque: Studies in a Point of View*. London: Frank Cass, 1967 (orig. publ. 1927).

Imbault Huart, Marie-José. 'L'Ecole Pratique de Dissection de Paris de 1750 à 1822: Ou, l'Influence du Concept de Médicine Pratique et de Médecine d'Observation dans l'Enseignement Médico-Chirurgical au XVIIIe Siècle et au Debut du XIXe Siècle', thèse, Université de Paris I, 1975.

Jacyna, L.S. 'Images of John Hunter in the Nineteenth Century', *History of Science* 21 (1983): 85–108.

Jaeckel, Gerhard. *Die Charité. Die Geschichte eines Weltzentrums der Medizin von 1710 bis zur Gegenwart*. Munich: Ullstein, 2001 (first publ. 1963).

Jarcho, Saul. 'Some Lost, Obsolete, or Discontinued Diseases: Serous Apoplexy, Incubus, and Retrocedent Ailments', *Transactions of the College of Physicians of Philadelphia* 49 (1980): 241–66.

Jarcho, Saul. 'Giovanni Battista Morgagni: His Interests, Ideas and Achievements', *Bulletin of the History of Medicine* 22 (1948): 503–27.

Jenty, Charles Nicholas. *A Course of Anatomico-Physiological Lectures on the Human Structure and Animal Œconomy; Interspersed with Various Critical Notes, Extracted from Memoirs, Transactions of Learned Societies, &c. and Pathological Observations, Deduced from the Dissection of Morbid Bodies. Including whatever is most Valuable in the Works of All the Most Eminent Professors on the Subjects. Particularly Winslow, Haller, Ruysch, Morgagn* [*sic*], *Monro, Heister. Illustrated with an Historical Compendium of the Rise, Progress, and Discoveries, which have been made in the Animal Œconomy, gradually Traced to the Present Time. To which are Prefixed, Two Essays on the Arts of Dissecting, Injecting and Making Anatomical Preparations*, 3 vols. London: James Rivington and James Fletcher, 1757.

Jenty, Charles Nicholas. *An Essay on the Demonstration of the Human Structure, Half as large as Nature, in Four Tables. From the Pictures Painted after Dissections, for that Purpose. Disposed in such a Manner, as to Represent gradually all the Capital Parts of the Human Body in their natural Situation, as they appear in Dissection when the Sanguiferous Vessels are Injected. Calculated to Convey a Clearer Artificial Idea of the Animal Œconomy than has hitherto appeared in any other Anatomical Figures*. London: Author, 1757.

Jocelyn, H. D., and B. P. Setchell, eds. *Renier De Graaf on the Human Reproductive Organs. An Annotated Translation of Tractatus De Virorum Organis Generationi Inservientibus (1668) and De Mulierum Organis Generationi Inservientibus Tractatus Novus (1672), Journal of Reproduction and Fertility* Supplement no. 17. Oxford: Blackwell Scientific Publications, 1972.

Johnstone, R. W. *William Smellie the Master of British Midwifery*. Edinburgh: E. & S. Livingstone, 1952.

Johnstone, William. *Roll of Commissioned Officers in the Medical Service of the British Army*. Aberdeen: Aberdeen University Press, 1917.

Kaiser, Wolfram, and Karl-Heinz Krosch. 'Zur Geschichte der Medizinischen Fakultät der Universität Halle im 18. Jahrhundert. Georg Daniel Coschwitz (1679–1729)', *Wissenschaftliche Zeitschrift der Martin-Luther-Universität Halle-Wittenberg. Mathematisch-Naturwissenschaftliche Reihe* 13 (1964): 617–816.

Kardel, Troels, ed. *Steno on Muscles. Introduction, Texts, Translations*, Transactions of the American Philosophical Society, vol. 84, part 1. Philadelphia: American Philosophical Society, 1994.

Kegel-Brinkgreve, E., and Luyendijk-Elshout, A. M., eds. *Boerhaave's Orations. Translated with Introduction and Notes*. Leiden: Leiden University Press, 1983.

Kemp, Martin, ed. *Dr William Hunter at the Royal Academy of Arts*. Glasgow: University of Glasgow Press, 1975.

Kerckhem, Johannes van, and Bernhard Siegfried Albinus, eds. *Hieronymi Fabricii ab Aquapendente Opera Omnia Anatomica et Physiologica. Una cum Prefatione Bernardi Siegfriedi Albini. Editio Novissima*. Leiden: Johannes van Kerckhem, 1738.

Keynes, Geoffrey. *The Life of William Harvey* (1966), 1978 repr. Oxford: Clarendon Press.

Klein, J. T. *Historiae Piscium Naturalis Promovendae Missus Primus De Lopillis Eorumque Numero in Cranis Piscium, Cum Praefatione de Piscium Auditu*, 5 vols. Danzig: Schreiber, 1740–49.

Knoeff, Rina. 'Moral Lessons of Perfection: A Comparison of Mennonite and Calvinist Motives in the Anatomical Atlasses of Bidloo and Albinus', in Ole Peter Grell and Andrew Cunningham, eds, *Medicine and Religion in Enlightenment Europe*, 121–43. Aldershot: Ashgate, 2007.

Knoeff, Rina. *Herman Boerhaave (1668–1738): Calvinist Chemist and Physician*. Amsterdam: Koninklijke Nederlandse Akademie van Wetenschappen, 2002.

Krivatsky, P. 'Le Blon's Anatomical Color Engravings', *Journal of the History of Medicine and Allied Sciences* 23 (1968): 153–8.

Kühn, C. Gottlob, ed. *Georgii Baglivi Opera Omnia Medica-Practica et Anatomica. Editionem Reliquis Omnibus Emendatiorem et Vita Auctoris Auctam. Curavit C. Gottl. Kühn*, 2 vols. Leizig: Leopold Vossius, 1827.

Kurpershoek, Ernest. *De Waag Op De Nieuwmarkt* [Amsterdam Anatomical Theatre]. Amsterdam: Stadsuitgeverij Amsterdam, 1994.

Lacépède, Bernard, and Georges Cuvier. *La Ménagerie du Muséum d'Histoire Naturelle, ou Les Animaux Vivants, Peints d'après Nature, sur Vélin, par le Citoyen Maréchal, Peintre du Muséum, et Gravés au Jardin des Plantes, avec l'Agrément de l'Adminstration, par le Citoyen Miger, Graveur, Membre de ci-devant Académie Royale de Peinture, avec une Note Descriptive et Historique pour Chaque Animal, par Les Citoyens Lacépède et Cuvier, Membres de l'Institut*. Paris: Chez Miger, Patris, Grandcher, Dentu, 1801.

Laissus, Yves. 'Le Jardin Du Roi [au XVIII Siècle]', in René Taton, ed., *Enseignement et Diffusion des Sciences en France au XVIIIe Siècle*, 287–341. Paris: Hermann, 1964.

Lamarck, Jean Baptiste. *Philosophie Zoologique*. Paris: Dentu, 1809.

Lamy, Edouard. *Les Cabinets d'Histoire Naturelle en France au XVIIIe Siècle et le Cabinet du Roi (1635–1793)*. Paris: Chez l'Auteur, 1931.

Lamy, Guillaume. *Discours Anatomiques de Monsieur Lamy, Docteur en Médecine de la Faculté de Paris. Revues et Augmentés de Toutes les plus Curieuses Découvertes des Anatomistes Modernes. Avec Plusieurs Lettres du Même Auteur, Et ses Reflexions sur ses Discours. Seconde Edition*. Brussels and Paris: Laurent d'Houry, 1685 (first publ. 1675).

Lancisi, Giovanni Maria. *De Aneurysmatibus. Opus Posthumum. Aneurysms: The Latin Text of Rome, 1745*, trans. Wilmer Cave Wright. New York: Macmillan, 1952.

Lancisi, Jean Marie (Giovanni Maria). 'Extract of a Letter from Jean Marie Lancisi, Prof. Anat. Rom. To Mr Bourdelot, Giving an Account of Mr Malpighi, the Circumstances of His Death, and What Was Found Remarkable at the Opening of his Body. Being Art. 1 of the 3d. Journal of Brunets Progres de la Medecine', *Philosophical Transactions of the Royal Society of London* 19 (1697): 467–71.

Lapauze, Henry [sic]. *Histoire de l'Académie de France à Rome*, 2 vols. Paris: Plon, 1934.

Lassus, Pierre. *Essai ou Discours Historique et Critique sur les Découvertes Faites en Anatomie par les Anciens et par les Modernes*. Paris: M. Lambert et F. J. Baudouin, 1783.

Latour, Bruno. *Science in Action: How to Follow Scientists and Engineers through Society*. Milton Keynes: Open University Press, 1987.

Lawrence, Christopher. 'Alexander Monro *Primus* and the Edinburgh Manner of Anatomy', *Bulletin of the History of Medicine* 62 (1988): 193–214.

Lawrence, John. *The History and Delineation of the Horse, in all his Varieties*. London: James Cundee and John Scott, 1809.

Le Cat, Claude Nicholas. 'An Account of Glasses of a New Contrivance, for Preserving Pieces of Anatomy or Natural History in Spirituous Liquors', *Philosophical Transactions of the Royal Society of London* 46 (1749–50): 6–8.

Le Cat, Claude Nicholas. 'Discours Prononcé par M. Le Cat à la Prémière Ouverture de ses Cours d'Anatomie et d'Opérations', *Bibliothèque françoise ou Histoire littéraire de la France* 25 (1737): 248–61.

Lennox-Boyd, Christopher, Rob Dixon and Tim Clayton. *George Stubbs: The Complete Engraved Works*. London: Stipple Publishing, 1989.

Leveling, Heinrich Palmaz, ed. *Anatomische Erklärung der Original-Figuren von Andreas Vesal, samt einer Anwendung der Winslowischen Zergliederungslehre in Sieben Büchern*. Ingolstadt: Attenkhouer, 1783.

Lieberkuhn, Johann Nathaniel. *Dissertationes Quatuor, Nimirum de Valvula Coli et Usu Processus Vermicularis. De Fabrica et Actione Villorum Intestinorum Tenuium Hominis. Sur les Moyens Propres à Decouvrir la Construction des Visceres. Description d'un Microscope Anatomique. Omnia Nunc Primum in Unum Collecta et Edita Cura et Studio Joannis Sheldon, Anatomes Praelectoris et Societatis Chirurgorum Londinensis Sodalis*. London: T. Cadell etc., 1782 (1744).

Lieutaud, Joseph. *Historia Anatomico-Medica, Sistens Numerosissima Cadaverum Humanorum Extispicia, Quibus in Apricum Venit Genuina Morborum Sedes; Horumque Reserantur Causae, vel Patent Effectus. Opus Quadripartitum, Cujus Liber Primus Recenset Laesiones Internas Abdominis. Secundus Exhibet Variam Stragem Pectoris. Tertius Prodit Diversam Labem Cerebri. Quartus Vero Vitia Externa Colligit. Recensuit et Suas Observationes Numero Plures Adjecit, Uberrimumque Indicem Nosologico Ordine Concinnavit Antonius Portal*, 4 books in 2 vols. Paris: Vincent, 1767.

Lieutaud, Joseph. *Elementa Physiologiae, Juxta Solertiora, Novissimaque Physicorum Experimenta, & Accuratiores Anatomicorum Observationes Concinnata* (1st edn 1746). Amsterdam: Fratres Detournes, 1749.

Lindberg, David C., and Ronald L. Numbers, eds. *When Science and Christianity Meet*. Chicago: Chicago University Press, 2003.

Lindeboom, G.A. *Herman Boerhaave: The Man and his Work*. London: Methuen, 1968.

Linebaugh, P. 'The Tyburn Riot against the Surgeons', in Douglas Hay, ed., *Albion's Fatal Tree: Crime and Society in 18th-Century England*, 65–117. London: Allen Lane, 1975.

Lister, Dr Martin. *A Journey to Paris in the Year 1698*. London: Jacob Tonson, 1699.

Loisel, Gustave. *Histoire des Ménageries de l'Antiquité à nos Jours*, 2 vols. Paris: Octave Doin et Fils; Henri Laurens, 1912.

Long, Esmond R. *A History of Pathology*. New York: Dover Publications, 1965.

Louis, A. *Éloges Lus dans les Séances Publiques de l'Académie Royale de Chirurgie de 1750 à 1792*. Paris: J.-B. Ballière et Fils, 1859.

Lovejoy, Arthur O. *The Great Chain of Being. A Study of the History of an Idea* (first publ. 1936). New York: Harper and Row, 1965.

Luchtmans, Pieter. *Oratio Inauguralis de Anatomicis Saeculi XVIII Incrementis, Publice Dicta Die XVII Martii MDCCLX. Quum Ordinariam Medicinae, Anatomes et Chirurgiae Professionem in Academia Trajectina* [= Utrecht] *Auspicaretur*, 2nd edn. Utrecht: Joannis Broedelet, 1760.

Lyon, John, and Philip R. Sloan, eds. *From Natural History to the History of Nature: Readings from Buffon and his Critics*. Notre Dame: University of Notre Dame Press, 1981.

Lyser, Michael. *The Art of Dissecting the Human Body, in a Plain, Easy, and Compendious Method. With the Several Instruments Necessary, Curiously Engraved; Wherein all the Different Parts of the Human Body are Described, but Particularly the Muscles, in a Full and Accurate Manner, and their Uses Clearly Explained. Translated from the Latin of Lyserus. By G. Thomson, M. D.* London: Joseph Davidson, 1740.

Lyser, Michael. *Culter Anatomicus. Hoc est Methodus brevis facilis ac perspicua artificiose et compendiose humana incidendi cadavera. Cum nonnullorum instrumentorum iconibus*. Copenhagen: Lamprecht, 1653.

Ma, Eikoh. 'Japan's Encounter with Western Medical Science: "The beginning of Dutch study", Being the Memoirs of an 18th-Century Doctor', *Bulletin of the History of Medicine* 33 (1959): 315–29.

Macdonald, Helen. *Human Remains: Dissection and its Histories*. New Haven, CT: Yale University Press, 2006.

Maehle, Andreas-Holger. 'The Ethical Discourse on Animal Experimentation, 1650–1900', in Andrew Wear, Johanna Geyer-Kordesch and Roger French, eds, *Doctors and Ethics: The Earlier Historical Settings of Professional Ethics*, 203–51. Amsterdam: Rodopi, 1993.

Maerker, Anna. '"Turpentine Hides Everything": Autonomy and Organization in Anatomical Model Production for the State in Late Eighteenth-Century Florence', *History of Science* 45 (2007): 257–86.

Maerker, Anna. 'The Anatomical Models of La Specola: Production, Uses, and Reception', *Nuncius* 21, special issue, 'Felice Fontana and his collections' (2006): 295–321.

Magendie, François. *Elementary Compendium of Physiology; for the Use of Students* [*Précis Élémentaire de Physiologie*], trans. E. Milligan. Edinburgh: John Carfrae and Son, 1826.

Malpighi, Marcello. 'Storie Anatomico-Mediche', in Luigi Belloni, ed., *Opera Scelte di Marcello Malpighi*. Turin: Unione Tipografico-Editrice Torinese, 1967.

Malpighi, Marcello. *Anatome Plantarum. Cui Subjungitur Appendix, Iteratas et Auctas Ejusdem Authoris de Ovo Incubato Observationes Continens. Regiae Societati, Londini ad Scientiam Naturalem Promovendam Institutae, Dicata*. London: John Martyn, 1675.

Manget, Jean Jacques, and Daniel Le Clerc, eds, and by John Nutt and John Morphew. *Bibliotheca Anatomica, Medica, Chirurgica, &c. Containing a Description of the Several Parts of the Body: Each Done by Some One or More Eminent Physician or Chirurgeon; with their Diseases and Cures. Wherein are not only All the Tracts of Use that are in the Second Edition of the Bibliotheca Anatomica, Lately Publish'd by Daniel Clericus and Jacob Mangetus, in Two Volumes in Folio, but an Addition also of near Double the Number of Other Curious Tracts, Which were either Omitted in the said Bibliotheca, or Have Been Publish'd since, Some of them Translated, Others Faithfully Abridg'd; Very Few of which were ever before in English. Illustrated with Several Hundred Figures, Done by the Best Artists, and from the Truest Designs; as will Appear from the Name to Each Figure*, 3 vols. London: Printed by John Nutt; and Sold by W. Lewis, Dan Browne, J. Pemberton, R. Knaplock, R. Wilkin, M. Atkins, W. Taylor, T. Hone, A. Bell and J. Morphew, 1711–14.

Manget, Jean Jacques, and Daniel Le Clerc, eds. *Bibliotheca Anatomica sive Recens in Anatomia Inventorum Thesaurus Locupletissimus, in quo Integra atque Absolutissima Totius Corporis Humani Descriptio, eiusdemque Œconomia e Praestantissimorum quorumque Anatomicorum Tractibus Singularibus, tum hactenus in Lucem Editis, tum etiam Ineditis, Concinnata Exhibetur. Adjecta est Partium Omnium Administratio Anatomica, Cum Variis earundem Praeparationibus Curiosissimus. Digesserunt, Tractatus Suppleverunt, Argumenta, Notulas, & Observationes Anatomico-Practicas Addiderunt Daniel Le Clerc & I. Jacobus Mangetus, M. M., D. D. Cum Indicibus Necessariis, Figurisque Æneis*, 2 vols. Geneva: Jean Antoine Chovet, 1685.

Marino, Luigi. 'Soemmering, Kant and the Organ of the Soul', in Stefano Poggi and Maurizio Bossi, eds, *Romanticism in Science: Science in Europe, 1790–1840*, 127–42. Dordrecht: Kluwer Academic Publishers, 1994.

Marland, Hilary. 'The "*Burgerlijke*" Midwife: The *Stadsvroedvrouw* of Eighteenth-Century Holland', in Marland, ed., *The Art of Midwifery: Early Modern Midwives in Europe*, 192–213. London: Routledge, 1993.

Martin, Julian. 'Sauvages's Nosology: Medical Enlightenment in Montpellier', in Andrew Cunningham and Roger French, eds, *The Medical Enlightenment of the Eighteenth Century*, 111–37. Cambridge: Cambridge University Press, 1990.

Martínez, Martín. *Anatomia completa del hombre, con todos los hallazgos, nuevas doctrinas y observaciones raras ... y muchas advertencias necessarias para la cirugia*. Madrid: P. del Castillo, 1728.

Martínez-Vidal, Àlvar, and José Pardo-Tomàs. 'Anatomical Theatres and the Teaching of Anatomy in Early Modern Spain', *Medical History* 49 (2005): 251–80.

Mascagni, Paolo. *Anatomia Universa XLIV Tabulis Aeneis Juxta Archetypum Hominis Adulti Accuratissime Repraesentata dehinc ab Excessu Auctoris Cura ac Studio Eq. Andreae Vaccà Berlinghieri, Jacobi Barzellotti et Joannis Rossini in Pisana Universitate Professorum Absoluta atque Edita*. Pisa: Nicholo Capurro, 1823.

Mascagni, Paolo. *Anatomia per Uso degli Studiosi di Scultura e Pittura. Opera Postuma di Paolo Mascagni* [edited by Bernardino and Aurelio Mascagni]. Florence: Dalla Tipografia di Giovanni Marenigh, 1816.

Mascagni, Paolo. *Vasorum Lymphaticorum Corporis Humani Historia et Ichnographia Auctore Paulo Mascagni in Regio Senarum Lyceo Publico Anatomes Professore*. Siena: Pazzini Carli, 1787.

Maulitz, Russell C. *Morbid Appearances: The Anatomy of Pathology in the Early Nineteenth Century*. Cambridge: Cambridge University Press, 1987.

Mayer, Joseph. *Memoirs of Thomas Dodd, William Upcott, and George Stubbs, R. A*. Liverpool: Joseph Mayer, F. S. A., 1879.

McMath, James. *The Expert Mid-Wife: A Treatise of the Diseases of Women with Child, and in Child-Bed: As also, of the Best Ways and Means of Help in Natural and Unnatural Labours*. Edinburgh: George Mosman, 1694.

Meijer, Miriam Claude. *Race and Aesthetics in the Anthropology of Petrus Camper (1722–1789)*. Amsterdam: Rodopi, 1999.

Messbarger, Rebecca. 'Waxing Poetic: Anna Morandi Manzolini's Anatomical Sculptures', *Configurations* 9 (2001): 65–97.

Meyer, Arthur William. *An Analysis of the De Generatione Animalium of William Harvey*. Stanford, CA: Stanford University Press, 1936.

Monro, Alexander (*Primus*). *Traité d'Ostéologie, Traduit de l'Anglois de M. Monro, Professeur d'Anatomie, et de la Société Royale d'Edimbourg: Où l'on a Ajouté des Planches en Taille-Douce, qui Représentent au Naturel tous les Os de l'Adulte et du Foetus, avec leurs Explications. Par M. Sue, Professeur et Démonstrateur d'Anatomie aux Ecoles Royales de Chirurgie, de l'Académie Royale de Peinture et de Sculpture, Censeur Royal, et Conseiller du Comité de l'Académie Royale de Chirurgie*, 2 vols. Paris: Guillaume Cavelier, 1759.

Monro, Alexander (*Primus*). *An Essay on Comparative Anatomy*. London: John Nourse, 1744.

Monro, Alexander (*Secundus*). *The Structure and Physiology of Fishes Explained and Compared with Those of Man and Other Animals. Illustrated with Figures*. Edinburgh: Charles Elliot, 1785.

Monro, Alexander (*Secundus*). *Observations, Anatomical and Physiological, Wherein Dr Hunter's Claim to Some Discoveries is Examined. With Figures*. Edinburgh: [For the Author], 1758.

Monro, Alexander (*Secundus*). *De Venis Lymphaticis Valvulosis et de earum inprimis Origine*. Berlin: Christianus Frideric. Henningius, 1757.

Monro, Alexander (*Secundus*). *Dissertatio Medica Inauguralis, De Testibus et de Semine in Variis Animalibus. Pro Gradu Doctoratus Eruditorum Examini Sujicit*. Edinburgh: G. Hamilton and J. Balfour, 1755.

Monro, Alexander (*Secundus*). 'A Description of the Seminal Vessels', *Essays and Observations, Physical and Literary. Read before a Society in Edinburgh, and Published by Them* 1 (1754): 396–402.

Monro, Donald. 'The Dissection of a Woman with Child; and Remarks on Gravid Uteri', *Essays and Observations, Physical and Literary. Read before a Society in Edinburgh, and Published by Them* 1 (1754): 403–25.

Monti, Maria Teresa, ed. *Albrecht Von Haller: Commentarius De Formatione Cordis in Ovo Incubato. Edizione Critica* [English translation by Helen Tooke], *Studia Halleriana VI*. Basel: Die Albrecht von Haller-Stiftung der Burgergemeinde Bern, 2000.

Moore, Wendy. *The Knife Man: The Extraordinary Life and Times of John Hunter, Father of Modern Surgery*. London: Bantam Press, 2005.

Morgagni, Giovanni Battista. 'Le Autobiografie', in A. Pazzini, ed., *Opera Posthuma. Ms. Laurenziano Fondo Ashburnhamian 227–159*. Rome: Istituto di Storia della Medicina dell'Università di Roma, 1964.

Morgagni, Giovanni Battista. *Autobiografia. Introduzione e Commento di Adalberto Pazzini, Trascrizione del Dr. L.B. Borri*. Forlì: Cassa dei Risparmi di Forlì, 1956.

Morgagni, Giovanni Battista. 'Animadversio I: Anatomes Tradendae Methodus, Ejusque Elementa', 1717. Reprinted in his *Opera Omnia, Vol. 2, Adversaria Anatomica Omnia*. Venice, 1762, 'Adversaria Anatomica Altera. Observationes Complectuntur Distributas in L. Animadversiones ad Librum I, *Theatri Anatomici* Cl. V. Jo. Jacobi Mangeti Sereniss. Regis Prussiae Archiatri', 32–33.

Morgagni, John Baptist [Giovanni Battista]. *The Seats and Causes of Diseases Investigated by Anatomy. In Five Books Containing a Great Variety of Dissections, with Remarks, to which are added Very Accurate and Copious Indexes of the Principal Things and Names Therein Contained*, trans. Benjamin Alexander, MD, 3 vols. London: A. Millar etc, 1769; 1960 facsimile edn *With a Preface, Introduction and New Translation of Five Letters by Paul Klemperer, M. D., Sc. D. (H. C)*. Haffner, New York.

Morrison, Venetia. *The Art of George Stubbs*. London: Quarto Publishing, 1989.

Myasita, Saburo. 'A Bibliography of the Dutch Medical Books Translated into Japanese', *Archives Internationales d'Histoire des Sciences* 25 (1975): 8–72.

N., T. 'The Lion and Horse. An Engraving from a Picture of the Late Mr. Stubbs', *The Sporting Magazine* 32 (1808): 155–7.

Newman, Thelma R. *Wax as Art Form*. London: Thomas Yoseloff, 1966.

Nicholls, Frank. *Syllabus Partium Corporis quae Urinae Excretione, Generatione, Sanguinis Circulationi, Respirationi, Vocificationi, Visui, Auditui, Nervis & Olfactui Inserviunt, Praelectionibus Aptatus Mox Habendis in Theatro Anatomico Chirurgorum Londinensium*. London, 1733.

Nicolson, Malcolm. 'Giovanni Battista Morgagni and Eighteenth-Century Physical Examination', in Christopher Lawrence, ed., *Medical Theory, Surgical Practice*, 101–34. London, 1992.

Notice des Animaux Vivants, Actuellement à la Ménagerie du Muséum National d'Histoire Naturelle, et une Description Historique sur la Manière de Vivre, et les Habitudes de chacun d'eux. Paris: La Ménagerie du Muséum, 1801.

Nourse, Edward. *Syllabus Totam Rem Anatomicam Humanam Complectens, et Praelectionibus Aptatus, Annuatim Habendis*. London: S. Palmer and J. Huggonson, 1729.

Nuck, Antonius. *Adenographia Curiosa et Uteri Foeminei Anatome Nova. Cum Epistola ad Amicum, de Inventis Novis*. Leiden: Jordanus Luchtmans, 1691.

Nyström, Gunnar. *Olaus Rudbeck's Anatomical Theatre in the 'Gustavianum'*. Uppsala: Almquist & Wiksells Boktryckeri AB, 1955.

Olmsted, J. M. D. *François Magendie: Pioneer in Experimental Physiology and Scientific Medicine in XIXth Century France*. New York: Schuman's, 1944.

Ottley, Drewry. 'The Life of John Hunter, F. R. S.', in James F. Palmer, ed., *The Works of John Hunter, F. R. S. In Four Volumes. Volume 1*, 3–198. London: Longman, Rees, Orme, Brown, Green and Longman, 1835.

Outram, Dorinda. 'Uncertain Legislator: Georges Cuvier's Laws of Nature in their Intellectual Context', *Journal of the History of Biology* 19 (1986): 323–68.

Outram, Dorinda. *Georges Cuvier: Vocation, Science and Authority in Post-Revolutionary France*. Manchester: Manchester University Press, 1984.

Paravicini-Bagliani, Agostino. *The Pope's Body*, trans. David S. Peterson. Chicago: Chicago University Press, 2000.

Park, Katherine. 'The Life of the Corpse: Division and Dissection in Late Medieval Europe', *Journal of the History of Medicine and Allied Sciences* 50 (1995): 111–32.

Park, Katherine. 'The Criminal and the Saintly Body: Autopsy and Dissection in Renaissance Italy', *Renaissance Quarterly* 47 (1994): 1–33.

Parker, Constance-Anne. *George Stubbs: Art, Animals and Anatomy*. London: J.A. Allen, 1984.

Parker, Constance-Anne. *Mr Stubbs the Horse Painter*. London: J.A. Allen and Co. Ltd, 1971.

Peachey, George C. *A Memoir of William and John Hunter*. Plymouth: William Brendon and Son Ltd for the Author, 1924.

Pecquet, Jean. *New Anatomical Experiments of John Pecquet of Deip. By which the Hitherto Unknown Receptacle of the Chyle, and the Transmission from thence to the Subclavial Veins by the Now Discovered Lacteal Chanels of the Thorax, is Plainly Made Appear in Brutes. As also an Anatomical Dissertation of the Motion of Blood and Chyle. Together with the Further Description of the Same Lacteal Chanels Newly Discovered in the Body of Man as Well as Brutes. Being an Anatomical Historie, Publickly Propos'd by Thomas Bartoline, Dr. and Reg. Professor Both in Physick and Anatomy, to Michael Lysere, Answering*. London: Octavian Pulleyn, 1653.

Perrault, Claude. *The Natural History of Animals, Containing the Anatomical Description of Several Creatures Dissected by the Royal Academy of Sciences at Paris. Wherein the Construction, Fabrick and Genuine Use of the Parts, are Exactly and Finely Delinated in Copper Plate, and the Whole Enriched with Many Curious Physical and no less Anatomical Remarks, Being One of the Most Considerable Productions of that Academy. Done into English by a Fellow of the Royal Society. To which is added an Account of the Measure of a Degree of a Great Circle of the Earth, Published by the Members of the Same Academy; English'd by R. W. Srs. Published by an Order of Council of the Royal Society* (first edn publ. 1688). London: Printed for R. Smith, 1702.

Perrault, Claude. *Mémoires pour Servir à l'Histoire Naturelle des Animaux*, 2 vols. Paris: L'Imprimerie Royale, 1671, 1676.

Pevsner, Nikolaus. *Academies of Art Past and Present*. Cambridge: Cambridge University Press, 1940.

Peyer, Johan Conrad. 'Exercitatio Anatomico-Medico de Glandulis Intestinorum Earumque Usu et Affectionibus. Cui Subjungitur Anatome Ventriculi Gallinacei', in *Parerga Anatomica et Medica Septem. Ratione ac Experientia Parentibus Concepta et Edita*. Amsterdam: Henricus Wetstentius, 1682.

Peyer, Johan Conrad. 'Methodus Historiarum Anatomico-Medicarum. Exemplo Ascitis Vitalium Organorum Vitio ex Pericardii Coalitu cum Corde Nati Illustrata', in *Parerga Anatomica et Medica Septem. Ratione ac Experientia Parentibus Concepta et Edita*. Amsterdam: Henricus Wetstentius, 1682.

Pignatti, Terisio. *Pietro Longhi. Paintings and Drawings. Complete Edition*, trans. Pamela Waley. London: Phaidon, 1969.

Platter, Thomas. *Journal of a Younger Brother: The Life of Thomas Platter as a Medical Student in Montpellier at the Close of the Sixteenth Century*, trans. and ed. Seàn Jennet. London: Frederick Muller Limited, 1963.

Poirer, Jean-Pierre. *Histoire des Femmes de Science en France du Moyen Age à la Révolution*. Paris: Editions Pygmalion/Gérard Watelet, 2002.

Pole, Thomas. *The Anatomical Instructor; or, an Illustration of the Modern and Most Approved Methods of Preparing and Preserving the Different Parts of the Human Body, and of Quadrupeds, by Injection, Corrosion, Maceration, Distention, Articulation, Modelling, &c. With a Variety of Copper-Plates*. London: The Author, and W. Darnton, 1790.

Portal, Antoine. *Cours d'Anatomie Médicale, ou Élémens de l'Anatomie de l'Homme, avec des Remarques Physiologiques et Pathologiques, et les Résultats de l'Observation sur le Siége* [*sic*] *et la Nature des Maladies, d'après l'Ouverture des Corps*, 5 vols. Paris: Baudouin, Imprimeur de l'Institut National, 1803.

Portal, Antoine. *Mémoires sur la Nature et le Traitment de Plusieurs Maladies*, 4 vols in 2. Paris: Chez Arthus Bertrand, 1800, 1808.

Portal, Antoine. *Histoire de l'Anatomie et de la Chirurgie, Contenant l'Origine et les Progrès de ces Sciences; Avec un Tableau Chronologique des Principales Découvertes, et un Catalogue des Ouvrages d'Anatomie et de Chirurgerie, des Mémoires Académiques, des Dissertations Insérées dans les Journaux, et de la Plupart des Theses qui ont été Soutenues dans les Facultés de Médecine de l'Europe*, 6 vols. Paris: P. Fr. Didot le Jeune, 1770–73.

Portal, Antoine, and Joseph Lieutaud. *Historia Anatomico-Medica*, 1767. *See under* Lieutaud.

Porter, Roy. *Flesh in the Age of Reason*. London: Allen Lane, 2003.

Porter, Roy. 'Willam Hunter: A Surgeon and a Gentleman', in W. F. Bynum and Roy Porter, eds, *William Hunter and the Eighteenth-Century Medical World*, 7–34. Cambridge: Cambridge University Press, 1985.

Price, Uvedale. *An Essay on the Picturesque, as Compared with the Sublime and Beautiful; and, on the Use of Studying Pictures, for the Purpose of Improving Real Landscape*. London: J. Robson, 1794.

Programmes des Cours d'Enseignement dans l'École de Santé de Montpellier. [n.p.] An III (1795).

Punt, H. *Bernard Siegfried Albinus (1697–1770). On 'Human Nature': Anatomical and Physiological Ideas in Eighteenth Century Leiden*, trans. T. van Winsen, M. Hannay and L. Wiemans. Amsterdam: B.M. Israël, 1983.

Rabl, Carl. *Geschichte der Anatomie an der Universität Leipzig. Mit 10 Tafeln*, vol. 7 of *Studien zur Geschichte der Medizin*. Leipzig: J.A. Barth, 1909.

Rau, J.J. *Oratio Inauguralis De Methode Anatomen Docendi et Discendi. Habita in Auditorio Majori, Cum ordinariam, Anatomiae, et Chirurgiae Professionem in alma Academia Lugduno-Batava ex Illustriss. Curatorum et Consulum Decreto auspicaretur. a.d. XXVI. Septembris. Anno MDCCXIII*. Leiden: Samuel Luchtmans, 1713.

Rehbock, Philip F. 'Transcendental Anatomy', in Andrew Cunningham and Nicholas Jardine, eds, *Romanticism and the Sciences*, 144–60. Cambridge: Cambridge University Press, 1990.

Richardson, Samuel. *Familiar Letters on Important Occasions, with an Introduction by Brian W. Down*. London: Routledge, 1928.

Richter, Gottfried. *Das Anatomische Theater. Abhandlungen zur Geschichte der Medizin und der Naturwissenschaften, Heft 16*. Berlin, 1936.

Ridley, Glynis. *Clara's Grand Tour: Travels with a Rhinoceros in 18th-century Europe*. London: Atlantic Books, 2005.

Riese, W. 'The 150th Anniversary of S. T. Soemmerrings's *Organ of the Soul*: The Reaction of his Contemporaries and its Significance Today', *Bulletin of the History of Medicine* 20 (1946): 310–21.

Robbins, Louise E. *Elephant Slaves and Pampered Parrots: Exotic Animals in Eighteenth-Century Paris*, in the series *Animals, History, Culture*, ed. Harriet Ritvo. Baltimore, MD: Johns Hopkins University Press, 2002.

Roberts, K. B., and J. D. W. Tomlinson. *The Fabric of the Body: European Traditions of Anatomical Illustration*. Oxford: Clarendon Press, 1992.

Rodari, Florian. *Anatomie de la Couleur. L'Invention de l'Estampe en Couleurs*. Paris: Bibliothèque Nationale de France, 1996.

Rodger, N. A. M. *The Wooden World: An Anatomy of the Georgian Navy*. London: Fontana, 1986.

Roe, S. A. *Matter, Life, and Generation: Eighteenth-Century Embryology and the Haller–Wolff Debate*. Cambridge: Cambridge University Press, 1981.

Roederer, Johann Georg. *Élémens de l'Art des Accouchements. Augmentés des Observations sur les Accouchemens Laborieux. A l'Usage des Étudians en Médecine et en Chirurgie. Traduits sur la Derniere Edition Par M****. Avec Figures*. Paris: P. Fr. Didot le Jeune, 1765.

Roederer, Johann Georg. *Icones Uteri Humani Observationibus Illustrati*. Göttingen: Vandenhoek, 1759.

Roederer, Johann Georg. *Elementa Artis Obstetriciae in Usum Praelectionum Academicarum*. Göttingen: Sumptibus Bibliopolii Acad. Privilegiati, 1753.

Roger, Jacques. *Buffon: A Life in Natural History*, trans. Sarah Lucille Bonnefoi. Ithaca, NY, and London: Cornell University Press, 1997.

Rookmaaker, L. C. 'Histoire du rhinocéros de Versailles', *Revue d'Histoire de Science* 36 (1983): 307–18.

Rosen, Richard L. 'The Academy of Sciences of the Institute of Bologna, 1690–1804', PhD thesis, Case Western Reserve University, 1971.

Rousseau, Jean-Jacques. *Reveries of the Solitary Walker*, trans. Peter France. Harmondsworth: Penguin, 1979.

Rudwick, Martin. 'The Emergence of a Visual Language for Geological Science 1760–1840', *History of Science* 14 (1976): 149–95.

Ruini, Carlo. *Anatomia del Cavallo, Infermita, et suoi Rimedii. Opera Nuova, Degna di Qualsivoglia Prencipe, et Cavaliere, et Molto Necessaria à Filosofi, Medici, Cavallerizzi, et Marescalchi. Del Sig. Carlo Ruini Senator Bolognese. Adornata di Bellissime Figure, le Quali Dimostrano Tutta l'Anatomia di Esso Cavallo. Divisa in Due Volumi. De Quali Questo Primo, in Cinque Libri Copiosamente Dichiara Tutte le Cose Appartenenti alla detta Anatomia. Con Due Bellissime Tavole, Una de' Capitoli & l'Altra delle Cose Notabili. Con Licenza De' Superiori, et Privilegio*, first publ. Bologna 1598; 4th edn, Venice: Fioravante Prati, 1618.

Russell, Kenneth Fitzpatrick. *British Anatomy, 1525–1800: A Bibliography of Works Published in Britain, America and on the Continent* (first publ. 1963), 2nd edn. Winchester: St Paul's Bibliographies, 1987.

Russell, Kenneth Fitzpatrick. 'The *Osteographia* of William Cheselden', *Bulletin of the History of Medicine* 28 (1954): 32–49.

Ruysch, Frederic. *Dilucidatio Valvularum in Vasis Lymphaticis et Lacteis 1665. Facsimile of the First Edition, with an Introduction by A.M. Luyendijk-Elshout*, in the series *Dutch Classics on History of Science*. Nieuwkoop (Holland), 1964.

Ruysch, Frederic. *Opera Omnia Anatomico-Medico-Chirurgica, Huc Usque Edita*, 2 vols. Amsterdam: Jansonius-Waesbergii, 1737.

Ruysch, Frederic. *Thesaurus Anatomicus Primus*, in *Opera Omnia Anatomico-Medico-Chirurgica, Huc Usque Edita*, 2 vols, vol. 2. Amsterdam: Jansonius-Waesbergii, 1737.

Ruysch, Frederic. *Responsio ad Dissertationem Epistolicam, Viri Doctissimi, Jo. Christoph. Bohlii, Quam, de Usu Novarum Cavae Propaginum in Systemate Chylopœo, nec non de Cortice Cerebri, Conscripsit*. Amsterdam: Jansonius-Waesbergii, 1727.

Ruysch, Frederic. *Responsio ad Godefridi Bidloi Libellum Quem Vindicias Inscripsit*. Amsterdam: Jansonius-Waesbergii, 1721.

Ruysch, Frederic. *Observationum Anatomico-Chirurgicarum Centuria: Accedit Catalogus Rariorum, quae in Museo Ruyschiano Asservantur*. Amsterdam: Boom, 1691.

Saban, Roger. 'Deux Anatomistes Monpelliérains Professeurs au Collège Royal de France et au Jardin du Roi: Ferrein et Portal', in *Actes du 110e Congrès National des Sociétés Savantes, Montpellier 1985*, 141–62. Paris: C. T. H. S., 1985.

Saban, Roger. 'La Dissection au Muséum de Rhinocéros de la Ménagerie de Versailles en 1793 et les "Vélins du Muséum"', *Comptes Rendus 108e Congrès National des Sociétés Savantes 1983, [section] Sciences* (1983): fasc. IV, pp. 33–40.

Sabatier, J.-C. *Recherches Historiques sur la Faculté de Médecine de Paris depuis son Origine Jusqu'à nos Jours*. Paris: J.-B. Bailliere, 1837.

Sachs, Michael. 'Die "Anatomische Tabellen" (1722) des Johann Adam Kulmus (1689–1745). Ein Lehrbuch für die (Wund-)Ärtzliche Ausbildung im Deutschen Sprachraum und in Japan', *Sudhoffs Archiv* 86 (2002): 69–85.

Sandifort, E. *Museum Anatomicum Academiae Lugduno-Batavae Descriptum ab Eduardo Sandifort*, 2 vols. Leiden: S. and J. Luchtmans, 1793.

Sandifort, Eduard. *Oratio de Circumspecto Cadaverum Examine Optimo Practicae Medicinae Adminiculo, Habita Publice, cum in Academia Lugduno-Batava Ordinariam Anatomes et Chirurgiae Professionem Auspicaretur, A.D. xxv Maji MDCCLXXII*. Leiden: Luchtmans, 1772.

Santorini, Giovanni Domenico. *Observationes anatomicae*. Leiden: G. Langerak, 1739 (first publ. 1724).

Schiebinger, Londa. 'Skeletons in the Closet: The First Illustrations of the Female Skeleton in Eighteenth-Century Anatomy', *Representations* 14 (1986): 42–82.

Schiller, Joseph. 'Les Laboratoires d'Anatomie et de Botanique à l'Académie des Sciences au XVIIè Siècle', *Revue d'Histoire des Sciences* 17 (1964): 97–114.

Schlumbohm, Jürgen. '"The Pregnant Women are Here for the Sake of the Teaching Institution": The Lying-in Hospital of Göttingen University, 1751–c.1830', *Social History of Medicine* 14 (2001): 59–78.

Schullian, Dorothy M., ed. *The Baglivi Correspondence from the Library of Sir William Osler*. Ithaca, NY, and London: Cornell University Press, 1974.

Schupbach, William. *The Paradox of Rembrandt's 'Anatomy of Dr Tulp'. Medical History, Supplement No. 2*. London: Wellcome Institute for the History of Medicine, 1982.

Semenzato, Camillo, ed. *The Anatomy Theatre [at Padua]: History and Restoration*. Padua: Università degli Studi di Padova, 1995.

Seneca. *Ad Lucilium Epistulae Morales*, trans. Richard M. Gummere. London and Cambridge, MA: William Heinemann Ltd, and Harvard University Press, 1967.

Sheldon, John. *The History of the Absorbent System, Part the First. Containing the Chylography, or Description of the Human Lacteal Vessels, with the Different Methods of Discovering, Injecting, and Preparing them, and the Instruments Used for these Purposes. Illustrated by Figures. By John Sheldon, Surgeon, F. R. S.* London: Printed for the Author, 1784.

Smellie, William. *A Collection of Cases. To Illustrate his Former Treatise, or First Volume, on that Subject*, 2 vols. London, 1754.

Smellie, William. *A Sett of Anatomical Tables, with Explanations, and an Abridgement of the Practice of Midwifery, with a View to Illustrate a Treatise on that Subject, and Collection of Cases*. London: n.p., 1754.

Smellie, William. *A Treatise on the Theory and Practice of Midwifery*. London, 1752.

Sömmering, Samuel Thomas. *Über das Organ der Seele*. Königsberg: Friedrich Nicolovius, 1796.

Sparrow, Walter Shaw. *British Sporting Artists from Barlow to Herring*. London: John Lane, The Bodley Head, 1922.

Staden, Heinrich von. 'Anatomy as Rhetoric: Galen on Dissection and Persuasion', *Journal of the History of Medicine and Allied Sciences* 50 (1995): 47–66.

Staden, Heinrich von. 'The Discovery of the Human Body: Human Dissection and its Cultural Contexts in Ancient Greece', *Yale Journal of Biology and Medicine* 65 (1992): 223–41.

Steinke, Hubert. *Irritating Experiments: Haller's Concept and the European Controversy on Irritability and Sensibility, 1750–90*. Amsterdam: Rodopi, 2005.

Steno, Nicolaus. *Discours de Monsieur Stenon sur l'Anatomie du Cerveau. A Messieurs de l'Assemblé, qui se fait chez Monsieur Thevenot*. Paris: Robert de Ninville, 1669. Facsimile edn Copenhagen 1950, by NYT Nordisk Forlag.

Stevenson, L.G. 'A Note on the Relation of Military Service to Licensing in the History of British Surgery', *Bulletin of the History of Medicine* 27 (1953): 420–27.

Stroup, Alice. *A Company of Scientists: Botany, Patronage, and Community at the Seventeenth-Century Parisian Royal Academy of Sciences*. Berkeley, CA: University of California Press, 1990.

Stroup, Alice. *Royal Funding of the Parisian Académie Royale des Sciences during the 1690s*, Transactions of the American Philosophical Society, vol. 77, part 4. Philadelphia, 1987.

Stubbs, George. *The Anatomy of the Horse. Including a Particular Description of the Bones, Cartilages, Muscles, Fascias, Ligaments, Nerves, Arteries, Veins, and Glands. In Eighteen Tables, All Done from Nature. By George Stubbs, Painter*. London: Printed by J. Purser, for the Author, 1766.

Stukeley, William. *An Essay Towards the Anatomy of the Elephant, from One Dissected at Fort St George Oct. 1715, and Another at London Oct. 1720*. Publ, in his *Of the Spleen, its Description and History, Uses and Diseases, Particularly the Vapors, with their Remedy. Being a Lecture Read at the Royal College of Physicians, London, 1722. To which is Added, Some Anatomical Observations in the Dissection of an Elephant*. London: For the Author, 1723.

Sue [Suë], Jean-Joseph, *fils*. *Recherches Physiologiques, et Expériences sur la Vitalité, lues à l'Institut National de France, le 12 Messidor, an V de la République. Suivies d'une Nouvelle Edition de son Opinion sur le Supplice de la Guillotine ou sur la Douleur qui survit à la Décolation*. Paris: L'Auteur, 1797 (An VI).

Sue [Suë], Jean-Joseph, *père*. *Anthropotomie, Ou l'Art d'Injecter, de Disséquer, d'Embaumer et de Conserver les Parties du Corps Humain &c.* (first publ. 1748, as part of his *Abrégé de l'Anatomie*). Paris: Chez l'Auteur et chez Cavellier, 1765.

Sugita, Genpaku. *Dawn of Western Science in Japan* [*Rangaku Kotohajime*]. *Translated by Ryozo Matsumoto, Supervised by Tomio Ogata*. Tokyo: Hokuseido Press, 1969.

Syllabus Musculorum Corporis Humani: Seu Ordo in Musculis Secandis Tenendus Corpore a Vertice Suspenso. In Sex Praelectiones Distinctus. In Usum Collegii Chirurgorum Londinensium. London, 1711.

Syllabus Musculorum Humani Corporis (Prout Dissectioni Sese Offerunt, una cum Actionibus suis), Praelectionibus Aptatus Mox Habendis in Theatro Chirurgorum Londinensium Anatomico. London, 1729.

Tassin, Leonard. *Les Adminstrations Anatomiques, et la Myologie*. Paris: Michel Vaugon, 1678.

Taylor, Douglass W. *The Monro Collection in the Medical Library of the University of Otago: A Descriptive Catalogue with Annotations and Introduction*. Dunedin: University of Otago Press, 1979.

Tenon, Jacques. 'Observations sur les Obstacles qui s'Opposent aux Progrès de l'Anatomie'. Paris: B.D. Pierres, 1785.

The Life of Sir Robert Christison, Bart. Edited by His Sons. Edinburgh: William Blackwood and Sons, 1885.

Thomas, H. Leigh. 'An Anatomical Description of a Male Rhinoceros', *Philosophical Transactions of the Royal Society of London* 91 (1801): 145–52.

Thornton, John L. *Jan Van Rymsdyk, Medical Artist of the Eighteenth Century*. Cambridge: Oleander Press, 1982.

Torlais, Jean. 'Le Collège Royal [au XVIII Siècle]', in René Taton, ed., *Enseignement et Diffusion des Sciences en France au XVIIIe Siècle*, 261–86. Paris: Hermann, 1964.

Tröhler, Ulrich. 'Tracing Emotions, Concepts and Realities in History: The Göttingen Collection of Perinatal Medicine', in Renato G. Mazzolini, ed., *Non-Verbal Communication in Science prior to 1900*, 345–73. Florence: Leo S. Olschki, 1993.

Tyson, Edward. *Orang-Outang, sive Homo Sylvestris: Or, the Anatomy of a Pygmie Compared with that of a Monkey, an Ape, and a Man. To which is added a Philological Essay Concerning the Pygmies, the Cynocephali, the Satyrs, and Sphinges of the Ancients. Wherein it will appear that they are all either Apes or Monkeys, and Not Men, as Formerly Pretended*. London: Thomas Bennet and Daniel Brown, 1699.

Tyson, Edward. *Phocœna, or the Anatomy of a Porpess, Dissected at Gresham Colledge: With a Preliminary Discourse Concerning Anatomy, and a Natural History of Animals*. London: Benj. Tooke, 1680.

Valentinus, Michael Bernhard. *Amphitheatrum Zootomicum, Tabulis Aeneis Quamplurimis Exhibens Historiam Animalium Anatomicam è Miscellaneis S. R. I. Academiae Naturae Curiosorum, Diariis Societatum Scientiarum Regiarum, Parisiensis, Anglicae et Prussiacae, Actis Hafniensibus et Lipsiensibus, Zootomiis Anatomicorum Celeberrimorum Aliisque Scriptis Rarioribus Collectam. Accedit Methodus Secandi Cadav. Humana, cum Enchiresibus Injiciendi Ceram, Hydrargyrum et Stannum in Vasa Sanguifera et Lymphatica, ut et Ars Dealbandi Ossa pro Sceletopœia, cum Osteologia, Tabulis Myologicis aliisque Mss. Rauianis Hactenus Summopere Expetitis, Accurante Variisque Notis et Figuris Illustrante Michaele Bernhardo Valentini, Archiattro et P. P. Gisseno*. Frankfurt: Heirs of the Zunnerians and Joh. Adam Jung, 1720.

Vartanian, Aram. *La Mettrie's L'Homme Machine: A Study in the Origins of an Idea*. Princeton, NJ: Princeton University Press, 1960.

Vater, Christian. *Physiologia Experimentis et Demonstrationibus Mechanicis, Chimicis et Anatomicis Illustrata et Confirmata*. Wittenberg: Kreusigii, 1701.

Vaughan, Louella. *"Improvements in the Art of Healing": William Heberden (1710–1801) and the Emergence of Modern Medicine in Eighteenth-Century England*, D.Phil. thesis, Oxford University, 2004.

Vesalius, Andreas. *De Humani Corporis Fabrica Libri Septem*. Basel: Froben, 1543.

Vesling, Johann. *The Anatomy of the Body of Man. Englished by Nich. Culpeper*. London: Peter Cole, 1653.

Vesling, Johann. *Syntagma Anatomicum*. Padua, 1641.

Vial De Sainbel, Charles. *Elements of the Veterinary Art, Containing an Essay on the Proportions of the Celebrated Eclipse; Six Lectures on the Art of Farriery, or the Art of Horse-Shoeing, and on the Diseases of the Foot. An Essay on the*

Grease, which Obtained the Prize Given by the Royal Society of Medicine. An Essay on the Glanders; and Observations on the Gripes. To which is Prefixed, a Short Account of his Life, 3rd edn. London: J. Wright, 1797.

Vicq d'Azyr, Félix. *Traité d'Anatomie et de Physiologie, avec des Planches Coloriées Représentant au Naturel les Divers Organes de l'Homme et des Animaux*. Paris: A l'Imprimerie de Franc. Amb. Dido l'Aîné, 1786, 1789.

Vines, Richard. *Article 1: On the Origin of the Veterinary Science in Great Britain*. London, 1836.

Visser, R.P.W. *The Zoological Work of Petrus Camper (1722–1789)*. Amsterdam: Rodopi, 1985.

Waddington, Ivan. 'The Struggle to Reform the Royal College of Physicians, 1767–1771: A Sociological Analysis', *Medical History* 17 (1973): 107–26.

Wall, Cecil. *The History of the* [*London*] *Surgeons' Company 1745–1800*. London: Hutchinson's Scientific and Technical Publications, 1937.

Wardrop, James. *The Works of Matthew Baillie, M.D. To which is Prefixed, an Account of his Life, Collected from Authentic Sources*, 2 vols. London: Longman, Hurst, Rees, Orme, Brown, and Green, 1825.

Watson, E.C. 'The Early Days of the Académie des Sciences as Portrayed in the Engravings of Sébastien Le Clerc', *Osiris* 7 (1939): 556–87.

Wear, Andrew. 'Harvey and the "Way of the Anatomists"', *History of Science* 21 (1983): 223–49.

Weber, Giorgio. *L'Anatomia Patologica di Lorenzo Bellini, Anatomico (1643–1704): In Appendice: Il Methodus Historiarum Anatomico-Medicarum (1678) di J.K. Peyer*. Accademia Toscana di Scienze e Lettere La Colombaria. Studi, 171. Florence: Leo S. Olschki, 1998.

Weber, Giorgio. *Aspetti poco noti della storia dell'anatomia patologica tra '600 e '700: William Harvey, Marcello Malpighi, Antonio Cocchi: verso Morgagni*. Florence: Leo S. Olschki Editore, 1997.

Wendler, Dietmar. '275 Jahre Theatrum Anatomicum Berolinense – ein Streifzug durch seine ältere Geschichte', *Charité-Annalen (Neue Folge)* 8 (1988): 230–37.

White, Andrew Dickson. *A History of the Warfare of Science with Theology in Christendom*, 2 vols. New York: D. Appleton and Co, 1896. Repr. 1960; New York: Dover Publications.

White, John Stuart. 'The 1653 English Edition of *De Motu Cordis*, Shown to be Harvey's Vernacular Original and Revealing Crucial Aspects of his Pre-Circulation Theory and its Connection to the Discovery of the Circulation of the Blood', *History and Philosophy of the Life Sciences* 21 (1999): 65–91.

Wilkie, J. S. 'Marcello Malpighi, 1628–1694', in Donald Hutchings, ed., *Late Seventeenth Century Scientists*. Oxford: Pergamon Press, 1969.

Williams, Raymond. *Keywords: A Vocabulary of Culture and Society*. London: Flamingo–Fontana, 1976, repr. 1981.

Williamson, Raymond. 'Sir Busick Harwood: A Reappraisal', *Medical History* 27 (1983): 423–33.

Willis, Thomas. *The Anatomy of the Brain and Nerves. Tercentenary Edition 1664–1964. Edited by William Feindel*, trans. S. Pordage, 2 vols. Montreal: McGill University Press, 1965.

Wilson, Adrian. *The Making of Man-Midwifery: Childbirth in England, 1660–1770*. London: UCL Press, 1995.

Wilson, Adrian. 'William Hunter and the Varieties of Man-Midwifery', in W. F. Bynum and Roy Porter, eds, *William Hunter and the Eighteenth-Century Medical World*, 343–69. Cambridge: Cambridge University Press, 1985.

Winslow, Jacques Bénigne [Jakob Winsløv]. *L'Autobiographie de J.B. Winslow Publiée par Vilhelm Maar*. Copenhagen and Paris: Vilhelm Tryde and Octave Dios et Fils, 1912.

Winslow, Jacques Bénigne [Jakob Winsløv]. *The Uncertainty of the Signs of Death, and the Danger of Precipitate Interments and Dissections, Demonstrated, I. From the Known Laws of the Animal Œconomy. II. From the Structure of the Parts of the Human Body. And III. From a Great Variety of Amusing and Well-Attested Instances of Persons who Have Returned to Life in Their Coffins, in Their Graces, under the Hands of the Surgeons, and after they had Remained Apparently Dead for a Considerable Time in the Water*. Dublin: George Faulkner, 1748.

Winslow, Jacques Bénigne [Jakob Winsløv]. *An Anatomical Exposition of the Structure of the Human Body*, 2 vols, trans. G. Douglas. London: N. Prevost, 1733.

Winslow, Jacques Bénigne [Jakob Winsløv]. *Exposition Anatomique de la Structure du Corps Humain*, 4 vols. Paris: Duprée and Dessesartz, 1732.

Wittenbergius, Burchardus. *Declaration de Monsr. Burchardus Wittenbergius, Docteur en Medecine, et Medecin de l'Hospital de Bruges, pour donner à connoistre à tout le monde la nouvelle dissection sans effusion de sang; et l'embaumement de Monsr. Louys de Bils, Escuyerm Seigneur de Coppensdamme, Boonem, Bailly de la ville d'Ardenbourg, &c.* Bruges: [n.p.], 1657.

Wright-St.Clair, R.E. *Doctors Monro: A Medical Saga*. London: The Wellcome Historical Medical Library, 1964.

Young, Robert M. *Mind, Brain and Adaptation in the Nineteenth Century: Cerebral Localisation and its Biological Context from Gall to Ferrier*. Oxford: Clarendon Press, 1970.

Young, Sidney. *The Annals of the Barber-Surgeons of London, Compiled from their Records and other Sources*. London: Blades, East and Blades, 1890.

Index

Abernethy, John, 141–2
absorbents, *see* lymphatics
Académie de chirurgie, Paris, 82, 146, 367
Académie de France à Rome, 266–7; illustration 268
Académie des sciences, Paris (post-Revolution), 371
Académie royale de peinture et de sculpture, Paris, 265–6
Académie royale des sciences, Paris, 101, 104, 106, 146, 234, 236, 244, 247, 296, illustrations of 298–9, 327, 367, 369; *Memoirs to aid the natural history of animals* 104
Accademia degli Inquieti, Bologna (later the Accademia delle Scienze dell'Istituto di Bologna), 207–8; anatomical wax museum 247–8
Accademia delle Scienze dell'Istituto di Bologna, 14
Ackerknecht, Erwin, 380
Adolph, Johannes Travgott, 46
Akenside, Mark, 79
Albertini, Hannibal, 212
Albinus, Bernhard Siegfried, 75, 78, 79, 82, 87, 90, 94, 123, 124, 160, 162, 173, 175, 184, 250, 264, 272, 275, 281, 307, 312n, 346, 347, 348, 349, 352
  Albinus method of portrayal 253–62
  illustrations 254–5, 258–61
Albinus, Bernhard, 46, 87
Albinus, Christian Bernard, 87
Albinus, Frederick, 87
Alexander, Benjamin, 217n
Alexandria, 18
Alfort veterinary school, 333, 341
Altdorf University, 30, 46
Amsterdam (*see also* Waag), 36, 48, 83, 111, 112, 171, 200, 202, 256, 314
Amsterdam Academy of Art, 267–8
Amsterdam anatomical theatre, *see* Waag
Amsterdam Athenaeum Illustre, 222, 284
Amsterdam, Collegium Medicum, 286
Amsterdam, Collegium Obstetricum, 171
Ancients, Antiquity, 3, 66, 71, 121, 206, 207, 269, 310
Andry, Nicolas, 109
angels, 328
Angers, 112
animal electricity, 14
animal, 'the animal' 140–2
Aquapendente, Fabricius ab, 85, 167, 168
Aristotle, xxi, 17, 48, 165, 167, 168, 169, 311, 318, 324
art academies 265–75
art students, 265–75
artificial anatomy, 145–6
Aselli, Gaspar, 69, 149, 277n
Atkinson, Charles, 345
Augsburg, 78
Augustine, Saint, 14
Avicenna, his *Canon*, 127–8

Bacon, Francis, and Baconians, 312, 373
Baglivi, Georgio, 2n, 6n, 7
Baillie, Mathew, 96–7, 219–22
  preparations illustrated 221, 380
Banister, John, 29, 48
Barcelona Royal College of Surgery, 44
  anatomical theatre 112, 143
  illustrated 44
Barclay, John, 367, 385
Barthholin, Caspar (Kaspar, the younger), 86–7
Bartholin, Caspar (Kaspar, the elder), 59, 85–6, 361
  his book Institutions of anatomy, 86
Bartholin, Thomas, 46n, 48, 56, 60, 80, 86, 106, 190, 200

his book Anatomy, from the Institutes of his father, 86
illustration 61,
Basel, Inselspitel, 95
Basle University, 86, 200
Basle, 200
Bassi, Laura, 143
Bauhin, Caspar, 85
Beccario, Jacob Bartholomew, 209
Bell, Charles, 385
Bellini, Lorenzo, 2n, 6n
Belon, 325
Bendetti, Alessandro, 47
Berger, J. G., 46
Berlin University, 30, 90, 119n
anatomical theatre illustrated 32, 60
Berlin veterinary school, 341
Bern, 81
Bernard, Claude, 156, 372, 374
Bernoulli, Jean, 95
Berrettini, Pietro (= Pietro da Cortona), 75
illustration 76
Bidloo, Govard, 157, 172, 175, 238n, 270, 282–3, 283–7
Biheron, Marie-Catherine, 145–6
Bils, Louis de, 233
Blair, Patrick, 305
illustration of elephant skeleton 307
Blancard, Stephen, 200
Blankaart (Blancard), Stephen, 111
Blasius, Gerard, 286, 314
illustration 316
Blon, Jakob (Jacob) Christophe le, 277
Blumenbach, J. F., 354–5
Boerhaave, 78, 89, 91, 94, 118, 139, 158, 160, 169, 171, 198
Institutes of 94
Bogarucci, Prospero, 78
Bohlius, Christopher, 287
Bologna hospitals, 203, 206, 207, 209
St Mary of Life 225
St Mary of Death 225, 250
Istituto degl' Innocenti, 250
Bologna University, 2, 7, 18, 25, 28, 30, 47, 52, 64, 83, 143, 207, 209
anatomical theatre 207, 209, 387
Bonet, Théophile, 190–4, 199, 213, 216, 217
Bonnet, Charles, 22, 23
Bontekoë, Cornelius, 69
Bossuet, Jacques-Bénigne, Bishop of Meaux, 105, 131
Bourquet, Louis, 118
Boyle, Robert 233–4
brain, 65–6
Brisbane, John, 272–3
Britain, British army, 84, 117; British navy, 84, 113–14, 117
British Lying-in Hospital, 181
Brosse, Gui (Guy) de la, 99
Browne, John, 51
Budapest veterinary school, 341
Buffon, Georges, 82, 166n, 305, 329–32, 333, 336, 340, 378
illustration 329 and 331
Burke, William, 231
Burmeister, Martha, 247
Burton, John, 173, 176, 345
Burton, Robert, xxiv
Byzantium, 18

Cabinet du Roi, Paris, 98, 332
Cambridge University, 18, 20, 29, 53, 70, 113, 173, 188–9, 337, 345
camera obscura, 262–5
Camper, Petrus (Peter, Pierre), 38, 46n, 90, 177, 187, 267–79, 281, 305, 346, 367, 352–3
illustration by 180
facial line 269
carnival, 308
Carpi, Berengario da, 23
Cartesian, *see* Descartes
Carus, Carl Gustav, 379
Cassebohm, Johann Friedrick, 60
Cat, *see* Le Cat
Catholic Church, 12–15
Cemetery of the Holy Innocents, Paris, 225
chain of being, 328
Charles II, King of England, 300, 305
Charleton, Walter, 30, 53
Cheselden, William, 41, 65, 88, 113, 118
method of portrayal (camera obscura) 262–5
illustrations from and of xxiv, 263–4
Chinese medical system, 363

Choulant, Ludwig, 81 n
Christianity, 361
Cicero, 126
circulation of the blood, 120, 126
City anatomist, in Amsterdam 112, in Venice, 112
Clara the rhinoceros, 246, 272, 307–8
  illustrations 309, 254–5
Clement XI, pope, 14, 73
Cleopatra, 216
Clift, William, 220, 333, 334
clinical medicine, *see* Paris hospital medicine
Cockburn, William, 277n
Coiter, Volcher, 325
Colbert, Jean-Baptiste, 266
Cole, F. J., 236–8, 311–12, 314
College (Royal College) of Physicians of London, 18, 20, 53, 79–80, 113, 188, 189, 217, 218, 224, 319
  anatomical theatre illustrated 33
Collège de France, 164, 218
Collège royal, Paris, 103–4, 106, 164, 218, 375
Collegium Carolinum, Kassel, 386
Collignon, Charles, 53, 119n
Collins, Samuel, 313–14, illustration 315
Colombo, Realdo, 64
Commission for Epidemics and Epizootics (French), 333
Comparative anatomy, 186, 310–40, 354–9, 375–9
Copenhagen University, 46n, 56, 57, 85–8, 106, 190
  anatomical theatre illustrated 49
Copenhagen veterinary school, 341
Copenhagen, 307
Cornell University, 13
Coro Anatomico (Bologna), 207
Corvisart, Jean-Nicholas, 380, 381–2
Coschwitz, Georg Daniel, 39, 41
Coventry, 319
Cowper, William, 238, 282–3, 314, 317
Cox, Thomas, 234
Creator, 17 (*see also* God, Maker etc), 52, 66, 166, 338
Cressé, Pierre, 54, 100, 120
Cruickshank, William, 158
Cullen, William, 96, 199
cumul, 106
Cureau de la Chambre, François, 99, 100
Cureau de la Chambre, Marin, 99
Cutler, Sir John, 30
Cuvier, Georges, 102, 238n, 305, 310, 311, 312, 334, 336, 340, 370, 371, 373, 376–9

Dalembert (D'Alembert), Jean Le Rond, 383n
Danzig, 307, 362
Daquin, Antoine, 144
Darwin, Charles, Darwinian, 310, 312, 314, 359, 378, 379, 384
Daubenton, Louis-Jean-Marie, 98, 308, 329, 332, 368, 375, 376n
Dauphin, 100, 105, 376
Dauphine, 144–5, 184
Delft, 150
Democritus, xxiv, 166
Descartes, René, and Cartesians, 53, 54, 66, 121, 158, 197, 295
Desnoües, Guillaume, 145, 247
Diderot, Denis, 22, 23n, 225
Dionis, Pierre, 50, 100–1, 119–21, 132, 144–5,151–152, 168, 169, 300
Dodart, Claude, 297
dogs, 135, 230, 262
  illustration 241, 263
Douglas, George, 158n
Douglas, James, 94, 96–7, 316–7
Draper, John William, 12
Drelincourt, Charles, 69
Dresden veterinary school, 341
Drummond, George 118
Dürer, Albrecht, 257n, 267
Duverney, Emmanuel-Maurice, 103
Duverney, Guichard Joseph, 3n, 44, 72, 78, 82, 86, 94n, 101, 102–3, 104, 199, 278, 300, 305
Duvernoy (Duvernoi), J. Georgius. (of Tübingen), 94, 305n

Écoles de Santé, Montpellier, 164, Paris, 146
écorché figures, 274–5; illustration 274

Edinburgh Corporation of Surgeon-Apothecaries, 34, 117, 121, 227–8
Edinburgh Town Council, 117–9, 123, 227–8
Edinburgh University, 83, 87–8, 154, 228, 324
  anatomical theatre 34
  illustration 35–6
  foundation of medical school 117–9
Edinburgh, 30, 367
elephant, 305–7
  illustration 306
Eliot, Robert, 117, 122
Elizabeth, Empress of Russia, 146
embalming, 233
embryology, 165
engraving, 275–8
Enlightenment, 383, 384
Ent, Sir George, 21n, 50n, 80, 166
Epicurus, Epicurean, 54, 55
Errard, Charles, the younger, 266–7, 269
Eustachius, Bartholomeus, 46, 73–5, 204, 209, 252, 270
  illustration 76,
experimental physiology, 150, 155, 156, 371–5, 385

Fabricius ab Aquapendente, Hieronymus, 64, 79, 367
Fabricius, P. C., 46
Fagon, Guy-Crescent, 101, 144
Fallopian tubes, 169
Fallopius, 367
Fallopius, Gabriel, 78, 367
Ferdinand II, Grand Duke of Tuscany, 66
Fernel, Jean, 56
Ferrara University, 30
Ferrein, Antoine, 103, 104, 106
Figlio, Karl, 385
Fire, Great, of London, 18, 30
First royal surgeon, France, 107, 108; *see also* Mareschal and Peyronie
Fischer, Johann Heinrich, 184
Florence, 66
  anatomical wax museum 'La Specola' 249–50
  hospital of Saint Matteo 273
Fontana, Felice, 249–50
Fontenelle, Bernard Le Bovier, de, 101
forceps, 171, 177
  illustration 180, 181
Foster, Sir Michael, 155
Foucault, Michel, 380
Foundling hospital, Paris, 172
Fourens, Pierre, 373–4
France, French army, 84; French navy, 84
Francus, Georgius, 46
Frankfurt an der Oder, 46, 57
Frederick I, King of the Prussians, 39
Frederick III, Elector of Brandenburg, 39
Frederick William I, Duke, 41
French Revolution, 162, 217, 303, 310, 341, 368, 370, 375, 382, 388
Fried, Johann Jakob 183, 184n

Galen, Galenic, Galenist, 12, 17, 18, 25, 26, 28, 29, 48, 56, 64, 73, 126, 160, 188, 191, 192, 197, 311, 314, 363
  his Ars medica 127
Galleria dell'Accademia di Belli Arti, Florence, 273
Galletti, Giuseppe, 249
Galvani, Luigi, 14, 52, 162
Gautier d'Agoty, Le Sieur (Fabien Gautier d'Agoty), 251, 275, 278
Gayant, Louis, 300
Gelfand, Toby, 107
generation, 165–85
Genga, Bernardino, 266–7, 269
Gesner, Conrad, 297
Getty Museum of British Art, Yale University, 357–9
Gild, Ignatius, 194, illustration by, 194
Gilpin, William, 271
Glaser, Henrick, 200
Glasgow University, 96, 252
Glisson, Francis, 20–22, capsule of 60, 69, 70, 188, 236, 318, 321, 324
God, 46, 48–55, 237, 312n, 328, 384, 389
  finger of 52
  hand of 319
  perfections of 54, 167
  see also Creator
Goldin, Grace, 111
Gornia, Giovanni Battista, 203

Göttingen University (the Georg-August-Universität, or Georgia Augusta), 46, 52, 81, 83, 89, 91, 95, 103, 159, 182, 201, 227
  anatomical theatre 160
  midwifery hospital illustrated 186
Göttingen, Royal Society of Sciences, 170n
Graaf, Regnier (Regnerus) de, 101, 150–1, 166, 236
grave-robbers (resurrection men), 108, 135, 137, 146, 228–9, 388
Greene, Ann, 5
Grégoire, father and son, teachers of midwifery in Paris, 89, 118, 176; the younger, 179, 184
Grew, Nehemiah, 318–24
  illustration 323
Groningen University, 352, 353
Gulston, Theodore, 188
Gunning, John, 386–7
Gustavianum (Uppsala), 34, 36
  anatomical theatre illustrated 37–8
Guy's Hospital, London, 357n

Halle University, 39
anatomical theatre illustrated 42, 198
Haller, Albrecht von, 45, 52, 62, 70, 72, 73, 81–2, 83, 89, 152–3, 157, 158, 159, 182, 187, 194, 196, 201–2, 227, 281, 295, 318, 367, 373, 383
  his education 91–5
  portrait of 92
  illustration of house and anatomical theatre of 93
  his *Elements of physiology* 93
Hamberger, Georg Erhard, 160–2
Hamilton, James, Jr., 217
Hanover veterinary school, 341
Hare, William, 231
Harvey, William, xxii, 26, 29, 50, 64, 69, 70, 79, 93, 123, 165, 166–8, 169, 189, 200, 236, 291, 367
Harwood, Sir Busick, 337–40
  illustration 339
Heberden, William, 217n
Heidelberg University, 46
Heister, Lorenz, 80–1, 163n
Helmstadt, Royal and Ducal Julius-University of, 46, 80
Heraclitus, 48
Hewson, William 119n, 146, 241, 279
Hippocrates, Hippocratic, xxiv, 46, 193, 194, 197
  his *Aphorisms* 127
history of animals, 325
Hoffmann, Friedrich, 41, 198
Hoffmann, Moritz, 46
Home, Everard, 333–4, 336
Hooke, Robert, 30, 320
Horne, Johannes van, and his Anatomical Cabinet, 88
horse, anatomy of, 340–54
  illustrated 343, 350–4
Hôtel Dieu, Paris, Salles d'accouchements, 171, 172
Houston, James, 171–2
Humphrey, Ozias, 342–9
Hunauld, François, 103
Hunauld, Pierre, 112
Hunter (William) versus Hunter (John) 291–3
Hunter, John, 2n, 82, 96–7, 137, 152, 156, 187, 202, 218, 219, 220, 241, 274–5, 295, 305, 333, 334–7, 340, 341, 355
  his anatomical course 139–42
  his museum 141–2, 235
Hunter, William, 17, 48, 62, 75n, 78, 82, 84–5, 90, 96–7, 103, 142, 146, 158, 159, 179, 181–2, 218, 219, 230, 237, 252–3
  proposed royal anatomical school 115
  illustrated 116
  his anatomical course 135–9
  his anatomical school 137–8
  on preparations 232–3, 275, 305, 367
  anatomical collection of 97
  professor at Royal Academy 270–2
  controversies 279–80, 287–93
  his dissecting room illustration 280

Ingolstadt University, 38, 78
  illustration 41
injections (water, mercury, wax), 158, 235–46

Innocent XII, pope, 1, 14
Institut national pour les sciences et les arts (Institut de France), 162, 369, 370, 371, 382
intellectual property 279–93
Italian hospital anatomical theatres, 111

Jacyna, L. S., 337
Japan, 361–6
Jardin royal, Paris, (Jardin des plantes; later Muséum d'histoire naturelle, q.v.) 44, 54, 78, 98–103, 113, 145, 151, 168, 199, 217, 218, 303, 329, 332, 375
  anatomical theatre 98
  illustration of 99
Jena University, 160–2
Jenner, Edward, 2n, 333
Jenty, Charles Nicholas, 5, 81, 237–8, 276, 278
Job, 253
Johnson, Dr., 313
Joseph II, Emperor, 227, 250–1; Josephinum 250
Josephinum, 250–1

Kerkham, Johannes van, 79
Kotsugahara, 362
Kulmus, Johann Adam, 361–6

L'Admiral, Jan, 277, 278
La Charité hospital, Paris, 265, 380
Lacépède, Étienne de, 376
lacteals, 120
Laennec, René Théophile, 380
Lairess. G. de, 283
Lamarck, Jean-Baptiste, 304, 376, 379
Lamb, Richard, 225
Lambertini, Prospero, 14, 247–8 (Benedict XIV, pope)
Lamy, Guillaume, 54–5
Lancisi, Giovanni Maria, 1, 2n, 5, 6, 7n, 73–5, 88, 204–6, 209, 266
Laocoön statue, 267; illustrations 267
Laplace, Pierre, 371, 372
Larber, Antonio, 127
Lassus, Pierre, 81
Latour, Bruno, 153
Laurens, André Du, 22
Lavoisier, Antoine-Laurent, 370
Lawrence, Christopher, 134
Lawrence, John, 354
Lawrence, William, 385
Le Cat, Claude-Nicholas, 65, 83, 89, 105, 234
  illustration 235
Le Clerc, Daniel, 67
Le Dran, François Antoine, 94
Leeuwenhoek, Antoni van, 168
Leiden University, 18, 30, 34, 83, 123–5, 157, 171, 173, 184, 222, 241, 253–61, 267, 295, 319, 345
  anatomical theatre 36, 57, 94
Leiden, 307
Leipzig University, 46
Lelli, Ercole, 248, 274
Leveling, Heinrich Palmaz, 78
Levret, André, 184
Lietaud, Joseph, 80, 82, 157, 196, 218
life classes, life school, 265–75
Linnaeus, Linnean, 330, 334
lion, anatomy of, illustrated, 302
Lister, Martin, 102
Liverpool, 344
London, 17, 18, 26, 41, 83
Longhi, Pietro, 308
  illustration by 309
Louis XIV of France, 100, 171, 172, 265, 296, 300, 303, 305
Louis XV of France, 303
Louis XVI of France, 303
Louvre, Paris, 297
Lower, Richard, 200, 295
Luchtmans, Peter, 80
lymphatic vessels and system, 138, 158, 241, 244, 281, 288–91
Lyons veterinary school, 341
Lyser, Michael, 56–9, 69, 89

Maastricht, royal hospital, 59
Macbride, David, 217
Mackenzie, Colin, 291–3
Madrid, anatomical theatre, 38, 112
  illustration 40
Magendie, François, 371–3, 384

Malpighi, Marcello, 1, 3, 5, 6, 69, 70, 73, 82, 86, 88, 156, 168, 197, 198, 203–4, 212, 311, 319
  portrait 4,
Manfredi, Eustachio, 207
Manfredi, Heraclito, 209
Manget, Jean Jacques, 67
manikin, 179
man-midwifery, 118, 165–85
Manzolini, Giovanni, 248
  illustration 249
Maréchal, Nicolas, 308
Mareschal, Georges, 108
Maria Theresa, Empress, 226
Martínez, Martin, 38
Mascagni, Paolo, 25n, 244, 273, 289
  illustration 245
masks, 143, 308–9
McGill, John, 117
Meckel, Johann Friedrich, the elder, 90
Mede, Joseph, 29
Mediavia, Nicholas, 210
menagerie, 104, 300, 303–4, 308, 329
  illustration 303, 376
meritocratic views, 85
Mertrud, Antoine-Louis-François, 98, 278, 308, 376, 377
Michelangelo, 64, 265, 356
microscope, 156, 168, 169, 203, 216, 285
Middlesex Hospital, 181
midwife, midwifery (cf. man-midwifery), 114, 167, 170, 171, 172, 179, 181
Mieg, J. Rudolf, 95
Moderns, 56, 69, 71, 73, 123, 207, 321
Mondini, Carlo, 387
Mondino, *see* Mundino
Monro family versus Hunters, 287–91
Monro, Alexander, *Primus*, 34, 80, 81, 82, 87, 96, 118–19, 135n, 176, 228, 229, 317, 367
  his anatomical course 133–5
  his anatomical encheireses (manual) 133
Monro, Alexander, *Secundus*, 34, 82n, 87, 90, 119, 134–5, 154, 175, 317, 324–5
  illustration 326, 367
  his thesis *On the testicles and semen in different animals* 289–91
Monro, Alexander, *Tertius*, 87–8, 90, 229
Monro, Donald, 89, 175–6
Monro, John, 88, 118
Montpellier University, 22, 143
Montpellier veterinary school, 341
Morand, Sauveur, 146
Morandi, Anna, 145, 248
  illustration 249
Morgagni, Giovanni Battista (Giambattista), 2, 72–3, 74, 79, 88, 125–31, 144, 155, 204, 206–16, 217, 218, 219, 225, 277, 281, 381–2
  portrait 211,
Münchausen. Baron Gerlach Adolph von, 91, 95
Mundino (Mundinus, Mondino), 25, 26, 28, 191, 363, 387
Murder Act of 1752, 225
Muséum d'histoire naturelle, 163, 217, 305, 308, 375, 377, 379

Napoleon, Emperor, 382, 384
National Assembly, 375
natural history, 325–40
natural theology, 54
Newgate prison, 225
Newton, Isaac, and Newtonians, 158, 197, 198, 372
Nicholls, Frank, 96, 176, 230, 237
Noah's Flood, 82
Northcote, William, 81
Nosology, 199
Nuck, Antoine, 240–1
  illustration 241

Oken, Lorenz, 337
orang-outan, anatomy of, 300
  illustrated 301
organisation, 383–6
Owen, Richard ('Professor'), 334, 336
Oxford University, 5, 18, 70, 113, 218
Oxford, 295

Pacchioni, Antonio, 205–6
Padua College of Physicians, 210

Padua hospitals, 206
Padua University, 18, 26, 28, 30, 83, 125–31, 210;; 34, 64, 72, 144, 167, 206
 anatomical theatre, illustrated 31
Padua veterinary school, 341
Padua, Hospital, 127
Paris College of Surgery, 108
Paris Faculty of Medicine, 54, 107, 189, 109–10
 anatomical theatre (Winslow's) illustrated 84
Paris hospital medicine, 380
Paris hospitals, 97, 190
 La Charité, 89, 112
 l'Hôtel Dieu, 89, 225
 Hôpital Général, 108
 Salpetrière, 108, 225
 Invalides, 112
Paris manner of teaching, 135–9, 230–1, 387
Paris surgeons, long gown 107, short gown 107
Paris University, 78, 104
Paris, 28, 83, 97, 171, 327
Parma, veterinary school, 341
pathology, 186–222
Paulli, Simon, 46n
Peachey, George, 113, 136, 230
Pecquet, Jean, 69, 74, 149–150
perfection of anatomy, 73, 82
Perrault, Claude, 297, 299, 300, 305, 327, 330
 illustration 302
Peter the Great, Czar, 240
Petit, Antoine, 82, 103, 332
Petit, Stephan Pourfour du, 110
Peyer, Johann Conrad, 69, 191, 194, 199, 200
Peyronie, François de la, 108
phantom, 179
Philosophical Transactions of the Royal Society, 65
physiocratic principles, 249
physiology, 93–4, 128, 138–9, 140–2, 156–65
Piccolomini (Piccolhominus), Alessandro, 3
picturesque theory of art, 271–3
Pidcock, Gilbert, 304, 308, 355
Pietro da Cortona (= Berrettini, Pietro), 75, illustration 76
Pitcairne, Archibald, 30, 122, 227
Platner, J. Z., 46n
Plato, 26
Platter, Thomas, 85, 143
Pole, John, 53
Pole, Thomas, 246
 illustrations 245, 296
population increase, 84
Portal, Antoine, 82, 102, 103, 106, 163, 164, 196, 217, 218, 368, 380
Porter, Roy, 85
post-mortem room, 382
Potsdam, 303
preparations, 231–5
principles of surgery, 140–2
printing, colour, 276–8
private anatomical teachers, 96, 387

Queen of France, 144–5
Quirinale Palace, 6

rabbits, 106, 277
Rackstrow's Museum of Anatomy and Curiosities, London, 247
Ramazzini, Bernadino, 212
Rau, Johannes (Jacobus), 89, 123–5,
 his anatomical collections 124–5, 222
Réamur, René-Antoine), 234
Redi, Francesco, 69
Redouté, Pierre-Joseph, 308
Reisch, Gregor, 26
 illustration 28
Rembrandt, 48
Rhinoceros, anatomy of, 46n, 307–9
Richardson, Samuel, 223
Riolan, Jean, 56, 189
Roederer, Johann Georg, 182–4
 illustrations by 185
Rome hospitals, 190, 204
 San Spirito 204, 266
Rome, 1, 6, 14, 17. 18, 30, 73, 75
Rome, University, *see* Sapienza
Rotterdam, 233
Rouen, 65, 83, 89, 234

public anatomy school 105
amphitheatre 105
Rousseau, Jean-Jacques, 25
Rowlandson, Thomas, 279–80
Royal Academy of Arts, London 181, 241, 270–2, 274–5, 356n
Royal Academy of Painting and Sculpture of Paris, 60
Royal Academy of Sciences, Paris, 199
royal anatomist, 104–6, 277
Royal Cabinet, Paris, *see* Cabinet Royal
Royal College of Anatomy, Paris, 25
Royal College of Physicians of London, *see* College of Physicians
Royal College of Physicians, Edinburgh, 34
Royal Society (of London), 64, 113, 234, 236, 282–3, 292–3, 299, 300, 305, 320, 321
museum 322
Rudbeck, Olaus (Olof), 34
Rudwick, Martin, 11
Ruini, Carlo, 297, 313, 342
illustration, 343
Russia, 389
Ruysch, Frederic, 36, 69, 81, 82, 88, 171, 202, 237–40, 275, 277, 282, 283–7, 295, 311
'Come and *see*', 239, 283–7
his anatomical museum 239–40
Rymsdyk (Riemsdyk), Jan Van, 177, 181

Sabbatier, Raphaël Bienvenu, 368
Saint Bel, Charles Vial de, 341
Saint Petersburg Academy of Sciences, 240
Saint-Côme, college of, 108
Saint-Hilaire, Geoffroy Étienne, 303, 337, 376, 377, 379
Sampson, Henry, 319
San Gregorio, Church of, Bologna, 7
Sandifort, Eduard, 222
Santo Spirito Hospital, Rome, 73
Santorini, Giovanni Domenico, 113, 209
Sapienza (Rome University), 30, 73, 204–5
Sassonia, Ercole, 194
Sauvages, Boissier de, 198
Savant of Jena, 161 (*see* Hamberger)
scale of nature, 328, 340
Schamberg, J. C., 46n
Schelhammer, Gunther, 69
Schiebinger, Londa, 146–7
Schönbrunn Palace, 303
Schools of Health, *see* Écoles de Santé
Schupbach, William, 48
Schweiger, Nicolaus, 200
secularisation, 388
Senac, Pierre, 218
Seneca, 67
sensibility and irritability, 152–3, 281
Shelburne, Lord, 335
Sheldon, John, 241–4
illustrations 243
Shogun, 361, 363
Siena University, 244
Sloane, Sir Hans, 305
Slop, Dr, 175
Smellie, William, 96, 175, 176,179, 181, 184, 276
illustrations 178 and 179,
Soemmering, Samuel Thomas, 306n, 383
Soldati, Francesco, 206
soul, 22, 121, 167, 198, 295, 321, 383–6
Spallanzi, abbé Lazarus, 82, 166n
Spanish hospital anatomical theatres, 112
Spella, Alessio, 1n
Spencer, Mary, 346, 347, 355–6, 357n, 359
Sperling, P. G., 46
St André, Nathaniel, 106, 277
St George's Hospital (London), 97, 219
St James's Park, London, 303, 305
Stahl, Georg Ernst, 198
Steno, Nicholaus (Niels Stensen), 65, 66, 67, 69, 88, 101, 105–6, 131–2
Sterne, Lawrence, 173
Strasbourg University, 184n
Strasbourg, Royal Hospital, 112
Stubbs, George, 173, 274, 275, 304, 308, 340–59
Stuckley, William, 305
Stuttgart, veterinary school, 341
Sue (Suë), Jean Joseph, the elder, 60, 62, 146, 162, 265
Sue (Suë), Jean-Joseph, the younger, 162–3, 274
Sugita, Genpaku, 361–6

Surgeons Company (later Incorporation, later Royal College) of London, 41, 113, 114, 139, 224–5, 337, 386
anatomical theatre illustrated 43, 48
Surgeons Hall 114
Swalve, Bernhard, 111
Swammerdam, Jan, 69, 86, 88, 101, 168, 237, 286, 327n
Swieten, Gerhard von, 226
Sydenham, Thomas, 187, 194, 201

Tamanti, Sebastiano, 206
Tassin, Leonard, 59
Tenon, Jacques, 25, 367, 368
Tertullian, 14
Thiroux d'Arconville, Marie-Geneviève-Charlotte, 146 7
Thomson, G., 57
Toft, Mary, 106, 277
Tokyo (Edo), 361, 365
Tortebat, François, 266
Tower of London, 305, 333
transfusion of blood, 236
Trevi fountain, Rome, 7
Tristram Shandy, 175
Tübingen, 94
Tulp, Dr Nicolaus, 48
Turgot, Ann-Robert Jacques, 333
Turin veterinary school, 341
Tyburn, 223, 274, 389
Tyson, Edward, 300, 316, 327, 328
illustration 301

Uppsala University, 34 (*see also* Gustavianum)
Utrecht, 80; University of Utrecht, 87
Utrecht, veterinary school, 341

Valentinus (Valentin), Michael, 59, 119n
Valsalva, Antonio Maria, xxii, 1, 88, 207, 208, 209, 212–16, 218, 367
Valverde, Juan, 23
Vandergucht, Gerard, 262
Varro (Terence), 130
Vater, Christian, 163n
Vatican, 267
Venice, 75, 112, 209, 308
Senate of 125
Riformati 125, 130
Verney, Joseph-Guichard du, *see* Duverney
Versailles, 104, 296, 300, 303, 305, 329, 376
Vesalius, xxii, 12, 13, 18, 23–4, 28, 29, 47, 50, 64, 67, 73, 74, 78, 89, 120, 132, 206, 252, 262, 275, 363, 365
Vesling, Johann, 128–9
Veterinary College of London, 341
Veterinary schools, 333, 341
Vicq d'Azyr, Félix, 55, 103, 308, 332–3, 367, 368
Vienna University, 194
Vienna, 226, 250, 303, 307
Vincent and Anastasius, Church of Saints, 1
Vinci, Leonardo da, 78, 265, 311, 356
Virgil, 129–30
Vivisection, 14, 23–4, 85, 295–6
Volpi, Antonio, 1, 125
Von Hagens, Professor, 14n

Waag anatomical theatre, Amsterdam, 36
illustration 39, 202
Walther, Augustin Frederick, 46, 46n
Wandelaar, Jan, 78, 253–62, 272, 275, 307, 348
Warsaw, 307
Weiss, Jacob Nicolaus, 46
Wharton, Thomas, 69, 93, 113n
White, Andrew Dickson, 12–3, 15
Wilkins, John, 320
Willis, Thomas, 200, 295
Windmill Street School (of anatomy), 62–3
Winslow, Jacques-Bénigne (Jakob Winsløv), 23n, 78–9, 82, 89, 94, 103, 106, 109, 115, 131–3, 158, 160, 367
Winslow anatomical theatre illustrated 110,
Wittenberg University, 46, 163n
women and anatomy 142–7
Worm, Ole, 86
worms, Humble Petition of, 229n

York County Infirmary, 345
York Infirmary, 173
York, 345

Zani, Valerio, 215
Zerbis, Gabriel de, 64
Zoo, zoological gardens, 304
Zoological Society of London 304
zoology, 376
zootomy, 325
Zumbo (Zummo), Gaetano Giulio, 247